Advances in Rotating Electric Machines

Advances in Rotating Electric Machines

Volume 2

Editor

Sérgio Cruz

MDPI • Basel • Beijing • Wuhan • Barcelona • Belgrade • Manchester • Tokyo • Cluj • Tianjin

Editor
Sérgio Cruz
University of Coimbra
Portugal

Editorial Office
MDPI
St. Alban-Anlage 66
4052 Basel, Switzerland

This is a reprint of articles from the Special Issue published online in the open access journal *Energies* (ISSN 1996-1073) (available at: https://www.mdpi.com/journal/energies/special_issues/Rotating_Electric_Machines).

For citation purposes, cite each article independently as indicated on the article page online and as indicated below:

LastName, A.A.; LastName, B.B.; LastName, C.C. Article Title. *Journal Name* **Year**, *Article Number*, Page Range.

Volume 2
ISBN 978-3-03936-840-2 (Hbk)
ISBN 978-3-03936-841-9 (PDF)

Volume 1-2
ISBN 978-3-03936-842-6 (Hbk)
ISBN 978-3-03936-843-3 (PDF)

Contents

About the Editor

Sérgio Cruz (Ph.D.) received his E.E. diploma and M.Sc. and Dr.Eng. degrees in electrical engineering from the University of Coimbra, Coimbra, Portugal, in 1994, 1999, and 2004, respectively. He is currently an Associate Professor and the Director of the Electric Machines Laboratory at the Department of Electrical and Computer Engineering, University of Coimbra. His teaching and research interests include power transformers, rotating electric machines, electric drives, and power electronic converters, with special emphasis on fault diagnosis, fault tolerance, and digital control.

energies

MDPI

Article

Electromagnetic Field Analysis and Design of an Efficient Outer Rotor Inductor in the Low-Speed Section for Driving Electric Vehicles

Myeong-Hwan Hwang [1,2], Hae-Sol Lee [2,3], Se-Hyeon Yang [1,2], Hyun-Rok Cha [1] and Sung-Jun Park [2,*]

1 EV Components and Materials Research and Development Group, Korea Institute of Industrial Technology, Gwangju 61012, Korea; han9215@kitech.re.kr (M.-H.H.); sehyun048@kitech.re.kr (S.-H.Y.); hrcha@kitech.re.kr (H.-R.C.)

2 Department of Electrical Engineering, Chonnam National University, Gwangju 61186, Korea; eddylee12@kitech.re.kr

3 Robotics and Virtual Engineering, Korea University of Science and Technology, Daejeon 34113, Korea

* Correspondence: sjpark1@jnu.ac.kr

Received: 4 November 2019; Accepted: 3 December 2019; Published: 4 December 2019

Abstract: Currently, the eco-friendly vehicle market is growing continuously. In the automobile industry, various electric vehicle models are being developed, and several technological innovations are being made. Certain limited vehicle types, such as passenger cars, are being converted to electric vehicles; moreover, a variety of small electric vehicles, including smart mobility vehicles, are being developed. The driving motor of an electric vehicle, e.g., a brushless Direct Current motor (BLDC), is one of the key components that determine its driving performance. However, since the recent hike in prices of the rare earth magnets used in BLDCs, the development of induction motor with lower cost and a simple product structure has become essential. Therefore, this study proposes an optimized design for an outer rotor induction motor with high efficiency in the low-speed section for electric vehicles. The motor designed in this study is efficient for speeds less than 1000 rpm, and our experimental results prove that the prototypes can provide up to 84.8% efficiency. This optimized motor is expected to have widespread application in the eco-friendly vehicle market.

Keywords: outer rotor inductor; electric vehicle; high-efficiency; eco-friendly

1. Introduction

The eco-friendly vehicle market is expanding, and various electric vehicle models and associated technologies are being developed. Moreover, strict regulations are imposed on the average CO_2 emissions and the particulate matter generation from internal combustion engine vehicles, such as gasoline- and diesel-fueled cars. As a result, electric vehicles are gaining acceptance in most countries, such as in Europe and the United States of America. Furthermore, limited vehicle types, such as passenger cars, are being converted into electric vehicles; moreover, various small electric vehicles, including smart mobility vehicles, are under development [1].

The driving motors in electric vehicles have a considerable influence on the driving performance, and are available in various configurations [2–5]. However, recently, the price of the rare earth magnets used in brushless DC motors (BLDCs) has rapidly increased, thus causing a rise in product prices, and therefore, decreasing their product price competitiveness. As a result, most of these vehicle manufacturers' profits are now being spent on material imports, causing a direct reduction in the profitability of their companies. Therefore, there is a growing technical need to replace BLDCs, and hence, developmental research on simply structured induction motors that are cheaper than BLDCs is indispensable [6–8].

Various approaches have been employed to improve the efficiency of induction motors; moreover, research and development is still ongoing. Typical methods include modeling using the finite element method [9,10], optimization using artificial neural networks (ANNs) coupled with the genetic algorithms (GAs) [11,12], development of the associated motor materials [13], and improvisation through magnetization of the barriers and slits [14].

Previously, Kumar et al. proposed a new approach to minimizing copper and iron losses and optimizing the efficiency of variable speed induction motor drives [15]. Their method was based on a simple induction motor field-oriented control model. However, they only used the conventional induction motor parameters, and as a result, some iron loss occurred. Additionally, Sakthivel and Subramanian proposed a new approach that utilizes particle swarm optimization (PSO) to evaluate the field efficiency of the induction motors by employing a model based on the modified induction motor equivalent circuits [16]. Similarly, Delgado et al. presented an optimization plan called "edge optimization," which is a simple recognition algorithm for induction motors, and has no derivative model. Their proposed approach relies on the hardware or software startup information of the motor for identification of all seven induction motor parameters, namely, the stator leakage inductance, rotor leakage inductance, stator resistance, rotor resistance, mutual inductance, mechanical inertia, and the friction coefficient [17].

Considering the parameters affecting the performance of induction motors, Faiz et al., using analytical statistics, demonstrated the negative impact of the power-supply unbalanced voltage on the efficiency of the induction motors, as well as the associated financial losses [18]. Similarly, Donolo et al. investigated the effects of voltage imbalance on the performance of induction motors. Using a sequence equivalent circuit, they determined the increase in losses of the induction motors and analyzed motors with open and closed rotor slots [19]. Jabr and Kar proposed experimental procedures for determining the mechanical parameters and the saturation characteristics [20], by employing an experimental procedure that facilitates easy measurement of the reactivity saturation characteristics of both the stator and the rotor. In addition, Rasouli et al. investigated possible induction motor parameter identification, with particular emphasis on the subset selection and reduction methods, such that the identification method could focus on the most important parameters [21]. Finally, Kostov et al. proposed an efficient approach to determine the equivalent circuit parameters of squirrel-cage induction motors based on genetic algorithms [22]. Moreover, when three data sets were used in their study, the maximum relative error of the estimated parameters with respect to the analysis values was found to be less than 1%.

In this study, shape optimization design for a low-cost, high-efficiency, outer rotor induction motor is performed based on electromagnetic field analysis and experimental observations. Therefore, an induction motor design with high efficiency, and hence, a high-power output in various speed ranges is proposed. This is achieved by targeting a low-speed and high-torque-section setup, to ensure that the motor provides high efficiency and high-power output at various driving speeds other than the rated speed. Note that an outer rotor induction motor is selected over an inner rotor induction motor, as the former has strong rotational inertia, a small rate of change of speed, and stronger structural characteristics compared to the latter. As the rated output and rated speed of the induction motor are determined in proportion to the frequency and number of poles, the basic design specifications are selected via a parameter-based formula.

The remainder of this paper is organized as follows. Section 2 presents the motor shape design and behavioral trends based on the induction motor design procedure. Through the understanding of the characteristics of outer rotor induction motors, the basic specifications are tuned by changing the number of poles. In Section 3, the feasibility of the basic design of the outer rotor induction motor is discussed using an electromagnetic field analysis assessment. Sections 4 and 5 report the production of prototypes of the optimally designed outer rotor induction motor and their overall characteristics. The design error is determined and minimized by comparing the electromagnetic field analysis values and the measured values. Finally, the conclusion is presented in Section 6.

2. Outer Rotor Induction Motor

An induction motor is a representative example of alternate current (AC) motors. Owing to the rotating magnetic field generated by the stator, an induction current is generated in the rotor of the electric conductors and a rotational torque corresponding to the slip is generated. Induction motors are divided into single-phase induction motors and three-phase induction motors, according to the type of AC power input; three-phase current is generally used, which can obtain a rotating magnetic field without any special techniques. As it does not step out like a synchronous motor, which is also an AC motor, it is considered suitable for loads with large torque fluctuations. However, it has difficultly in controlling the rotational speed, owing to the principle of obtaining torque through slip. Nevertheless, because the rotational speed can be freely controlled by the inverter circuit, thanks to the development of power electronics, this problem can be considered nearly solved.

One advantage of the outer rotor induction motor is that the generated torque increases with an increase in the cross-sectional area of the permanent magnet, and its structure is advantageous for prevention of scattering of the magnet. However, it is difficult to rotate at high speeds due to problems with mechanical stability and the possibility of demagnetization. Nevertheless, the outer rotor induction motor has numerous advantages in mechanical performance. First, to accommodate the stator, the outer rotor is larger than the rotor of the inner rotor induction motor. Owing to an increased rotor size, the inertia increases, thereby reducing the torque ripple and the cogging torque and providing smooth and stable operation, even at lower speeds. Second, it normally generates a higher torque compared to similar sized inner rotor induction motors. The torque is a magnetic force multiplied by the air gap radius (magnitude of magnetic flux), which is related to the output of the induction motor. For the same induction motor diameter, the outer rotor induction motor has a larger air gap area than that of the inner rotor induction motor, and larger air gaps can produce higher power. Third, a larger radius of the air gap also results in an increased "lever arm" for torque generation. In an outer rotor induction motor, the larger the diameter of the rotor, the higher the number of poles the rotor can accommodate, which further increases the magnetic flux.

The outer rotor induction motors have shorter accumulation compared to the inner rotor induction motors with similar performance characteristics. Their smaller size and a higher torque production facilitate their application as in-wheel induction motors in electric vehicles and propellers in remote-controlled model drones. In the case of high-precision applications, such as optical drives, their smooth and consistent speeds have an advantage over other induction motor types. Moreover, in various load applications, such as industrial power tools, pumps, fans, and blowers, the high inertia of outer rotor induction motors can reduce the load changes and provide a stable output torque. Due to the specific design advantages of outer rotor induction motors, they are commonly applied in fans and blowers. The outer rotor can serve as a hub for the fan or blower impeller. It provides a compact case and acts as a heat sink for the impeller to rotate, which facilitates induction motor cooling.

3. Outer Rotor Induction Motor Design Outer Rotor Induction Motor

To design the outer rotor induction motor for application in electric vehicles, the basic design specifications were determined considering the driving characteristics of electric vehicles. Then, the equivalent circuit method and the finite element method were primarily applied to the design process. The typical characteristics of the outer rotor induction motor were identified. After identifying the performance problems through prototype production, the design was optimized to make it more similar to the prototype by adjusting the number of poles.

3.1. Basic Design Specification Selection

An electric vehicle motor has a limited battery capacity; therefore, the input voltage was limited to 48 V. For low-speed operation, the rated speed was limited to 1000 rpm or less. The design targets were a 1.2-kW rated output, 80% efficiency, 12 Nm torque, and an output density of 0.5 kW/kg or more.

3.2. Electrical Steel Selection

There are two types of electrical steel sheets: Oriented and non-grain oriented sheets. The former is an electrical steel sheet in which the crystal magnetization is aligned in the rolling direction and is mainly used in transformers or reactors. A non-grain oriented electrical steel sheet exhibits uniform magnetic properties in the rolling and other directions, and is mainly used as a material for electric motor cores. Non-grain oriented electrical steel sheets are further divided into two types, depending on their thickness: 0.5 and 0.35 mm. There are 10 types of 0.5 mm sheets, which range from 50PN370 to 50PN1650 depending on the silicon content. There are five types of 0.35 mm sheets, ranging from 35PN320 to 35PN560, depending on the silicon content. As the first two digits of the label represent the iron plate thickness and the last three digits represent the iron loss value, it is advantageous to use a thinner steel plate to reduce the Eddy current losses, and to select a material with a lower iron loss.

The electrical steel used in this study was S08 0.35T, 35PN230 according to the POSCO standards. This was a non-grain oriented electrical steel sheet with a density of 7.6 kg/dm^3, a maximum iron loss of 2.3 W/kg, a minimum magnetic flux density of 1.62 T, and a space factor exceeding 95%.

3.3. Number of Poles and Slot Combination

The number of slots in the squirrel-cage rotor must be carefully determined considering the number of stator slots. In addition to the noise generated during operation or starting, the starting torque causes a significant change in the rotor position, and the primary cause of abnormal torque is the mismatch between the stator and rotor slot combinations.

For three poles, the number of grooves in the stator is preferably a multiple of three. A multiple of 6, 12 or 18 may be selected for 2, 4 or 6 poles, respectively, as detailed in Table 1. It is common for the number of rotor slots to exceed the number of stator slots. Furthermore, designing a difference of more than 20% between the number of stator and rotor slots can reduce the motor noise and the leakage reactance. Moreover, the higher the number of grooves, the better the output, maximum torque, efficiency, and the power factor. However, this setup also reduces the coil space factor; therefore, appropriate values must be selected.

Table 1. Stator and rotor combination according to pole number.

Number of Poles	Number of Stator Slots	Number of Rotor Slots	Number of Poles Number	Number of Stator Slots	Number of Rotor Slots
2	24	28, 16, 22	6	36	42, 48, 54, 30
	36	24, 28, 48, 16		54	72, 88, 48
	48	40, 52		72	90, 88, 84, 50
4	36	24, 40, 42, 60	8	36	48
	48	60, 84, 56, 44		48	72, 60
	60	72, 48, 84, 44		72	96

For this study, the number of poles for the design was selected as 6, which could yield the same torque at lower frequency compared to 8 poles, as detailed in Table 2. For high output and high efficiency, the number of stator and rotor slots was 72 and 88, respectively.

Table 2. Rated speed and torque depending on pole number.

8 Poles	6 Poles
Rotation speed: 1000 rpm $N = \frac{120 \times f}{P} = \frac{120 \times 66.6}{8} = 1000$ rpm Torque: 11.4 Nm $T = \frac{974 \times Power}{RPM} = kgf \times 9.8 = $ Nm $= \frac{974 \times 1.2}{1000} = 1.1688 \times 9.8 = 11.4$ Nm	Rotation speed: 1000 rpm $N = \frac{120 \times f}{P} = \frac{120 \times 50}{6} = 1000$ rpm Torque: 11.4 Nm $T = \frac{974 \times Power}{RPM} = kgf \times 9.8 = $ Nm $= \frac{974 \times 1.2}{1000} = 1.1688 \times 9.8 = 11.4$ Nm

3.4. Motor Dimensions

In order to design an outer rotor motor in this study, it was necessary to determine the outer diameter of the rotor. First, we considered mounting of the motor on an electric bike, with its volume being equivalent to D^2L, which is proportional to the motor output and torque and which determines the motor thermal stability. The detailed dimensions of the stator and the rotor of motor were determined through application of the design procedure. The relevant formulae and associated values in millimeters (mm) are presented in Table 3.

Table 3. Rotor and stator dimensions in mm.

Rotor Core		Stator Core	
Outer diameter	216	Outer diameter	187.4
Inner diameter	188	Inner diameter	146
Stack	20	Stack	20
Air gap	0.3		
w20, d20 ≥ 0.8–1.2(d20 ≥ w20) Slot opening w 20 = 0.8 Slot face width w 20 = 0.8 Tooth thickness = $\frac{0.95 \times S1 \times t_0}{S2}$ $= \frac{0.95 \times 72 \times 3}{88} = 2.5$ Core back $= 0.95 \times h_0$ $= 0.95 \times 8.1 = 7.7$ Slot depth = Rotor outer radius − Rotor inner radius − Core back = 108 − 94 − 7.7 = 6.3		Slot open = 2–3 mm ⇒ Selected as 2 mm considering the windings Slot face width = 0.8 mm Tooth thickness = 3 mm Core back $= \frac{1.05 \times D}{k_e \times P} \times \frac{B_0}{B_{CO}}$ $= \frac{1.05 \times 146}{0.95 \times 6} \times 0.3 = 8.1$ Core space factor = 0.9–0.95 Slot depth $= \frac{[D_0 - (2h_0 + D)]}{2}$ $= \frac{187.4 - (2 \times 8.1 + 146)}{2} = 12.6$	

4. Analysis of Induction Motor Characteristics via Electromagnetic Field Analysis

4.1. Slot Combination

The stator and rotor slot combination to yield the optimal conditions with regards to the output, efficiency, torque, and the weight was derived according to the pole number fluctuations with 16 poles, as demonstrated in Figure 1. With a higher number of slots, the output and the torque were found to improve, as shown in Table 4. Moreover, higher efficiencies were obtained in a certain range. Considering the overall characteristics of the motor, the optimal characteristics could be obtained with a combination of 90 stator slots and 124 rotor slots, as indicated by the blue box in Figure 1. Table 4 lists the motor characteristic data values along with the torque values.

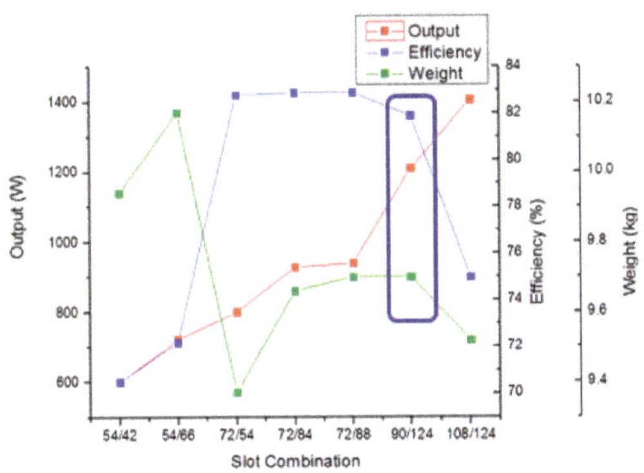

Figure 1. Graphs of motor characteristics according to slot combination.

Table 4. Motor characteristics according to slot combination.

	Output (W)	Efficiency (%)	Torque (Nm)	Weight (kg)
54/42	598	70.5	4.5	9.94
54/66	722	72.2	5.5	10.17
72/54	800	82.8	6.1	9.37
72/84	927	82.9	7.1	9.66
72/88	939	82.9	7.2	9.7
90/124	1210	81.9	9.3	9.7
108/124	1406	75	10.8	9.52

4.2. Stator Slot Shape

Figure 2a depicts a stator slot opening. With an increase in the stator slot opening, the output increases, and the efficiency decreases, as shown in Figure 2b. In Figure 2a, SOAng is the stator slot opening angle, and SO-S is the slot opening. TW_S is the width of the stator rotor, and SD_S is the slot depth of the stator. Note that the coil thickness for winding should be considered, i.e., only the slot opening with dimensions larger than the winding thickness should be selected.

Furthermore, as the stator tooth thickness increases, as shown in Figure 2c, the output and efficiency increase, as well because it is in proportion to the thickness. However, such an increase in tooth thickness renders the motor predominantly iron-based, which increases its weight.

Finally, as the stator slot height increases, the overall output and efficiency decrease (Figure 2d). However, for this study, it was necessary to increase the winding area to accommodate as many windings as possible. Therefore, it was essential to secure a certain slot area, even if it decreased the output and efficiency slightly.

4.3. Rotor Bar Shape

Considering the rotor shape design (Figure 3a), with increases in the bar thickness and the depth, as shown in Figure 3b,c, respectively, the motor output and efficiency increases as well. In Figure 3a, SetBack is the depth of the rotor slot opening, and SD-R is the slot depth of the rotor. Furthermore, SO-R is the width of the rotor slot opening. Accordingly, the slot area becomes wider, which yields increased power output due to the increased amount of aluminum.

An increase in the rotor slot opening corresponds to an increased power output (Figure 3d); however, the efficiency is inversely related. Therefore, the slot opening of the bar should be set to the optimal point in terms of output and efficiency.

4.4. Lamination

As shown in Figure 4, increasing the lamination increases the efficiency, and above a certain level of lamination, it reduces the power output of the induction motor. After the efficiency exceeds a certain level, the graph exhibits a saturation-like behavior, showing a very slight change. In this study, considering the output density, an optimum level of 30 mm was selected, which yields the highest efficiency compared to weight.

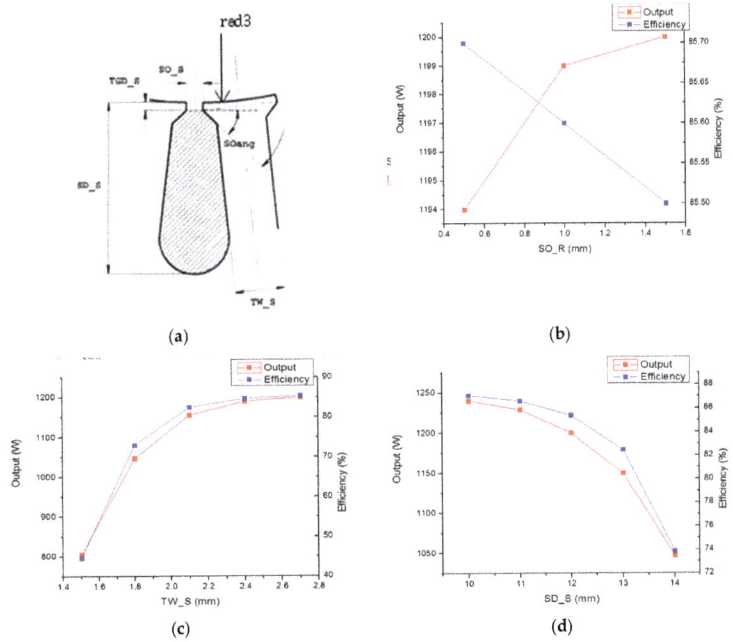

Figure 2. Graphs of motor characteristics according to stator slot shape: (**a**) Stator slot shape; graphs of output and efficiency according to (**b**) slot opening, (**c**) tooth thickness and (**d**) slot height.

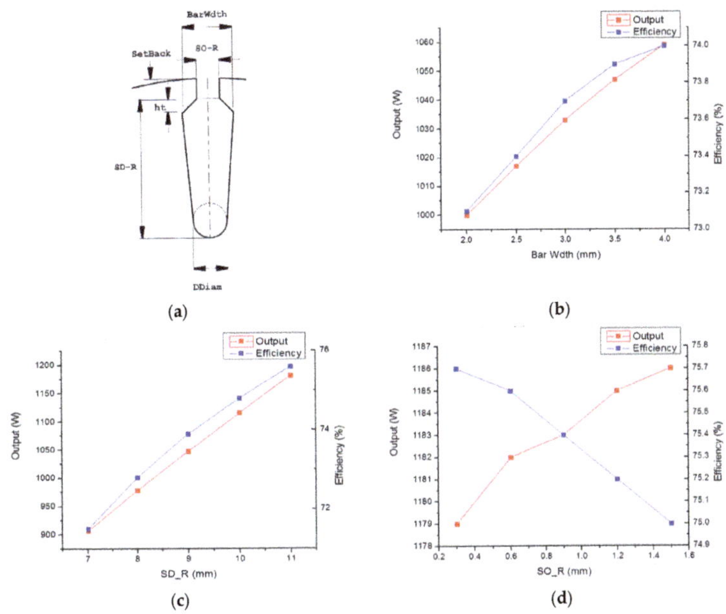

Figure 3. Graphs of motor characteristics according to rotor bar shape: (**a**) Rotor bar shape; graphs of power output and efficiency according to (**b**) rotor bar thickness, (**c**) rotor bar height and (**d**) rotor slot opening.

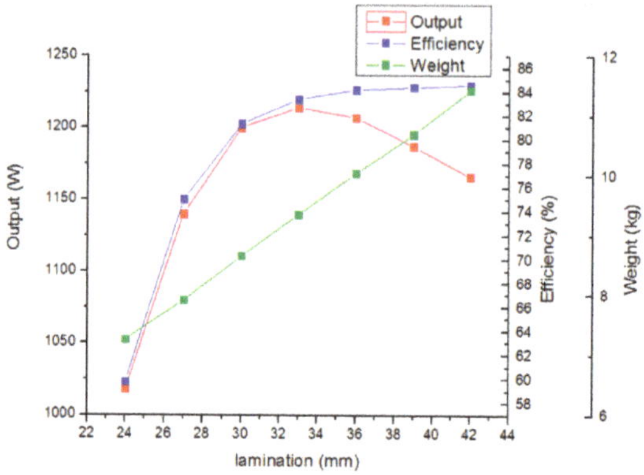

Figure 4. Graphs of motor characteristics according to lamination.

4.5. Wire Diameter and Turn Count

For winding optimization, it is necessary to determine the wire diameter and turn count by considering the stator's space factor. In this study, the optimal point was derived by fixing this space factor to be less than 45%, and through adjustment of the wire diameter and the turn count. Examination of the turn count trend revealed that the output, efficiency, torque, and the current density tend to decrease as the turn count increases, as shown in Figure 5. The point at which the desired output relative to the efficiency was obtained was considered the optimal point. The characteristics according to the turn count are listed in Table 5.

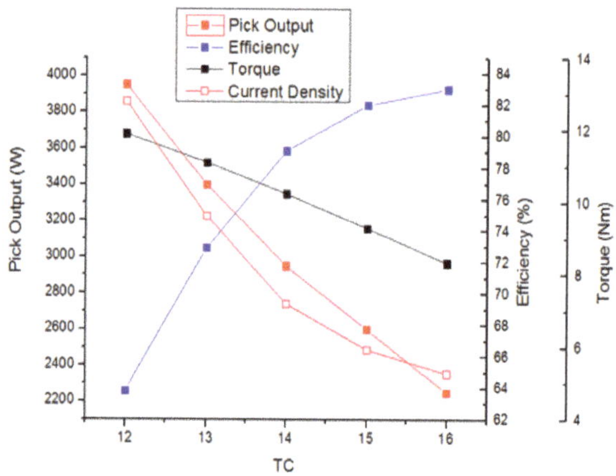

Figure 5. Graphs of motor characteristics according to turn count.

Table 5. Characteristics according to turn count.

	Turn Count	Wire Diameter (mm)	Reel Count	Space Factor (%)	Rated Output (W)	Efficiency (%)	Torque (Nm)	Current Density (A/mm^2)
1	12	0.5	4	42.2	1561	63.8	11.9	12.8
2	13	0.55	3	41.5	1447	72.9	11.11	9.62
3	14	0.35	7	42.2	1336	79.1	10.25	7.21
4	15	0.45	4	42.7	1210	82	9.29	5.94
5	16	0.5	3	42.2	1086	83	8.33	5.28

4.6. End Ring Design

Due to the use of distributed windings, the coil end rings had a greater height than that obtained for concentrated windings. Therefore, when the rotor was die-cast, it was necessary to produce a rotor end ring of similar height to increase the output and efficiency. Figure 6 depicts the structure of such a rotor end ring. Generally speaking, however, if the end ring height increases continuously, this may result in an increase in the overall motor weight, thus yielding a reduced output density. In this study, the windings were set as fixed components and the motor characteristics were observed with the variation in height of the rotor end ring. When this height exceeded a certain level, the efficiency curve became saturated (Figure 7). The point with the best motor characteristics relative to the weight was identified as the optimal point.

Table 6 shows the motor characteristics according to the motor end ring structure. The output efficiency weight was analyzed when the end ring of the motor was increased from 10 mm to 15 mm. The motor end ring is 80.8% efficient at 10 mm thick and 81.3% at max. 15 mm.

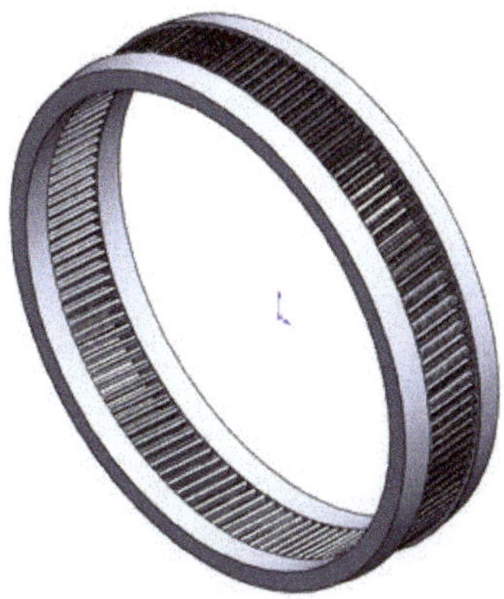

Figure 6. Rotor end ring structure of outer rotor induction motor.

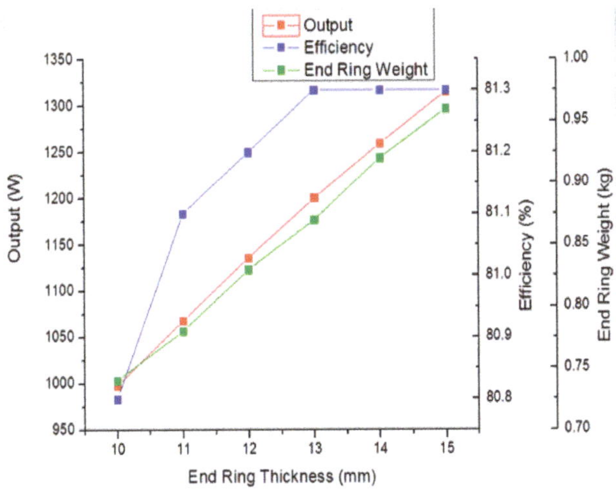

Figure 7. Graphs of motor characteristics according to rotor end ring structure.

Table 6. Motor characteristics according to rotor end ring structure.

Rotor End Ring Thickness (mm)	Rated Output (W)	Efficiency (%)	Weight of the Rotor End Ring (kg)
10	998	80.8	0.74
11	1068	81.1	0.78
12	1136	81.2	0.83
13	1200	81.3	0.87
14	1259	81.3	0.92
15	1315	81.3	0.96

4.7. Summary of Trend Analysis

As the design optimization progressed, the influence of each item on the motor characteristics was determined and the motor was reconstructed. Table 7 summarizes these trends and elucidates the characteristics of the outer rotor induction motor. The rated output, efficiency, and weight, are directly related to the output density, which are indicated by arrows. The factors with considerable changes are indicated by red arrows.

Table 7. Trend analysis of motor characteristics.

With Increasing Value of	Output	Efficiency	Weight
Lamination	↓	↑	↑
Turn Count	↓	↑	↑
Wire Diameter	↓	↑	↑
Throw	↓	↑	↑
Rotor End Ring Thickness	↑	↑	↑
Pole Number	↓	↑	↑
Stator Slot	↑	↑	↓
Rotor Slot	↑	↑	↑
Stator Outer Diameter	↑	↑	↑
Stator Inner Diameter	↑	↑	↑
Rotor Outer Diameter	↑	↑	↑
Rotor Inner Diameter	↓	↑	↓
Stator Slot Opening	↑	↓	↓
Stator Tooth Thickness	↑	↑	↑
Stator Slot Height	↓	↓	↓
Rotor Bar Thickness	↑	↑	↑
Rotor Bar Height	↑	↑	↑
Rotor Slot Opening	↑	↓	↑

5. Fabrication of Outer Rotor Induction Motor Prototype and Performance Evaluation Device

Based on the design data, an induction motor prototype was fabricated to allow comparison of the electromagnetic field analysis results with actual experimental data. The device used for evaluating the performance of the fabricated inductor (speed, torque, efficiency, etc.) is also discussed in this section.

5.1. Fabrication of Outer Rotor Inductor

Figure 8 shows three-dimensional (3D) modeling images and photographs of the actual outer rotor inductor and the assembled outer rotor inductor. In detail, Figure 8a,b shows a 3D drawing of the inductor stator and rotor, respectively, while Figure 8c shows the winding method for the inductor stator. Figure 8d shows the actual stator fabricated based on the drawing, which was manufactured using a 35PN230 non-grain oriented electrical steel sheet with 90 slots. Figure 8e shows the inductor rotor core, which was manufactured using the same material as the stator. Figure 8f shows the coil winding of the stator. The turn count was 15 and the wire diameter was 0.45 mm. The winding was produced with four reels and 6 poles in parallel. Finally, Figure 8g shows the assembled fabricated outer rotor inductor.

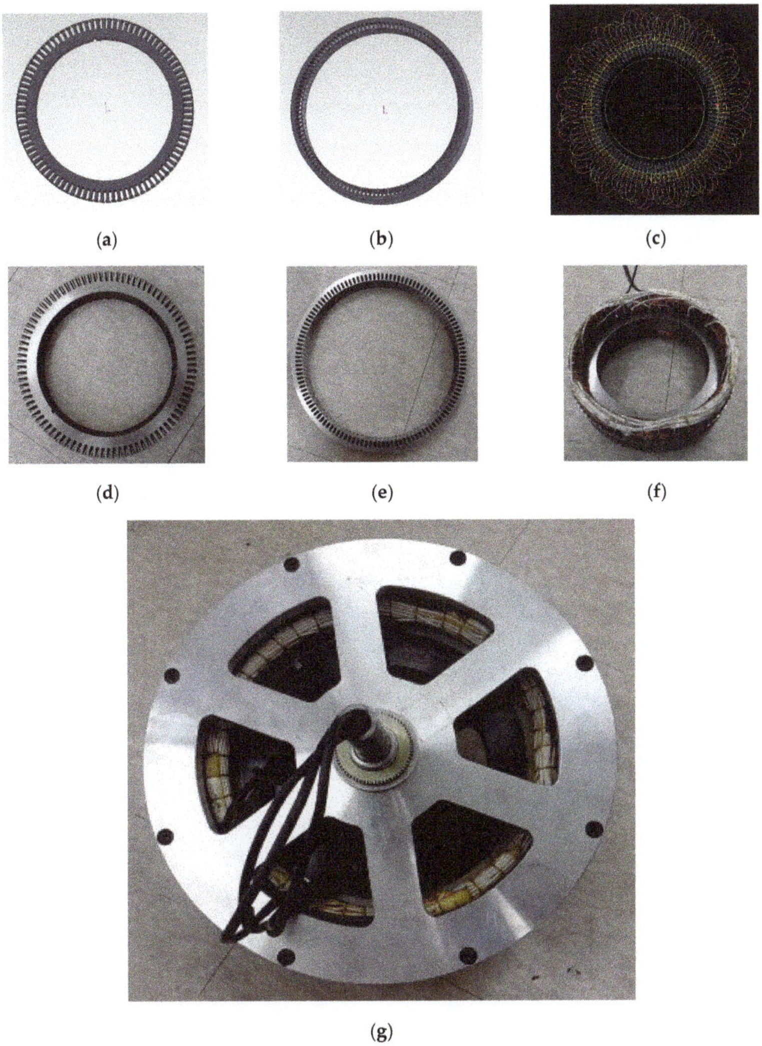

Figure 8. Three-dimensional (3D) modeling images and photographs of actual fabricated components of the outer rotor (induction motor and assembled device: 3D drawing of (**a**) stator and (**b**) rotor; (**c**) two-dimensional (2D) drawing of winding; produced (**d**) stator, (**e**) rotor, (**f**) winding and (**g**) assembled outer rotor inductor.

5.2. Outer Rotor Inductor Performance Evaluation Device

Figure 9a shows the inductor test equipment. The test equipment employed in this part of the study was a dynamo system produced by Dr. Staiger Mohilo & Co. GmbH (Lorch, Germany). The dynamo system was equipped with a servo motor to simulate load conditions, a torque sensor to measure the motor torque, and a power analyzer that could assess a variety of electrical conditions. Therefore, this device could measure the overall performance of the motor. Figure 9b shows the motor configuration installed into the test equipment. Measurements were performed under test conditions of 25 °C and 50% humidity. A certain frequency was applied according to the number of poles. To perform measurements in the 1000 rpm section, which was the rated speed section, the efficiency, power, torque, and power factor were measured by adjusting the load from 0 to 800 rpm.

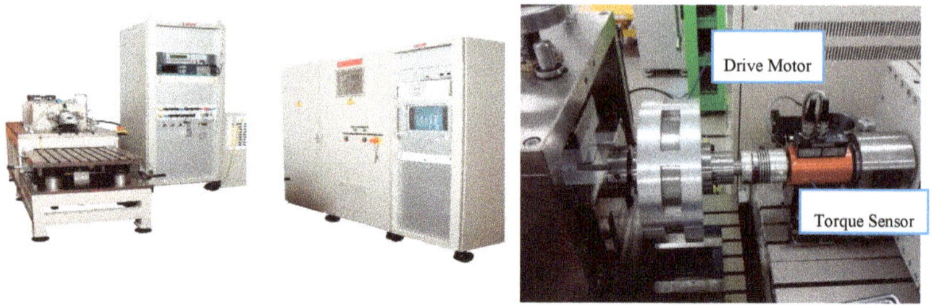

(**a**) (**b**)

Figure 9. Inductor test equipment and test device configuration: (**a**) Induction motor test equipment, (**b**) test device configuration.

6. Inductor Performance Evaluation

In accordance with the design procedure for the outer rotor induction motor, the overall shape design and operating trends were first identified using the electromagnetic field analysis design tool described above. An actual motor having the shape-optimized model was then fabricated and tested.

A six-pole model was selected as the initial model for production and performance evaluation. Comparison of the electromagnetic field analysis and actual measurement results revealed that the efficiency of the latter was at least 30% lower. To improve the efficiency, the frequency was increased in order to accelerate the induction motor. Thus, the design was optimized by varying the frequency by increasing the number of poles; as a result, the difference between the electromagnetic field analysis and measurement results was found to be reduced.

6.1. 6-Pole Model Performance Evaluation

The motor constructed using the shape-optimized model was utilized and the motor efficiency, torque, and output were tested by varying the voltage with respect to the frequency. To obtain the rated speed of 1000 rpm with 6 poles, a frequency of 55 Hz was required, and the efficiency was measured to be 53% under those conditions. Experiments performed with varying frequencies and voltage values revealed that a higher frequency corresponds to higher efficiency, but lower output and torque. In addition, the driving voltage was adjusted based on 48 V, and was found to be proportional to the output and torque, but inversely proportional to the efficiency, which is contrary to the results obtained for frequency. These results are presented in Figure 10 and Table 8.

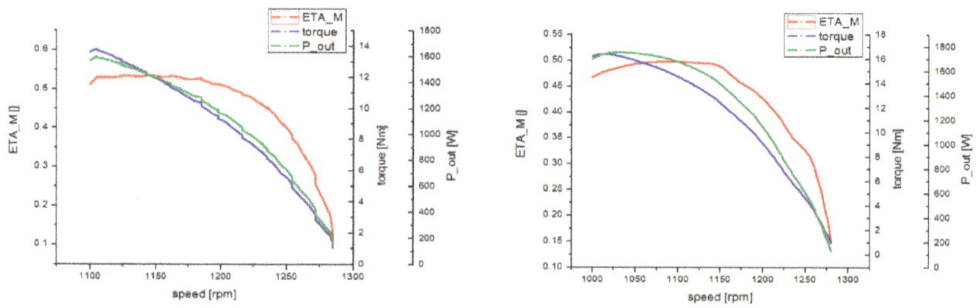

(**a**) (**b**)

Figure 10. *Cont.*

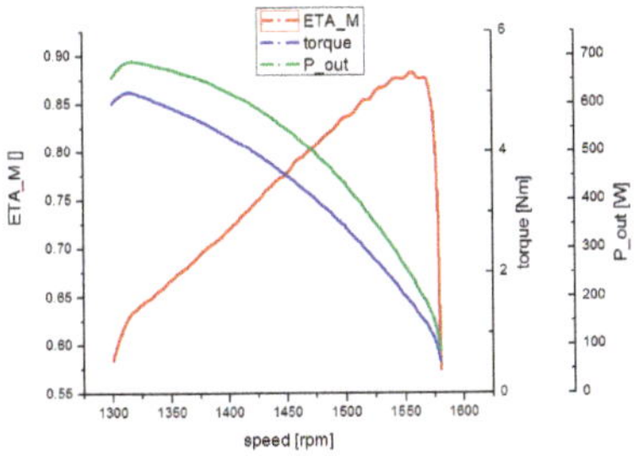

(c)

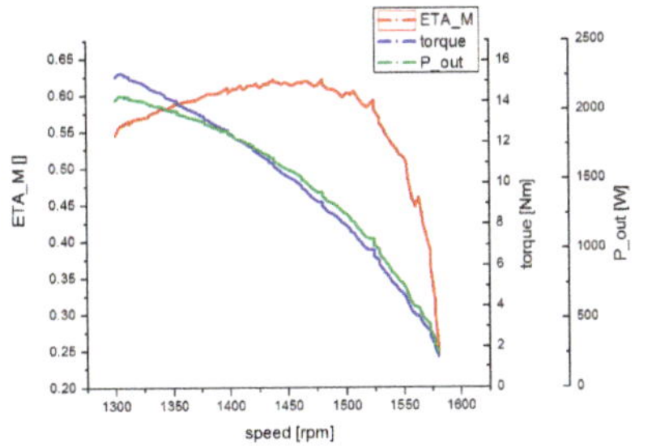

(d)

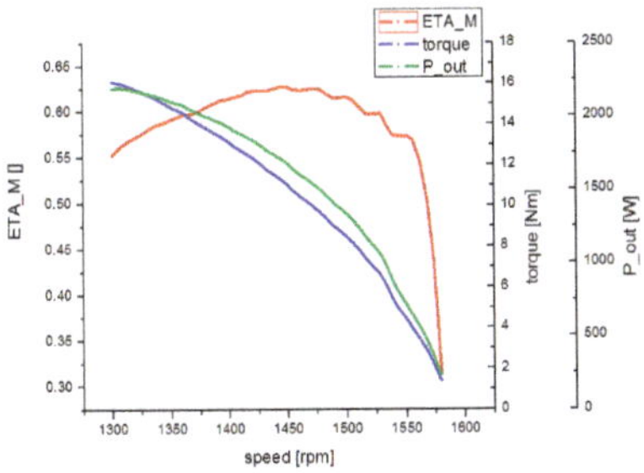

(e)

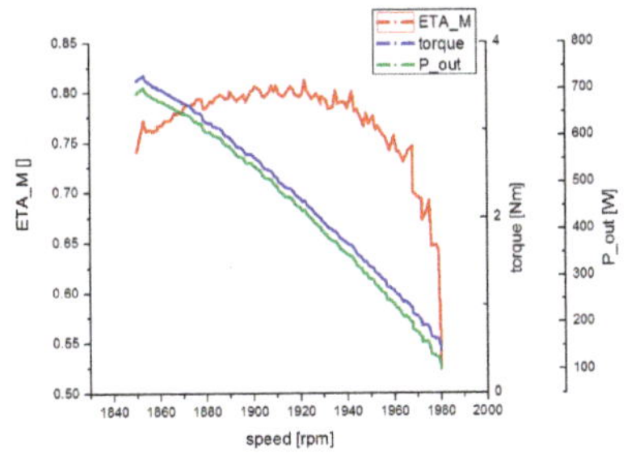

(f)

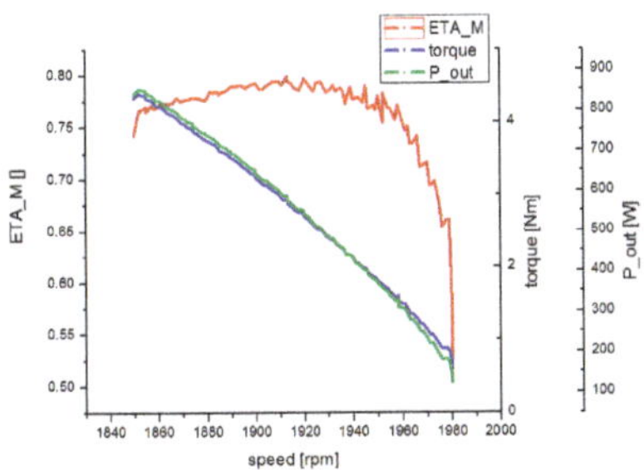

(g)

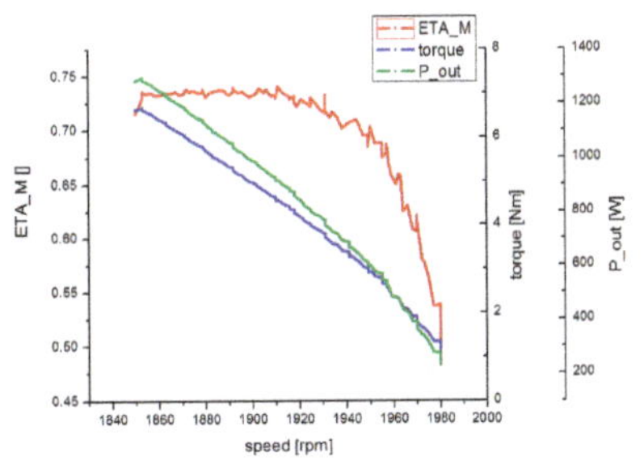

(h)

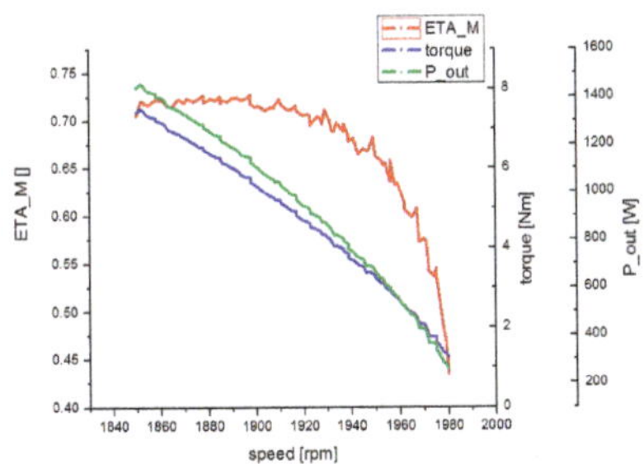

(i)

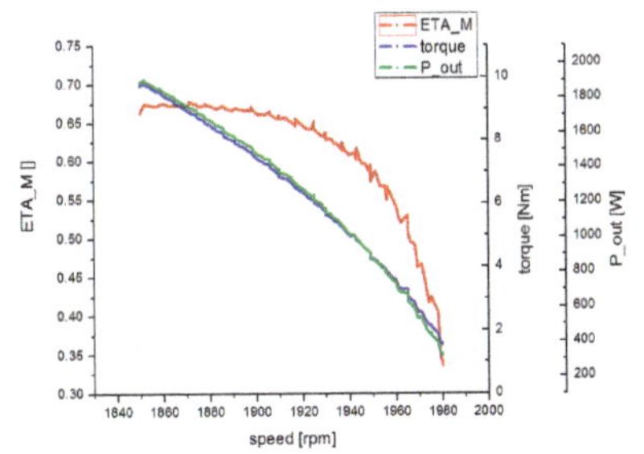

(j)

Figure 10. *Cont.*

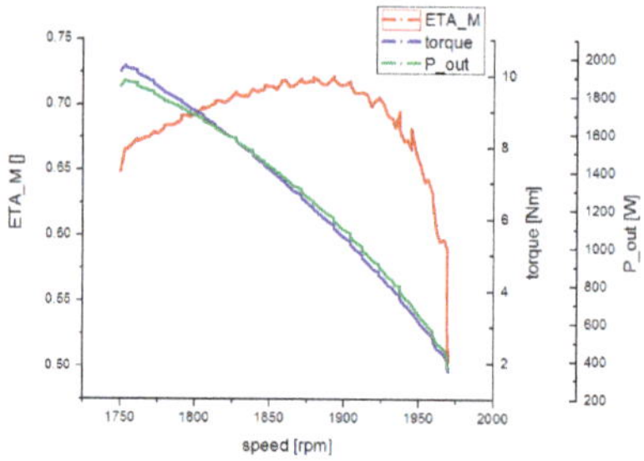

(k)

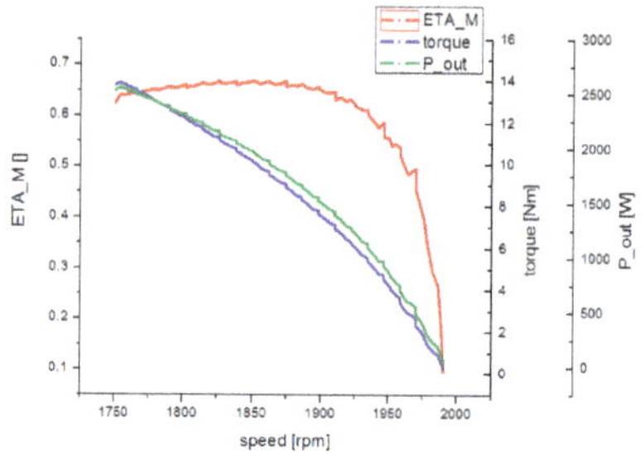

(l)

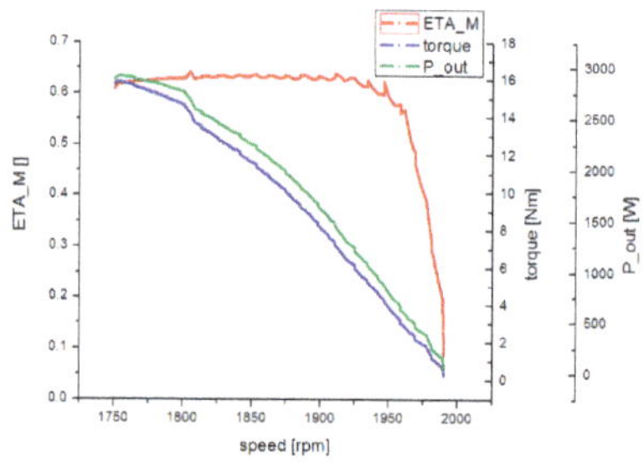

(m)

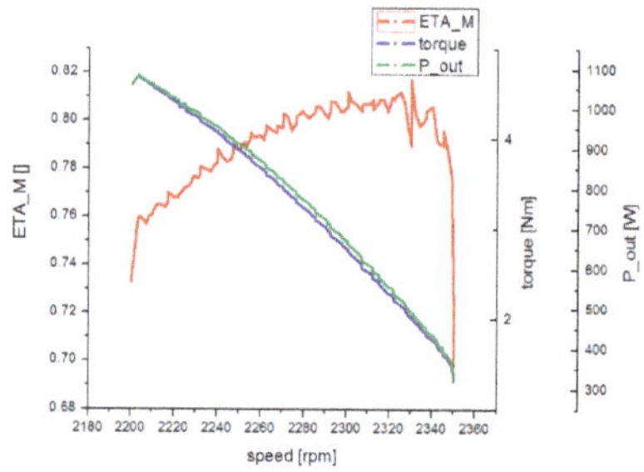

(n)

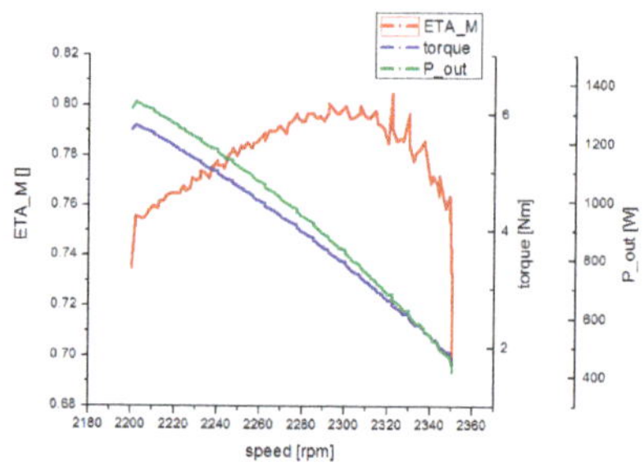

(o)

Figure 10. Graphs of motor characteristics according to voltage and frequency for the outer rotor inductor with 6 poles: Voltage_frequency of (**a–o**): 41.7 V_55 Hz, 51.9 V_65 Hz, 25.3 V_80 Hz, 47.2 V_80 Hz, 51.9 V_80 Hz, 31.3 V_100 Hz, 34.4 V_100 Hz, 43.2 V_100 Hz, 46.7 V_100 Hz, 51.9 V_100 Hz, 54.1 V_100 Hz, 62.3 V_100 Hz, 69.2 V_100 Hz and 36.6 V_120 Hz, respectively.

Table 8. Motor characteristics according to voltage and frequency for the outer rotor inductor with 6 poles.

Voltage (V)	Frequency (Hz)	Maximum Efficiency (%)	Speed at Maximum-Efficiency Point (rpm)	Output at Maximum-Efficiency Point (W)	Torque at Maximum-Efficiency Point (Nm)
41.7	55	53	1010	1516.8	12.8
51.9	65	49.8	1076	1723.7	15.3
25.3	80	88.2	1555	245.4	1.5
47.2	80	62.7	1444	1711.7	11.3
51.9	80	62.2	1438	1654.2	11
31.3	100	81.2	1922	442.3	2.2
34.4	100	79.9	1913	587.7	2.9
43.2	100	74	1892	1049.9	5.3
46.7	100	72.8	1898	1148	5.8
51.9	100	72.2	1893	1160.1	5.9
54.1	100	67.8	1871	1747.6	8.9
62.3	100	66.8	1826	2165.1	11.3
69.2	100	63.7	1853	2238.3	11.5
36.3	120	81.2	2300	666.5	2.8
41	120	80.1	2292	874.8	3.6

6.2. 10-pole Model Performance Evaluation

For the 6-pole design and measurement, improved efficiency with increased frequency was confirmed. While maintaining the rated speed, consistent at 1000 rpm, the pole number of the fabricated motor was changed to 10. The subsequent experiment was conducted in the same manner as the experiment with 6 poles. Based on the rated speed, 90 Hz frequency was applied for 10 poles. The efficiency was found to be increased compared to the 6-pole model, with a confirmed maximum of 77%. This result validates the approach of changing the number of poles to improve the efficiency. The detailed results are shown in Figure 11 and Table 9.

Table 9. Motor characteristics according to voltage and frequency for outer rotor inductor with 10 poles.

Voltage (V)	Frequency (Hz)	Maximum Efficiency (%)	Speed at Maximum-Efficiency Point (rpm)	Output at Maximum-Efficiency Point (W)	Torque at Maximum-Efficiency Point (Nm)
36.67	90	77	1027	1197.9	11.14
46.71	90	76.6	1028	1182.8	11
52.76	90	72.9	1011	1797.1	16.9
34.6	100	80.6	1165	454	3.72
46.71	100	79	1153	1029	8.5
46.36	110	80.7	1267	1052.8	7.9
49.65	110	80.3	1265	1229.6	9.28
52.76	110	79.9	1269	1286.1	9.37

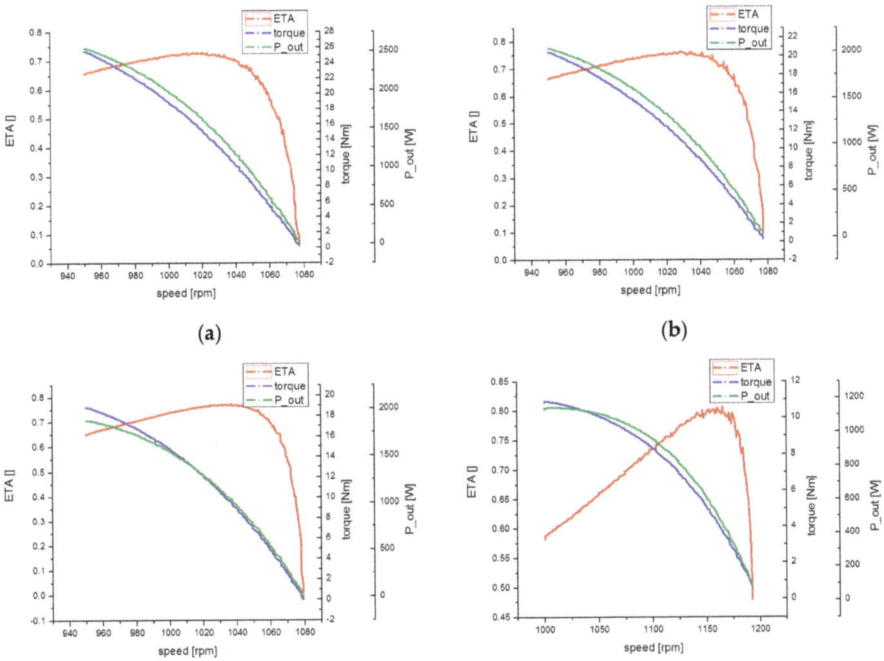

(a) (b) (c) (d)

Figure 11. *Cont.*

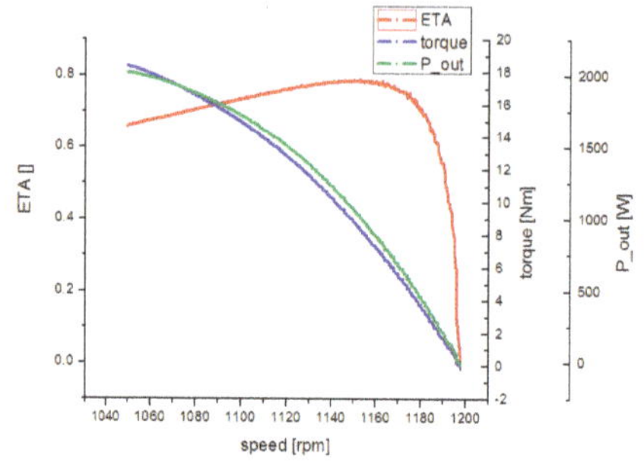

(e)

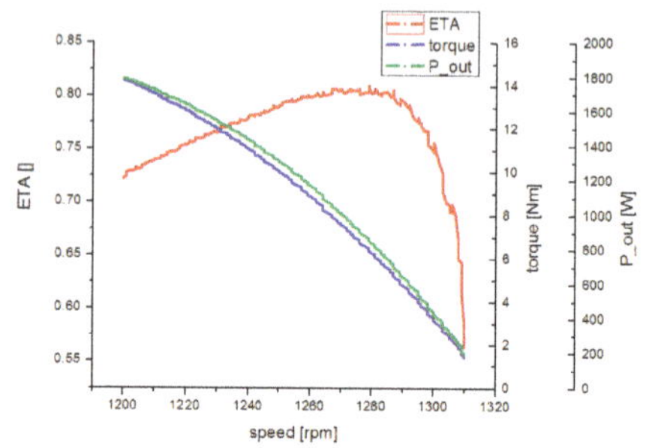

(f)

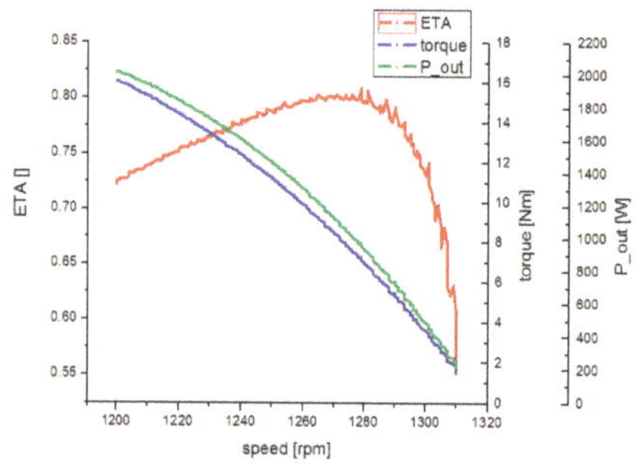

(g)

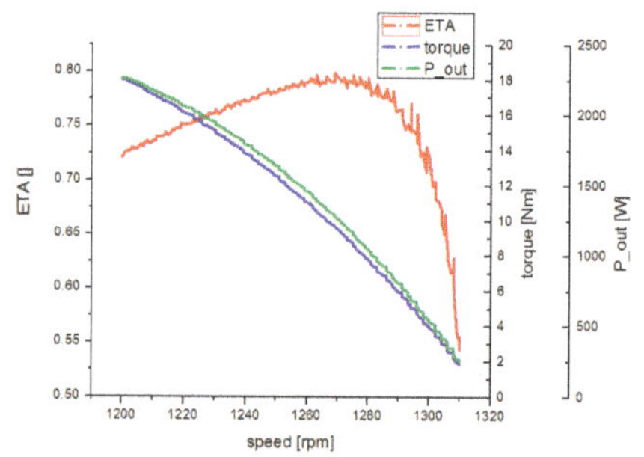

(h)

Figure 11. Graphs of motor characteristics according to voltage and frequency for outer rotor inductor with 10 poles: Voltage_frequency of (**a–h**): 36.7 V_90 Hz, 46.7 V_90 Hz, 52.7 V_90 Hz, 34.6 V_100 Hz, 46.7 V_100 Hz, 46.3 V_110 Hz, 49.6 V_110 Hz and 52.7 V_110 Hz, respectively.

6.3. 16-pole Model Performance Evaluation

To improve the efficiency of the induction motor, the number of poles of the motor was increased to 16 poles during fabrication. In terms of the experimental method, this study evaluated the performance of the outer rotor inductor in the same manner as the 10-pole model. Figure 12 shows graphs of the motor characteristics of the inductor according to the number of poles and voltage frequencies; the experimental results for each frequency are shown. The experiment was conducted within a frequency range of 126–140 Hz, and a maximum efficiency of 84.8% was experimentally confirmed. With 16 poles, the frequency for the rated speed was 140 Hz, and in terms of the measured values, at 33.2 V, the efficiency was 82.7%, the output was 1.2 kW, and torque was 11.4 Nm. As such, the design through the electromagnetic field analysis and the results of the actual manufactured outer rotor inductor were found to be consistent. In the case of a lesser number of poles of the induction motor, the coil end extended, causing extensive copper loss, and therefore, degrading the efficiency. Through this experiment, this study verified the method of improving the efficiency of a large induction motor by increasing the number of poles. The detailed results are presented in Table 10.

Table 10. Motor characteristics according to voltage and frequency for outer rotor inductor with 16 poles.

Voltage (V)	Frequency (Hz)	Maximum Efficiency (%)	Speed at Maximum-Efficiency Point (rpm)	Output at Maximum-Efficiency Point (W)	Torque at Maximum-Efficiency Point (Nm)
32.35	126	81.6	914	1189	12.4
35.12	126	79.7	912	1405	14.7
33.21	137	82.9	999	1090	10.4
33	138	83	1007	1059	10
23.87	140	84.8	1027	452	4.2
33.2	140	82.7	996	1192.6	11.4

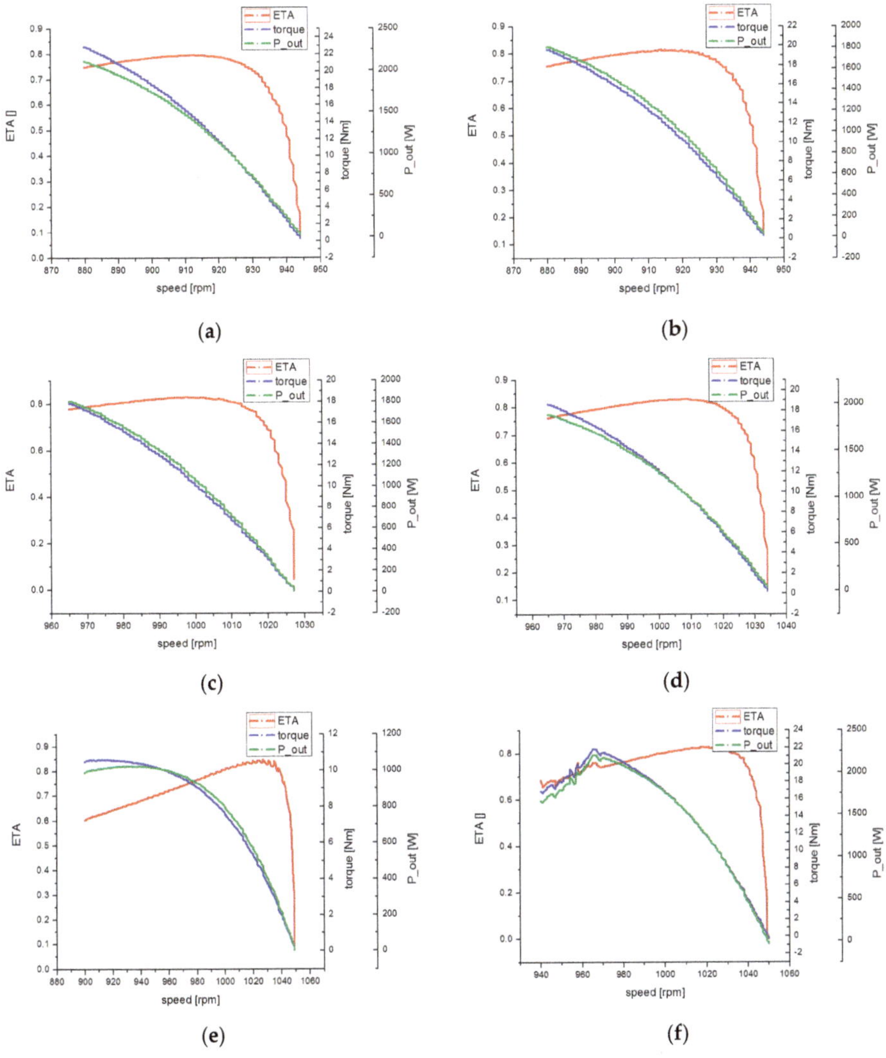

Figure 12. Graphs of motor characteristics according to voltage and frequency of outer rotor inductor with 10 poles: Voltage_frequency of (**a**–**f**): 32.3 V_126 Hz, 35 V_126 Hz, 33.2 V_137 Hz, 33 V_138 Hz, 23.9 V_140 Hz and 33.2V_140 Hz, respectively.

7. Conclusions

In this paper, the basic theory and design procedure of an outer rotor induction motor were described; moreover, a drive motor shape was designed using the equivalent circuit method and the finite element method. Based on the basic shape design, an optimized design was developed by adjusting the number of poles.

For effective application of an induction motor to an electric vehicle, a design yielding improved efficiency and power density, which are the most critical factors, is required. In this study, the design variables that affect the efficiency and power density of an outer rotor induction motor were classified. Furthermore, by analyzing the trends appearing through variable adjustments, the effects of each variable on the efficiency and output were determined.

An outer rotor induction motor has a large outer diameter and short axial length. Further, additional poles are known to be required, compared to those of a typical inner rotor induction motor. This is because the coil pitch is increased according to the employed distributed winding method. Moreover, the amount of available windings was reduced, owing to the larger outer diameter; as a result, there was a longer coil pitch and a reduced number of available windings. This aspect may have yielded the measurement results obtained in this study for the outer rotor induction motor with 6 poles. Superior results were obtained when the motor was fabricated with a higher number of poles. This is because the narrower pole spacing reduced the coil pitch, significantly increasing the usable range of the coil compared to the six-pole model; hence, the efficiency measured in the actual performance evaluation matched the analytical efficiency value. Use of a pancake-shaped winding in the motor, with a large outer diameter and a short axial length, appeared to be effective. This approach was also found to be more advantageous in terms of output density, because the maximum output can be increased with application of the driving voltage when the motor characteristics were obtained at a voltage lower than the driving voltage.

Thus, to achieve the maximum performance of an electric vehicle with an outer rotor induction motor, characteristics such as high torque of the motor parts in the low-speed range, high efficiency in the operating range, and a relatively small battery to allow low-voltage operation and to reduce the overall weight are required. To realize these characteristics, a pancake-shaped outer rotor induction motor, which is different than that of an inner rotor induction motor, was designed in this study. Furthermore, the important design variables were derived through shape optimization design and production of many prototypes. The motor was then optimized to meet the performance indicated by the electromagnetic field analysis. The performance evaluation device confirmed that the test results were similar to the designed results, and the validity of the design was verified. In the future, this outer rotor induction motor design is expected to be widely applied to electric vehicles and bicycles.

Author Contributions: Conceptualization, M.-H.H. and H.-S.L.; data curation, H.-S.L.; formal analysis, S.-H.Y.; methodology, M.-H.H.; supervision, S.-J.P.; validation, H.-R.C. and M.-H.H.; visualization, H.-S.L.; writing—original draft, M.-H.H.; writing—review and editing, S.-J.P.

Funding: This study has been conducted with the support of the Korea Institute of Energy Technology Evaluation and Planning as "Developing image big data based real time detection system for detecting defective module applied to solar power plant (KETEP 20183010014230).

Conflicts of Interest: The authors declare no conflict of interest.

References

1. Wang, S.; Zhao, X.; Yu, Q.; Zhang, S.; Shi, P.; Yu, M. Research on strategy of the stability control system of dual-motor drive electric vehicle. In Proceedings of the 2019 IEEE International Symposium on Circuits and Systems (ISCAS), Sapporo, Japan, 26–29 May 2019; pp. 1–5.
2. Akbaba, M. Energy conservation by using energy efficient electric motors. *Appl. Energy* **1999**, *64*, 149–158. [CrossRef]
3. Mathaba, T.; Xia, X. Optimal and energy efficient operation of conveyor belt systems with downhill conveyors. *Energy Effic.* **2017**, *10*, 405–417. [CrossRef]
4. Han, J.; Yun, S.J. An analysis of the electricity consumption reduction potential of electric motors in the South Korean manufacturing sector. *Energy Effic.* **2015**, *8*, 1035–1047. [CrossRef]
5. Marinakis, V.; Doukas, H.; Karakosta, C.; Psarras, J. An integrated system for buildings' energy-efficient automation: Application in the tertiary sector. *Appl. Energy* **2013**, *101*, 6–14. [CrossRef]
6. Chua, K.J.; Chou, S.K.; Yang, W.M.; Yan, J. Achieving better energy-efficient air conditioning—A review of technologies and strategies. *Appl. Energy* **2013**, *104*, 87–104. [CrossRef]
7. Sun, L.; Zhang, N. Design, implementation and characterization of a novel bidirectional energy conversion system on DC motor drive using super-capacitors. *Appl. Energy* **2015**, *153*, 101–111. [CrossRef]
8. Tahan, M.; Tsoutsanis, E.; Muhammad, M.; Karim, Z.A. Performance-based health monitoring, diagnostics and prognostics for condition-based maintenance of gas turbines: A review. *Appl. Energy* **2017**, *198*, 122–144. [CrossRef]

9. Sundaram, M.; Navaneethan, P. On the influence of stator slot shape on the energy conservation associated with the submersible induction motors. *Am. J. Appl. Sci.* **2011**, *8*, 393–399. [CrossRef]
10. Sujitjorn, S.; Areerak, K.L. Numerical approach to loss minimization in an induction motor. *Appl. Energy* **2004**, *79*, 87–96. [CrossRef]
11. Cunkas, M. Intelligent design of induction motors by multiobjective fuzzy genetic algorithm. *J. Intell. Manuf.* **2010**, *21*, 393–402. [CrossRef]
12. Kentli, F. A survey on design optimization studies of induction motors during the last decade. *J. Electr. Electron. Eng.* **2009**, *9*, 969–975.
13. Manoharan, S.; Devarajan, N.; Deivasahayam, S.M.; Ranganathan, G. Review on efficiency improvement in squirrel cage induction motor by using DCR technology. *J. Electr. Eng.* **2009**, *60*, 227–236.
14. Aho, T.; Nerg, J.; Pyrhönen, J. The effect of the number of rotor slits on the performance characteristics of medium-speed solid rotor induction motor. In Proceedings of the 3rd IET International Conference on Power Electronics, Machines and Drives (PEMD 2006), Dublin, Ireland, 4–6 April 2006; pp. 515–519.
15. Kumar, K.R.; Sakthibala, D.; Palaniswami, S. Efficiency optimization of induction motor drive using soft computing techniques. *Int. J. Comput. Appl.* **2010**, *3*, 6–12. [CrossRef]
16. Sakthivel, V.P.; Subramanian, S. On-site efficiency evaluation of three-phase induction motor based on particle swarm optimization. *Energy* **2011**, *36*, 1713–1720. [CrossRef]
17. Campos-Delgado, D.U.; Arce-Santana, E.R.; Espinoza-Trejo, D.R. Edge optimization for parameter identification of induction motors. *IET Electr. Power Appl.* **2011**, *5*, 668–675. [CrossRef]
18. Faiz, J.; Ebrahimpour, H.; Pillay, P. Influence of unbalanced voltage supply on efficiency of three phase squirrel cage induction motor and economic analysis. *Energy Convers. Manag.* **2006**, *47*, 289–302. [CrossRef]
19. Donolo, P.; Bossio, G.; De Angelo, C. Analysis of voltage unbalance effects on induction motors with open and closed slots. *Energy Convers. Manag.* **2011**, *52*, 2024–2030. [CrossRef]
20. Jabr, H.M.; Kar, N.C. Leakage flux saturation effects on the transient performance of wound-rotor induction motors. *Electr. Power Syst. Res.* **2008**, *78*, 1280–1289. [CrossRef]
21. Rasouli, M.; Westwick, D.T.; Rosehart, W.D. Reducing induction motor identified parameters using a nonlinear lasso method. *Electr. Power Syst. Res.* **2012**, *88*, 1–8. [CrossRef]
22. Kostov, I.; Spasov, V.; Rangelova, V. Application of genetic algorithms for determining the parameters of induction motors. *Tech. Gaz.* **2009**, *16*, 49–53.

energies

MDPI

Article

Optimal Design of PMSM Based on Automated Finite Element Analysis and Metamodeling

Yong-Min You

Department of Automotive Engineering, Honam University, Gwangju 62399, Korea; ym.you@honam.ac.kr; Tel.: +82-62-940-5499

Received: 30 October 2019; Accepted: 4 December 2019; Published: 9 December 2019

Abstract: To obtain accurate optimal design results in electric machines, the finite element analysis (FEA) technique should be used; however, it is time-consuming. In addition, when the design of experiments (DOE) is conducted in the optimal design process, mechanical design, analysis, and post process must be performed for each design point, which requires a significant amount of design cost and time. This study proposes an automated DOE procedure through linkage between an FEA program and optimal design program to perform DOE easily and accurately. Parametric modeling was developed for the FEA model for automation, the files required for automation were generated using the macro function, and the interface between the FEA and optimal design program was established. Shape optimization was performed on permanent magnet synchronous motors (PMSMs) for small electric vehicles to maximize torque while maintaining efficiency, torque ripple, and total harmonic distortion of the back EMF using the built-in automation program. Fifty FEAs were performed for the experimental points selected by optimal Latin hypercube design and their results were analyzed by screening. Eleven metamodels were created for each output variable using the DOE results and root mean squared error tests were conducted to evaluate the predictive performance of the metamodels. The optimization design based on metamodels was conducted using the hybrid metaheuristic algorithm to determine the global optimum. The optimum design results showed that the average torque was improved by 2.5% in comparison to the initial model, while satisfying all constraints. Finally, the optimal design results were verified by FEA. Consequently, it was found that the proposed optimal design method can be useful for improving the performance of PMSM as well as reducing design cost and time.

Keywords: automation; finite element analysis; PMSM; DOE; optimization; metamodeling

---

1. Introduction

The necessity of eco-friendly vehicles has been highlighted owing to environmental pollution and depletion of fossil fuels. Global electric car stocks are growing rapidly, crossing the 3 million vehicle threshold in 2017. The estimated demand for electric vehicles by 2030 is 100 to 140 million. The core of an electric vehicle is the electric powertrain, which consists of a traction motor, a reduction drive, an inverter, and a power delivery module. Permanent magnet synchronous motors (PMSMs) have been mainly used as a traction motor for electric vehicles because they have high efficiency and high output power density characteristics.

Several studies have been conducted on the PMSMs used in electric vehicles that require various characteristics such as torque, efficiency, and harmonic distortion (THD). Optimal design is essential to satisfy the various design requirements of PMSM at once. Optimal design is a method of finding the values of design variables to obtain an optimal solution within a range of constraints. The optimal design for PMSMs is created by combining design methods such as the analytical model, magnetic equivalent circuits (MEC) model, and finite element analysis (FEA) with optimal design algorithms [1–8]. First of all, there are studies on optimal design using the analytical model [1,2]. In [1], the optimal design of a

PMSM based on the magnetic field analytical model was determined. The objective function used in that study consisted of efficiency, electrical time constant, and mechanical time constant. The experimental results showed that the efficiency increased by 1%. To minimize torque ripple, a novel analytical solution of a PMSM was proposed [2]. The stator current was optimized considering magnetic saturation using an analytical expression. The following are studies on optimization using the MEC model [3,4]. In [3] it was reported that the MEC optimization method combined with an optimization algorithm can optimize the volume and energy loss of a PMSM. A novel MEC model of a PMSM to obtain the maximum efficiency, minimum weight, and price was developed [4]. K-means clustering algorithm was utilized to obtain the best solution out of the eight clusters. Finally, some research on optimization combined with FEA have been published [5–7]. The work in [5] performed multi-objective shape optimization of a PMSM based on FEA and particle swarm optimization algorithm. Five rotor topologies were compared, aimed at efficiency, flux-wakening rate, and price. The work in [6] proposed an optimization process of a PMSM to optimize the weight, output power, and suitability. It performed shape optimization of permanent magnets and rotor core using FEA with the fuzzy inference system strategy. Using a novel memetic algorithm, an optimal design based on FEA to minimize torque ripple in a PMSM was created [7]. In [8], multi-physics and multi-objective optimization of a PMSM based on FEA and analytical magnetic model were studied. Although the FEA optimization method combined with optimization algorithm has the highest accuracy, it has high computational cost [5].

There are two main ways to optimize design variables: To combine the optimal algorithm with design methods directly and combine the optimal algorithm with the metamodel from the results of design of experiments (DOE). Metamodel is a mathematical model that approximates the relationships between design variables and responses. DOE is an application of statistics aimed at designing experimental methods and analyzing the results to identify relationships between design variables and responses. First, directly connecting the optimal algorithm with the design methods can determine the best solution more clearly [9]. However, this method takes a long time to optimize and it is difficult to predict the design time. Additionally, if the formulation of the optimal design is wrong, it is difficult to find the best solution. In the case of optimization by creating metamodels using DOE results, it is possible to predict the optimal results by analyzing the sensitivity between design variables and target goals. In addition, the time taken for the optimization design is clear. However, the number of DOE and test points must be selected properly, and the metamodel must be made correctly. Meanwhile, non-automated DOE requires a lot of effort and time because mechanical modeling and analysis must be performed as many times as DOE. Although a large number of DOEs are required to achieve good optimal design results, it takes a significant amount of effort and time. The work in [10] reported optimization results using response surface methodology combined with metamodels from the DOE results. To produce DOE results, a total of 15 models were made and 15 FEAs were conducted. The study in [11] optimized a PMSM by combining an optimal algorithm and metamodel, i.e., the genetic algorithm and the Kriging model, based on DOE. In that study, to obtain the DOE results, several models had to be designed and FEAs were required.

The novelty of this distinguishes it from previous studies for the following reasons: First, optimal design can be easily processed based on a novel automated DOE procedure based on FEA, so it can be done faster and more accurately. In general, DOE by FEA consists of modeling process using CAD tools, analysis condition setting process for FEA, FEA process, and post process for extracting and organizing results. To obtain a reliable optimal design result, a large number of DOE have to be carried out. However, the conventional method of manually performing the process was complicated and time consuming, and thus the number of DOE was limited [10–18]. However, using the automated DOE process proposed in this study, not only can the DOE be easier but also the number of DOE can be dramatically increased, resulting in high reliability of the optimal design result. The proposed automated design method is expected to reduce the design cost and time. Moreover, it can be used to find the optimal solution for various design problems as well as PMSMs. In addition, since the proposed procedure is based on commercial tools, it has a ripple effect that can easily apply optimal design in academia and industry.

Most of the previous studies have been applied to optimal design using metamodel generated in one way. There have been a lot of optimizations recently using a single metamodeling technique such as Kriging and response surface method [10–18]. However, since a suitable metamodel is different according to each design problem and condition, it is necessary to select the best metamodel through accuracy evaluation after generating several metamodels. This is because the accuracy of the metamodel must be high to obtain good optimal design results. In this study, metamodels of objective functions and constraints are generated in 11 ways, and the most accurate metamodels are selected through the root mean squared error (RMSE) test, respectively.

In this study, shape optimization is performed for a PMSM to maximize the torque while maintaining efficiency, torque ripple and THD in the back electromotive force (EMF). First, the design target specification of a PMSM for small electric vehicles is established, and the characteristics of the initial model are analyzed using FEA. To improve the accuracy of the design results, DOE is performed using FEA. After the creation of metamodels using the DOE results, the optimal values are obtained by the optimal algorithm. The optimal Latin hypercube design (OLHD) technique [19] is applied for the DOE, and the appropriate DOE number and test point number are selected to produce accurate metamodels. To perform DOE easily and accurately, this study proposes an automated DOE procedure through linkage between an FEA program and an optimal design program. Using the DOE results, the relationship between the design and output variables are analyzed by screening. To generate an accurate metamodel, the RMSE tests are performed on eleven metamodels for each output variable, and the best metamodels are selected for each output variable. Optimization based on metamodels is performed, and the global optimization algorithm hybrid metaheuristic algorithm (HMA) [20] is utilized as the optimal algorithm. The overall process of this study is represented in a flowchart, as shown in Figure 1.

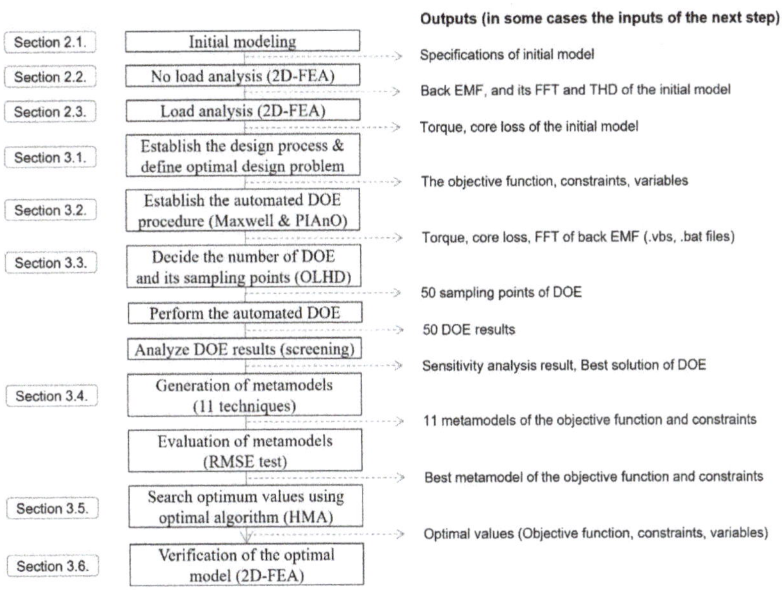

Figure 1. Flowchart of the overall research procedure.

2. Finite Element Analysis

The target specifications are determined by referring to the Renault's Twizy with a torque of 57 N·m and an output power of 13 kW at 2100 rpm. In this study, a PMSM is selected as the design model, and the target output power is 15 kW which should satisfy 60 N·m at 2387 rpm.

2.1. Initial Model

Figure 2 and Table 1 show the structure and specifications of a 15 kW PMSM for a small electric vehicle, respectively. The PMSM has 8 poles, 36 slots, and distributed winding.

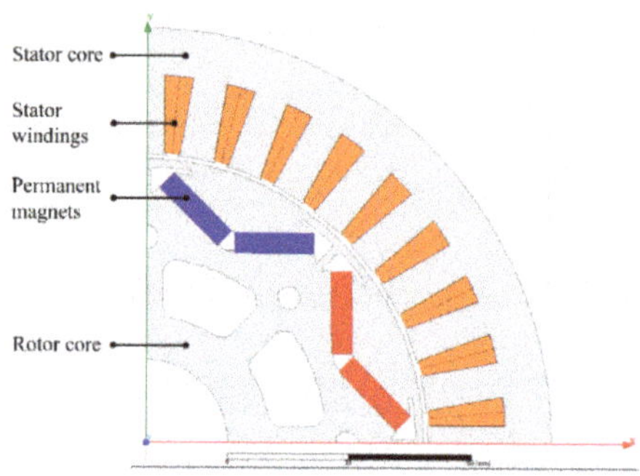

Figure 2. Structure of initial permanent magnet synchronous motor (PMSM) model (1/4 model).

Table 1. Specifications of analysis model.

Items		Unit	Value
Required specification	Max. output power	kW	15
	Max. torque	N·m	60
	Rated speed	rpm	2387
Electrical	Continuous current	A_{rms}	146
	Current phase angle	°	25
	Winding specification	-	ø 0.9, 11 turns (3 strand)
Mechanical	No. of poles and slots	ea	8/36
	Outer diameter of stator	mm	202
	Inner diameter of stator	mm	140
	Outer diameter of rotor	mm	138
	Inner diameter of rotor	mm	40
	Air-gap	mm	1
	Lamination	mm	45
Thermal	Reference temperature	°C	20

The electromagnetic, mechanical and thermal properties of the 35PN210 core material are shown in Table 2. Core loss is the sum of hysteresis loss, eddy-current loss, and excess loss, and is calculated by Equation (1). The core loss varies with frequency, but the analysis is based on 60 Hz.

$$P_c = K_h f (B_m)^2 + K_c (f B_m)^2 + K_e (f B_m)^{1.5} \quad (1)$$

where P_c is the core loss, K_h is the hysteresis loss coefficient, K_c is the eddy current loss coefficient, K_e is the excess loss coefficient, f is the frequency, and B_m is the amplitude of the alternating flux component.

The electromagnetic, mechanical and thermal properties of permanent magnets are shown in Table 3. V-shaped N38UH grade NdFeB are applied to concentrate the magnetic flux. The magnetic flux density and coercivity decreased with increasing temperature, but the analysis is conducted at 20 °C.

Table 2. Properties of electrical steel.

Items		Unit	Value
Grade (Manufacturer)		-	35PN210 (POSCO)
Electromagnetic	Anisotropy type	-	Isotropic
	Flux density at 2.5 kA/m	T	1.56
	Flux density at 5 kA/m	T	1.65
	Conductivity	S/m	1,694,915
	Frequency	Hz	60
	Hysteresis loss coefficient	w/m^3	85.0512
	Eddy-current loss coefficient	w/m^3	0.34153
	excess loss coefficient	w/m^3	4.94904
Mechanical	Thickness	mm	0.35
	Mass density	kg/m^3	7600
Thermal	Reference temperature	°C	20

Table 3. Properties of permanent magnets.

	Items	Unit	Value
	Grade	-	N38UH
Electromagnetic	Magnetizing direction	-	Parallel
	Permeability type	-	Anisotropic
	Relative permeability	-	1.05
	Residual induction	T	1.23
	Coercivity	kA/m	−932.193
	Bulk conductivity	S/m	625,000
Mechanical	Thickness	mm	5
	Length	mm	20
	Mass density	kg/m^3	7650
Thermal	Reference temperature	°C	20
	Reversible temp. coefficient of induction	%/°C	−0.12
	Reversible temp. coefficient of coercivity	%/°C	−0.465

2.2. No Load Analysis

Characteristic analysis of the initial model is performed by FEA under the no load condition without current excitation. When the rotor of the initial model rotates at the rated speed, back EMF is induced in the stator winding. Since the back EMF simulation is performed while rotating the rotor under the no load condition, the equivalent circuit when the PMSM operates as a generator should be considered, as shown in Figure 3. The voltage equation of the equivalent circuit is shown in Equation (2). However, since no current flows in the armature winding under the no load condition, the terminal voltage and the no load EMF are the same. The back EMF of phase A can be obtained by Equation (3), and the analysis result by FEA is illustrated in Figure 4a.

$$\dot{V} = \dot{E_0} - (j(X_a + X_l) + R_a)\dot{I_a}\ [\mathrm{V}] \tag{2}$$

where $\dot{V}$ is the terminal voltage, $\dot{E_0}$ is the no load EMF, X_a is the armature reation reactance, X_l is the leakage reactance, R_a is the armature resistance and I_a. is the load current.

$$E_a = N\varnothing\omega cos\omega t\ [\mathrm{V}] \tag{3}$$

where E_a. is the back EMF of phase A, N is the number of the amarture turns, $\varnothing$ is the magnetic flux and ω is the electrical angular velocity.

THD is an important factor in the electrical equipment and power systems. THD can be obtained by adding the harmonic components to the fundamental wave components of voltage or current as shown in Equation (4) [21]. A higher THD increases the core loss in electric machines, which can reduce the efficiency and generate excessive heat. The harmonic analysis result of the back EMF waveform is shown in Figure 4b, and the THD of the back EMF calculated by Equation (4) is 3.52%.

$$V_{\mathrm{THD}} = \frac{\sqrt{{V_2}^2 + {V_3}^2 + {V_4}^2 + \cdots + {V_n}^2}}{V_1} \times 100 \quad (4)$$

where V_{THD} is the THD of the back EMF, V_1 is the RMS voltage of the fundamental frequency and V_n is the RMS voltage of nth harmonic.

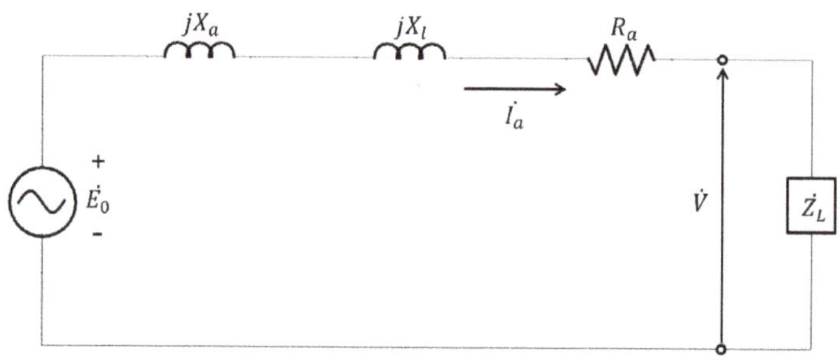

Figure 3. The equivalent circuit of PMSM (generator mode).

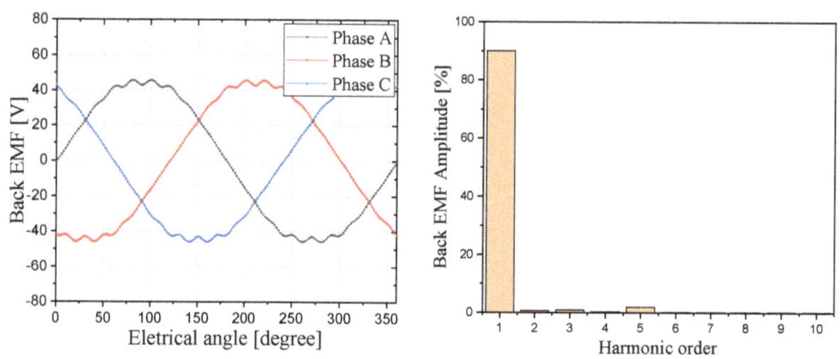

(**a**) (**b**)

Figure 4. The back EMF of initial model under the no load condition: (**a**) Waveform; (**b**) Harmonic.

2.3. Load Analysis

Target specifications are 15 kW and 60 N·m at 2387 rpm as shown in Table 1. To perform load analysis, a current condition satisfying the target torque of 60 N·m at 2387 rpm should be found. Through static torque analysis, the current condition is determined as the RMS value of 146 A and phase angle of 25°. As shown in Figure 5a, the average torque is 59.95 N·m and the torque ripple is 5.09% of torque. The core loss is interpreted as shown in Figure 5b, and total losses are the sum of the core loss and copper loss. The output power of the motor is calculated as the product of torque and angular velocity, and the efficiency can be calculated from the output power and total losses of the motor. The efficiency of the initial model is calculated to be 91.42%.

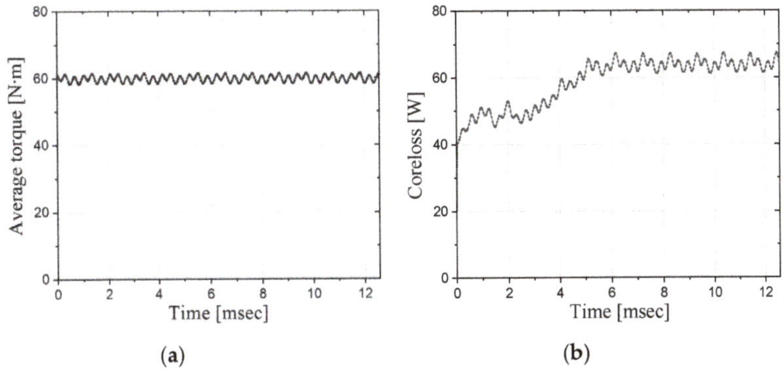

Figure 5. Torque and core loss of the initial model under the load condition: (**a**) Torque; (**b**) Core loss.

3. Design Optimization

3.1. Design Process

The design optimization process for maximizing the average torque, while maintaining THD of the back EMF, efficiency, and torque ripple, is shown in Figure 6. The objective function, constraints, thermal condition, and design variables are established, as described in Equations (5)–(9) and Figure 7. To improve the average torque, the average torque is set as both an objective function and a constraint, with a goal of 2% improvement over the initial model. THD of the back EMF and the efficiency are set as constraints to maintain the same level as the initial model. The torque ripple is set below 10%, which is an acceptable level as a traction motor for electric vehicles [22]. To improve the accuracy of the design results, DOE is performed using FEA. Because DOE using FEA requires a significant amount of time and effort, interworking is conducted between the FEA and optimal design programs, which are ANSYS Maxwell and PIAnO, respectively, to automatically perform DOE. When the analysis and extraction of results for one experiment is finished, the values of the design variables are automatically changed to perform the FEA at the next DOE point. From the DOE results, sensitivity analysis between design variables and output variables is conducted using screening, and each metamodel for output variables is generated. RMSE test was conducted to evaluate the predictive performance of the metamodels, and the best metamodel is selected for each output variable. Based on the selected metamodels, the optimal values are obtained using the HMA.

Objective function

$$\text{Maximize the average torque} \tag{5}$$

Mechanical Constraints

$$\begin{gathered}\text{Average torque} \geq 61.152\ \text{N·m}\\ \text{Torque ripple} \leq 10\ \%\end{gathered} \tag{6}$$

Electrical Constraints

$$\begin{gathered}\text{THD of the back EMF} \leq 3.414\ \%\\ \text{Efficiency} \geq 91.42\ \%\end{gathered} \tag{7}$$

Thermal condition

$$\text{Reference temperature} = 20\ \text{C}^{\circ} \tag{8}$$

Design variables (based on the value of the initial model)

$$\begin{gathered} -4\text{ mm} \leq \text{DV1 (Barrier length)} \leq 10\text{ mm} \\ -1.0\text{ mm} \leq \text{DV2 (Rib thickness)} \leq 0.5\text{ mm} \\ -1.0\text{ mm} \leq \text{DV3 (Teeth width)} \leq 0\text{ mm} \\ 0\text{ mm} \leq \text{DV4 (Teeth thickness)} \leq 1.0\text{ mm} \\ 0\text{ mm} \leq \text{DV5 (Barrier gap)} \leq 2.0\text{ mm} \end{gathered} \quad (9)$$

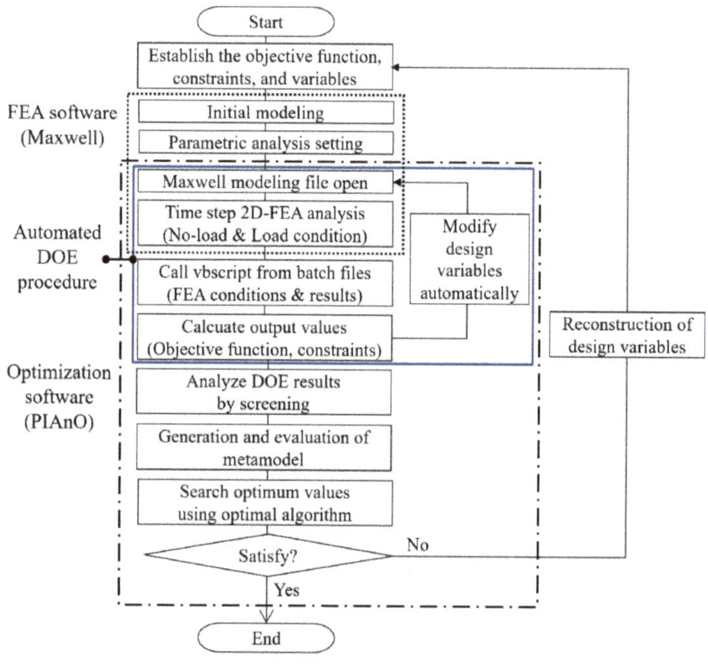

Figure 6. Optimization design process.

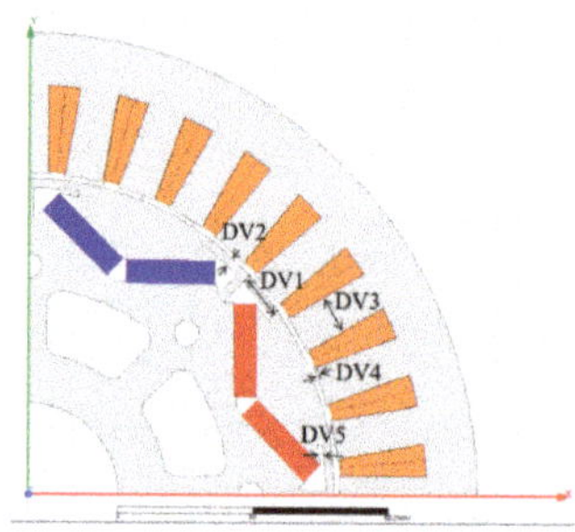

Figure 7. Shape design variables.

The mechanical constraints of the barrier length are set from a range that facilitates the flow of magnetic flux to a range that inhibits the flow of magnetic flux very much, as shown in Figure 8a. Rib thickness is set to be at least 1 mm in consideration of workability at manufacture and mechanical rigidity at high speed, as shown in Figure 8b. Figure 8c shows the mechanical constraints of the teeth width, and the range is set so that the slot is smaller than the initial model and maintains the proper fill factor. The range of teeth thickness is set up to reduce the saturation of the magnetic flux at the tip of the teeth and maintain the proper fill factor, as shown in Figure 8d. Barrier gap size affects the formation of magnetic flux and the motor performance since the permanent magnet position also changes. Therefore, its mechanical constraints are set as shown in Figure 8e.

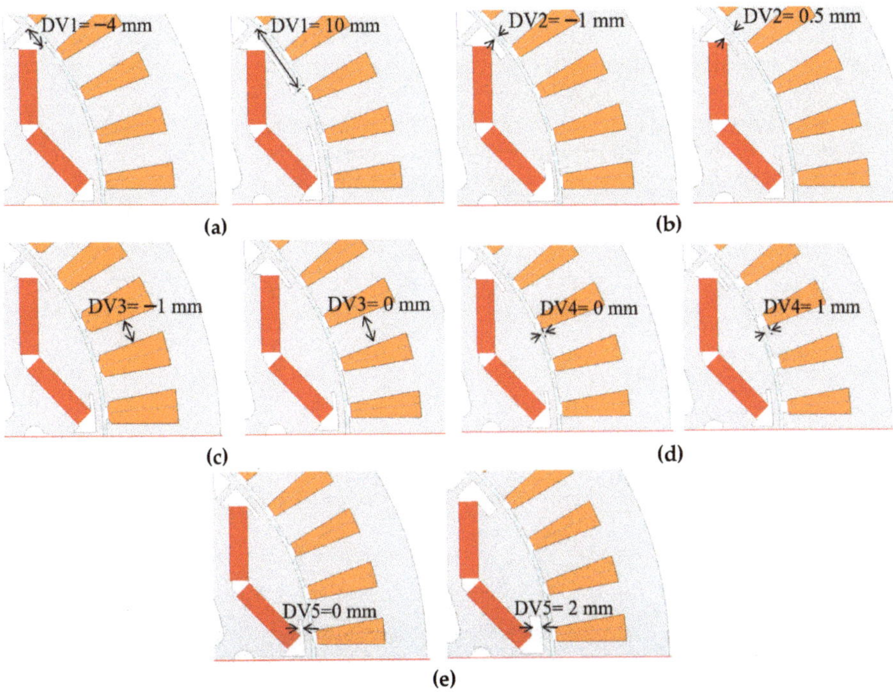

Figure 8. Mechanical constraints of design variables: (**a**) DV1 (barrier length); (**b**) DV2 (rib thickness); (**c**) DV3 (teeth width); (**d**) DV4 (teeth thickness); (**e**) DV5 (barrier gap). Notes: The values of the design variables are relative to the values of the initial model.

3.2. Automated DOE Procedure

To perform DOE, the shape of design variable should be changed. However, when DOE is processed manually, the shape of each model is drawn using the CAD tool. Next, the designed shape should be imported into the FEA program and the FEA should be performed for each model. After FEA, the post process is required to calculate the desired result. As manual DOE requires a lot of effort and time, this study suggests the automation of the DOE process. First, the Maxwell's parametric sweep setup function is used to change the shape of design variables without using the CAD tool. The use of this function can change the shape of the FEA model by inputting numerical values in the Maxwell program. Next, Maxwell's Macro function is used to perform DOE using PIAnO, an optimal design program. As the design variables change, the vbscript and batch files are created to automatically change the shape of the FEA model. In addition, vbscript and batch files are generated for FEA under the no load and load conditions for each experiment. Vbscript and batch files also are generated to output and quantify the torque, core loss and FFT analysis results of the back EMF obtained through FEA. The files created through the Macro function are shown in Figure 9. Next, the interface configuration between Maxwell and PIAnO for automation is shown in Figure 10. To process DOE in PIAnO, the files created in Maxwell are imported as shown in Figure 10a, and the script for calculating the output variables is shown in Figure 10b.

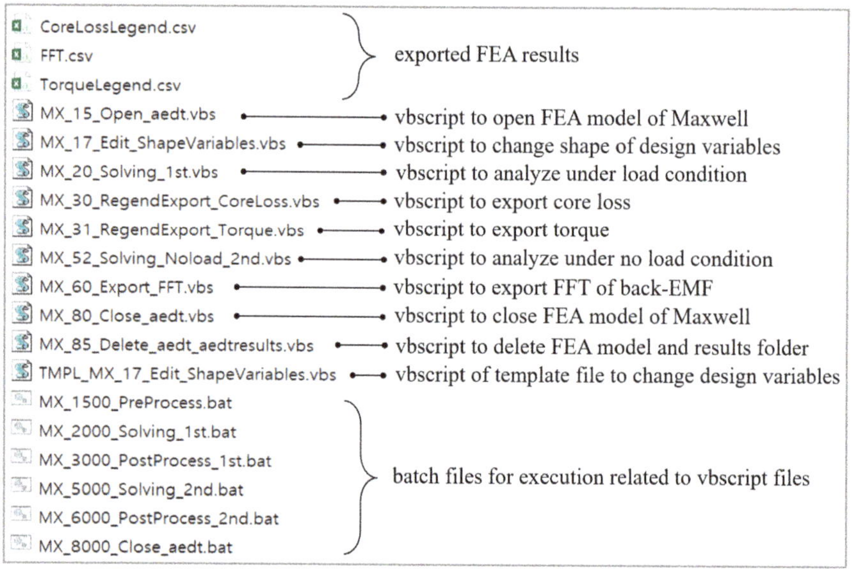

Figure 9. Vbscript, batch, and output files using the Macro function of Maxwell.

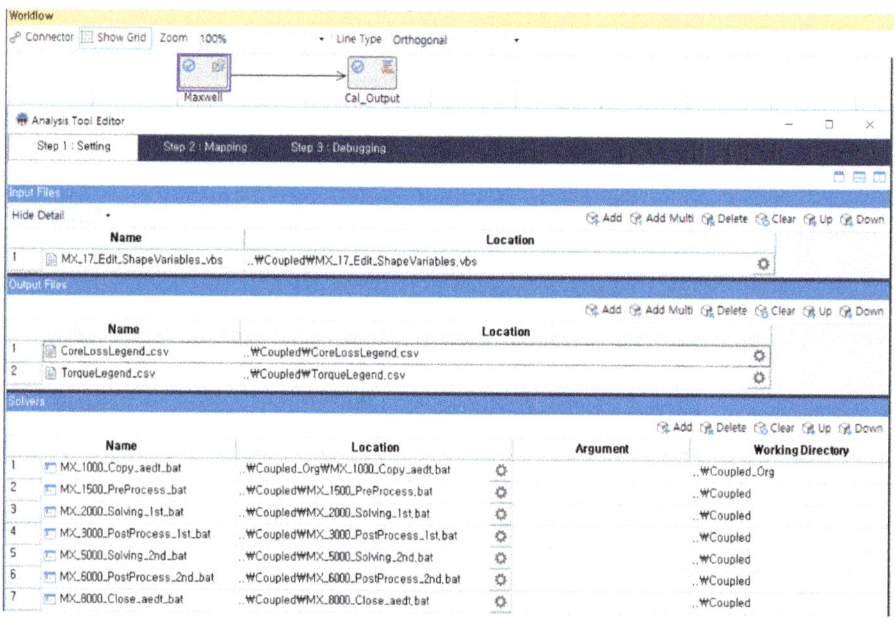

(a)

Figure 10. *Cont.*

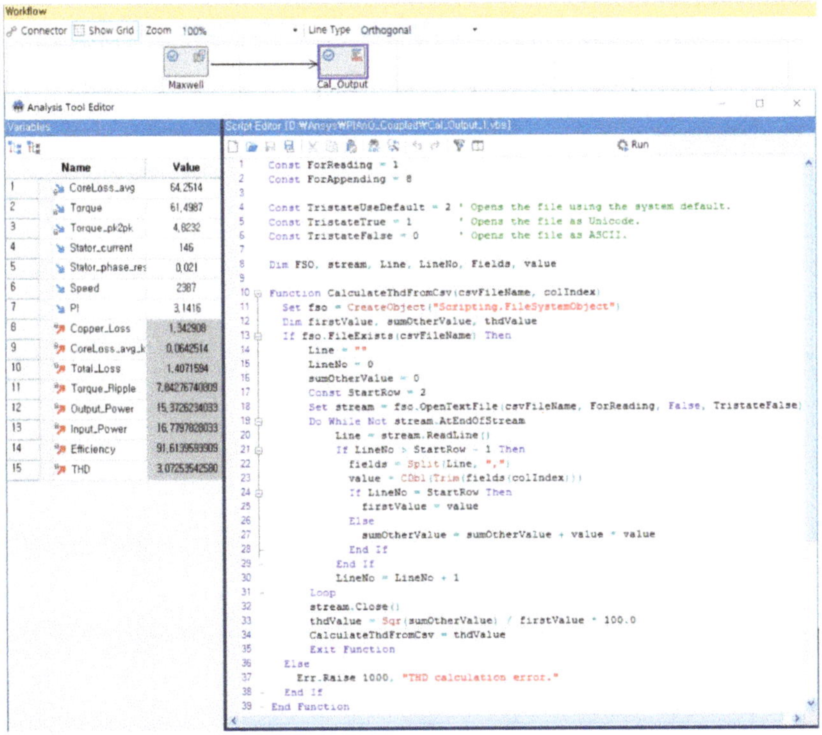

(b)

Figure 10. Interface configuration with Maxwell using PIAnO: (**a**) Interface setting; (**b**) Script for calculating output variables.

3.3. Design of Experiment

The number of experiments and the number of test points are determined in three steps [23]. First, the number of experiments should be selected according to the number of design variables. When the number of design variables is ten or less, the number of experimental points is determined by Equation (10),

$$\text{nEXP} > 1.5 \times \text{nSAT} = 1.5 \times \frac{(\text{nDV} + 1) \times (\text{nDV} + 2)}{2} \tag{10}$$

where nEXP is the number of DOE, nSAT is the number of saturation, and nDV is the number of design variables.

Next, the number of DOEs that can be used as test points for evaluating the accuracy of the metamodel should be secured by Equation (11):

$$\text{nEXP} > \min\left[\frac{(\text{nDV} + 1) \times (\text{nDV} + 2)}{2},\ 10 \times \text{nDV}\right] + (5 \times \text{nDV}) \tag{11}$$

Because five design variables are used in this study, the number of DOE should be more than 46 by Equations (10) and (11). Therefore, the number of DOE is determined to be 50, which is a multiple of the design variables. If fifty experiments are manually operated, a significant amount of effort and time would be required. However, in this study, automation is implemented so that DOE can be easily developed and design cost can be reduced. Finally, the number of test points for evaluating the accuracy of the metamodel is determined to be five by Equation (12),

$$\text{nEXP\_ts} > \min[\text{nEXP} \times 10\%,\ 10 \times \text{nDV}] \quad (12)$$

where nEXP_ts is the number of test points.

The OLHD technique is applied to determine the sampling point of the DOE. OLHD is a type of DACE sampling technique developed for computational experiments. In the computer experiment, because there are no random errors, only the bias error should be considered and the test point should be spread evenly inside the design area. OLHD improves the space-filling property by using the optimum conditions and spreads the test points evenly; thus, even if there are several test points, they can be selected efficiently. DOE for fifty test points selected by OLHD is easily performed using the automated program. Sensitivity analysis is conducted to analyze the correlation between the design variables and design results. Figure 11 shows that the barrier length has the highest impact on the output variables among the five design variables. However, as shown in Table 4, even the most optimal experimental point among the 50 experiments does not satisfy the constraints. Therefore, metamodeling based on DOE results is conducted.

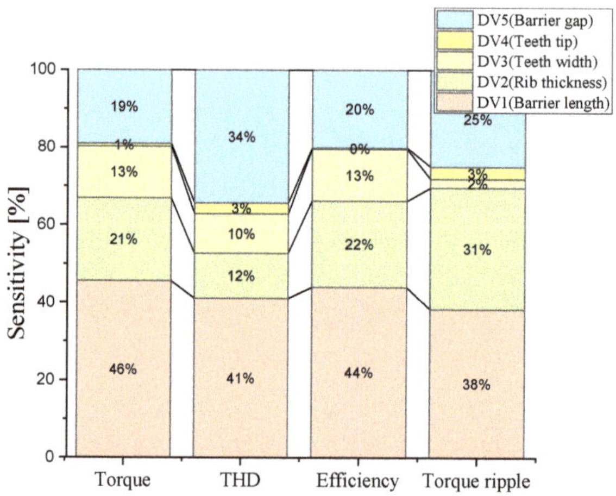

Figure 11. Sensitivity analysis using screening.

Table 4. DOE results.

Items		Unit	Initial Model	Best Solution of DOE
Design variables	Barrier length	mm	0	−0.642
	Rib thickness	mm	0	0.37
	Teeth width	mm	0	−0.34
	Teeth thickness	mm	0	0.337
	Barrier gap	mm	0	0.76
Design results	Average Torque	N·m	59.95	59.9
	THD of the back EMF	%	3.414	3.245
	Efficiency	%	91.42	91.41
	Torque ripple	%	5.086	4.131

3.4. Metamodeling

Five test points are selected to evaluate the metamodel, and eleven metamodels are generated for each output variable. The metamodel can be classified into a regression model and an interpolation model. The regression model, i.e., polynomial regression (PR), radial basis function regression (RBFr), ensemble of decision trees (EDT), and multi-layer perceptron (MLP), smoothens the noise data because they do not pass through the test points exactly. Therefore, this model is useful for real experiments with random errors. PR allows free choice of regression terms [24]. RBF is easy to design and generalize, and has strong tolerance to input noise [25]. EDT is advantageous for expressing nonlinearity in large amounts of data. MLP is type of deep learning algorithm and has the advantage of being able to represent the nonlinear relationships between input and output variables [26]. In contrast, the interpolation model, i.e., Kriging and radial basis function interpolation (RBFi), is well suited for function approximation using analytical results without random errors because it passes through the test points exactly. The estimated equation of the Kriging model was defined to eliminate bias and thereby minimize error variance [27]. Thus, a numerically robust model is provided. RBFi was first popularized in the machine learning community and has been used in computer graphics [28].

The accuracy of the metamodel is a very important factor in the optimal design using metamodel [12]. This is because the predictive performance of the metamodel affects the reliability of the optimal design. Most of the existing studies have been metamodeled by a single method such as Kriging and RSM, and the accuracy evaluation has not been performed [10–18]. In this study, however, metamodels for the objective function and constraints are generated in 11 ways provided by PIAnO, and the best metamodels are selected, respectively, by comparing the RMSE test results to evaluate the metamodel accuracy. The predictive performance of the metamodel is evaluated by the RMSE test and is calculated by Equation (13) [23],

$$\mathrm{RMSE} = \sqrt{\frac{1}{\mathrm{nEXP\_ts}} \sum_{i=1}^{\mathrm{nEXP\_ts}} \left[y(X_i) - \hat{y}(X_i) \right]^2} \tag{13}$$

where $y(X_i)$ is the value of the real function and $\hat{y}(X_i)$ is the value of the metamodel.

Through the RMSE test, the predictive performances of the metamodels are evaluated for the output variables. The RMSE test showed the best predictive performance of RBFr as a metamodel of the average torque as shown in Table 5. Similarly, the RMSE tests are conducted on the metamodel for efficiency, torque ripple, and THD of the back EMF. Based on the test results, the metamodels with the best predictive performance for each output variable are selected for use in the optimal design, as shown in Table 6.

Table 5. RMSE test results of metamodels for the objective function.

Rank	Metamodel	RMSE Test Value
1	RBFr	0.34648496
2	RBFi	0.38436657
3	PR (Backward stepwise regression)	0.40021405
4	PR (Simple cubic model)	0.44883739
5	PR (Full quadratic model)	0.50425469
6	PR (Forward stepwise regression)	0.68032589
7	PR (Linear model)	1.15518494
8	PR (Simple quadratic model)	1.21540869
9	Kriging	1.30829623
10	EDT	1.32713329
11	MLP	1.38523885

Table 6. Selected metamodels of the output variable by RMSE test.

Output Variable	Metamodel	RMSE Test Value
Average torque	RBFr	0.346485
THD of the back EMF	Kriging	1.276249
Efficiency	MLP	0.025028
Torque ripple	Kriging	0.228137

3.5. Design Optimization Based on Metamodel

The HMA, a global optimization algorithm, is used for the optimal design based on the metamodel. The HMA was proposed in 2016 by Park [20]. HMA can determine the global optimum faster than other global optimizers owing to the combined advantages of improved constrained differential evolution and modified cuckoo search.

The optimum design results predicted from the metamodel based HMA are shown in Table 7 and verified through FEA. The predicted results showed that the average torque, THD of the back EMF, efficiency and torque ripple results are similar to the FEA results. Therefore, the automated DOE procedure and the generation and evaluation of the metamodel were verified. The average torque of the optimal model was 2.5% better than the initial model, and the torque ripple increased slightly, as shown in Figure 12. THD of the back EMF and efficiency set by the constraints were slightly improved. Although the torque ripple of the optimal model is 7.822%, it is very acceptable as a traction motor for electric vehicles [22].

Table 7. Optimization results.

Items		Unit	Initial Model (FEA)	Optimal Model (Predicted)	Optimal Model (FEA)
Design variables	Barrier length	mm	0	−0.356	−0.356
	Rib thickness	mm	0	−0.893	−0.893
	Teeth width	mm	0	−0.236	−0.236
	Teeth thickness	mm	0	0.269	0.269
	Barrier gap	mm	0	1.388	1.388
Design results	Average torque	N·m	59.95	61.03	61.43
	THD of the back EMF	%	3.414	3.424	3.065
	Efficiency	%	91.42	91.57	91.61
	Torque ripple	%	5.086	7.258	7.822

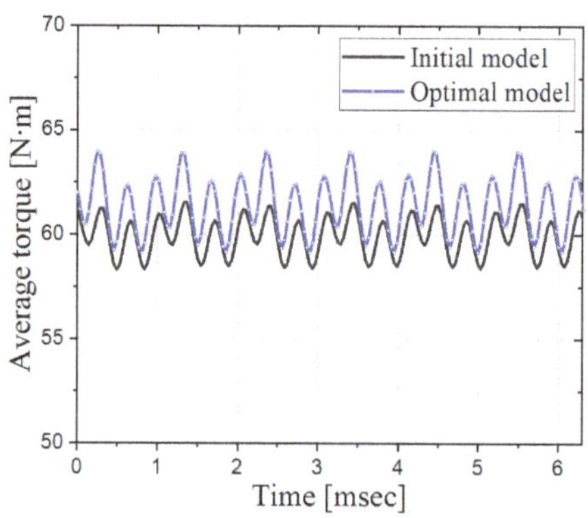

Figure 12. Torque waveforms.

3.6. Consideration of Optimal Design Results

Figures 13 and 14 show the flux distributfion and flux density of the initial and optimal models, respectively. In comparison to the initial model, the rib thickness and barrier length of the optimal model were reduced, and the barrier gap was increased. As the rib thickness decreased, the unnecessary flux flow between the north pole and south pole through the rotor rib was reduced. In addition, the flux flow was smoothly improved owing to the reduction in barrier length, and consequently, more flux passed through the stator core. The improvements in the magnetic flux flow and the change in reluctance can be considered to be the cause of the increase in the back EMF and torque [29]. Owing to the improvements of in the flux flow, the back EMF of the optimum model was 35.0 V, which showed an improvement of 7.4% in comparison to 32.6 V of the initial model, as shown in Figure 15. In addition, owing to the sinusoidal improvement in the waveform of the back EMF of the optimal model, the THD was slightly improved from 3.414% to 3.065%.

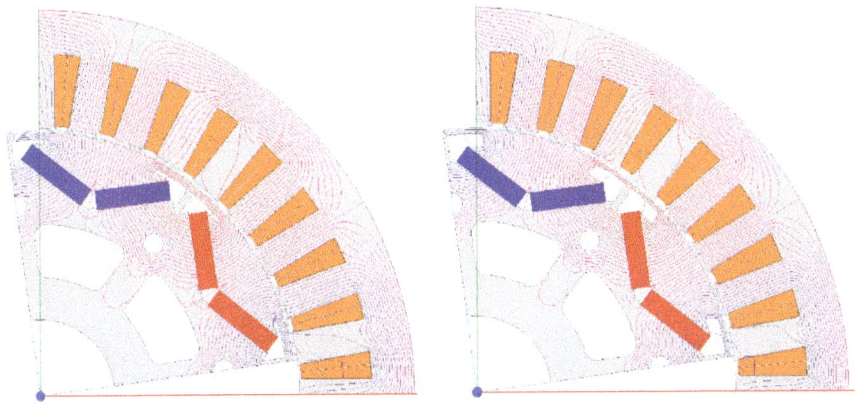

(a) (b)

Figure 13. Flux distribution under no load condition: (**a**) Initial model; (**b**) Optimal model.

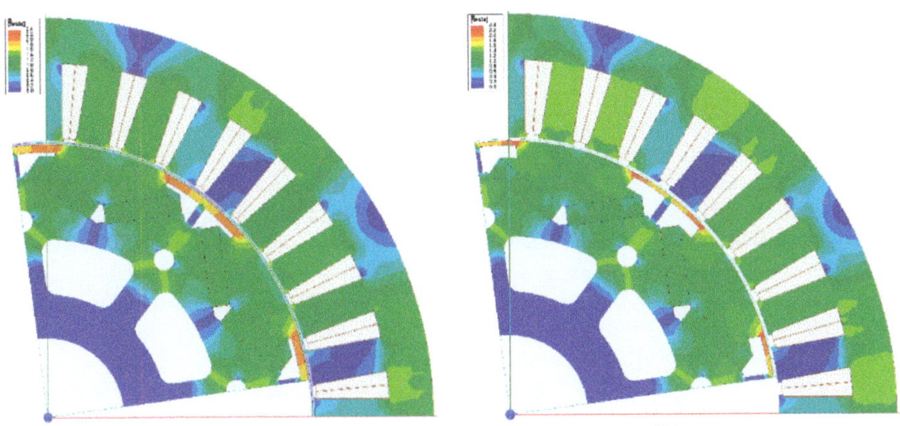

(a) (b)

Figure 14. Flux density under no load condition: (**a**) Initial model; (**b**) Optimal model.

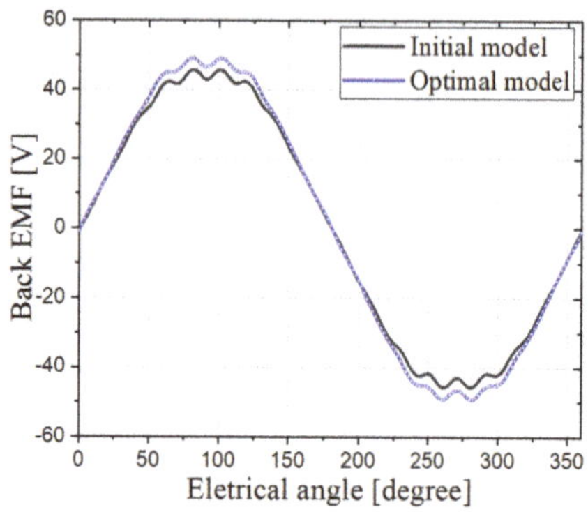

Figure 15. Back EMF waveforms.

4. Conclusions

This paper presented shape optimization of a PMSM for small electric vehicles to maximize torque while maintaining efficiency, torque ripple and THD of the back EMF. To improve the accuracy of the optimal design results, DOE was performed using FEA. This study proposed an automated DOE procedure through linkage between an FEA and optimal design programs to perform DOE easily and accurately. Parametric modeling was performed for the FEA model to change the shape variables automatically, and automation-related files were created using Maxwell's Macro function. In addition, an interface was established to link the FEA program with PIAnO, an optimal design program. Using the built-in automation program, 50 FEAs for the experimental points selected by OLHD were easily performed. From the DOE results, the relationship between the design and output variables was analyzed by screening. Among the five design variables, the barrier length was found to have the greatest effect on the output variables. Eleven metamodels were created for each output variable and RMSE test was conducted to evaluate the predictive performance of the metamodels. Consequently, the metamodels with the best predictive performance for each output variable were selected. Finally, the optimization design based on the metamodel was determined using the HMA to find the global optimum. The objective average torque improved by 2.5% over the initial model while satisfying all the constraints. The optimal design results were finally verified by FEA.

The proposed automated design method is expected to reduce design cost and time. Moreover, it can be used to find the optimal solution for various design problems as well as PMSMs. By following the procedure given below, the proposed optimal design method can be applied to any type of motor without any special constraints. First, in order to change the shape of the optimum design variable automatically, the dimension of the optimal design variable should be set using Maxwell's parametric sweep setup function. Next, determine the values that you want to extract from Maxwell and create vbscript and batch files to extract them. Finally, an interface setting must be performed to accommodate Maxwell's output values in an optimization program called PIAnO.

Optimization of multi-physics systems by simulation takes significant computing time for each simulation run, and its process depends on numerous runs, making it difficult and expensive [30]. However, using the automated DOE procedure suggested in this study can reduce design cost and time, so I think multi-physics analysis is possible in the near future. In the next project, I will consider multi-physics analysis that takes into account the mechanical and thermal properties.

Funding: This study was supported by research fund from Honam University, 2017 and the National Research Foundation of Korea (NRF). Grant funded by the Korea government (MSIT). (No. NRF-2018R1C1B5046117).

Acknowledgments: The author express gratitude to PIDOTECH and FRONTIS for their technical support.

Conflicts of Interest: The authors declare no conflict of interest.

Abbreviations

Acronym	Descriptor
DOE	Design of Experiments
EDT	Ensemble of Decision Trees
FEA	Finite Element Analysis
HMA	Hybrid Metaheuristic Algorithm
MEC	Magnetic Equivalent Circuits
MLP	Multi-layer Perceptron
OLHD	Optimal Latin Hypercube Design
PMSM	Permanent Magnet Synchronous Motor
PR	Polynomial Regression
RBFi	Radial Basis Function Interpolation
RBFr	Radial Basis Function Regression
RMSE	Root Mean Squared Error
THD	Total Harmonic Distortion

References

1. Guo, H.; Tian, W.; Xaiofeng, D. Multi-objective Optimal Design of Permanent Magnet Synchronous Motor for High Efficiency and High Dynamic Performance. *IEEE Access* **2018**, *6*, 23568–23581.
2. Feng, G.; Lai, C.; Kar, N.C. An Analytical Solution to Optimal Stator Current Design for PMSM Torque Ripple Minimization with Minimal Machine Losses. *IEEE Trans. Ind. Electron.* **2017**, *64*, 7655–7665. [CrossRef]
3. Dang, L.; Bernard, N.; Bracikowski, N.; Berthiau, G. Design Optimization with Flux Weakening of High-Speed PMSM for Electrical Vehicle Considering the Driving Cycle. *IEEE Trans. Ind. Electron.* **2017**, *64*, 9834–9843. [CrossRef]
4. Ilka, R.; Alinejad-Beromi, Y.; Yaghobi, H. Techno-economic Design Optimisation of an Interior Permanent-Magnet Synchronous Motor by the Multi-Objective Approach. *IET Electr. Power Appl.* **2018**, *12*, 972–978. [CrossRef]
5. Song, T.; Zhang, Z.; Liu, H.; Hu, W. Multi-objective Optimisation Design and Performance Comparison of Permanent Magnet Synchronous Motor for EVs based on FEA. *IET Electr. Power Appl.* **2019**, *13*, 1157–1166. [CrossRef]
6. Krasopoulos, C.T.; Beniakar, M.E.; Kladas, A.G. Multicriteria PM Motor Design based on ANFIS Evaluation of EV Driving Cycle Efficiency. *IEEE Trans. Electr.* **2018**, *4*, 525–535. [CrossRef]
7. Lee, J.H.; Kim, J.W.; Song, J.Y.; Kim, Y.J.; Jung, S.Y. A Novel Memetic Algorithm Using Modified Particle Swarm Optimization and Mesh Adaptive Direct Search for PMSM Design. *IEEE Trans. Magn.* **2016**, *52*, 7001604. [CrossRef]
8. Zhao, W.; Wang, X.; Gerada, C.; Zhang, H.; Liu, C.; Wang, Y. Multi-Physics and Multi-Objective Optimization of a High Speed PMSM for High Performance Application. *IEEE Trans. Magn.* **2018**, *54*, 8106405. [CrossRef]
9. Kim, S.W.; Kang, K.B.; Yoon, K.C.; Choi, D.H. Design Optimization of an Angular Contact Ball Bearing for the Main Shaft of a Grinder. *Mech. Mach. Theory* **2016**, *104*, 278–302. [CrossRef]
10. Chung, S.U.; Kim, J.W.; Chun, Y.D.; Woo, B.C.; Hong, D.K. Fractional Slot Concentrated Winding PMSM with Consequent Pole Rotor for a Low-Speed Direct Drive Reduction of Rare Earth Permanent Magnet. *IEEE Trans. Energy Convers.* **2015**, *30*, 103–109. [CrossRef]
11. You, Y.M.; Chung, D.W. Optimal Design of a Permanent Magnet Synchronous Motor to Improve Torque and Demagnetization Characteristics. *J. Magn.* **2017**, *22*, 423–429. [CrossRef]
12. Kang, G.J.; Park, C.H.; Choi, D.H. Metamodel-based Design Optimization of Injection Molding Process Variables and Gates of an Automotive Glove Box for Enhancing its Quality. *J. Mech. Sci. Technol.* **2016**, *30*, 1723–1732. [CrossRef]
13. Zhang, B.S.; Song, B.W.; Mao, Z.Y.; Tian, W.L.; Li, B.Y.; Li, B. Novel Parametric Modeling Method and Optimal Design for Savonius Wind Turbines. *Energies* **2017**, *10*, 301. [CrossRef]
14. You, Y.M.; Lipo, T.A.; Kwon, B.I. Optimal Design of a Grid connected to Rotor Type Doubly Fed Induction Generator for Wind Turbine Systems. *Energies IEEE Trans. Magn.* **2012**, *48*, 3124–3127. [CrossRef]

15. You, Y.M.; Jung, D.W. A Study on Performance Improvement of Polygon Mirror Scanner Motor. *J. Electr. Eng. Technol.* **2019**, *14*, 745–755. [CrossRef]
16. Kwon, J.W.; Lee, J.H.; Zhao, W.L.; Kwon, B.I. Flux-Switching Permanent Magnet Machine with Phase-Group Concentrated-Coil Windings and Cogging Torque Reduction Technique. *Energies* **2018**, *11*, 2758. [CrossRef]
17. Chai, W.P.; Lipo, T.A.; Kwon, B.I. Design and Optimization of a Novel Wound Field Synchronous Machine for Torque Performance Enhancement. *Energies* **2018**, *11*, 2111. [CrossRef]
18. Li, Y.K.; Song, B.W.; Mao, Z.Y.; Tian, W.L. Analysis and Optimization of the Electromagnetic Performance of a Novel Stator Modular Ring Drive Thruster Motor. *Energies* **2018**, *11*, 1598. [CrossRef]
19. Butler, N.A. Optimal and Orthogonal Latin Hypercube Designs for Computer Experiments. *Biometrika* **2001**, *88*, 847–857. [CrossRef]
20. Park, K.B. An Efficient Hybrid Metaheuristic Algorithm for Solving Constrained Global Optimization Problems. Ph.D. Thesis, Hanyang University, Seoul, Korea, February 2016.
21. Narayanan, G.; Ranganathan, V.T. Analytical Evaluation of Harmonic Distortion in PWM AC Drives Using the Notion of Stator Flux Ripple. *IEEE Trans. Power Elctron.* **2005**, *20*, 466–474. [CrossRef]
22. Chau, K.T. *Electric Vehicle Machines and Drives: Design, Analysis and Application*, 1st ed.; Wiley-IEEE: Hoboken, NJ, USA, 2015; p. 243.
23. PIDOTECH Inc. *PIAnO User's Manuals and Tutorials*; PIDOTECH Inc.: Seoul, Korea, 2019.
24. Myers, R.H.; Montgomery, D.C. *Response Surface Methodology: Process and Product Optimization Using Designed Experiments*, 2nd ed.; John Wiley & Sons: New York, NY, USA, 1995.
25. Orr, M.J.L. *Introduction to Radial Basis Functions Networks*; Edinburgh University: Edinburgh, UK, 1996.
26. Zhang, L.; Tian, F. Performance Study of Multilayer Perceptrons in a Low-Cost Electronic Nose. *IEEE Trans. Instrum. Meas.* **2014**, *36*, 1670–1679. [CrossRef]
27. Sacks, J.; Welch, W.J.; Mitchell, T.J.; Wynn, H.P. Design and Analysis of Computer Experiments. *Stat. Sci.* **1989**, *4*, 409–435. [CrossRef]
28. Anjyo, K.; Lewis, J.P. RBF Interpolation and Gaussian Process Regression through an RKHS Formulation. *J. Math Ind.* **2011**, *3*, 63–71.
29. Du, X.; Liu, G.; Chen, Q.; Xu, G.; Xu, M.; Fan, X. Optimal Design of an Inset PM Motor With Assisted Barriers and Magnet Shifting for Improvement of Torque Characteristics. *IEEE Trans. Magn.* **2017**, *53*, 8109204. [CrossRef]
30. Robert, F.; Bensetti, M.; Santos, V.D.; Dufour, L.; Dessante, P. Multiphysics Modeling and Optimization of a Compact Actuation System. *IEEE Trans. Ind. Electron.* **2017**, *64*, 8626–8634. [CrossRef]

energies

MDPI

Article

Study on Line-Start Permanent Magnet Assistance Synchronous Reluctance Motor for Improving Efficiency and Power Factor

Hyunwoo Kim [1], Yeji Park [1], Huai-Cong Liu [2], Pil-Wan Han [3] and Ju Lee [1,*]

1 Department of Electrical Engineering, Hanyang University, Seoul 04763, Korea; khw7481@hanyang.ac.kr (H.K.); yejipark@hanyang.ac.kr (Y.P.)
2 Hyundai Transys, Hwaseong 18280, Korea; hcliu@hyundai-transys.com
3 Electric Machines and Drives Research Center, Korea Electrotechnology Research Institute, Changwon 51543, Korea; pwhan@keri.re.kr
* Correspondence: julee@hanyang.ac.kr; Tel.: +82-2220-0342

Received: 12 December 2019; Accepted: 9 January 2020; Published: 13 January 2020

Abstract: In order to improve the efficiency, a line-start synchronous reluctance motor (LS-SynRM) is studied as an alternative to an induction motor (IM). However, because of the saliency characteristic of SynRM, LS-SynRM have a limited power factor. Therefore, to improve the efficiency and power factor of electric motors, we propose a line-start permanent magnet assistance synchronous reluctance motor (LS-PMA-SynRM) with permanent magnets inserted into LS-SynRM. IM and LS-SynRM are selected as reference models, whose performances are analyzed and compared with that of LS-PMA-SynRM using a finite element analysis. The performance of LS-PMA-SynRM is analyzed considering the position and length of its permanent magnet, as well as its manufacture. The final model of LS-PMA-SynRM is designed for improving the efficiency and power factor of electric motors compared with LS-SynRM. To verify the finite element analysis (FEA) result, the final model is manufactured, experiments are conducted, and the performance of LS-PMA-SynRM is verified.

Keywords: efficiency; finite element analysis; line-start synchronous reluctance motor; permanent magnet; power factor

---

1. Introduction

As the electrical machine industry is developing, electric energy consumption has been increasing. In industrial applications, electric motors account for energy consumption between 35% and 40% [1–3]. If the efficiency of the electric motor is low, environmental problems such as emission of greenhouse gases are induced. Electrical motors are not a unique problem regarding energy consumption and efficiency. There are also more complex systems such as modern data centers, where there are a lot of electrical motors (and highly energy consumption), because of environment control, cooling, air ventilating, and so on [4,5]. In order to solve these environmental problems, the minimum energy performance standards (MEPS) are enacted by regulating motor efficiency. According to MEPS, the energy efficiency of industrial motors is standardized by power range (0.75–150 kW), and efficiency classes are defined from IE1 to IE4. Recently, the industrial motors have been required to satisfy the IE4 class efficiency according to the strength of MEPS [6].

Induction motors (IMs) account for 70% of industrial motors because of their simple structure and low manufacture cost [1–3]. However, IMs have the secondary copper loss that limits the improvement in their efficiency [7–10]. Therefore, to replace IMs, new types of motor are being studied to improve the efficiency such as line-start synchronous reluctance motors (LS-SynRMs). As LS-SynRMs are operated at a synchronous speed, they do not exhibit secondary copper loss [9,10]. Furthermore, unlike the synchronous motor, this machine does not require an inverter to reach the synchronous speed. Therefore, because the inverter is not necessary, the system cost of these machines is lower compared with synchronous motors and efficiency of LS-SynRMs is better compared with IMs [11–13]. Therefore, LS-SynRM has received attention as a better alternative to IM as an industrial motor [14,15].

The characteristics of LS-SynRM are the same as those of SynRM. Therefore, LS-SynRMs can improve the efficiency of electric motors; however, the power factor is reduced compared with that of IMs because the saliency characteristic [16–18]. However, for the industrial machines, the power factor is also an important factor because the reactive power is affected by the generated power of a generator; further, power factor affects electrical cost [19]. Therefore, according to IEC 60034-1, there is a specification for the power factor of the industrial motor. The permanent magnet synchronous reluctance motor (PMA-SynRMs) with ferrite magnet inserted into SynRM has higher efficiency and power factor compared with SynRMs [20,21]. Therefore, we proposed a line start permanent magnet assistance synchronous reluctance motor (LS-PMA-SynRM) for improving the efficiency and power factor of LS-SynRMs [22,23].

When designing an electrical machine, a magnetic equivalent circuit (MEC) or a finite element analysis (FEA) is used [24–31]. MEC is constructed using the magnetic resistance and magneto motive force considering the magnetic flux paths. By applying a circuit theory, the equation of MEC is solved and the magnetic flux density is simply calculated. However, in a complex rotor structure such as SynRM, it is difficult to analyze the motor because MEC is not easy to construct [30]. On the other hand, in FEA, the solution region is discretized into finite elements using mesh generators and the solution of the governing equation for each element is solved using numerical method [32]. As the solution is based on the element, FEA can be applied to various electrical machine structures [28–31]. Therefore, FEA is powerful numerical method to analyze LS-PMA-SynRM having the complex rotor structure.

In this study, the characteristics of LS-PMA-SynRM are analyzed to improve the efficiency and power factor using FEA. Based on the mathematical model, the characteristics of LS-SynRM and LS-PMA-SynRM are compared. Furthermore, the power factor is analyzed according to the magnetic flux of the permanent magnet. To compare the characteristics of LS-PMA-SynRM, a 5.5 kW IM and LS-SynRM are selected as the reference models. The LS-PMA-SynRM is designed based on the LS-SynRM. The characteristics of LS-PMA-SyRM are then analyzed considering the position and length of the permanent magnet. The final model is designed to maximize the efficiency and power factor. To verify the FEA result, the final model is manufactured and tested for the characteristics of LS-PMA-SynRM.

This paper is organized as follows. In Section 2, the characteristics of LS-SynRM and LS-PMA-SynRM are discussed with respect to the efficiency and power factor. In Section 3, the characteristics of the reference models (IM and LS-SynRM) are analyzed and compared. In Section 4, the performance of LS-PMA-SynRM is analyzed considering the position and length of the permanent magnet. In Section 5, to verify the FEA result, experiments are conducted, and the results obtained for IM and LS-PMA-SynRM are compared. Finally, Section 5 presents the conclusion.

2. Characteristics of LS-SynRM and LS-PMA-SynRM

The operating principle of LS-machines is classified into asynchronous and synchronous speed. Therefore, the subsections are discussed with the characteristic of LS-machines according to the operating speed. In Section 2.1, the principle of LS-machines is discussed at asynchronous speed. In Section 2.2, the main performances, such as efficiency and power factor, are determined at synchronous speed.

2.1. Asynchronous Operation

In asynchronous speed, the rotor current is generated by the slip speed that is the relative speed between the synchronous speed and the rotor speed. Similar to the operation principle of IMs, the rotor current provides the required magnetic torque for the speed to approach synchronous speed. This magnetic torque is expressed as follows and is the same for IMs [33]:

$$T_c = K_T \Phi_m I_2 \cos \phi, \tag{1}$$

where T_c is the torque generated by squirrel-cage bar, K_T is the torque constant, Φ_m is the mutual flux, I_2 is rotor current, and ϕ is the power factor angle.

2.2. Synchronous Operation

2.2.1. Efficiency

In synchronous speed, the operation principle of LS-machines is the same as that for synchronous machines. The efficiency of the synchronous machines is determined based on the mechanical power of motor, core loss, and copper loss as follows:

$$\eta = \frac{P_{out}}{P_{in}} = \frac{P_{out}}{P_{out} + P_{loss}} = \frac{T_e \omega_e}{T_e \omega_e + P_{copper} + P_{core}}, \tag{2}$$

where η is efficiency, P_{out} is mechanical power, P_{in} is electrical input power, P_{loss} is total loss of motor, T_e is torque of motor, ω_e is synchronous speed, P_{copper} is copper loss, and P_{core} is core loss.

Equation (2) indicates that if the power is the same, the efficiency is determined by the losses of LS machines. The main losses of LS machines are classified into core loss and copper loss. The core loss is dependent on the magnetic flux density of the core and frequency. However, the magnetic flux density of the core is determined when the motor is designed, and the frequency is determined by the synchronous speed of LS machines. Therefore, there is a limitation when reducing core loss. In contrast, the copper loss is dependent on the resistance and current. If the torque per current is increased, the copper loss is reduced, and the efficiency can be increased. Therefore, the design method for improving the torque per current is important to improve the efficiency.

When comparing LS-SynRMs and LS-PMA-SynRMs, the LS-PMA-SynRMs generate the additional magnetic torque. Therefore, because the torque per current of LS-PMA-SynRM is larger than LS-SynRM, the efficiency of LS-PMA-SynRM can be improved. The torque of LS-PMA-SynRM is as follows:

$$T_e = \frac{3}{2}\frac{P}{2}[\lambda_a i_q + (L_d - L_q) i_d i_q], \tag{3}$$

where T_e is torque of motor, P is pole, λ_a is the flux linkage, i_d and i_q are dq-axis current, and L_d and L_q are dq-axis inductance.

2.2.2. Power Factor

The power factor is determined by the phase angle difference between the voltage and the current. This phase angle difference can be determined from the dq-axis vector diagram. Figure 1a,c,e shows the structure of an interior permanent magnet synchronous motor (IPMSM), LS-SynRM, and LS-PMA-SynRM to define the dq-axis vector diagram. In general, because the d-axis is defined as the main flux axis, the d-axis of IPMSM is the magnetizing direction of a permanent magnet and the d-axis of the LS-SynRM is a segment through which the main magnetic flux passes. In the case of LS-PMA-SynRM, the d-axis can be defined as the segment because the permanent magnet inserted into LS-SynRM is only used as the assistant torque. Therefore, in Figure 1e, the magnetic flux of the permanent magnet is generated along the q-axis. Figure 1b,d,f shows the vector diagram of IPMSM, LS-SynRM, and LS-PMA-SynRM. The magnetic flux of the permanent magnet is the d-axis in Figure 1b. In contrast, the magnetic flux of the permanent magnet is negative along the q-axis in Figure 1f because the path of main flux is segments.

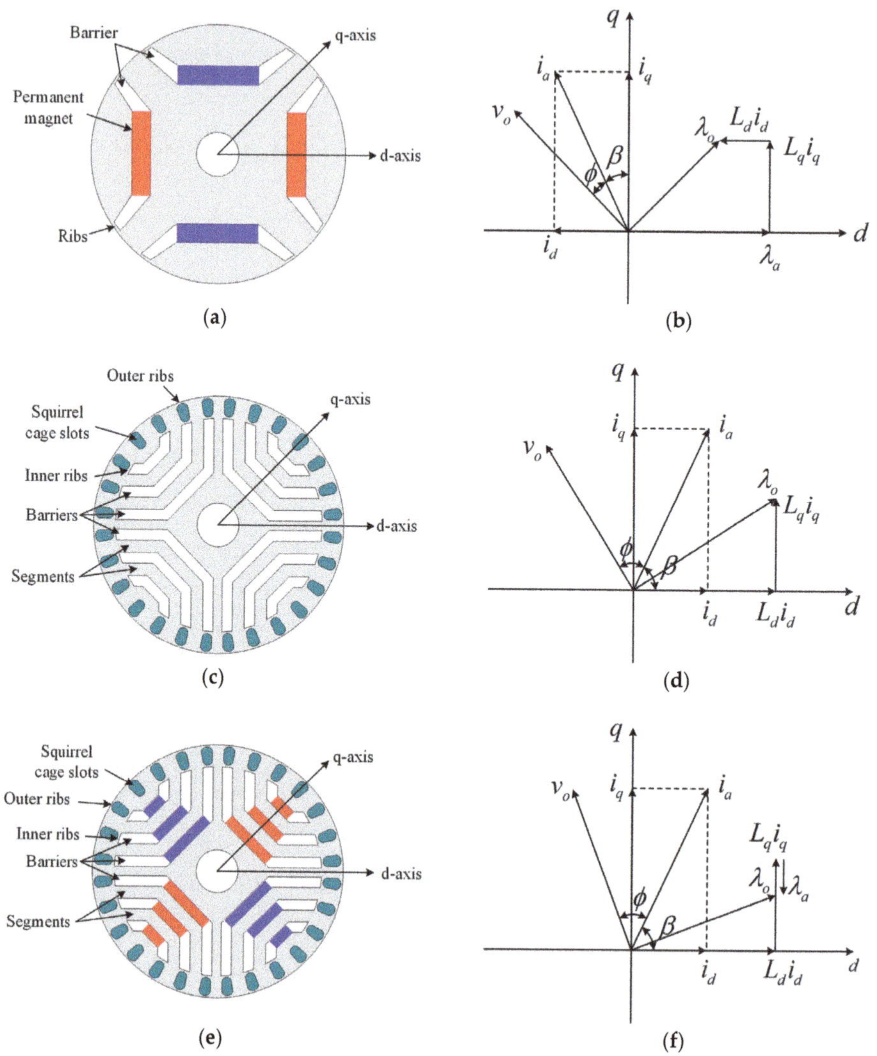

Figure 1. Structure of a line-start synchronous reluctance motor (LS-SynRM) and vector diagram (**a**) structure of an interior permanent magnet synchronous motor (IPMSM) (**b**) vector diagram of an IPMSM, (**c**) structure of a LS-SynRM, (**d**) vector diagram of a LS-SynRM, (**e**) structure of a line-start permanent magnet assistance synchronous reluctance motor (LS-PMA-SynRM), (**f**) vector diagram of a LS-PMA-SynRM.

LS-PMA-SynRMs with ferrite permanent magnet inserted into SynRM have the flux linkage owing to the permanent magnet. This flux linkage is affected by the vector diagram and the power factor is also affected. Figure 2 shows the vector diagram of LS-PMA-SynRM according to flux linkage owing to the permanent magnet (λ_a). In Figure 2, the difference between the voltage and the current phase is decreased according to the increasing flux linkage. Therefore, the larger the flux linkage, the larger the power factor. As a result, LS-PMA-SynRM can improve the power factor and efficiency of electric motors compared with LS-SynRM.

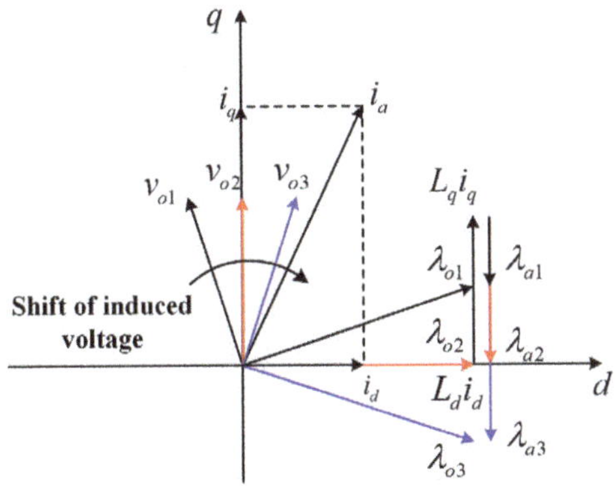

Figure 2. The vector diagram of LS-PMA-SynRM according to flux linkage by permanent magnet.

3. FEA Models of Reference Models

In this section, reference models (IM and LS-SynRM) are discussed in terms of specification and FEA model. In Section 3.1, the specification of reference models is introduced. In Section 3.2, the materials of motor are discussed for FEA. In Section 3.3, the performances of reference modes are analyzed by FEA.

3.1. Specification of Reference Models

In order to compare the efficiency and power factor of LS-PMA-SynRM, the 5.5 kW 6-pole IM and LS-SynRM are selected as reference motors. Figure 3 shows the 2D FEA model of reference motors. The stators of IM and LS-SynRM have the same specification but the rotor is different. The number of squirrel-cage bars of the reference motors is the same, 42 bars. Therefore, considering the number of squirrel-cage bars, the number of barriers is designed to be three [8]. The detail specifications of the reference motors are shown in Table 1.

Table 1. Specification of reference models.

Item		Value	Unit
Rated	Output Power	5.5	kW
	Input voltage	380	V
	Input frequency	60	Hz
Stator	Number of slots	54	-
	Stator outer diameter	220	mm
	Stator inner diameter	145	mm
Rotor	Number of slots	42	-
	Rotor outer diameter	144.2	mm
	Rotor inner diameter	42	mm
Number of poles		6	-
Stack length		170	mm
Airgap length		0.3	mm
Rib length		0.4	mm

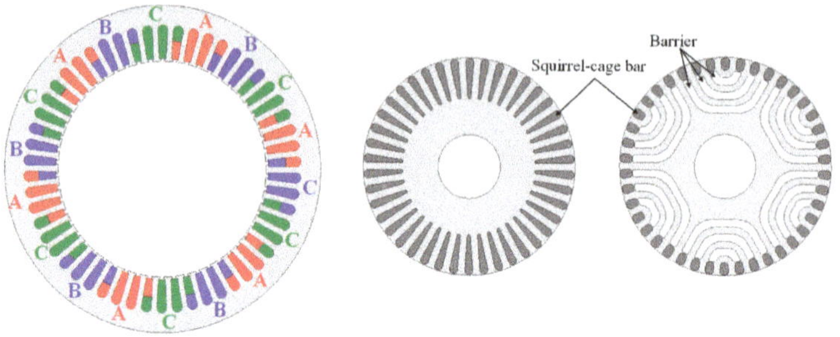

(a) (b) (c)

Figure 3. Reference models: (**a**) stator, (**b**) rotor of induction motor (IM), (**c**) rotor of LS-SynRM.

3.2. Material for FEA

Table 2 shows the material of the stator, rotor, winding, and the squirrel-cage bar. In industrial application, the cost of manufacture is the important factor because of price competitiveness. Considering the cost and core loss, 50PN470 (S18) has been mainly used as soft magnetic material of the stator and rotor in industrial applications. Figure 4 shows the B-H curve and core loss in 60 Hz of 50PN470. In general, the copper can improve the efficiency more than aluminum because of the high conductivity of copper as shown in Table 2. However, because the metal must be melted in the die-casting process, the aluminum with low melting point is mainly used as the material of die-casting. Therefore, considering the manufacture and copper loss, the material of winding and squirrel-cage bar is copper and aluminum, respectively.

Table 2. The material of each parts for finite element analysis (FEA).

Item	Material	Conductivity	Unit
Stator Rotor	50PN470 (S18)	2.38×10^6	S/m
Winding	Copper	5.8×10^7	S/m
Squirrel-cage bar	Aluminum	3.6×10^7	S/m

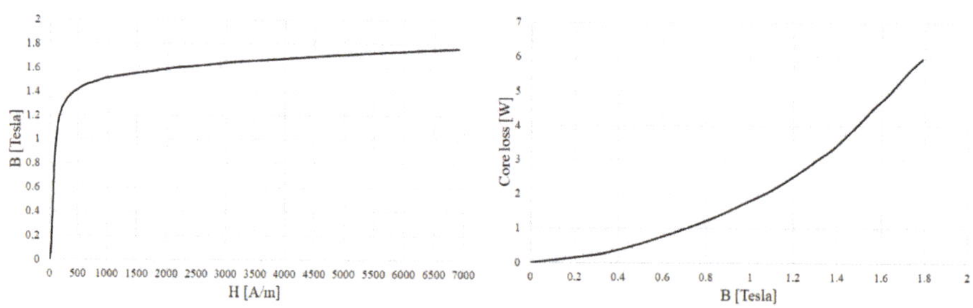

(a) (b)

Figure 4. Characteristic of 35PN230: (**a**) B-H curve, (**b**) core loss.

3.3. FEA Result

To analyze the performance of the reference models, ANSYS MAXWELL 18.2 version was used. The time-step FEA was used to consider the mechanical and electromagnetic transient analysis [31,33]. Figure 5 shows the magnetic field density of the reference motors using FEA and Table 3 shows the characteristics of the reference motors. In Table 3, the core loss of IM is higher than that of LS-SynRM because the magnetic flux density of IM is higher than that of LS-SynRM in Figure 5. Furthermore, IMs have a slip that represents the difference between the synchronous speed and the rotor speed. This slip induces an electromotive force in the squirrel-cage bar and generates the rotor copper loss that accounts for 31% of the total losses in Table 3. On other hand, for the stator copper loss, the current of LS-SynRM is higher than that of IM because the power factor is low under same output power and input voltage. As a result, the efficiency of IM is higher than that of LS-SynRM considering the total loss.

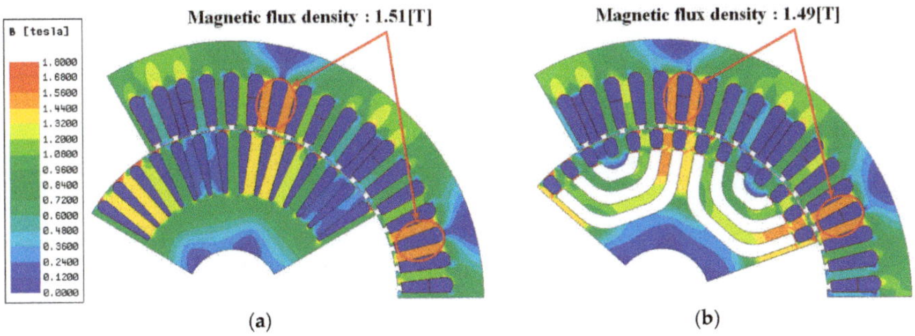

Figure 5. Electromagnetic analysis result of reference models: (**a**) IM, (**b**) LS-SynRM.

Table 3. Characteristics of reference models using FEA.

Item	Value		Unit
	IM	LS-SynRM	
Power	5.5	5.5	kW
Speed	1170.8	1200	rpm
Torque	48.1	45.5	Nm
Current	11.55	12.88	A
Core loss	94.36	90.12	W
Stator copper loss	214.35	266.69	W
Rotor copper loss	143.8	25.39	W
Total loss	452.55	382.21	W
Efficiency	91.46	92.46	%
Power factor	80.14	70.39	-

4. Design of LS-PMA-SynRM

In this section, the performance of LS-PMA-SynRM is analyzed by FEA according to the position and length of permanent magnet. In Section 4.1, LS-PMA-SynRM is analyzed according to the position of magnet considering the manufacture. In Section 4.2, LS-PMA-SynRM is analyzed according to the length of the permanent magnet in the position obtained in Section 4.1.

To analyze the characteristics of LS-PMA-SynRM, the ferrite permanent magnet was inserted into the barriers of LS-SynRM. As LS-SynRM has the end-ring structure, as shown in Figure 6, the position of the magnet is limited. From Figure 6, the position magnet can be inserted into only first and second barriers. Therefore, the efficiency and power factor of LS-PMA-SynRM were analyzed considering the position of the magnet. Furthermore, the length of the magnet was designed to maximize the efficiency and power factor. Table 4 shows the ferrite permanent magnet data to analyze the LS-PMA-SynRM.

Table 4. Permanent magnet data.

Item	Value	Unit
Residual flux density	0.39	T
Coercive force	3.703	kOe

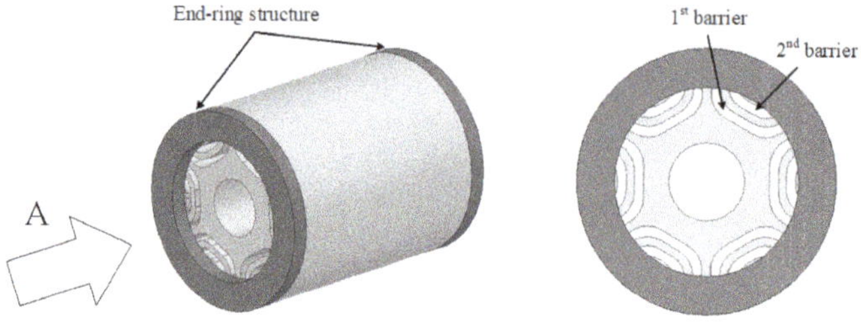

(a) (b)

Figure 6. Rotor structure of LS-SynRM with end-ring (**a**) rotor and (**b**) view of A.

4.1. Position of Magnet

Figure 7 shows the FEA models for analysis according to the position of the magnet. To analyze the characteristic of the LS-PMA-SynRM considering the position of the magnet, the length of the magnet is the same as 7 mm. Moreover, considering the end-ring structure, the position of the magnet is inserted into the first and second barrier. Figure 8 shows the magnetic flux density using FEA, and Figure 8 shows the loss analysis, efficiency, and power factor under the same output power conditions. In Figure 8, because the magnetic flux density of model 2 is larger than models 1 and 3, the core loss of model 2 is the larger in Figure 9. Figure 10 shows the magnetic flux density of air gap. As the magnetic flux density of model 3 is larger than other models, the efficiency and power factor are better, as shown in Figure 9b. Therefore, the model 3 was selected as the base model to analyze the characteristics of LS-PMA-SynRM according to the length of the magnet.

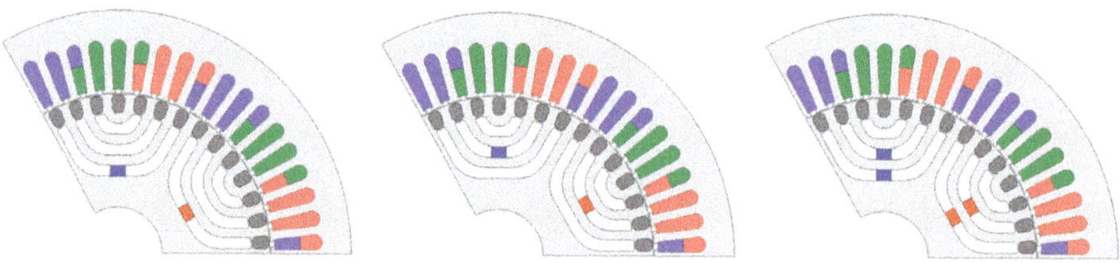

(a) (b) (c)

Figure 7. FEA model for analysis according to the position of magnet in (**a**) model 1, (**b**) model 2, and (**c**) model 3.

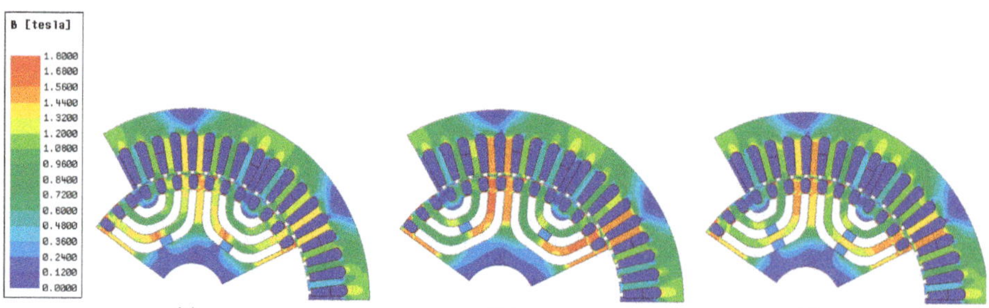

(a) (b) (c)

Figure 8. Electromagnetic analysis result using FEA according to the position of magnet in (**a**) model 1, (**b**) model 2, and (**c**) model 3.

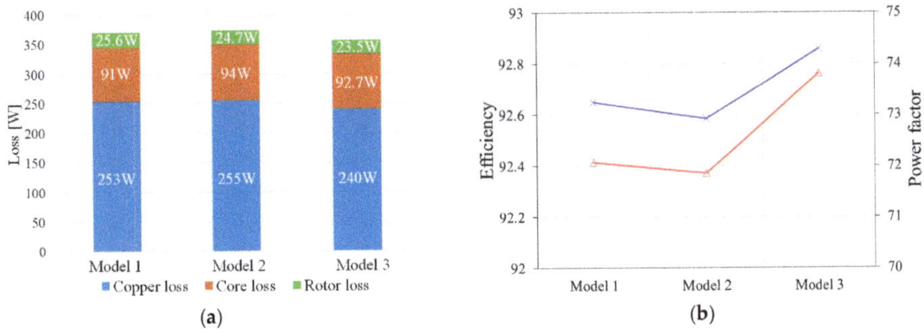

Figure 9. Electromagnetic analysis result using FEA according to the position of magnet: (**a**) loss, (**b**) efficiency and power factor.

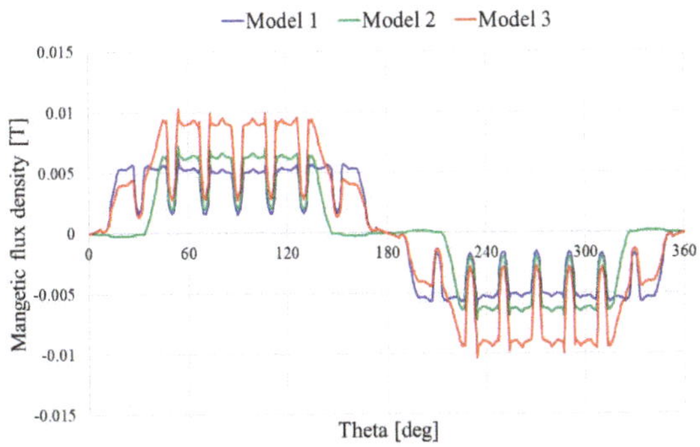

Figure 10. Magnetic flux density in air gap of analysis models.

4.2. Length of Magnet

The characteristics of LS-PMA-SynRM were analyzed considering the length of the magnet. Figure 11 shows the design parameter of LS-PMA-SynRM. The lengths of the magnets of the first and second barriers are defined as W_{m1} and W_{m2}, respectively. Considering the length of the barriers and their manufacturer, the parameter ranges were selected. The length of the first magnet ranges from 1 to 17 mm and the length of second magnet from 0.5 to 8.5 mm. Figure 12 shows the FEA analysis result for parameters (W_{m1}, W_{m2}). The longer is the magnet, the higher the efficiency and power factor. Based on the FEA result, the final model was designed to maximize the efficiency and power factor.

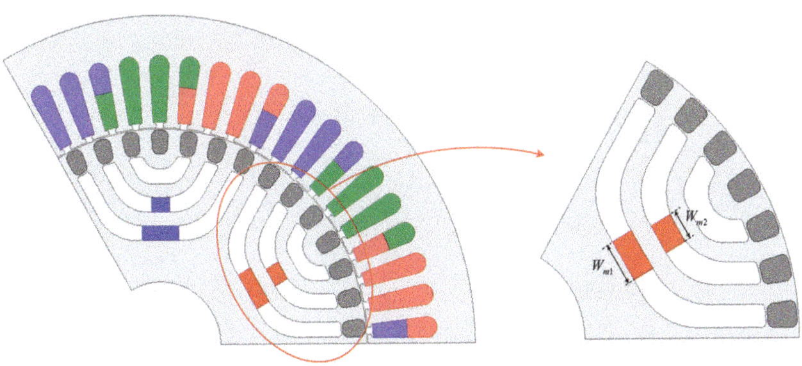

Figure 11. Parameter for analysis of the characteristic according to length of magnet.

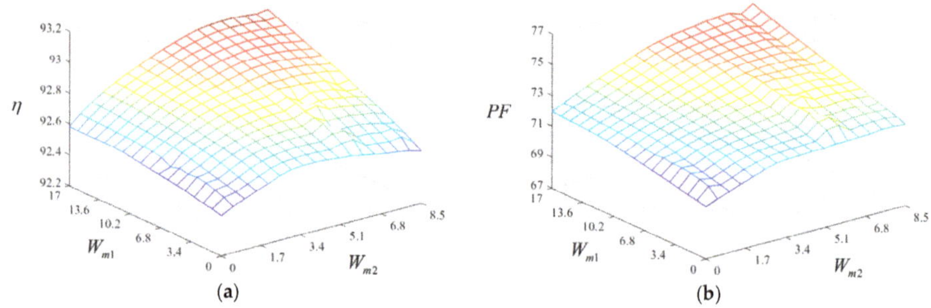

Figure 12. FEA analysis result according to length of magnet: (**a**) efficiency, (**b**) power factor.

4.3. Final Model of LS-PMA-SynRM

In Figure 12, considering the efficiency and power factor, the final model was designed so that the length of the first magnet was 17 mm and that of the second magnet was 8.5 mm. Figure 13a shows the final model of LS-PMA-SynRM and Figure 13b shows the magnetic flux density using FEA. Table 5 shows the FEA result for LS-PMA-SynRM. Compared with Table 3, the efficiency is improved by approximately 5.6% and the power factor is improved by approximately 5.2% compared with LS-SynRM. Based on the FEA result, the permanent magnet can be used to improve the efficiency and power factor based on the analysis result.

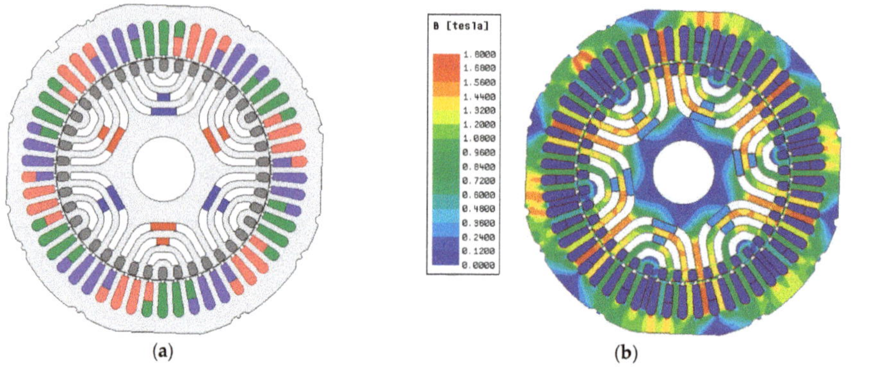

Figure 13. The final model of LS-PMA-SynRM: (**a**) FEA model, (**b**) magnetic flux density.

Table 5. FEA result of LS-PMA-SynRM.

Item	Value	Unit
Power	5.5	kW
Speed	1200	rpm
Torque	44	Nm
Current	11.98	A
Core loss	93.79	W
Stator copper loss	229.04	W
Rotor copper loss	24.84	W
Total loss	347.68	W
Efficiency	93	%
Power factor	75.5	-

5. Verification

The final models of LS-PMA-SynRM and reference model of IM were manufactured to verify the FEA result. In Section 5.1, the manufacture of LS-PMA-SynRM is discussed considering the die-casting process and permanent magnet. In Section 5.2, the experiments of IM and LS-PMA-SynRM are conducted and compared with FEA and experiment result according to load. In Section 5.3, the main advantages are discussed in comparison with previous literature.

5.1. Manufacture

Figure 14a,b shows the rotor and stator of the manufactured LS-PMA-SynRM. In Figure 14a, there is a bridge in the barrier as a support structure for the permanent magnet. In the die-casting process, the high pressures lead to aluminum leaking into the barriers. To prevent this problem, thin electrical steel sheets are placed on both ends of the rotor, as shown Figure 14c. Therefore, the electrical steel sheet is eliminated to insert the permanent magnet, as shown Figure 14d. When the die-casting is performed, the temperature is high, and this can lead to thermal demagnetization. Furthermore, because of the squirrel-cage bar slot, magnetization is difficult to attain. Therefore, the permanent magnet is assembled after magnetization.

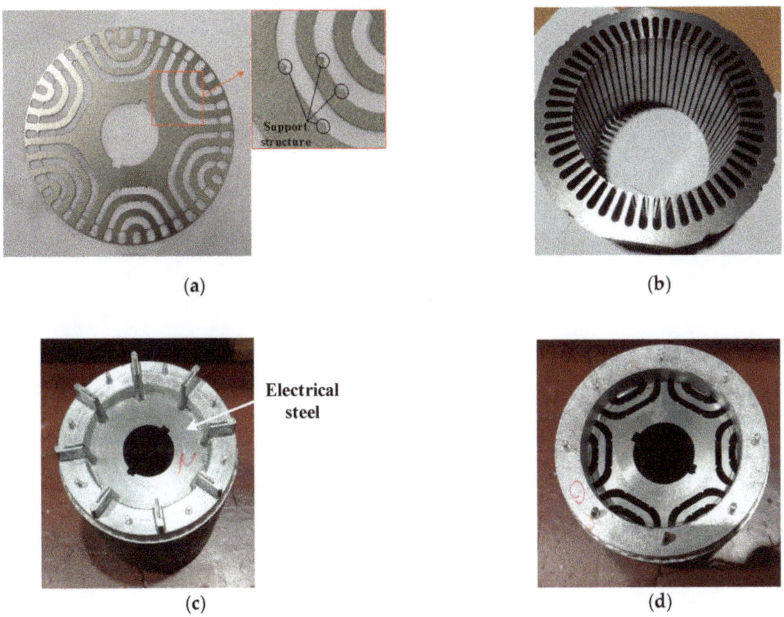

Figure 14. Manufacture LS-PMA-SynRM: (**a**) rotor core, (**b**) stator, (**c**) rotor before removing the electrical steel, and (**d**) rotor after removing the electrical steel.

5.2. Experiment Result

Experiments were conducted to verify the efficiency and power factor of IM and the final LS-PMA-SynRM. Figure 15 shows the dynamometer motor and experiment environment. The performance of the test motor was calculated using the power analyzer that is Yokogawa's WT1803E model. In addition, the temperature logger was used for the temperature saturation test. The temperature was measured through the temperature sensor and the measured temperature was recorded in the temperature logger as shown Figure 15. The dynamometer motor is the IM and allows the test motor to rotate at a synchronous speed. Through the V/f control, the dynamometer motor runs up and reaches synchronous speed. The test motor is also operated at a synchronous speed and the voltage is applied to the test motor. The torque control is used to maintain the 5.5 kW output power in the dynamometer motor. The current sensor measures the current in the test motor, and the torque and speed are measured using the dynamometer motor. In the power analyzer, the input power, output power, efficiency, and power factor are calculated based on the measured voltage, current, torque, and speed. Figure 16 shows the efficiency of IM and LS-PMA-SynRM according to different loads. The performances of IM and LS-PMA-SynRM agree well with the results obtained from the FEA and experiments. Table 6 shows the experiment result of IM and LS-PMA-SynRM. Compared with the FEA result in Tables 3 and 5, the efficiency of IM and LS-PMA-SynRM is decreased by approximately 0.3% and the power factors are decreased by 6% and 4%, respectively.

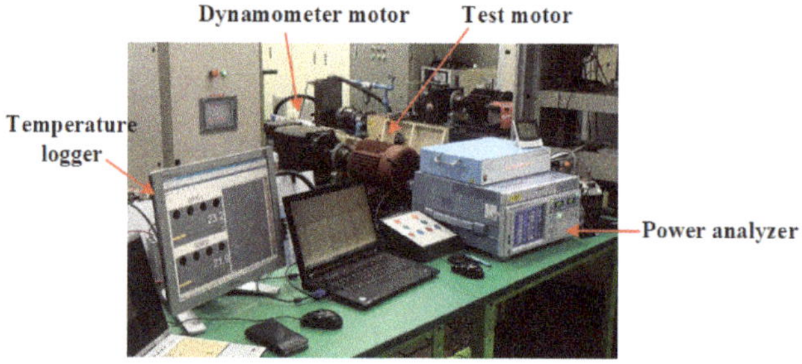

Figure 15. Experiment environment and test dynamometer.

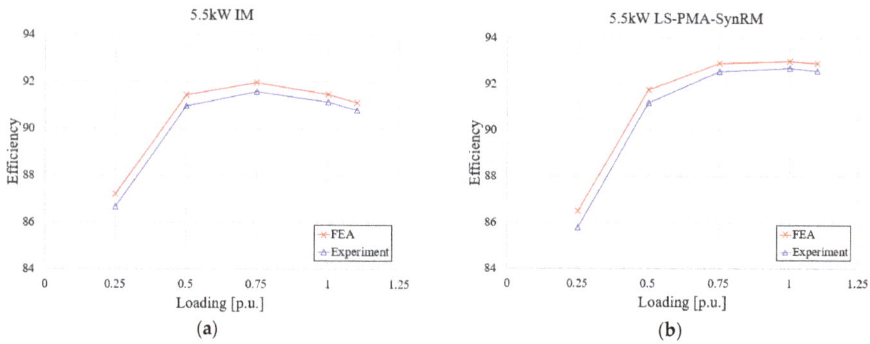

Figure 16. Experiment result of manufactured motors: (**a**) IM, (**b**) LS-PMA-SynRM.

Table 6. Experiment result of IM and LS-PMA-SynRM using the power analyzer.

Item	Value		Unit
	IM	**LS-PMA-SynRM**	
Power	5.5	5.5	kW
Speed	1175	1200	rpm
Torque	44.7	43.7	Nm
Current	12.3	12.6	A
Core loss	106	105.9	W
Stator copper loss	249.5	257.7	W
Rotor copper loss	124.6	0	W
Total loss	535.3	433.75	W
Efficiency	91.13	92.69	%
Power factor	74.5	71.2	-

5.3. Discussion

In previous literature, LS-SynRM have been studied as alternatives to IM due to high efficiency. However, because of the saliency characteristic of SynRM, LS-SynRM have a limited power factor. Referring to IEC 60034-1, not only the efficiency but also power factor is the important factor in industrial electrical machines. Therefore, the power factor of LS-SynRM must be improved to replace IM. In general, due to magnetic flux by permanent magnet, the power factor of PMA-SynRM is better than SynRM. However, because of the thermal demagnetization of ferrite magnet in the die-casting process, there is no research on LS-PMA-SynRM. This study discussed the design and analysis of LS-PMA-SynRM according to the position and length of permanent magnet considering the manufacture such as die-casting. In addition, the efficiency and power factor of LS-SynRM were verified by comparison with FEA and experiment. As a result, the efficiency and power factor were improved compared with LS-SynRM.

6. Conclusions

This study analyzed the efficiency and power factor of LS-PMA-SynRM considering the position and length of its permanent magnet. In addition, IM and LS-SynRM were analyzed to verify the performance of LS-PMA-SynRM under the condition of the same stator specification. Considering the die-casting and manufacturing processes, the permanent magnet was assembled after magnetization. Furthermore, the position of the permanent magnet was selected considering the end-ring. Further, considering this position, the performance of the LS-PMA-SynRM was analyzed and the base model was selected. Based on this model, the analysis parameter was selected, and the efficiency and power factor of electric motors was analyzed according to the length of the permanent magnet. The final model was designed using FEA. To verify the FEA result, LS-PMA-SynRM was manufactured and experiments were conducted. As a result, the FEA and experiment result of IM and LS-PMA-SynRM agree well, and the performance of LS-PMA-SynRM is improved compared with LS-SynRM. This study provided the design and manufacture of LS-PMA-SynRM as the alternative to IM. However, the power factor of LS-PMA-SynRM was still lower than IM. Therefore, the design of barrier, such as the length, angle, and thickness of barrier, must be optimized. In addition, the position and length of the permanent magnet according to design of barrier must be optimized to improve the efficiency and power factor. This requires further investigation about optimal design.

Author Contributions: H.K. conducted an analysis to improve the efficiency and power factor of electric motors. Y.P. investigated the research trend obtained using LS-SynRM. H.-C.L. reviewed the FEA result and guided this paper. P.-W.H. funded the research. J.L. verified and supervised the research. All authors have read and agreed to the published version of the manuscript.

Funding: This research was funded by the Energy Efficiency & Resources of the Korea Institute of Energy Technology Evaluation and Planning (KETEP) grant funded by the Korea government Ministry of Knowledge Economy (No. 2018201010633A) and in part by the Human Resources Program in Energy Technology of the Korea Institute of Energy Technology Evaluation and Planning (KETEP), granted financial resource from the Ministry of Trade, Industry & Energy, Republic of Korea (No. 20174030201750).

Conflicts of Interest: The authors declare no conflict of interest.

References

1. Kersten, A.; Liu, Y.; Pehrman, D.; Thiringer, T. Rotor Design of Line-Start Synchronous Reluctance Machine with Round Bars. *IEEE Ind. Appl.* **2019**, *55*, 3685–3696. [CrossRef]
2. Almeida, A.T.D.; Ferreira, F.J.T.E.; Baoming, G. Beyond Induction Motors-Technology Trends to Move Up Efficiency. *IEEE Ind. Appl.* **2014**, *50*, 2103–2114. [CrossRef]
3. Almeida, A.T.D.; Ferreira, F.J.T.E.; Fong, J.A.C. Standards for Efficiency of Electric Motor. *IEEE Ind. Appl. Mag.* **2011**, *17*, 12–19. [CrossRef]
4. Matko, V.; Brezovec, B. Improved Data Center Energy Efficiency and Availability with Multilayer Node Event Processign. *Energies* **2018**, *11*, 2478. [CrossRef]

5. Jia, M.; Srinivasan, R.S.; Raheem, A.A. From occupancy to occupant behavior: An analytical survey of data acquisition technology, modeling methodologies and simulation coupling mechanisms for building energy efficiency. *Renew. Sustan. Energy Rev.* **2017**, *68*, 525–540. [CrossRef]
6. Ferreira, F.J.T.E.; Baoming, G.; Almeida, A.T.D. Reliability and Operation of High-Efficiency Induction Motors. *IEEE Ind. Appl.* **2016**, *52*, 4628–4637. [CrossRef]
7. Xie, Y.; Pi, C.; Li, Z. Study on Design and Vibration Reduction Optimization of High Starting Torque Induction Motor. *Energies* **2019**, *12*, 1263. [CrossRef]
8. Rafajdus, P.; Hrabovcova, V.; Lehocky, P.; Makys, P.; Holub, F. Effect of Saturation on Field Oriented Control of the New Designed Reluctance Synchronous Motor. *Energies* **2018**, *11*, 3223. [CrossRef]
9. Aguba, V.; Muteba, M.; Nicolae, D.V. Transient Analysis of a Start-up Synchronous Reluctance Motor with Symmetrical Distributed Rotor Cage Bars. In Proceedings of the 2017 IEEE AFRICON, Cape Town, South Africa, 18–20 September 2017.
10. Liu, H.C.; Lee, J. Optimum Design of an IE4 Line-Start Synchronous Reluctance Motor Considering Manufacturing Process Loss. *IEEE Trans. Ind. Electron.* **2018**, *65*, 3104–3114. [CrossRef]
11. Lee, J.K.; Jung, D.H.; Lim, J.; Lee, K.D.; Lee, J. A Study on the Synchronous Reluctance Motor Design for High Torque by Using RSM. *IEEE Trans. Magn.* **2018**, *54*, 8103005. [CrossRef]
12. Ozcelik, N.G.; Dogru, U.E.; Imeryuz, M.; Ergene, L.T. Synchronous Reluctance Motor vs. Induction Motor at Low-Power Industrial Applications: Design and Comparison. *Energies* **2019**, *12*, 2190. [CrossRef]
13. Tampio, J.; Kansakangas, T.; Suuriniemi, S.; Kolehmainen, J.; Kettunen, L.; Ikaheimo, J. Analysis of Direct-On-Line Synchronous Reluctance Machine Start-Up Using a Magnetic Field Decomposition. *IEEE Trans. Ind. Appl.* **2017**, *53*, 1852–1859. [CrossRef]
14. Boroujeni, S.T.; Bianchi, N.; Alberti, L. Fast Estimation of Line-Start Reluctance Machine Parameters by Finite Element Analysis. *IEEE Trans. Energy Convers.* **2011**, *26*, 1–8. [CrossRef]
15. Liu, H.C.; Hong, H.S.; Cho, S.; Lee, J.; Jin, C.S. Bubbles and Blisters Impact on Diecasting Cage to the Designs and Operations of Line-Start Synchronous Reluctance Motors. *IEEE Trans. Magn.* **2017**, *53*, 8202504. [CrossRef]
16. Zhao, W.; Xing, F.; Wang, X.; Lipo, T.A.; Kwon, B.-I. Design and Analysis of a Novel PM-Assisted Synchronous Reluctance Machine with Axially Integrated Magnets by the Finite-Element Mehtod. *IEEE Trans. Magn.* **2017**, *53*, 8104104. [CrossRef]
17. Joo, K.-J.; Kim, I.-G.; Lee, J.; Go, S.-C. Robust Speed Sensorless Control to Estimated Error for PMa-SynRM. *IEEE Trans. Magn.* **2017**, *53*, 8102604. [CrossRef]
18. Kim, W.-H.; Kim, K.-S.; Kim, S.-J.; Kang, D.-W.; Go, S.-C.; Chun, Y.-D.; Lee, J. Optimal PM Design of PMA-SynRM for Wide Constant-Power Operation and Torque Ripple Reduction. *IEEE Trans. Magn.* **2009**, *45*, 4660–4663. [CrossRef]
19. Almeida, A.T.D.; Ferrera, F.J.T.E.; Duarte, A.Q. Technical and Economcal Considerations on Super High-Efficiency Three-Phase Motors. *IEEE Trans. Ind. Appl.* **2014**, *50*, 1274–1285. [CrossRef]
20. Kong, Y.; Lin, M.; Yin, M.; Hau, L. Rotor Structure on Reducing Demagnetization of Magnet and Torque Ripple in a PMa-synRM With Ferrite Permanent Magnet. *IEEE Trans. Magn.* **2018**, *54*, 8108705. [CrossRef]
21. Mingardi, D.; Bianchi, N. Line-Start PM-Assisted Synchronous Motor Design, Optimization, and Tests. *IEEE Trans. Ind. Electron.* **2017**, *64*, 9739–9747. [CrossRef]
22. Jung, D.-H.; Kwak, Y.; Lee, J.; Jin, C.-S. Study on the Optimal Design of PMa-SynRM Loading Ratio for Achievement of Ultrapremium Eifficieny. *IEEE Trans. Magn.* **2017**, *53*, 8001904. [CrossRef]
23. Liu, H.-C.; Joo, K.-J.; Oh, Y.J.; Lee, H.-J.; Seol, H.-S.; Jin, C.-S.; Kim, W.-H.; Lee, J. Optimal Design of an Ultra-Premium-Eifficiency PMA-Synchronous Reluctance Motor with the Winding Method and Stator Parameters to Reduce Flu Leakage and Minimize Torque Pulsations. *IEEE Trans. Magn.* **2018**, *54*, 8207505. [CrossRef]
24. Wang, X.; Zhu, C.; Zhang, R.; Tang, R.; Song-Yop, H. Performance Analysis of Single-Phase Induction Motor Based on Voltage Source Complex Finite-Element Analysis. *IEEE Trans. Magn.* **2006**, *42*, 587–590. [CrossRef]
25. Lee, S.-H.; Kwon, S.; Lee, J.-J.; Hong, J.-P. Characteristic Analysis of Claw-Pole Machine Using Improved Equivalent Magnetic Circuit. *IEEE Trans. Magn.* **2009**, *45*, 4570–4573.
26. Shimotani, T.; Sato, T.; Igarashi, H. Fast Finite-Element Analysis of Motors Using Block Model Order Reduction. *IEEE Trans. Magn.* **2016**, *52*, 7207004. [CrossRef]

27. Lee, J.-J.; Lee, J.; Kim, K.-S. Design of a WFSM for an Electric Vehicle Based on a Nonlinear Magnetic Equivalent Circuit. *IEEE Trans. Appl. Supercond.* **2018**, *28*, 5206304. [CrossRef]
28. Lin, I.-H.; Hsieh, M.-F.; Kuo, H.-F.; Tsai, M.-C. Improved Accuracy for Performance Evaluation of Synchronous Reluctance Motor. *IEEE Trans. Magn.* **2015**, *51*, 8113404. [CrossRef]
29. Lee, B.-H.; Hong, J.-P.; Lee, J.-H. Optimum Design Criteria for Maximum Torque and Efficiency of a Line-Start Permanent-Magnet Motor Using Response Surface Methodology and Finite Element Method. *IEEE Trans. Magn.* **2012**, *48*, 863–866. [CrossRef]
30. Li, N.; Zhu, J.; Lin, M.; Yang, G.; Kong, Y.; Hau, L. Analysis of Axial Field Flux-Switching Memory Machines Based on 3-D Magnetic Equivalent Circuit Network Considering Magnetic Hysteresis. *IEEE Trans. Magn.* **2019**, *55*, 7203104. [CrossRef]
31. Zhang, Y.; Chau, K.T.; Zhang, D.; Liu, C. A Finite Element-Analytical Method for Electromagnetic Field Analysis of Electric Machines with Free Rotation. *IEEE Trans. Magn.* **2006**, *42*, 3392–3394. [CrossRef]
32. Sadiku, M.N.O. A Simple Introduction to Finite Element Analysis of Electromagnetic Problems. *IEEE Trans. Magn.* **1989**, *32*, 85–93. [CrossRef]
33. Liu, H.C.; Seol, H.S.; Kim, J.Y.; Lee, J. Design and Analysis of an IE4 Class Line-Start Synchronous Reluctance Motor Considering Total Loss and Starting Performance. *J. Electron. Mater.* **2019**, *48*, 1386–1394. [CrossRef]

energies

MDPI

Article

Improvement of the Thermal and Mechanical Strength of the Starting Cage of Double-Cage Induction Motors

Jan Mróz [1,*] and Wojciech Poprawski [2]

1 The Faculty of Electrical and Computer Engineering, Rzeszow University of Technology, Aleja Powstańców Warszawy 12, 35-959 Rzeszów, Poland

2 Independent Researcher, 35-959 Rzeszów, Poland; w.poprawski@hotmail.com

* Correspondence: janmroz@prz.edu.pl; Tel.: +48-178651360

Received: 25 October 2019; Accepted: 27 November 2019; Published: 29 November 2019

Abstract: This article discusses the thermal and mechanical exposure of the starting cage of a double-cage induction motor rotor during start-up. Damage to the starting cage is the most common cause of failure of a double-cage winding during long start-ups. It has been indicated that the end region of the double-cage winding is a key area in the search for a more damage-resistant solution. Among the available studies on improving the mechanical strength of double-cage windings, which typically focuses on improving the cooling system, modifying the shape of the slots, or altering the bar material, a new concept of improving the mechanical strength through the modification of the structure of the end region has appeared. This is achieved by applying sleeves onto the ends of the starting cage bars, which helps to reduce the temperature of the connection between the starting bars and the end rings. A simulation of the temperature field of a double-cage induction motor with this new design is performed and discussed in this paper. It has been confirmed that the new design solution effectively improves the mechanical strength of the starting cage, making it less prone to damage caused by thermal stresses.

Keywords: double-cage induction motor; improvement of motor reliability; cage winding constructions; direct start-up; coupled electromagnetic-thermal model

---

1. Introduction

In a cage induction motor with an emergency locked rotor, or under long starting conditions, the element most vulnerable to damage is the cage winding of the rotor. Thermal exposure is particularly high in double-cage rotor motors owing to the relatively low thermal capacity of the starting cage bars. Such motors are typically found in drives requiring high starting torque, which translates directly into higher losses in the starting cage. Examples of damage to the starting cage in double-cage motors are displayed in Figures 1 and 2. The most common cause of double-cage winding failure is damage to the starting cage, while the working cage remains functional. This is a defect that is difficult to detect in its initial phase. There are numerous publications presenting new approaches to detecting starting cage failures in a double-cage motor. Thus, [1] presents a method of detecting outer cage damage in double squirrel cage induction motors. This diagnostic method relies on a discrete wavelet transform optimised for sensitive detection under transient operating conditions. Reference [2] presents a complete on-line condition monitoring system designed to detect incipient broken rotor bar faults in a double-cage induction motor using the stator current signature. It is based on successful combination of one of the latest variants of wavelet techniques, the recursive stationary wavelet packet transform and a tool widely used in quality control, the statistical process control in order to deal with several challenges in the continuous monitoring of the incipient fault. In [3] the possibility of using the stator phase current waveform as a diagnostic signal to detect faults in a double squirrel cage induction motor is discussed. On the basis of tests conducted on a double cage motor prototype, conditions were formulated for distinguishing faults in cages of both medium and high power machines.

Figure 1. Damaged rotor cage, view after cutting off the end ring of the starting cage—visible melting of the starting cage bar.

Figure 2. Damaged rotor cage—visible sheared bars of the starting cage.

An important, developing issue is diagnosing large induction motors fed by an inverter. Reference [4] presents the use of a fuzzy-based statistical feature extraction from the air gap disturbances for diagnosing broken rotor bars in large induction motors fed by line or an inverter. The method is based on the analysis of the magnetic flux density variation in a Hall Effect Sensor installed between two stator slots of the motor. Reference [5] provides a diagnosis of broken rotor bars in field oriented controlled double cage induction motors, based on current and vibration signature analysis techniques.

Designers and constructors are making attempts to make the double-cage winding more resistant to damage occurring during long start-ups. One of the lines of action is to improve the cooling system. In [6] the cooling performance of axial fans with forward-swept and inclined blades and a structure with low ventilation resistance in large-capacity open-type motors is studied.

Steps are taken for thermal analysis to be more deeply regarded in electric machine design. In [7] some of these problems are discussed and advice is provided as to dealing with them when developing algorithms for inclusion in design software. In [8], a design of a 115 kW squirrel cage induction motor for electric vehicle applications is presented. In the design procedure, initially, an analytical design of the electric vehicle motor is performed depending on specific design criteria. Then, the analytical design is verified by means of finite element analysis. Reference [9] presents the coupled fluid-thermal analysis for an induction motor with healthy and broken bar rotors. Much attention was paid to developing the fluid model on the basis of the computational fluid dynamic theory.

Another area of work aimed at improving the fault resistance of the double cage winding and the deep bar rotor to the effects of a prolonged start-up focuses on changing the cage winding material and modifying the shape of rotor slots. In [10] three double-cage induction motors have been simulated and their electromagnetic characteristics compared. The study is carried out using finite element method (FEM) analysis. Reference [11] presents dynamic modelling of a series of induction motor squirrel cages with different shapes of rotor deep bars, taking into account the skin effect. Reference [12] focuses on designing and optimizing an induction motor with a lower cost and high performance. The starting torque of the induction motor, which is an important aspect in traction applications, has been improved by applying a rotor with a double cage. Analytical modelling is carried out and it is validated by means of FEM analysis. In [13] a new design of the rotor bar which allows to improve the starting torque without decreasing motor efficiency is presented. Reference [14] deals with the influence of the shape of the cage on double cage induction motor's parameters, mainly the values of starting torque, breakdown torque, as well as the power factor, efficiency and starting current. The optimal shape of the rotor slot has been identified for the maximum size of the relative starting torque and minimum size of the relative starting current while maintaining a favorable power factor and efficiency of the motor.

Another area of work undertaken by constructors to improve the fault resistance of the cage winding are special motor structure designs. In [15] an investigation on the design of a high-power induction motor with special constraints is presented. Direct online start-up and pull-up torque of high value are the two imposed requirements. The proposed solution advances a new rotor structure with two different rotor cages. Reference [16] presents the elaborate design procedure for a double rotor double cage motor. The two rotors can run independently, at an equal or unequal speed, depending on their individual loading.

In [17,18], using mathematical models to consider the electromagnetic and thermal interrelations, the author demonstrated that the end region of the cage is a key location in the search for construction solutions more resistant to damage. These were likely the inspiration for the construction of the starting cage of a double-cage motor as discussed in [19], which, according to the authors, allows for a reduction of the temperature of the end region of the starting cage bars. This is achieved by applying sleeves onto the ends of the starting cage bars. Because [19] does not present any test results for a motor with a double-cage winding constructed in such a manner, it is necessary to test the effectivity of this new design. The aim of the present paper is to perform a simulation of the temperature field of a double-cage induction motor with the new construction solution of the starting cage, during start-up with a locked rotor. The results are compared with those of a motor with a starting cage of conventional structure. All results presented in this paper have been achieved through a simulation and should be verified through experimental research in the future. The problem of experimental research on high-powered motors is related to a number of logistic activities, as they typically must be performed under industrial conditions. This entails high experimental costs. Therefore, there are relatively few publications addressing experimental research related to the heating of high-power cage induction motors [20–22].

2. Method of Analysing the Temperature Field of a Double-Cage Induction Motor

The analysis of thermal and electromagnetic phenomena occurring in an induction motor is commonly conducted by means of professional software available on the market. There are many publications on this issue [6,8,9,12,23]. A considerable limitation, particularly at an early stage of work on the presented problem, is the cost and a long calculation time for 3D issues, hence the choice of faster and less expensive methods which have been partially verified through experimental research [17,24]. Reference [17] presents a mathematical model and corresponding 3D simulation model that allows the determination of the temperature field of a double-cage motor in transient electromechanical states. It uses a heat network created by the control volume method [25]. The model considers mutual electromagnetic and thermal dependencies. Because the dynamics of electromechanical phenomena in electric machines are considerably greater than those of thermal phenomena, an electromechanical problem can be solved at a specific moment and for a given temperature field. In the next time step, a new temperature field can be calculated using the solution of the electromechanical problem from the previous time step. This procedure is presented in the block diagram in Figure 3.

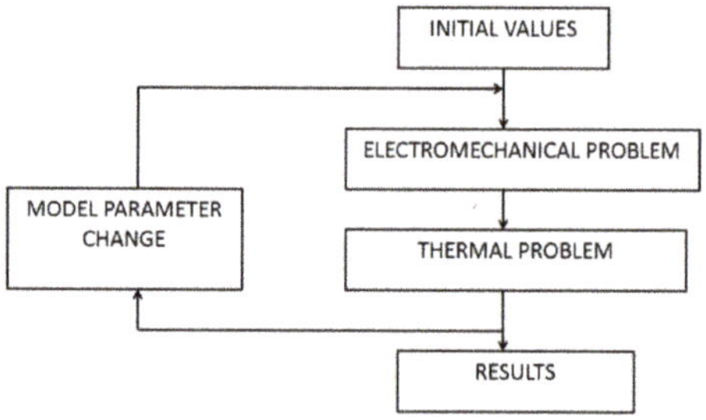

Figure 3. Algorithm for calculating rotor thermal field.

The motor model used in calculations is designed for simulations of short operation periods (operating with a locked rotor or during start-up). Research shows that under those operating conditions the heating of the stator components may be examined regardless of the rotor heating [17,24]. The operation periods analysed here are too short for the stator to exert a noticeable influence on the rotor heating and vice versa. Therefore, the heat exchange between the stator and the rotor can be omitted in the analysis. The temperature field is calculated for the area indicated in Figure 4, which encompasses a half of the rotor's length and half of the rotor's slot pitch. The analysed area has three types of boundary conditions: the Dirichlet boundary condition for the A_1 surface, the Neumann condition for the A_2, A_3, and A_4 surfaces, and the Newton condition for the areas that are in direct contact with a cooling fluid.

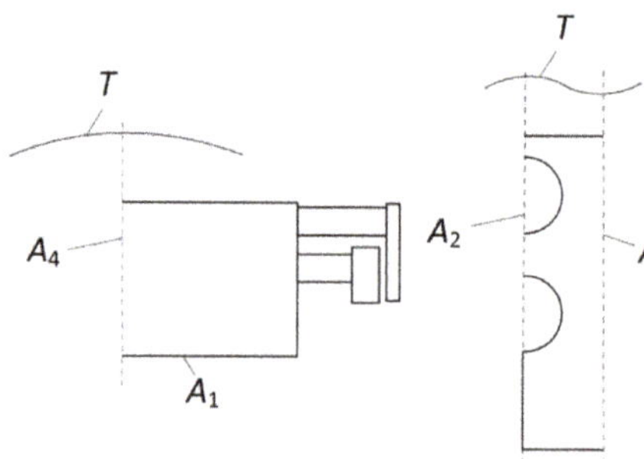

Figure 4. Analysed area of double-cage motor's rotor.

In the control volume method, the analysed object is subdivided into a number of homogeneous elements. The energy balance is calculated for the entire system. If it is assumed that in the analysed object, the separate control element with volume $V_{c(i)}$ is sufficiently small that the temperature in its entire area is the same and has the value $T_{(i)}$, and that the point heat source $\dot{q}_{v(i)}$ is located at the centre of gravity of this element, then the transient temperature field in a motor analysed using the heat network created by the control volume method [25] is described by the following system of Equation (1) together with the initial condition (2):

$$\sum_j \Lambda_{(i,j)} T_{(j)} - \left(\sum_m \Lambda_{a(i,m)} + \sum_j \Lambda_{(i,j)} \right) T_{(i)} + \sum_m \Lambda_{a(i,m)} T_{a(i)} + V_{c(i)} \dot{q}_{v(i)} = V_{c(i)} c_{(i)} \rho_{(i)} \frac{dT_{(i)}}{dt}, \quad (1)$$

where $i = 1, 2, \ldots, z_n$ is the number of elements into which the considered area is divided, z_n is the number of area elements, $\Lambda_{(i,j)}$ is the thermal conductivity for the heat flowing from node i to node j, $\Lambda_{a(i,m)}$ is the thermal conductivity for the heat flowing from node i to the external surface m and the heat transferred from surface m, c is the specific heat, ρ is the density, t is the time, $\dot{q}_{v(i)}$ is the volumetric density of the heat sources, and T is the temperature:

$$T(\mathbf{r}, t)\big|_{t=0} = T_0(\mathbf{r}), \quad (2)$$

where T_0 is the initial temperature and **r** is the positional vector describing the position of the element in question. To determine $\dot{q}_v$ in the area of the windings, it is necessary to calculate the distribution of the current density $J(\mathbf{r}, t)$. Then:

$$\dot{q}_v = J^2(\mathbf{r}, t) \frac{1}{\gamma(T)}, \quad (3)$$

where γ-conductivity.

For a double-cage induction motor with a soldered cage, the equations describing the transient electromechanical state in a two-axis coordinate system rotating at the speed ω_x have the form [26]:

$$\mathbf{U} = \frac{d}{dt}\Psi + \mathbf{\Omega}\Psi + \mathbf{RI}, \quad (4)$$

$$\Psi = \mathbf{LI}, \quad (5)$$

$$J_m \frac{1}{p} \frac{d\omega}{dt} = p\mathrm{Re}\left(j\underline{\Psi}_1 \underline{I}_1^*\right) - T_L \quad (6)$$

where $\mathbf{U} = \left[\underline{U}_1, 0, 0\right]^T$, $\mathbf{I} = \left[\underline{I}_1, \underline{I}_{2(1)}, \underline{I}_{2(2)}\right]^T$, and $\Psi = \left[\underline{\Psi}_1, \underline{\Psi}_{2(1)}, \underline{\Psi}_{2(2)}\right]^T$ are the voltages, currents, and linkage fluxes, respectively, $\mathbf{\Omega} = \mathrm{diag}[j\omega_x, j(\omega_x - \omega), j(\omega_x - \omega)]^T$, **R** and **L** are the resistance and inductance matrices, ω is the electric rotor angular velocity, U_1 is the amplitude of the supply voltage, $\underline{I}_1$, $\underline{I}_{2(1)}$, and $\underline{I}_{2(2)}$ are the stator and rotor two-axis vector currents (complexor), respectively, J_m is the moment of inertia, T_L is the load torque, and p is the number of pole pairs. Knowledge of the currents flowing in the motor windings allows the determination the volumetric density of heat sources $\dot{q}_V$ necessary to determine the temperature field of the motor winding. This method allows the calculation of the temperature field for the double-cage induction motor with the parameters presented in Table 1.

Table 1. Chosen motor parameters.

Parameter	Value
Rated power, kW	320
Rated voltage, V	6000
Rated frequency, Hz	50
Number of pole pairs	3
Air gap, mm	1.6
Stator diameter, m	0.888/0.600
Rotor diameter, m	0.597/0.369
Core length, m	0.56
Number of stator/rotor slots	72/58
Class of insulation	F
Ratio of coil span to pole pitch	10/12
Number of stator—winding turns per phase	240
Stator—winding resistance, Ω	0.648

The winding of the cage consists of bars short-circuited with copper end rings. The bars of the starting cage are made of brass, whereas the bars of the working cage are made of copper. The dimensions of the rotor and stator slots are shown in Figure 5.

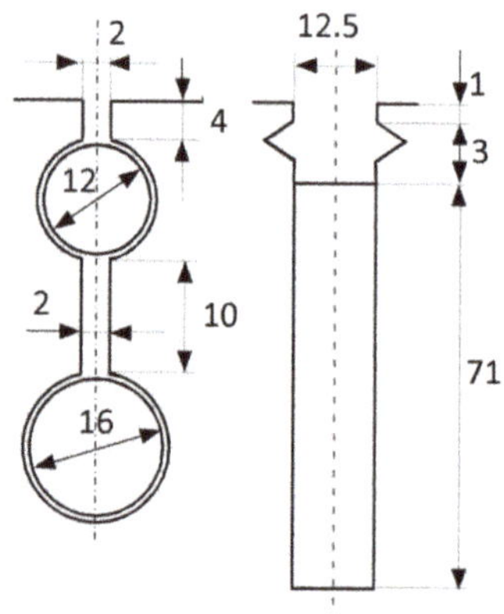

Figure 5. The shape and dimensions of rotor and stator slots of double-cage motor with a welded cage.

The heat transfer coefficient from the areas in direct contact with cooling air was calculated on the basis of paper [18]. It provides an equation for calculating the equivalent heat transfer coefficient through natural convection and radiation, in the following form:

$$\alpha = \nu_p K_p (T + T_0)\left(T^2 + T_0^2\right) + C_k \sqrt{\varphi}(T - T_0)^{0.25}, \quad (7)$$

where: ν_p is the emissivity coefficient of the surface, K_p is the Boltzman constant, T is the surface temperature, T_0 is the temperature of the air surrounding the surface, C_k is the coefficient included within the limits (2.79 … 3.39) W/m^{2}K, ϕ is the relative air humidity. For T = 300 °C, T_0 = 20 °C, ϕ = 0.9, ν_p = 0.85 (varnished surface), C_k = 3 W/m^{2}K$^{1.25}$ the heat transfer coefficient is α = 13.2 W/m^{2}K. For a rotating rotor (forced convection):

$$\alpha_v = \alpha\left(1 + k\sqrt{v}\right), \quad (8)$$

where v is the speed of the cooling air thrown over the surface, k the coefficient included within the range 0.5 ... 1.3, in the analysis $k = 1$ was assumed for the end region. For the rated speed: $\alpha_v = 58\ W/m^2K$.

Thermal conductivity values: for the bars of the working cage 372 W/m·K, for the end rings of the working cage 372 W/m·K, for the bars of the starting cage 103 W/m·K, for the end rings of the starting cage 372 W/m·K. For the rotor core—50 W/m·K (in axial direction), 10 W/m·K (in radial direction). A slight change due to the temperature of the thermal conductivity coefficient was omitted (within the expected temperature range within 20 ... 400 °C).

3. Influence of the End Region Structure of the Double-Cage Winding on Mechanical Stresses

In a double-cage motor with a soldered cage, the mechanical stresses associated with the skin effect in the rotor bars are not as important as in a deep bar motor. In a double-cage winding, the forces due to the thermal expansion of the end ring and centrifugal forces originating from the mass of the end region of the winding are more significant. The associated stresses can be dangerous, especially during long start-up periods, when the end rings experience maximum heat.

In double-cage motors with a soldered winding, the end rings are typically moved away from the core. There are many construction solutions for soldered bar connections with rings. Insert connections, as displayed in Figure 6, are frequently used.

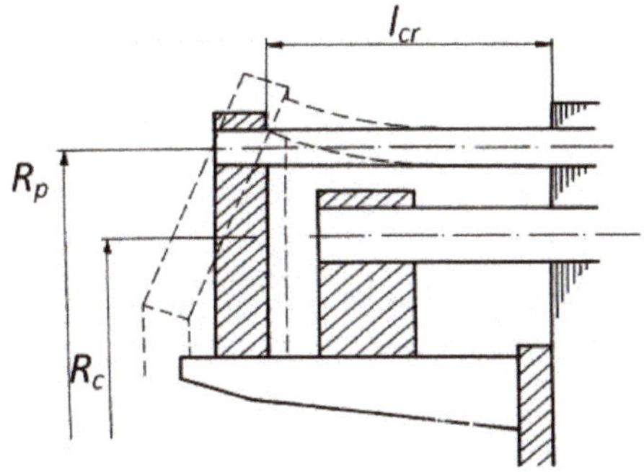

Figure 6. Insert connections in double-cage motor.

The length of the overhang of the bars beyond the core is limited mainly by implementation considerations. An excessive increase of the overhang is meaningless owing to the length of the entire machine. This problem was discussed, among others, in [18]. The model of the phenomena occurring in the external heated part, rotating at a constant winding speed, assumes that the bar is rigidly fixed in the packet, and a radial force and circular-symmetric moment act in the location of the rigid connection of the bar to the ring. They cause the displacement of the external part of the cage as indicated by the dashed line in Figure 6. During the motor start-up, the centrifugal forces and forces originating from the thermal deformations of the ring act together on the cage bars.

Using this model, simulation calculations were conducted for a double-cage induction motor with parameters presented in Table 1. The results are presented in Figure 7 as the total stress value in the bars during a prolonged motor start-up.

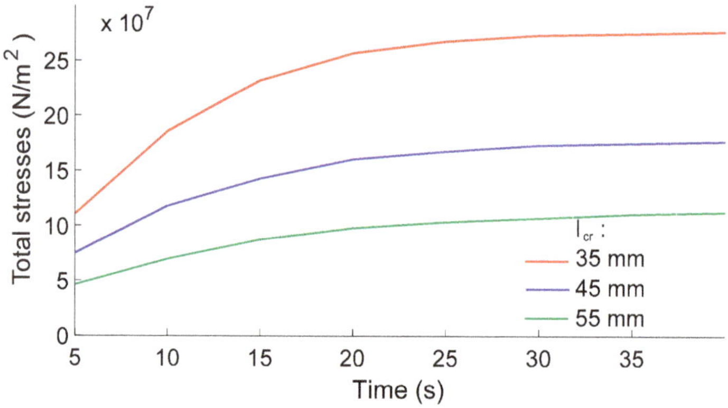

Figure 7. Stresses in starting-cage bar during motor start-up.

As can be observed in Figure 7, a change in the outward extension of the bars within the limits of 23% around the value of l_{cr} = 45 mm causes a significant change in the total stresses when the bar emerges from the core during a long-term startup. When reducing the size of l_{cr}, the stress increases by approximately 56%; when the value of l_{cr} increases, the stresses decrease by approximately 36%. In double-cage motors with bars of a circular cross section, the highest temperature occurs in the end part of the cage. The natural method to avoid a large unevenness in temperature distribution along the bar would be to eliminate the areas with elevated temperature from the cage structure, i.e., to shorten the bars. However, the reduction of this overhang leads to a significant increase in the mechanical stresses in the bars due to the thermal deformations of the end ring.

4. Possibilities of Equalising the Temperature Distribution along the Axis of the Motor Starting Cage Bar

In certain cases, the temperature in the end region of the cage is sufficiently high to cause a loss of elasticity of the material, whereas in the remainder of the winding, the temperature does not significantly reduce the mechanical strength of the bar material. These highly heated end regions of the rotor winding determine the durability and reliability of the entire motor, despite the fact that their share in the volume of the cage is small. Therefore, we must attempt to obtain a more uniformly heated structure. This applies, in particular, to work in conditions of prolonged start-up or operation with a locked rotor. This issue is discussed in [18], where the possibilities of equalising the temperature distribution along the axis of the starting cage bar of a double-cage motor were considered. The influence of the slot clearance was examined (Figure 8) for a double-cage induction motor with parameters presented in Table 1. The conditions of the heat transfer from these parts to the temperature field of the cage, in the state of working with the rotor locked, were also examined (Figure 9).

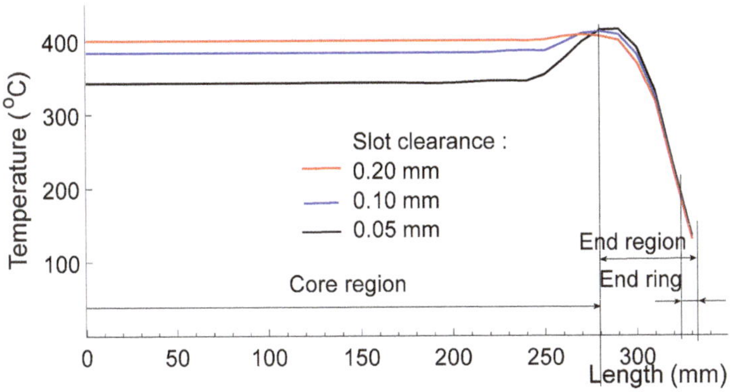

Figure 8. Influence of slot clearance on temperature distribution along starting-cage bar.

Figure 8 displays the temperature distribution along the starting cage bar with different fits of the bar to the slot (slot clearance 50, 100 or 200 μm). From the calculation results displayed in Figure 8, it can be observed that improving the fit of the bar to the slot leads to a significant reduction in the bar temperature in the core part; however, the temperature remains high in the end region and this costly treatment does not eliminate the cage damage discussed above. Because only somewhat exceeding a certain temperature (dependent on the bar material) decreases the tensile strength sharply, even a marginal reduction in the temperature of the end regions is beneficial from the point of view of its durability. To obtain similar conditions for heat exchange in the end region and slot part of the cage, an equality of thermal resistance should be ensured by means of heat transfer between the external part of the bar and the surrounding air, and between the cage bar and the rotor core. For the motor considered, the value of the heat transfer coefficient in the external parts of the cage was calculated, providing heat transfer conditions similar to those in the slot part, i.e., 254 W/m^{2}K. The temperature distribution along the bars of the starting cage after 12 s of operation of the motor with the rotor locked supplied with the rated voltage is displayed in Figure 9. It is not possible to obtain a coefficient of this value in ordinary construction solutions because the values of this coefficient encountered in practice are many times smaller than required.

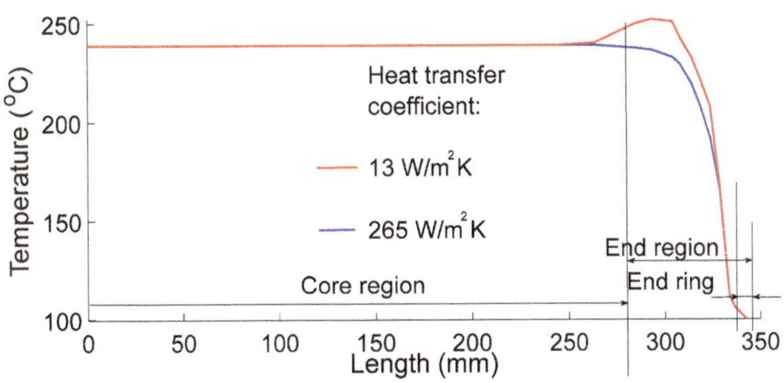

Figure 9. Influence of heat of transfer coefficient from end region of cage to temperature distribution along bar axis.

5. New Solution for the Construction of a Starting Cage of a Double-Cage Motor

The presented review of the results of calculations of the temperature of the double-cage winding indicates that in the end region of the cage construction, the possibility of further increasing the resistance of the cage to the related effects of working with the locked rotor and during a prolonged start-up should be investigated. The end region of the starting cage is most exposed to destructive activities under these working conditions, and it is necessary to investigate the possibility of increasing the resistance of the double-cage winding to the effects of a prolonged start-up.

The authors of [19] presented a proposal for an innovative construction solution of the starting cage that allows a limitation of the temperature of the end region of the bars. The starting cage displayed in Figure 10, based on this solution, is characterised by sleeves (3) of the same material as the bars, applied to the ends of all the bars (1) protruding from the core. The sleeves (3) adjacent to the end rings (2) are preferably permanently connected to the rings using a hard solder, welding, or sealing.

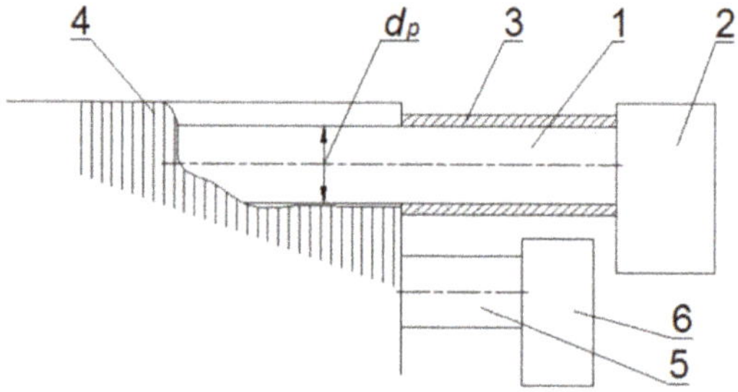

Figure 10. Innovative construction solution of starting cage: 1—starting cage bar, 2—end ring of starting cage, 3—sleeve, 4—core, 5—working cage bar, 6—end ring of working cage, d_p—diameter of starting bar.

The method of building the starting cage is based on the fact that after inserting the bars (1) into the slots, the sleeves (3) are heated to a temperature at which the inner diameter of the sleeve is larger than the diameter of the bar d_p. Before the sleeves are heated, the inner diameter of the sleeve d_r is less than the diameter of the rotor cage bar. Then, the hot sleeves (3) are applied to the ends of the bars (1). Finally, to the ends of the bars (1), the end rings (2) are inserted and joined with the bars and sleeves with hard solder, welding, or sealing.

The sleeves (3) superimposed on the ends of the bars (1) increase the cross section and thermal capacity of the external part of the bars (1); thus, in the ends of the bars (1), the current density is reduced and the temperature decreases. Hence, a starting cage prepared in this manner is characterised by greater start-up durability.

6. Simulation Tests of the Temperature Field of a Double-Cage Motor with a New Design Solution

Using Equations (1)–(6), simulation tests were performed for a double-cage motor with the parameters presented in Table 1, with an emergency locked rotor ($\omega = 0$). The calculations were performed for both a motor with a starting cage of conventional structure and a double-cage induction motor with the new construction solution for the starting cage presented in [19].

Figure 11 displays the temperature distribution along the centre axis of the starting cage bar for the classic solution and with 1.8 mm thick sleeves made of brass, similar to the starting cage bars. Figure 12 indicates the influence of the thickness of the sleeves (1.8 mm and 1.0 mm) on the distribution of the temperature along the axis of the starting cage bar. Owing to the new construction of the end region of the starting cage, locations previously threatened by overheating are largely eliminated. For the solution proposed in [19], the temperature of the starting bar in the region of the outward reach of the bars is considerably less than for the classical solution. Increasing the thickness of the sleeves further reduces this temperature (Figure 12).

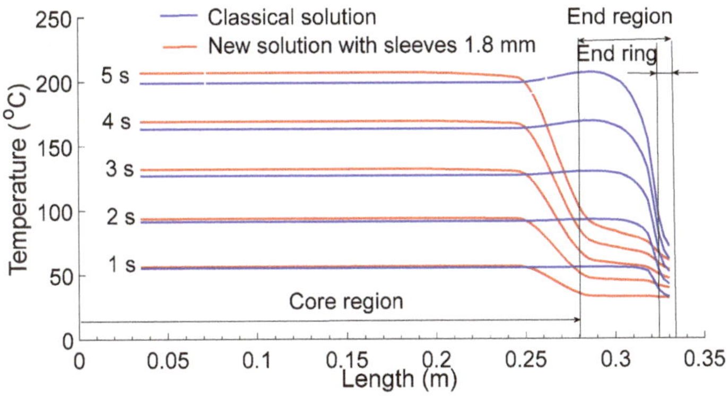

Figure 11. Distribution of temperature along axis of starting-cage bar for classic solution and with sleeves.

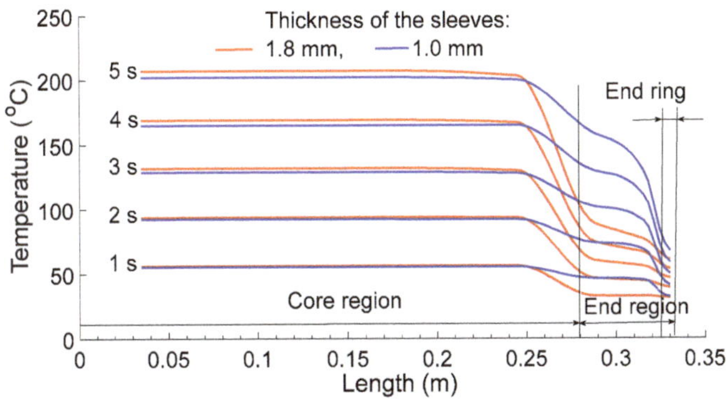

Figure 12. Influence of thickness of sleeves on temperature distribution along axis of starting-cage bar.

7. Conclusions

The startup process of the induction cage motor is one of the most important phases of the drive operation. During the start-up an electrical motor can be subject to severe electrical and thermal loads. These loads, despite their relatively short durations, significantly affect the motor's lifetime and reliability. The heat generated during a prolonged start-up causes mechanical stresses which may damage the motor cage. By using the double-squirrel-cage soldered winding, large values of starting torques are achieved. This is, however, at the cost of non-uniform heating of the cage, along with the tendency of the cage bars to overheat. A large amount of heat is released during start-up in the bars of the motor's starting cage. The highest temperature is observed in the end region of the starting cage bar. There is a significant difference in the axial temperature distribution in the bars owing to the different types of heat exchange in the core region and the end region of the bars. The limitation of the slot clearance to improve heat transfer into the core does not change the fact that the temperature of the end region of the bars remains high. Moreover, reducing the distance between the end rings and the core is disadvantageous owing to the increase in the bending stress in the bars from the thermal expansion of the rings. Excessive heating of the end region of the starting cage during start-up can lead to motor failure.

The application of sleeves onto the ends of the starting cage bars can significantly reduce the temperature of the connection of the starting cage bars with the rings and thus increase the resistance of the rotor starting cage to thermal exposure during motor start-up. The conducted simulation tests demonstrated that thanks to the new construction of the end region of the starting cage, locations previously threatened by overheating were largely eliminated. However, there arise additional technological difficulties associated with the process of setting the sleeves onto the bars of the starting cage.

The heating of the motor windings is a transient phenomenon which is closely related to the transient electromechanical process. Both processes are mutually related since the distribution of heat sources depends on the temperature of motor windings. The results of the analysis can be massively improved by applying detailed numeric calculation methods (FEM analysis). A considerable limitation, particularly at an early stage of work on the presented problem, is the cost and a long calculation time for 3D coupled problems, hence the choice of faster and less expensive methods which have been partially verified through experimental research.

In view of the relatively limited number of publications in the area of costly experimental research involving high-power motors, it is necessary to plan and conduct such research. This applies in particular to investigating energy aspects in transient states for verifying the simulation models and the further improvement of cage winding constructions.

Author Contributions: Conceptualization, J.M. and W.P.; methodology, J.M.; software, J.M.; formal analysis, J.M. and W.P.; investigation, J.M. and W.P.; resources, J.M.; writing—original draft preparation, J.M.; writing—review and editing, J.M. and W.P.; visualization, J.M.; supervision, J.M.; project administration, J.M.; funding acquisition, J.M.

Funding: This work is financed in part by the statutory funds of the Department of Electrodynamics and Electrical Machine Systems, Rzeszow University of Technology and in the part by Polish Ministry of Science and Higher Education under the program "Regional Initiative of Excellence" in 2019–2022. Project number 027/RID/2018/19, amount granted 11 999 900 PLN.

Conflicts of Interest: The authors declare no conflict of interest.

References

1. Gritli, Y.; Lee, S.B.; Filippetti, F.; Zarri, L. Advanced diagnosis of outer cage damage in double-squirrel-cage induction motors under time-varying conditions based on wavelet analysis. *IEEE Trans. Ind. Appl.* **2014**, *50*, 1791–1800. [CrossRef]
2. Hmida, M.A.; Braham, A. An on-line condition monitoring system for incipient fault detection in double-cage induction motor. *IEEE Trans. Instrum. Meas.* **2018**, *67*, 1850–1858. [CrossRef]
3. Tulicki, J.; Weireb, K.; Sulowicz, M. The possibility of distinguishing rotor cage bar faults in double squirrel cage induction motors on the basis of the stator current signal. In Proceedings of the 2017 International Symposium on Electrical Machines (SME), Naleczow, Poland, 18–21 June 2017.
4. Dias, C.G.; da Silva, L.C.; Chabu, I.E. Fuzzy-based statistical feature extraction for detecting broken rotor bars in line-fed and inverter-fed induction motors. *Energies* **2019**, *12*, 2381. [CrossRef]
5. Gritli, Y.; Tommaso, O.D.I.; Miceli, R.; Filippeti, F.; Rossi, C. Vibration signature analysis for rotor broken bar diagnosis in double cage induction motor drives. In Proceedings of the 4th International Conference on Power Engineering, Energy and Electrical Drives, Istanbul, Turkey, 13–17 May 2013; pp. 1814–1820.
6. Nakahama, T.; Biswas, D.; Kawano, K.; Ishibashi, F. Improved cooling performance of large motors using fans. *IEEE Trans. Energy Convers.* **2006**, *21*, 324–331. [CrossRef]
7. Staton, D.; Boglietti, A.; Cavagnino, A. Solving the more difficult aspects of electric motor thermal analysis in small and medium size industrial induction motors. *IEEE Trans. Energy Convers.* **2005**, *20*, 620–628. [CrossRef]
8. Ulu, C.; Korman, O.; Komurgoz, G. Electromagnetic and thermal design/analysis of an induction motor for electric vehicles. In Proceedings of the 2017 8th International Conference on Mechanical and Aerospace Engineering (ICMAE), Prague, Czech Republic, 22–27 July 2017.
9. Xie, Y.; Guo, J.; Chen, P.; Li, Z. Coupled fluid-thermal analysis for induction motors with broken bars operating under the rated load. *Energies* **2018**, *11*, 2024. [CrossRef]
10. Gyftakis, K.N.; Athanasopoulos, D.; Kappatou, J. Study of double cage induction motors with different rotor bar materials. In Proceedings of the 20th International Conference on Electrical Machines (ICEM), Marseille, France, 2–5 September 2012; pp. 1450–1456.

11. Maddi, Z.; Aouzellag, D. Dynamic modelling of induction motor squirrel cage for different shapes of rotor deep bars with estimation of the skin effect. *Prog. Electromagn. Res.* **2017**, *59*, 147–160. [CrossRef]
12. Sundaram, M.; Mohanraj, M.; Varunraj, P.; Kumar, T.D.; Sharma, S. FEA based electromagnetic analysis of induction motor rotor bars with improved starting torque for traction applications. In Proceedings of the International Conference on Automatic Control, Mechatronics and Industrial Engineering (ACMIE), Suzhou, China, 29–31 October 2018.
13. Lee, H.J.; Im, S.H.; Um, D.Y.; Park, G.S. A design of rotor bar for improving starting torque by analyzing rotor resistance and reactance in squirrel cage induction motor. *IEEE Trans. Magn.* **2017**, *99*, 1–4. [CrossRef]
14. Orsag, O.; Rusnok, S.; Sobota, P.; Kacor, P. Influence of rotor slot on the parameters of induction motor. In Proceedings of the 2017 IEEE International Conference on Environment and Electrical Engineering and 2017 IEEE Industrial and Commercial Power Systems Europe (EEEIC/I&CPS Europe), Milan, Italy, 6–9 June 2017.
15. Livadaru, L.; Simion, A.; Munteanu, A.; Cojan, M.; Dabija, O. Dual cage high power induction motor with direct start-up design and FEM analysis. *Adv. Electr. Comput. Eng.* **2013**, *13*, 55–58. [CrossRef]
16. Sinha, S.; Deb, N.K.; Biswas, S.K. The design and its verification of the double rotor double cage induction motor. *J. Inst. Eng. (India) Ser. B* **2017**, *98*, 107–113. [CrossRef]
17. Mróz, J. *The Analysis of Coupled Electromechanical and Thermal Problems in Transient States of Double-Cage Induction Motors*; Publishing House Rzeszow University of Technology: Rzeszow, Poland, 2013.
18. Mróz, J. Possibilities of equalizing the temperature distribution in the bars of the starting cage along the axis of a double cage induction motor. *Przegląd Elektrotech.* **1998**, *12*, 314–317.
19. Poprawski, W.; Wolnik, T. Innovative design of double squirrel cage induction motor for high start frequency operation. *Electr. Mach. Trans. J. Inst. Electr. Drives Mach. KOMEL* **2016**, *111*, 41–44.
20. Krok, R. Influence of work environment on thermal state of electric mine motors. *Arch. Electr. Eng.* **2011**, *60*, 357–370. [CrossRef]
21. Al'Akayshee, Q.; Staton, D.A. 1150 hp motor design, electromagnetic and thermal analysis. In Proceedings of the 15th International Conference on Electrical Machines (ICEM), Bruges, Belgium, 25–28 August 2002.
22. Dymond, J.H.; Ong, R.; Stranges, N. Instrumentation, testing, and analysis of electric machine rotor steady-state heating. *IEEE Trans. Ind. Appl.* **2002**, *38*, 1661–1667. [CrossRef]
23. Naderi, P. Modified magnetic-equivalent-circuit approach for various faults studying in saturable double-cage-induction machines. *IET Electr. Power Appl.* **2017**, *11*, 1224–1234. [CrossRef]
24. Mróz, J. The model of double-cage induction motor for the analysis of thermal fields in transient operations. *Arch. Electr. Eng.* **2017**, *66*, 397–408. [CrossRef]
25. Taler, J.; Duda, P. *Solving Straight and Reverse Problems of Heat Conduction*; WNT: Warsaw, Poland, 2003.
26. Paszek, W. *Transient States of AC Electric Machines*; WNT: Warsaw, Poland, 1986.

energies

MDPI

Article

Multi-Physics Tool for Electrical Machine Sizing

Yerai Moreno [1,*], Gaizka Almandoz [1], Aritz Egea [1], Patxi Madina [1] and Ana Julia Escalada [2]

[1] Faculty of Engineering, University of Mondragon, 20500 Mondragon, Spain; galmandoz@mondragon.edu (G.A.); aegea@mondragon.edu (A.E.); pmadina@mondragon.edu (P.M.);
[2] ORONA Elevator Innovation Centre, 20120 Hernani, Spain; ajescalada@orona-group.com
* Correspondence: yerai.moreno@alumni.mondragon.edu

Received: 17 February 2020; Accepted: 26 March 2020; Published: 2 April 2020

Abstract: Society is turning to electrification to reduce air pollution, increasing electric machine demand. For industrial mass production, a detailed design of one machine is usually done first, then a design of similar machines, but different ratings are reached by geometry scaling. This design process may be highly time-consuming, so, in this paper, a new sizing method is proposed to reduce this time, maintaining accuracy. It is based on magnetic flux and thermal maps, both linked with an algorithm so that the sizing process of an electrical machine can be carried out in less than one minute. The magnetic flux maps are obtained by Finite Element Analysis (FEA) and the thermal maps are obtained by analytical models based on Lumped Parameter Circuits (LPC), applying a time-efficient procedure. The proposed methodology is validated in a real case study, sizing 10 different industrial machines. Then, the accuracy of the sizing tool is validated performing the experimental test over the 10 machines. A very good agreement is achieved between the experimental results and the performances calculated by the sizing tools, as the maximum error is around 5%.

Keywords: sizing methodology; electrical machines; thermal model; electromagnetic model; permanent magnet

---

1. Introduction

Society is turning to electrification in transportation and industrial processes to reduce emissions, embrace alternative energy, and increase efficiency. In this regard, permanent magnet synchronous machines (PMSM) are mostly used for traction applications, due to their high power density and high efficiencies [1].

However, the main drawback of PMSM and electrical machines in general is their complex and time-consuming design process, as different targets must be reached optimizing various parameters that are cross-coupled. Usually, the final objective of the design is to minimize the cost of the machine while maximizing its efficiency. To fulfill these objectives coping with the growing machine demand, the design process must be improved.

Once the design requirements are defined, there are two designing scenarios, one where the machine is designed from scratch and another one where a 2D computer-aided design plane (2D CAD) is selected from the ones previously designed and the machine is sized setting the appropriate stack length and number of turns for the specific application. Different magnets and sheet types can be also chosen. Usually, in a company with a wide range of electric machines, the second scenario could be more usual, making it more competitive in the market, as the process is faster than designing a machine from zero. In addition, keeping the same electric sheets and the same motor concept reduces the manufacturing cost considerably, as the same manufacturing process can be used for building a wide range of electrical machines.

In this paper, the second scenario is analyzed, where the stator and rotor sheets are selected from a 2D plane database, and after the machine is sized. In Figure 1a, the general process is shown.

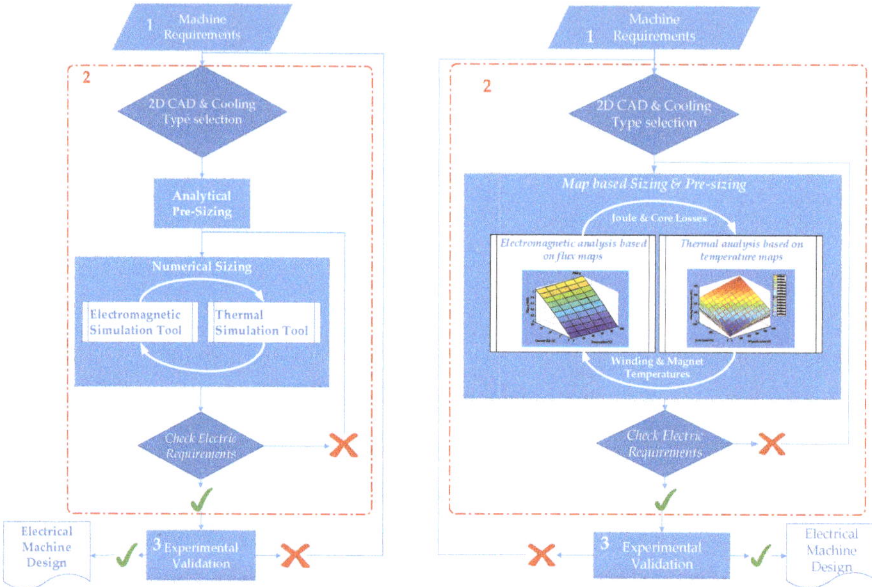

(**a**) (**b**)

Figure 1. Description of the sizing methodology using multi-physics tool. (**a**) classical Method; (**b**) integration of the sizing multi-physics tool in the classical method.

Firstly, machine requirements must be defined (1) including working currents, voltages, application temperature, output torque, and speed. These technical requirements come from the customer functional specifications that must be precise, as the design is optimum for the application.

Secondly, the machine is sized (2). The optimum 2D plane and cooling type must be selected for the application. Then, an analytical pre-sizing is made, to have an approximate number of turns, and stack length, together with the wire section. Next, the main step of the sizing process starts, using electromagnetic simulation tools together with thermal ones, to make a more accurate calculation of the machine. Lastly, the obtained performances are automatically checked, and if the machine does not comply the requirements, changes are made to the stack length or the number of turns, making the calculation again, until the design fulfills the requirements.

Finally, the machine is validated experimentally in (3) to check that the machine performance is equal to the simulated one. If the bench test is correct, the sizing of the machine is achieved; otherwise, the process must start again from the beginning.

As mentioned before, PMSM is the most used electric machines for traction, and one of their main characteristics is the nonlinearity, as well as their temperature dependences in the generated magnetic flux. They can also get demagnetized if they continuously suffer thermal overload [1–3]. This is why electromagnetic simulations must be linked to thermal ones, making them more accurate as a whole. However, linking the two simulations causes a longer iterative process, increasing the computational and design time.

To avoid this, a map interpolation is proposed, to obtain magnet and copper temperature based on Joule and Core losses. It will obtain the steady-state temperatures by a fast iteration taking into account loss change with temperature. Finally, working temperature performances can be obtained with a new magnetic flux interpolation and some simple calculations.

With respect to electromagnetic simulations, there are different tools to make it, with diverse accuracy and time consumption ranks—for instance, analytic equations, lumped parameter models, Fourier series models, and Finite Element Analysis (FEA).

The fastest way of sizing a machine is using fundamental machine sizing equations as shown in [4,5]. This analytical method is mostly used for preliminary machine sizing, in the first stages of the process due to its speed. Nevertheless, it does not provide a high accuracy compared to FEA solutions.

Another analytical method is the one based on Fourier series. In this case, the accuracy and computational time depend on the number of spatial harmonics considered in the different machine regions. The higher the harmonic orders are, the higher the accuracy is, but the longer time the computation takes. If magnetic saturation plays an important role in the designed machine, this method should be avoided using FEA, where local magnetic saturation may be considered [6].

With the aim to be more accurate, Lumped-Parameter (LP) models are developed. This method takes into account magnetic saturation and it can be more accurate than other analytical methods, still being faster than FEA. However, the more accurate is the result, the slower is the calculation, so the equilibrium between speed and accuracy is the key [1].

FEA is considered the standard tool for electric machine analysis, as it has a detailed magnetic field solution, including saturation, providing an accurate result of the machine performance according to the density of the mesh. Despite this, FEA needs a massive computational effort and time consumption [7].

In order to reduce FEA computation time maintaining its accuracy, different methods are found in the literature that mix initial FEA simulations with different scaling methods to size different machines. In [8,9], dimensional and number of turns scaling techniques are used after a base machine FEA simulation is carried out. This simulation generates magnetic flux, loss, and torque maps that are used to generate other machine models with the mentioned scaling techniques. For instance, second-order polynomial functions are used in [3] to describe magnetic flux linkage variations respect to current, using FEA models just to calculate parameters of the functions. Then, combined with other analytical models, other machines can be sized based on the reference. Moreover, in [10], FEA is also combined with analytical models to obtain a rapid tool of induction machine mapping in dq axes. The results of these methods are rather accurate, as they are based on FEA models.

As mentioned before, the thermal model must also be simulated, to evaluate working temperatures and their distribution along the machine, linking it to the electromagnetic model. As in electromagnetic calculations, there are various methods with different performances such as FEA, LP models, and Computational Fluid Dynamic (CFD), shown in [11,12].

As in the previous case, the objective is to simulate as accurately and as quickly as possible. The speediest way is to simulate the machine with LP thermal networks as they are quite accurate as shown in [4,13], but they can become as slow as FEA if many nodes are introduced. Therefore, some reduced node models are found in literature, reducing computation time considerably, maintaining good accuracy [14,15].

Gaining accuracy, there is CFD software, used for modelling cooling systems, calculating flow rates, and heat transfer. The main asset is that it can be used to predict the flow in complex regions, such as around the end windings, with great precision. Moreover, the data obtained from CFD can be used to improve analytical algorithms in analytic thermal networks. However, its big disadvantage is its huge computation needs and time consumption, making it unsuitable for fast sizing process, but might be used in big machines with a high cost of prototypes [11,12].

In thermal modelling, FEA is used to accurately calculate the conduction heat transfer in complex geometric shapes, such as heat transfer through strands of copper in a slot. Nevertheless, it has an important limitation as the software uses analytical/empirical-based algorithms for convection boundaries, exactly as in the lumped circuit analysis. As a result, the accuracy is dependent on the same factors for the thermal network; just making a difference when the solid component conduction must be calculated precisely. Some authors used a 3D reduced-order FEA model to reduce computational time maintaining a rather good accuracy [2,11].

Taking into account all the simulation methods found in literature, magnetic flux and temperature maps are used in the proposed sizing method. Magnetic flux maps are obtained from FEA simulations and temperature maps are generated from lumped parameter network simulations. In this way, a fast and accurate method can be defined.

In this article, a fast sizing method is proposed. It sizes a machine in about one minute, maintaining the accuracy of the Finite element models. The major novelty of this method is the coupling between thermal and electromagnetic fields. The sizing is done by an iterative algorithm.

Finally, the proposed method is validated in a real case study, sizing 10 industrial machines used for people transportation systems.

This paper is organized as follows: Section 2 describes the proposed method and explains the theoretical base of the method. The initial simulation process and the sizing algorithm are presented. Section 3 compares the obtained data with 10 industrial machine experimental results to validate the algorithm and the proposed process. In Section 4, the novelty of the work is presented, with the obtained results and their conclusions.

2. Description of the Proposed Sizing Procedure

In this section, the procedure for obtaining the magnetic flux maps and the thermal maps is explained. As it is shown in Figure 2, this procedure consists of two different simulations. Magnetic flux maps are obtained by electromagnetic FEA simulations performed using Altair Flux® (Troy, MI 48083, USA), whereas the temperature maps are obtained by thermal simulations carried out using Motor-CAD® (Wrexham LL13 7YT, UK).

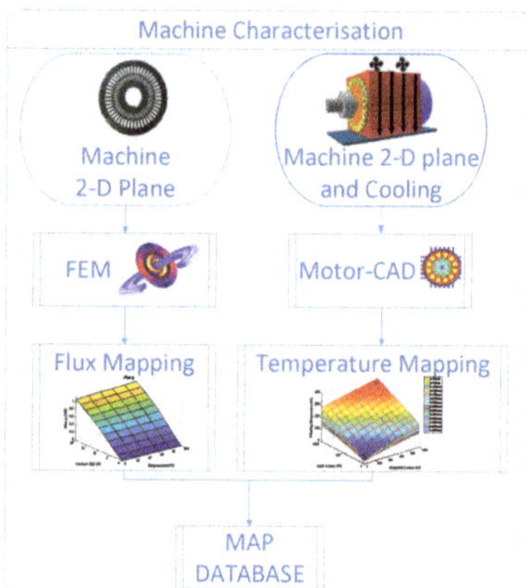

Figure 2. Map database creation.

Concerning the magnetic flux maps, d-q axis magnetic flux is computed as a function of d-q axis currents and magnet temperature. Regarding the temperature maps, two maps are obtained as well—one map for the average winding temperature and another one for the average magnet temperature. Both maps are computed as a function of the Joule and magnetic losses, accounting for different stack lengths.

These maps make up a database in which different magnetic circuit 2D geometries and cooling types are included. This way, during the sizing process of electrical machines, different magnetic flux and temperature maps are used according to the chosen 2D geometry of the lamination and the cooling solution. In the next sections, more details are given about the map creation procedure.

2.1. *Magnetic Flux Map Creation*

D-Q axis magnetic flux maps are obtained performing FEA magneto-static simulations. As the electrical machine analysis involves magnetic and electric domains, the magnetic circuit of the motor is coupled to the electric circuit.

In case the machine geometry is simple, FEA 2D simulations could be suitable. However, complex motor geometries might require FEA 3D simulations. The proposed procedure in this paper is suitable for both cases, FEA 2D and FEA 3D.

The objective is to generate magnetic flux maps depending on current (in d-q axis) and temperature, so $\varphi_q\left(i_q, i_d, T_{magnet}\right)$ and $\varphi_d\left(i_q, i_d, T_{magnet}\right)$ are obtained. As the map depends on three variables, magnet temperature, d, and q currents, many simulations must be performed varying these three variables.

The first step is to define the currents and temperature variation ranges to generate the solving scenario for the model, as shown in Figure 3.

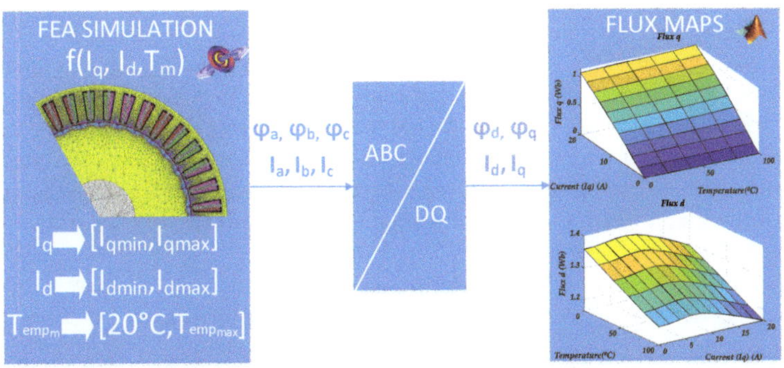

Figure 3. Magnetic flux map generation process.

One criterion for defining the currents' variation range could be the saturation of the magnetic material. For instance, for a given N, the maximum supplying current value could be defined according to the linearity of the q axis magnetic flux as a function of the q axis current, from 0 A up to the point at which the linearity of the q axis magnetic flux decrease in a given value. The effect of the magnet temperature is accounted for changing the remanence value of the magnets field. The maximum value of the remanence value can be set according to the standard data-sheets of commercial magnets. For instance, nowadays, the maximum remanence field that can be found in the market for Neodymium magnets is around 1.43 T (N52M from Baker Magnetics (5692 Elson, The Netherlands)). Concerning the minimum remanence value, it could be defined also considering standard properties of the magnets in the market. For example, a suitable criterion could be to consider the minimum remanence value at room temperature around 0.98 T (N25 BH from Baker Magnetics) and compute the remanence value at the maximum working temperature of the same magnet, about 240 °C. Applying this criterion, the magnet remanence value should be varied in the range of 0.74 T–1.43 T.

It is also important to define NI properly, to obtain data in the machine working range. If a working point of a machine exceeds the maximum value of the map range, the map will not be suitable for sizing the desired machine because it would need to extrapolate and extrapolation may generate incorrect results.

Flux® software performs the electromagnetic analysis solving Maxwell's equations with a magnetic vector potential, finally solving Equation (1) by finite element methods:

$$\nabla \times \left(v_0 \left[v_r \right] \nabla \times \vec{A} \right) + \left[\sigma \right] \left(\frac{\partial \vec{A}}{\partial t} + \nabla V \right) = 0 \tag{1}$$

where $[v_r]$ is the tensor of the reluctivity of the medium, v_0 is the reluctivity of the vacuum, $\overrightarrow{A}$ is the magnetic vector complex potential, $[\sigma]$ is the tensor of the conductivity of the medium, and V is the electric scalar potential.

Once the solving scenario is defined, the simulation is carried out to get $\varphi_a \left(i_q, i_d, T_{magnet} \right), \varphi_b \left(i_q, i_d, T_{magnet} \right)$ and $\varphi_c \left(i_q, i_d, T_{magnet} \right)$. Then, these variables are post-processed to get $\varphi_q \left(i_q, i_d, T_{magnet} \right)$ and $\varphi_d \left(i_q, i_d, T_{magnet} \right)$. This will be done with Clark–Park transformation [16], as shown in the second step of Figure 3. Finally, the obtained magnetic fluxes shown in Figure 3 will be saved in a magnetic flux map database for future use.

The resolution of the maps must be properly chosen as it might affect the accuracy of the final results given by the sizing tool. Defining at least 10 computation points in the variation range of each variable could be a criterion. This leads to at least 1000 different simulations to be performed by FEA. Another key point affecting the accuracy is the resolution of the simulations. A criterion could be to consider at least 100 points in a single electric period, so 100 points are performed for each simulation, leading to a total amount of 100,000 simulation points. Using an average computer (16 GB RAM, 64 bytes—3.41 GHz Microprocessor), the solving of a single point could take around 5 s, leading to a total simulation period of five days. It might not be too much considering that the magnetic flux maps are obtained once and then no more FEA simulations are required for a given magnetic circuit geometry. Nevertheless, in case the data-sheet must be made-up with many different magnetic circuit geometries, this task could take a lot. Thus, in this paper, a proposal is presented to reduce the computation load of the magnetic flux mapping process.

As shown in Figure 4, during one full electric period of φ_a, φ_b, and φ_c, there are six φ_d periods, so it is enough to simulate 1/6 of the period to obtain φ_d, reducing significantly the computation time.

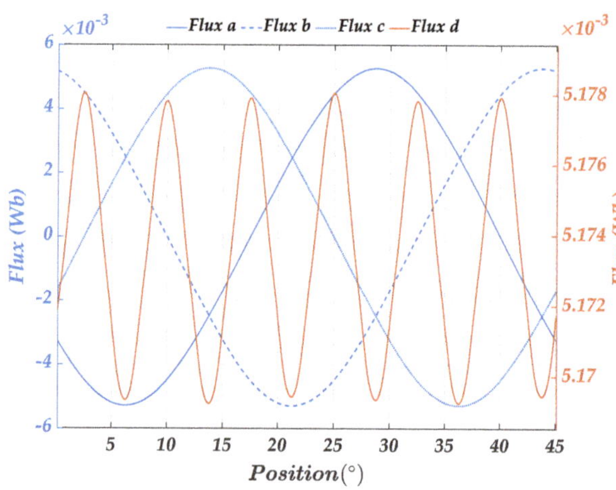

Figure 4. Magnetic fluxes during a full electrical period.

2.2. Temperature Map Generation

As mentioned in Section 1, there are different methods and software to model the thermal performance of electrical machines. In this paper, the commercial software Motor-CAD® is used to obtain the required temperature maps.

Motor-CAD® uses a three-dimensional lumped circuit model that can be used to calculate the steady-state and transient thermal characteristics of several motor types. One of the most complex aspects of motor thermal analysis is the prediction of Convection Heat Transfer mainly relating to the outer surface of the motor, but also for the internal air-gap. An estimation is made by the software using natural and forced convection correlations. Radiation Heat Transfer is also modelled in Motor-CAD.

The process for obtaining these maps is shown in Figure 5. First, the model is defined, using the stack 2D plane and the desired cooling system, generating the thermal network.

Average winding temperature and average magnet temperature maps are obtained as a function of the Joule and Magnetic losses and accounting for different stack lengths. The variation range for the losses and the stack length must be properly established to assure that all thermal situations demanded during the sizing process are covered by the maps.

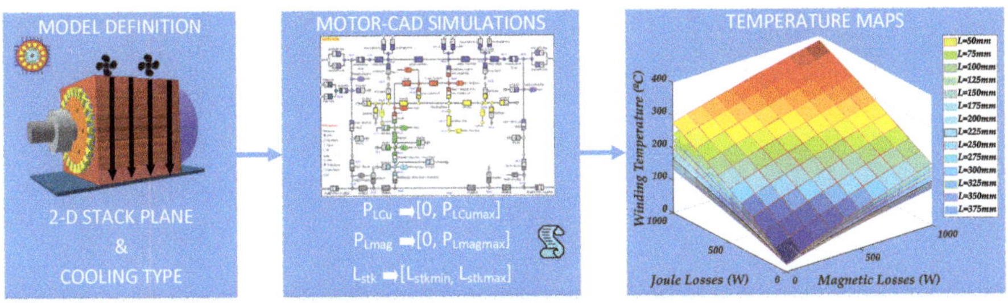

Figure 5. Temperature map generation.

The variation range and the resolution of the stack length will be defined by the user.

The variation range of the losses could be established accounting for the winding temperature. For instance, it does not make sense to consider losses that raise the temperature above the maximum limit for the maximum stack length (keeping the losses constant, the smaller the stack length is, the higher the temperatures are. Thus, for small stack lengths, the temperatures might lead above the maximum limit).

This way, the variation range of the losses and the stack length should be defined for every particular case. The model is simulated for each defined point of power losses and length combination. One of these simulations is done for each cooling type.

Finally, temperature maps are created. $T_{\text{magnet}}\left(P_{\text{LCu}}, P_{\text{Lmag}}\right)$ and $T_{\text{Cu}}\left(P_{\text{LCu}}, P_{\text{Lmag}}\right)$ maps are shown in next Section 3, Section 3.2. Each layer represents one stack length, from the smallest machine in the upper layer, to the longest in the lower (and coldest) layer.

2.3. *Sizing Method*

In this section, the proposed sizing process of electrical machines is described. As it is shown in Figure 6, before the sizing process begins, the design requirements must be defined. Then, the cooling solution and the 2D magnetic circuit geometry must be chosen. Once these two elements are chosen, the corresponding magnetic flux and temperature maps are uploaded to the sizing program. As it can be appreciated in Figure 6, the sizing process consists of three main stages:

- STAGE 1: DEFINITION OF L & Z. At the first step, preliminary values for the number of turns and the stack length are estimated. Then, if after the calculations the checking is not correct, L & Z will be recalculated. Depending on the obtained performances, L & Z values will increase or decrease.
- STAGE 2: ELECTROMAGNETIC & THERMAL ANALYSIS. With the defined L & Z, dq magnetic fluxes are obtained by the interpolation on the magnetic flux maps. Then, machine performances

are calculated, such as Joule and magnetic losses. These losses will be used to obtain the working temperature of the winding and the magnets by interpolating in the temperature maps. With the working temperatures, performances are calculated again, based on a new magnetic flux map interpolation. At the end of this process, performances of the machine are obtained at ambient and working temperatures.

- STAGE 3: AUTOMATIC CHECKING. The performances are checked and, depending on the results, the process is finished or a new iteration is started returning to Stage 1.

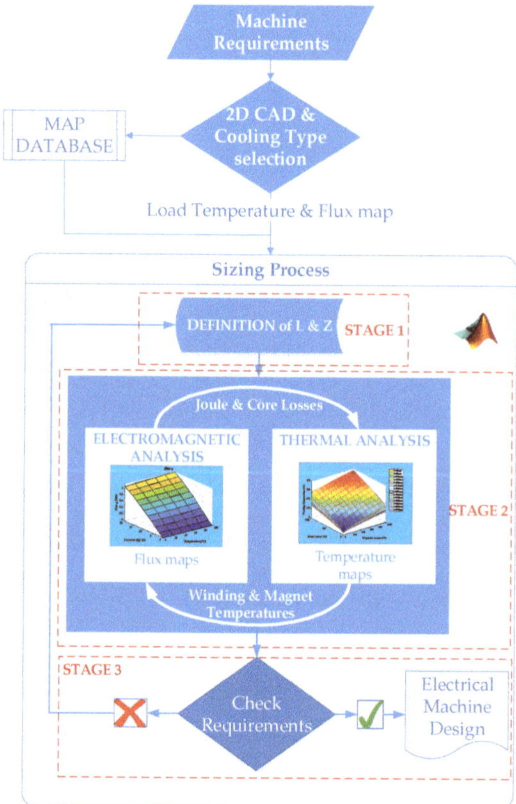

Figure 6. Sizing algorithm.

2.3.1. Stage 1: Definition of *L* & *Z*

Once the main design requirements are defined, the number of turns and the stack length are pre-calculated considering the torque Vs current requirement, and the voltage limitation, applying the next well known torque and voltage analytical Equations (2):

$$\begin{cases} T_{\mathrm{nom}} = 3p\varphi_{\mathrm{d}}LNI_{\mathrm{n}} \\ V_{\mathrm{max}}^2 = \left(-\varphi_{\mathrm{q}} + N^2Lk_{\mathrm{ov}}w_{\mathrm{e}}I_{\mathrm{n}}\right)^2 + \left(2\rho N^2 n_{\mathrm{cap}} I_{\mathrm{n}} \frac{L+L_{\mathrm{end}}}{A_{\mathrm{r}}k_{\mathrm{f}}} + \varphi_{\mathrm{d}}NLw_{\mathrm{e}}\right)^2 \end{cases} \quad (2)$$

where T_{nom} is the nominal torque (Nm), p is the machine pole pairs, φ_{d} is magnetic flux d (Wb), φ_{q} is magnetic flux q (Wb), L is the machine stack length (m), N is winding number of turns, I_{n} is the desired nominal current (A), V_{max} is the maximum allowed voltage (V), k_{ov} is the overlapping factor,

w_e is the machine rotational speed (rad/s), ρ is the wire resistivity (Ωm), n_{cap} is the number of winding layers, L_{end} is the end-winding length (m), A_r is the copper wire area (m^2), and k_f is the filling factor.

Solving the equation system (2), the initial length and number of turns are estimated, providing an accurate starting point of the iterative loop, instead of traditional iteration starting from a particular point every time. Overlapping and filling factor values for the estimation are fixed, but they could be adjusted depending on the number of turns and the final chosen wire section.

2.3.2. Stage 2: Electromagnetic & Thermal Analysis

Once initial number of turns (N_{ini}) and initial stack length (L_{ini}) are defined, it is time to interpolate in the magnetic flux map. At the first iteration, ambient temperature is taken for the interpolation. Nominal current is set to obtain the required torque, interpolating and obtaining φ_d and φ_q.

The interpolated value at a query point is based on linear interpolation of the values at neighbouring grid points in each respective dimension. This method is accurate enough if the map resolution is properly defined. Extrapolation is not recommended as it may result in false values of magnetic flux or temperature in case of thermal maps.

For resistance calculation, standard values of wire diameter are tabulated for each number of turns, to obtain a suitable filling factor depending on the chosen winding type, p.e. around 0.42 for distributed windings and around 0.5 for concentrated windings. The overlapping factor is defined with an experimentally adjusted curve. With this data and motor geometry, end-winding length (L_{end}) is calculated with (3), finally obtaining winding resistance (R_{Cu}) with (4), where ρ is copper resistivity at 20 °C. Moreover, q axis inductance is calculated by definition in (5):

$$L_{end} = \frac{\frac{\pi}{2}\left(\frac{D\pi}{Q_s} + w_d\right) * k_{ov}}{1000} \quad (3)$$

where D is stator diameter, Q_s is the number of slots, w_d is the slot width, and k_{ov} is the overlapping factor:

$$R_{Cu} = \frac{\rho(L_{end} + 2L)(1 + 0.0039(T_{Cu} - 20))}{A_r} \quad (4)$$

$$L_q = \frac{\varphi_q}{i_q} \quad (5)$$

Then, voltage is calculated, as shown in (6), where V_d and V_q are dq voltages, and L_σ is the leakage inductance:

$$\begin{cases} V_d = R_{Cu} i_d - L_\sigma w_e i_q - \varphi_q w_e \\ V_q = R_{Cu} i_q + L_\sigma w_e i_d + \varphi_d w_e \end{cases} \quad (6)$$

Finally, losses are calculated. Joule losses are calculated by Joule's law (7), while Core losses are calculated with the Bertotti's Model (8) shown in [17]:

$$P_{LCu} = 3 R_{Cu} I^2 \quad (7)$$

$$\begin{cases} p_{Fe} = p_h + p_c + p_e = k_h f B_s^\alpha + \sum_i k_c f^2 B_{si}^2 + \sum_i k_e f^{1.5} B_{si}^{1.5} \\ P_{Fe} = k_a p_{Fe} W_m \end{cases} \quad (8)$$

where p_{Fe} is the core loss per weight, p_h is the hysteresis loss, p_c is the eddy current loss, p_e is the excess loss, k_h is the hysteresis loss coefficient, k_c is the eddy current loss coefficient, k_e is the excess loss coefficient, α is an Steinmetz coefficient, f is the frequency, B_{si} is the ith harmonic amplitude of the stator magnetic flux density, k_a is the empirical coefficient, and W_m is the weight of motor.

Once Joule and Core losses are obtained at working temperature, they are sent to the thermal analysis. Then, the first interpolation can be made in the temperature maps, obtaining winding and magnet temperatures. Nevertheless, these temperatures are not the steady-state ones; as with temperature change, losses also change. To obtain the steady-state working temperatures, Joule losses are updated with temperature, as the resistance varies with temperature. The thermal analysis block will make this iteration until the steady-state losses and temperatures are obtained, taking about 16 iterations.

Once steady-state temperatures are obtained, they are returned to the electromagnetic analysis block, so the performances are obtained at working temperatures.

2.3.3. Stage 3: Automatic Checking

After obtaining the electrical performances at room and working temperatures, they must be checked, and, if they fulfil all the requirements, the sizing process will be finished, generating a favourable machine design report. Otherwise, the design parameters are changed and the calculus is addressed again. Figure 7 shows which parameters are checked, and the actions adopted (in STAGE 1) if they are not fulfilled. In the figure, L+ refers to increasing the machine length in one step, and Z+ or Z− means increasing or decreasing conductors in each slot.

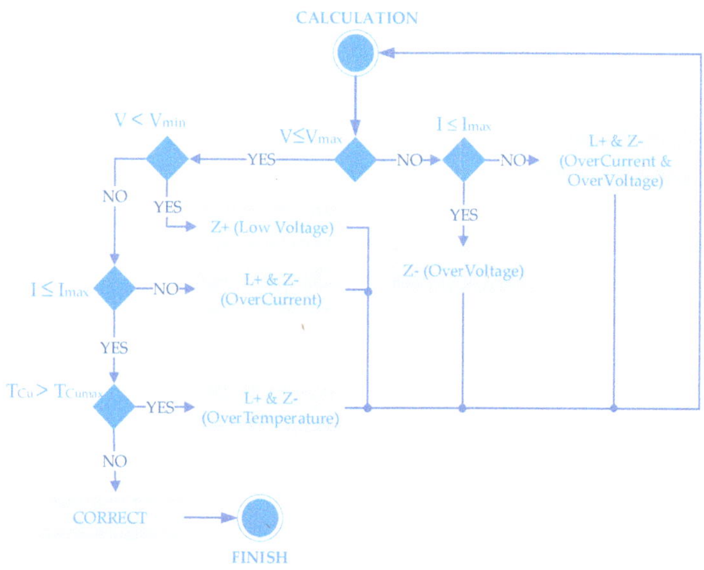

Figure 7. Automatic checking process after each iteration.

Checking minimum voltage is interesting, but it is not mandatory as some machines will not be able to fulfill both maximum and minimum voltages.

This checking is made automatically with the proposed algorithm taking into account designer specifications as the minimum and maximum voltages, maximum current, and maximum temperatures.

These requirements are set by the designer at the beginning of the process. The maximum voltage and current usually are limited by the inverter or the grid and the winding temperature normally is limited by the material or the machine class. If just one of the parameters does not comply, changes are made and another iteration is made, checking the four parameters again at its end. If all parameters are fulfilled, this is the optimum length and number of turns for the machine so the report is favourable, ending the process.

A Graphic User Interface (GUI) is designed to implement the proposed method in an easy and user-friendly way, to save time and effort when sizing machines. This GUI is composed of a database containing magnetic flux and temperature maps, an interface to choose those maps and introduce design requirements, a calculation core to implementing the proposed algorithm and a report generator to show the results after the results are automatically checked.

3. Case Study: Sizing of PMSM for People Transportation

In this section, the proposed improved sizing methodology is implemented in a real case study. The objective is to validate the multi-physics tool sizing several commercial PMSM for people transport application.

3.1. Description of the Machines

These machines are based on conventional topology comprising 36 slots in the stator and 30 poles in the rotor (Qs36p15). In total, ten different machines have been sized by the proposed method and tested experimentally. Their performances are shown in Table 1.

Table 1. Main machine performances

Motor ID	Speed [rpm]	Nominal Torque [Nm]	Power [kW]	Stack Length [mm]
M1	200	100	2.1	100
M2	400	100	4.2	100
M3	200	200	4.2	125
M4	400	200	8.4	125
M5	200	300	6.3	175
M6	400	300	12.6	175
M7	200	400	8.4	225
M8	400	400	16.7	225
M9	200	500	10.5	300
M10	400	500	21	300

In Figure 8, the stator and the rotor corresponding to one of the tested motors, and the ID 5 are shown.

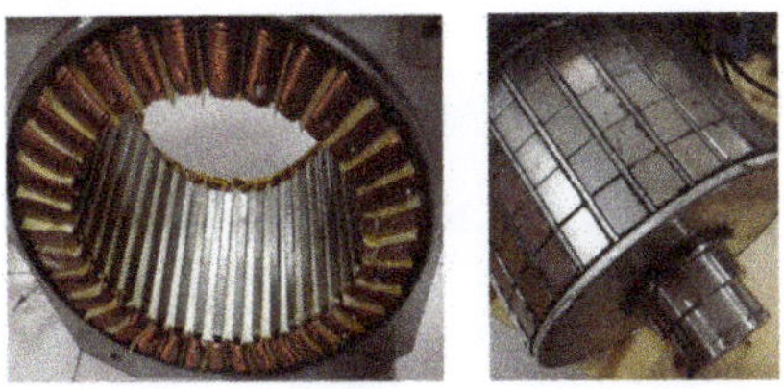

Figure 8. Rotor and Stator of one motor experimentally tested (Motor ID 5).

3.2. Map Creation

All of the tested machines have the same 2D magnetic circuit shown in Figure 9. The active length and the number of turns per phase are adjusted to fulfill the requirements of each application.

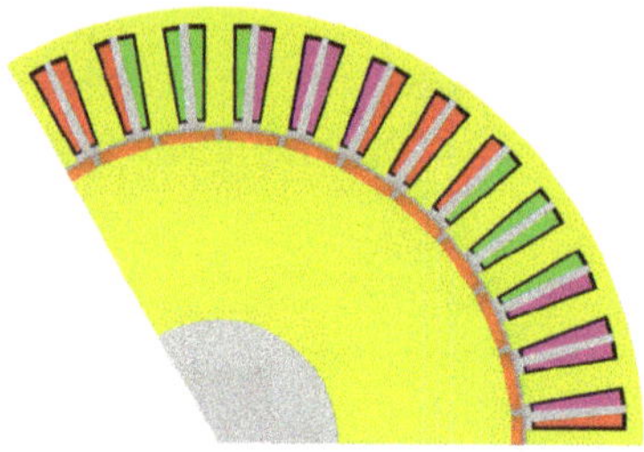

Figure 9. Geometry of the sized machines (Qs36p15).

In this case, the tested motors have surface-mounted permanent magnets and concentrated windings. In the future, the multi-physics sizing tool can be used with interior magnet motors or distributed windings to broaden the validation.

As mentioned in Section 2, magnetic flux maps must be created for each 2D geometry, generating the maps shown in Figure 10. Figure 10a shows magnetic flux in the d axis, whereas Figure 10b shows magnetic flux in the q axis.

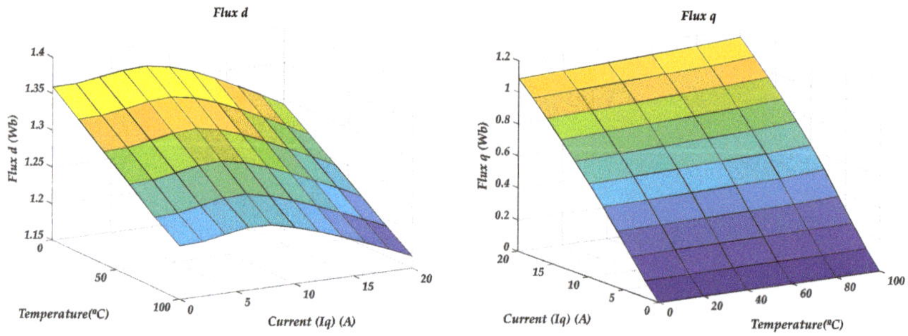

(a) (b)

Figure 10. Generated magnetic flux maps: magnetic flux vs. Current and Temperature for the Qs36p15 motor. (**a**) D magnetic flux map; (**b**) Q magnetic flux map.

As it can be appreciated in Figure 10, the temperature mainly affects to the d-axis magnetic flux because the remanent field of the magnets decreases linearly as the temperature increases. On the contrary, the q-axis magnetic flux changes slightly with the temperature. In this case, these small variations are due to changes in the saturation of the magnetic circuit. The magnetic flux created by the magnets changes with the temperature leading to variations in the saturation.

Ideally, in universal d-q axis models of the PMSM, it is commonly considered that the d-axis magnetic flux depends on the magnet flux and the d-axis current, while the q-axis magnetic flux depends only on the q-axis current. Nevertheless, there might be a fairly cross-coupling effect between the d-q axis depending on the saturation level of the motor, which might lead to relevant errors in the final results. In this case, this cross-coupling effect is clearly appreciated in Figure 10a as the d-axis magnetic flux changes with the q-axis current. In addition, this relationship is not linear, which makes it more difficult to model. Interpolating the magnetic flux maps, as it is done in the proposed tool, all these nonlinearities are taken into account, making it possible to achieve accurate results to some extent.

With respect to thermal maps, the Q36p15 motor model is shown in Figure 11. This model is used to create the maps, with the geometry and the selected cooling system—natural convection in this case.

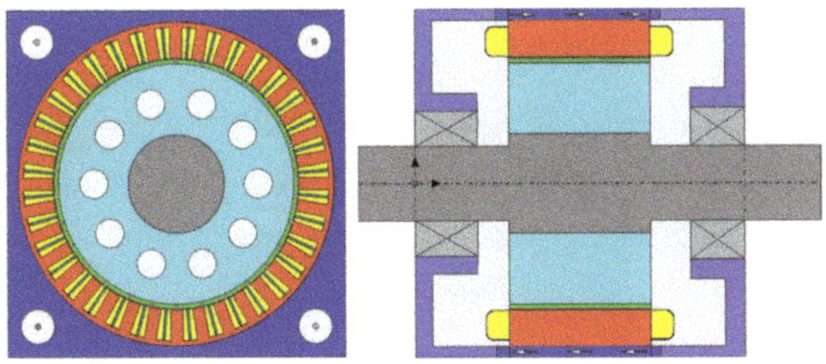

Figure 11. Motor-CAD® Model of the sized machines.

As mentioned in Section 2, temperature maps must also be created for each 2D geometry and cooling type, generating the maps shown in Figure 12. Figure 12a shows the mean winding temperature, whereas Figure 12b shows the mean magnet temperature. As it can be seen, shorter machines get warmer easier, and they will tolerate lower losses. In conclusion, if a machine exceeds the maximum desired temperature of the winding, or the demagnetization temperature of the magnets, a longer machine may be chosen.

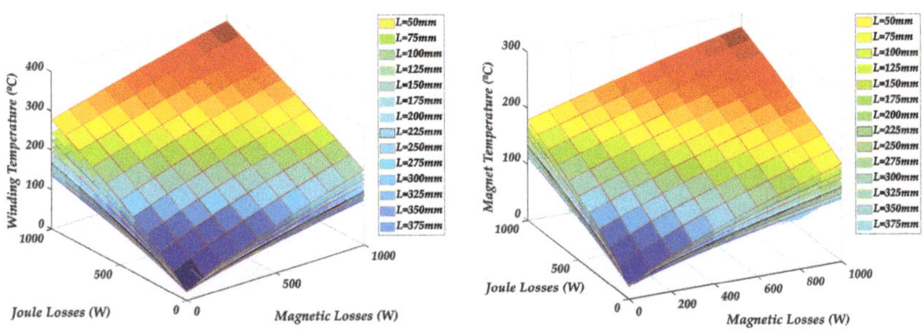

(**a**) (**b**)

Figure 12. Generated Temperature maps: Temperature vs. Magnet Losses, Copper Losses and Stack length in the naturally cooled machine with Qs36p15 configuration. (**a**) winding temperature map; (**b**) magnet temperature map.

3.3. Sizing

In Figure 13, sizing results are shown. It can be seen that most of the machines have the same EMF constant, as it was expected. In addition, voltage trends are very similar between machines, although some of them reach slightly higher voltage values. The figure also shows that machines are more efficient at nominal currents than at low and high currents.

With respect to time consumption, sizing a machine with the proposed sizing tool takes less than one minute for each machine. Taking into account that sizing a machine with the classical method can take about 8 h, the time reduction is considerable. Naturally, obtaining magnetic flux and temperature maps takes time, but these calculations are carried out only once, so this time is paid off when some machines are sized.

With respect to the mapping process, the computation of a temperature map with a resolution of 13 stack length values, 15 Joule loss values, and 15 Magnetic loss values could take around 2 h. This leads to a 3D matrix with 13 × 15 × 15 dimensions. Magnetic flux map creation for six temperatures varying i_q with 10 values over a full electric cycle takes approximately 6 h, but, applying the 1/6 reduction mentioned in Section 2, the consumed time is reduced in a 83%, leading to a 6 × 10 2D Matrix.

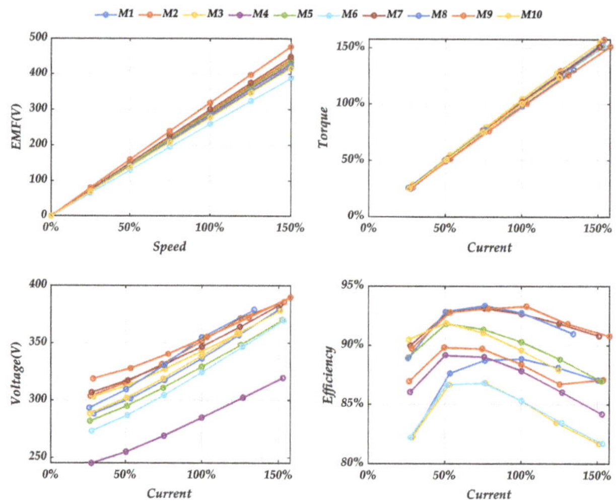

Figure 13. EMF, Torque, Voltage, and Efficiency vs. Current calculated by the Sizing Tool during the sizing process of 10 machines.

3.4. Experimental Validation

As the final step of the designing process, all sized machines are prototyped and experimentally tested to validate the designs. The test bench is shown in Figure 14.

Figure 14. Picture of the test bench consisting of the load motor, the motor under test (MUT), and the torque transducer.

In Figure 15, the main experimental performances, such as electromotive force, torque, supplying voltage, and efficiency are shown for all built prototypes.

As illustrated in the upper-left of Figure 15, speed and EMF are proportional and most of the machines have a similar k_e. In the bottom-left side, voltage is plotted against the current, and the majority of the machines follow a similar trend, although the values differ slightly. With respect to the efficiency, on the bottom-right side, it can be seen that efficiency decreases in low and high currents, and it increases in medium and nominal currents.

To validate the accuracy of the proposed sizing tool, the obtained results during the sizing process are compared to the experimental measurements. In Figure 16, the difference between sizing tool results and experimental results is plotted for the 10 sized motors.

It can be seen that the maximum difference in the EMF is about 1%, whereas, in the supplying voltage, it is about 2%. Nevertheless, the mean error is about 0.2% and 0.5%, respectively. With respect to the torque, the maximum difference is about 5% at low currents, mainly due to the uncertainty in the measurements at low currents. Nevertheless, it must be emphasized that the error is rather small as it is below 1.5%.

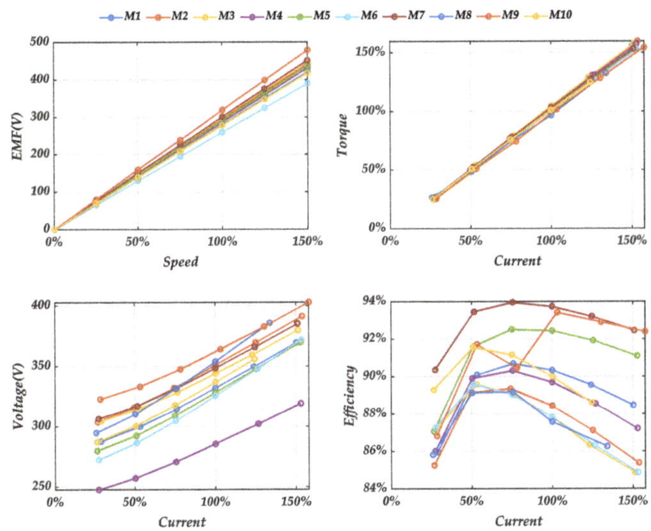

Figure 15. Experimental results of EMF, Torque, Voltage, and Efficiency vs. Current measured during the validation tests of the 10 sized motors.

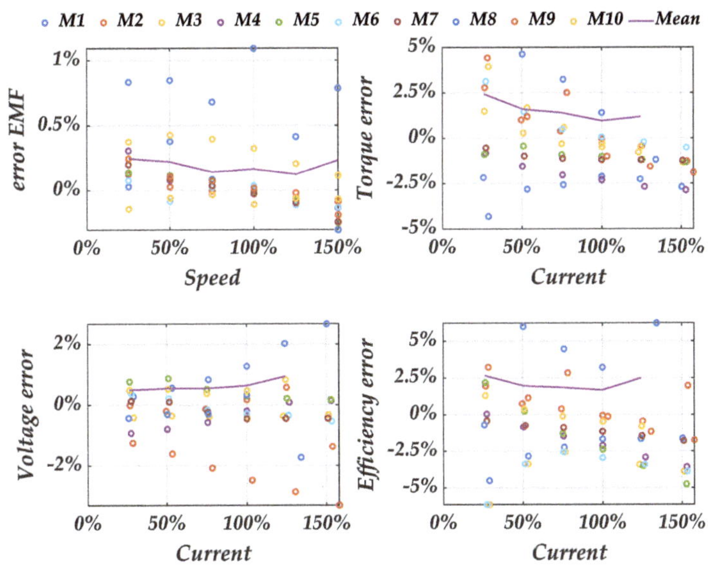

Figure 16. Error of the sizing tool computed as the difference in percentage between the experimental measurements and the results given by the sizing tool; error computed for the 10 sized motors.

Then, in Figure 17, the results of a given machine, the M7 (see Table 1), are shown for a close-up view. As illustrated in Figure 17, measurements agree with the sizing tool results with a small difference at some points.

Concerning the efficiency, higher differences can be observed. This could be due to many different factors—on one hand, due to the uncertainty in the calculation of iron losses, mechanical losses, and stray losses; on the other hand, due to the possible measuring errors in the torque transducer and Voltage/Current probes. Even these errors might not be very significant, as it is shown in the torque comparison; for instance, the accumulation of all of them could justify the differences in the efficiency. In any case, it must be stated that these differences are not very significant, as the error is very low, below 6%.

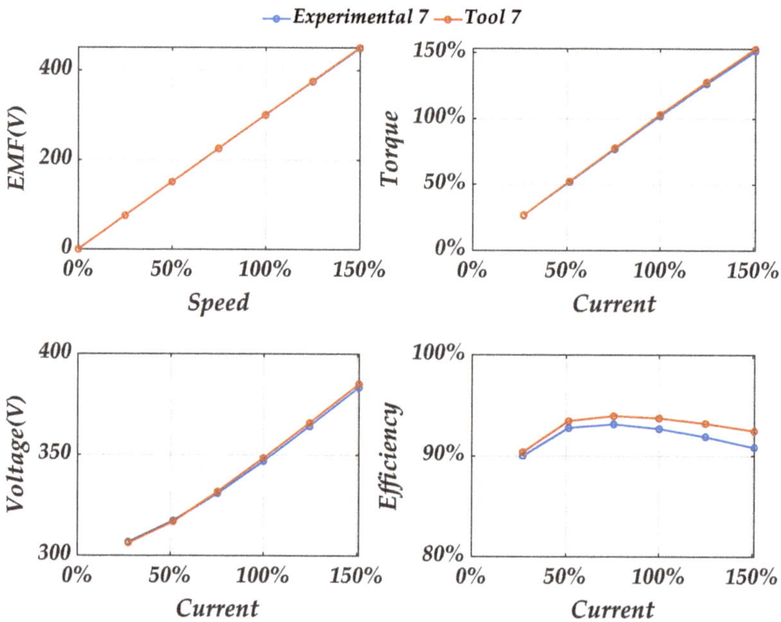

Figure 17. Comparison between experimental and calculated results in the M7 motor.

4. Conclusions

In this paper, a PM machines sizing methodology is developed. The proposed methodology, based on the coupling of magnetic flux and temperature maps, has been put into practice in a real case study. This methodology improves the competitiveness of ten industrial motors, reducing the design time and, consequently, the resources needed for that design. As a result, all motors have been sized accomplishing a very good trade-off between cost and required performances.

This procedure enables to perform sizing in a faster way, using less computational resources. Using magnetic flux and temperature maps enables achieving very good accuracy. As the influence of the temperature is considered, the accomplished results are more realistic. It must be remarked that specialized software is only needed for Map creation, and the sizing algorithm can be run on any computer.

Moreover, a faster procedure is described to obtain magnetic flux maps at different temperatures, just simulating 1/6 portion of the electric period, instead of considering the overall electric period. This method makes it possible to reduce the mapping process time in 83%.

Taking the overall results into account, the proposed sizing method fulfills the desired objectives of time reduction and accuracy in the sizing process, coupling electromagnetic with thermal effects, and sizing machines in less than a minute and with an error below 6%.

Author Contributions: G.A., A.E., and A.J.E. conceptualized the research; A.E. and Y.M. developed the methodology and the software; P.M. conceived and performed the experiments; P.M. and G.A. analyzed the data; Y.M. wrote the original draft. All authors have read and agreed to the published version of the manuscript.

Funding: This research received no external funding.

Conflicts of Interest: The authors declare no conflict of interest.

Abbreviations

The following abbreviations and symbols are used in this manuscript:

CAD	Computer Aided Design
CFD	Computational Fluid Dynamics
FEA	Finite-Element analysis
GUI	Graphic User Interface
LP	Lumped Parameter
LPC	Lumped Parameter Circuit
PMSM	Permanent Magnet Synchronous Machines

List of symbols

$\vec{A}$	Magnetic vector complexpotential (Wb/m)	p_{Fe}	Core loss per weight (W/kg)
A_r	Copper wire Area (m^2)	p_h	Hysteresis loss per weight (W/kg)
α	Steinmetz coefficient	P_{LCu}	Joule Losses (W)
B_{si}	i^{th} harmonic amplitude of the stator flux density (T)	P_{Lmag}	Magnetic Losses (W)
D	Stator Diameter (m)	$\varphi_{d,q}$	Magnetic Flux of d,q axis (Wb)
f	Frequency (Hz)	φ_a	Magnetic Flux A (Wb)
$i_{d,q}$	d,q axis current (A)	φ_b	Magnetic Flux B (Wb)
I_n	Desired nominal current (A)	φ_c	Magnetic Flux C (Wb)
k_a	Empirical coefficient	Q_s	Number of slots in the stator
k_c	Eddy current loss coefficient	R_{Cu}	Winding Resistance (Ω)
k_e	Excess loss coefficient	ρ	Wire resistivity (Ωm)
k_f	Filling factor	$[\sigma]$	Tensor of the conductivity of the medium (S)
k_h	Hysteresis loss coefficient	T_{Cu}	Copper Temperature (° C)
k_{ov}	Overlapping factor	T_{magnet}	Magnet Temperature (° C)
L	Machine stack length (m)	T_{nom}	Nominal Torque (Nm)
L_{end}	End-Winding length (m)	V	Electric scalar potential (V)
L_{ini}	Estimated initial machine stack length (m)	$[v_0]$	Reluctivity of the vacuum (m/H)
L_σ	Leakage inductance (H)	$[v_r]$	Tensor of the reluctivity of the medium
L_q	Q axis inductance (H)	$V_{d,q}$	d,q axis voltage (V)
N	Winding number of turns	V_{max}	Maximum allowed voltage (V)
N_{ini}	Estimated initial winding number of turns	w_e	Machine rotational speed (rad/s)
n_{cap}	Winding layers	W_m	Weight of motor (kg)
p	Machine pole pairs	w_s	Slot width (m)
p_c	Eddy current loss per weight (W/kg)	Z	Number of conductors per slot
p_e	Excess loss per weight (W/kg)		

References

1. Hoang, K.D.; Atallah, K. Rapid sizing concept of interior permanent magnet machine for traction applications. *J. Eng.* **2019**, *2019*, 3956–3961. [CrossRef]
2. Zhou, K.; Pries, J.; Hofmann, H. Computationally Efficient 3D Finite-Element-Based Dynamic Thermal Models of Electric Machines. *IEEE Trans. Transp. Electrif.* **2015**, *1*, 138–149. [CrossRef]

3. Goss, J.; Wrobel, R.; Mellor, P.; Staton, D. The design of AC permanent magnet motors for electric vehicles: A design methodology. In Proceedings of the 2013 IEEE International Electric Machines and Drives Conference (IEMDC 2013), Chicago, IL, USA, 12–15 May 2013; pp. 871–878. [CrossRef]
4. Xie, P.; Ramanathan, R.; Vakil, G.; Gerada, C. Simplified analytical machine sizing for surface mounted permanent magnet machines. In Proceedings of the 2019 IEEE International Electric Machines and Drives Conference, IEMDC 2019, San Diego, CA, USA, 12–15 May 2019; pp. 751–757. [CrossRef]
5. Wang, Y.; Bonfante, M.; Bianchi, N.; Petrella, R. Scalability of Synchronous Reluctance Machines Considering Thermal Performance. In Proceedings of the 2019 IEEE Energy Conversion Congress and Exposition (ECCE), Baltimore, MD, USA, 29 September–3 October 2019; pp. 1701–1707.
6. Tiegna, H.; Amara, Y.; Barakat, G. Overview of analytical models of permanent magnet electrical machines for analysis and design purposes. *Math. Comput. Simul.* **2013**, *90*, 162–177. [CrossRef]
7. Amrhein, M.; Krein, P.T. Magnetic equivalent circuit simulations of electrical machines for design purposes. In Proceedings of the IEEE Electric Ship Technologies Symposium (ESTS 2007), Arlington, VA, USA, 21–23 May 2007; pp. 254–260. [CrossRef]
8. Zhou, K.; Ivanco, A.; Filipi, Z.; Hofmann, H. Finite-Element-Based Computationally Efficient Scalable Electric Machine Model Suitable for Electrified Powertrain Simulation and Optimization. *IEEE Trans. Ind. Appl.* **2015**, *51*, 4435–4445. [CrossRef]
9. Ramakrishnan, K.; Stipetic, S.; Gobbi, M.; Mastinu, G. Optimal Sizing of Traction Motors Using Scalable Electric Machine Model. *IEEE Trans. Transp. Electrif.* **2018**, *4*, 314–321. [CrossRef]
10. Carbonieri, M.; Bianchi, N.; Alberti, L. Induction Motor Mapping Using Rotor Field-Oriented Analysis Technique. In Proceedings of the 2019 IEEE Energy Conversion Congress and Exposition (ECCE), Baltimore, MD, USA, 29 September–3 October 2019; pp. 2321–2328.
11. Boglietti, A.; Cavagnino, A.; Staton, D.; Shanel, M.; Mueller, M.; Mejuto, C. Evolution and modern approaches for thermal analysis of electrical machines. *IEEE Trans. Ind. Electron.* **2009**, *56*, 871–882. [CrossRef]
12. Huang, Z.; Márquez-Fernández, F.J.; Loayza, Y.; Reinap, A.; Alaküla, M. Dynamic thermal modeling and application of electrical machine in hybrid drives. In Proceedings of the 2014 International Conference on Electrical Machines, ICEM 2014, Berlin, Germany, 2–5 September 2014; Institute of Electrical and Electronics Engineers Inc.: Piscataway, NJ, USA, 2014; pp. 2158–2164. [CrossRef]
13. Li, J.; Ni, J.; Xu, X. *Lumped Parameter Transient Thermal Model of Motor Considering Temperature and Flow Rate of Cooling Water*; SAE Technical Papers; SAE International: Warrendale, PA, USA, 2019; Volume 2019. [CrossRef]
14. Kral, C.; Haumer, A.; Lee, S.B. A practical thermal model for the estimation of permanent magnet and stator winding temperatures. *IEEE Trans. Power Electron.* **2014**, *29*, 455–464. [CrossRef]
15. Sciascera, C.; Giangrande, P.; Papini, L.; Gerada, C.; Galea, M. Analytical thermal model for fast stator winding temperature prediction. *IEEE Trans. Ind. Electron.* **2017**, *64*, 6116–6126. [CrossRef]
16. Mondal, D.; Chakrabarti, A.; Sengupta, A.; Mondal, D.; Chakrabarti, A.; Sengupta, A. Fundamental Models of Synchronous Machine. In *Power System Small Signal Stability Analysis and Control*; Academic Press: Cambridge, MA, USA, 2014; Chapter 2, pp. 15–40. [CrossRef]
17. Yanli, F.; Chengning, Z. Analytical Calculation for Predicting the Core Loss of Surface-Mounted Permanent Magnet Machine. *Energy Procedia* **2017**, *105*, 2119–2124. [CrossRef]

energies

MDPI

Article

Enhanced Flexible Algorithm for the Optimization of Slot Filling Factors in Electrical Machines †

Armin Dietz [1], Antonino Oscar Di Tommaso [2,*], Fabrizio Marignetti [3], Rosario Miceli [2] and Claudio Nevoloso [2]

1 Technische Hochschule Nürnberg, Institut ELSYS, 90489 Nuremberg, Germany; armin.dietz@th-nuernberg.de
2 Dipartimento di Ingegneria, University of Palermo, viale delle Scienze, Parco d'Orleans, 90128 Palermo, Italy; rosario.miceli@unipa.it (R.M.); claudio.nevoloso@unipa.it (C.N.)
3 Department of electrical and Information Engineering (DIEI), University of Cassino and South Lazio, via. G. Di Biasio, 43, 03043 Cassino, Italy; marignetti@unicas.it
* Correspondence: antoninooscarditommaso@unipa.it

† This paper is an extended version of our paper published in: Caruso, M.; Di Tommaso, A.O.; Miceli, R.; Nevoloso, C. Algorithmic Approach for Slot Filling Factors Determination in Electrical Machines. In Proceedings of the 2018 7th International Conference on Renewable Energy Research and Applications (ICRERA), Paris, France, 14–17 October 2018; pp. 1489–1494.

Received: 14 January 2020; Accepted: 24 February 2020; Published: 26 February 2020

Abstract: The continuous development in the field of industrial automation and electric mobility has led to the need for more efficient electrical machines with a high power density. The improvement of electrical machines' slot filling factors is one of the measures to satisfy these requirements. In recent years, this topic has aroused greater interest in the industrial sector, since the evolution of the winding technological manufacturing processes allows an economically sustainable realization of ordered winding arrangements, rather than random ones. Moreover, the manufacture of electrical machines' windings must be preceded by an accurate design phase in which it is possible to evaluate the maximum slot filling factor obtainable for a given wire shape and for its dimensions. For this purpose, this paper presents an algorithmic approach for the evaluation of maximum slot filling factors in electrical machines under an ideal geometric premise. In particular, this algorithm has a greater degree of flexibility with respect to the algorithm approaches found in the literature, since the study has been extended to round, rectangular and hexagonal wire sections. Furthermore, the slot filling factor calculation was carried out both for standard and non-standard slots. The algorithmic approach proposed can be considered as an additional useful tool for the fast design of electrical machine windings.

Keywords: electrical motors; sot filling factor; optimization algorithm; windings; magnetic wire; filling factor optimization

1. Introduction

The development of more and more efficient electrical machines has become a topic of interest for various industrial sectors, such as automation or electric traction. In particular, high efficiency, high power density and cost-effective manufacturing are required in the automotive industry [1–3]. One possible solution to meet these requirements is to optimize the copper filling factor of stator winding [4–11]. In particular, a high copper filling factor involves a more rational and efficient use of copper with economic benefits and improved energy savings. Therefore, the optimization of the slot filling factor is a key focus in winding technology. The improvement of the slot filling factor depends mainly on the winding pattern schemes and the adopted winding manufacturing process.

The simplest type of winding pattern is so-called "random winding". In this case, the random winding process is sustainable for mass production and it is characterized by low manufacturing requirements. The main advantage is represented by its high production speed, while the disadvantages consist, generically, of lower filling factors. The highest possible filling factor is achieved by the "orthocyclic winding pattern" for round wires. This winding pattern presents a high packing density of wire, but the winding process is more complex and, therefore, more costly than the random winding process. In particular, the orthocyclic winding process requires very high manufacturing requirements to obtain an ordered wire positioning within each slot. Another possible winding pattern is the "layer winding", where the wires are uniformly arranged in layers [12]. This winding pattern allows us to obtain higher filling factors than those of the random winding pattern. The choice of typology winding pattern depends on the functions and the design requirements of the electrical machine. Therefore, orthocyclic (or ordered) winding structures are appropriate for high power density applications.

In the past, the realization of commercial solutions that allow the economically sustainable manufacturing of distributed windings with ordered structures was very difficult due to the high economic burdens. The automated winding process technologies available for making distributed windings are: insertion winding technology, flyer winding technology and needle winding technology [13]. In recent years, these winding process technologies have undergone a great technological evolution that has allowed us to reduce manufacturing costs and winding process times with an ordered structure [13,14]. In [12], a new and innovative needle winding method that allows shifted layer winding structures for distributed round wire applications is described, thus significantly increasing the copper filling factor. In this regard, it is of considerable interest to accurately evaluate the slot filling factor obtainable during the design phase of electrical machines. In the previous century, this task was performed by means of manual graphic analysis or by testing the stator of the electrical machine during the pre-production phase. In order to carry out this process, the resources in terms of money, time and technical staff are not indifferent [15–17]. Therefore, a preliminary analysis of the maximum possible value of filling factor is an important step forward for the design of electrical machine windings for a given slot, wire shape and set of dimensions.

In the scientific literature, there are several definitions of slot filling factors and, generally, an electrical slot filling factor f_{cu} and a mechanical slot filling factor f_{me} are defined. In this study, the electrical slot filling factor f_{cu} is given by the ratio between the total wire copper cross-section (N_w A_{cu}) and the total slot cross-section A_{slot}. Furthermore, the mechanical slot filling factor f_{me} is defined as the ratio between the total wire cross-section (N_w A_w) and the effective slot area A_{eff}, which is defined as the difference between the total slot cross-section A_{slot} and the area occupied by slot insulation. The two filling factors can be described by the following mathematical equations:

$$f_{cu} = \frac{N_w \, A_{cu}}{A_{slot}} \tag{1}$$

$$f_{me} = \frac{N_w \, A_w}{A_{eff}} \tag{2}$$

where N_w is the number of wires contained within a slot, A_{cu} is the maximum copper cross-section of a single wire (useful cross-section) and A_w is the maximum cross-section of a single wire (including its insulation layer). For the purpose of this work, the electrical slot filling factor values are considered and discussed. This paper proposes an algorithmic approach for the preliminary determination of the slot filling factors. In particular, the goal of the present paper is to extend the work done in [18,19] with reference to various types of wires (this work also considers hexagonal cross-sections) and slots. Moreover, the identification of the best arrangement of wires within the slot for each slot type is suggested, with a technique aimed at the reduction of computation times. In fact, the algorithmic approach described in the previous papers calculates the slot filling factors only for circular and rectangular wire shapes. This algorithmic approach requires the definition of the coordinates of the first wire, that can be useful for the investigation of singular cases where the goal is the evaluation of the slot filling factors for a specific wire's positioning. In fact, to research the best wire arrangement and, consequently, the highest obtainable values of the slot filling factors, the algorithm requires several executions of the procedure to obtain several values of the first wire position. Since each execution of the algorithm requires a few seconds, the time consumed can be significant.

However, the enhanced algorithm, here described, allows the determination of the maximum slot filling factor for a given slot, wire shape and set of geometrical dimensions. In particular, the algorithm procedure is designed in order to systematically investigate a large number of cases defined for several values of the coordinates of the first wire in order to optimize the computation time and find the best arrangement inside the slot, with an improved accuracy. In more detail, with a given cross-section geometry and given wire dimensions, the algorithm allows the calculation of the maximum number of wires that can be placed inside the slot. In this work, both wire and slot insulation are considered because, together, they cause a reduction of the useful slot area. In detail, with respect to the previous work, three different wire geometries, namely round, rectangular and hexagonal, are taken into consideration and the results are discussed. Since the use of hexagonal wires can present an innovative character in the design of electrical machines, attention has been paid to the comparative study of the filling factors obtainable with wires that have both circular and hexagonal cross-sections, by considering the same cross-section area.

This paper is structured as follows: Section 2 describes the state-of-the-art optimization algorithms of the slot filling factors, Section 3 describes the algorithmic approach proposed and Section 4 illustrates several cases of studies that have been carried out and the relevant obtained results.

2. State-of-the-Art Slot Filling Factor Optimization Algorithms

At the present time, the scientific literature does not include many works regarding the optimization of the slot filling factors with algorithmic approaches. Furthermore, the algorithmic approaches found in the literature refer to the case where the wire has a round cross-section. The optimal winding pattern is invariant with respect to the axis perpendicular to the stator cross-section. Therefore, the determination of the optimal winding pattern can be carried out with a bi-dimensional approach. A family of widespread algorithms is that of the orthocyclic windings algorithms (OWA). In [20], the results of this kind of algorithm are discussed. A study was carried out on the variation in the value of the mechanical and electrical slot filling factors when the radius of the circular conductor varies. In the first phase, the algorithm sets the coordinates of the center (x_i; y_i) of the first wire:

$$w_i = \begin{pmatrix} x_i \\ y_i \end{pmatrix}. \tag{3}$$

Then, the wires of the same layer are plotted and the distance between the centers of two adjacent wires is equal to the diameter of the wire d. Once the first layer is finished, the wire coordinates of the upper layer are determined through the following relation:

$$w_j = \begin{pmatrix} \cos\alpha \\ \sin\alpha \end{pmatrix} \cdot d + w_i \tag{4}$$

with α equal to 60° or 120° (Figure 1).

The algorithm ends when it is not possible to find a new wire position whose area intercepts the slot contour or its insulation contour; then, the slot is full.

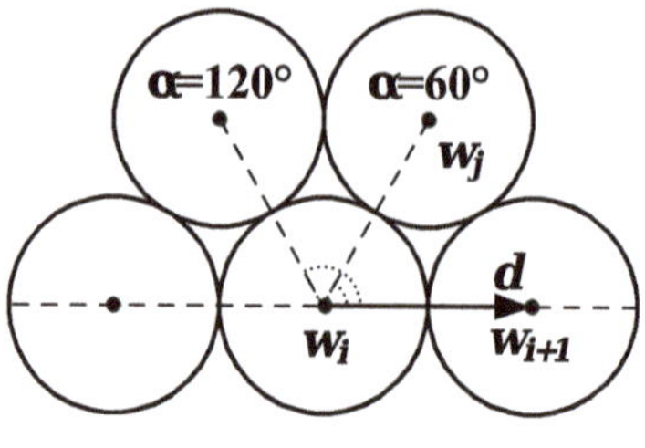

Figure 1. Round wires with orthocyclic distribution.

Another algorithm approach, named the filling factor estimation algorithm (FFEA), is described in [15]. The goal of this algorithm is to reduce the design time of the winding by estimating the filling factors. In this case, the first wire is placed centrally at the slot bottom. The next wire is placed by checking all valid positions along a circle around the first wire, with a radius equal to the distance $d_w + t_w$, where d_w and t_w are the diameter of the wire and the safety distance or the minimum gap between two wires, respectively. For each new wire positioned, the algorithm checks whether there are overlaps with the contour of the slot, taking into account the presence of the slot insulation. If the new position is valid, the algorithm places the new neighbor wire and continues going around until all positions have been tried. The procedure ends when it is not possible to position wires that do not overlap the slot profile. The result of this algorithm is a random positioning of the wires forming the winding. Since the value of the determined filling factor depends on the initial position of the first wire, the procedure is repeated for several different initial positions.

Another interesting approach is described in [20]. This algorithm is based on the basic rules of the two algorithms described above. This algorithm is called the needle winding simulation algorithm (NWSA) because the objective is to optimize the positioning of each wire in order to simulate the process of needle winding. In this case, for the positioning of wire, a cost function is taken into consideration, accounting for the position and the downward force acting on the wires. In the algorithm, there are constraints to avoid overlapping between wires and between the wire and the slot profile. Optimal positioning is achieved by identifying the global minimum of the cost function. Authors use a genetic algorithm to search the global minimum in order to reduce convergence times.

Beyond the approaches described above, there are the dense packing algorithms, also called the wire inflation algorithms (WIA). The goal of these algorithms is to find the maximum radius of spheres within a given boundary and their corresponding coordinates when the number of spheres is fixed. In [21], an algorithm is described that simulates a system of billiard spheres within a limited space, whose radius is made to grow until it reaches a state whereby it becomes a blockade in the system. This algorithm approach is used for the dense packing of spheres in circular, triangular and hexagonal spaces [22–24]. This algorithm performs an event based on the physical simulation of a billiard system where the coordinates, the number and the speed of the spheres are set in advance as input data. Events are represented by the collision of each sphere with other spheres or with the borders of the system. Depending on the event type (either sphere–obstacle or sphere–sphere interaction) an elastic impact is performed, and the new velocities are calculated. At the increasing radii of the spheres, an eventual jamming occurs, resulting in a dense packing. For the purposes of maximizing the slot filling factor, the objective is to maximize the number of magnetic wires, with given radius, within the slot. In this sense, in [20], the authors, in order to make a comparison with the NWSA, have modified the described approach. In particular, the wires are understood to be charged particles exerting a Coulomb force on each other, leading to the movement of the spheres during the simulation. The results of the comparison are widely discussed.

3. Proposed Enhanced Algorithmic Approach

The basic rules of the algorithm proposed in this paper are mainly inspired by the approach proposed in [15] and the treatment of the winding patterns with ordered structures in detail [18]. The algorithm is based on a general approach that is valid for different stator slots and for different wire shapes. In detail, for round and hexagonal wires, the algorithm focuses on the windings, with orthocyclic arrangement of the wires within the slot. Moreover, the algorithm has been designed so as to be able to define some critical aspects, such as the possible presence of slot insulation, the positioning of the wire parallel to the flank or to the bottom of the slot, any safety distances between the wires forming the winding and a large number of the possible coordinates of the first wire. In this way, the algorithm presents a high degree of flexibility that makes investigating the maximum slot filling factors obtainable in many cases of study. The algorithm has been implemented in the Matlab environment and it is described in detail below.

3.1. Slot Geometrical Features Definition

Compared to the works found in the literature, where simplified models of slot have been taken into account, in this work, both a standard slot (STSL) model and the non-standard (NSTSL) ones are used. The slot profiles are defined with the presence of an insulation sheet and, therefore, its thickness d_{ins} is taken into account. In general, the slot insulation sheet is pre-folded and does not substantially modify the slot cross-section profile available for the positioning of the wires. The geometries and the related contour data of a standard slot, typically used for wires, and a non-standard one are shown in Figure 2a,b [16], respectively.

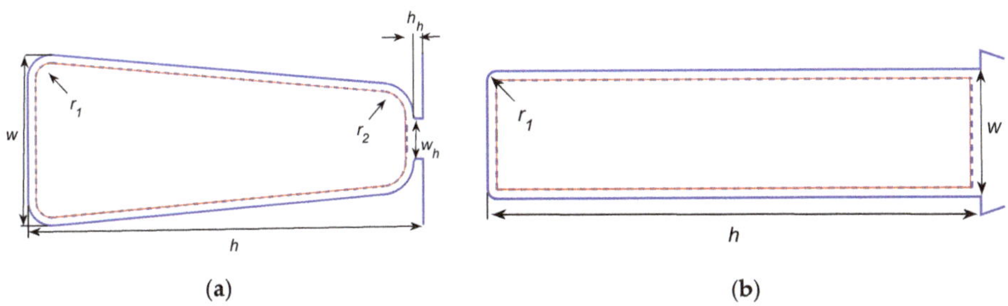

Figure 2. Cross-section of the standard slot (**a**) and non-standard slot (**b**) with contour parameters.

Generally, the slot profiles are supplied with CAD drawings by the manufacturers and, therefore, the geometrical parameters are easily determined. The dimensions of each slot are known and are reported in Tables 1 and 2.

Table 1. Standard slot geometric parameters.

Parameter	Value (mm)
w	10.360
h	24.930
w_h	2.500
h_h	0.500
r_1	1.500
r_2	1.971
d_{ins}	0.500

Table 2. Non-standard slot geometric parameters.

Parameter	Value (mm)
w	10.360
h	24.930
r_1	1.500
d_{ins}	0.500

The slot profiles are defined in the xy reference frame by means of characteristic points connected by lines and arcs. The implementation of the geometric model of the slot profile is carried out as described in [18]. The cross-sections of slots are calculated in a numeric way by the trapezium rule in the same way as [15]. In particular, the slot areas are divided into different sections whose characteristic points are known (Figure 3a,b). The calculated cross-sections show a maximum deviation of less than one percent from the cross-section given by the specifications.

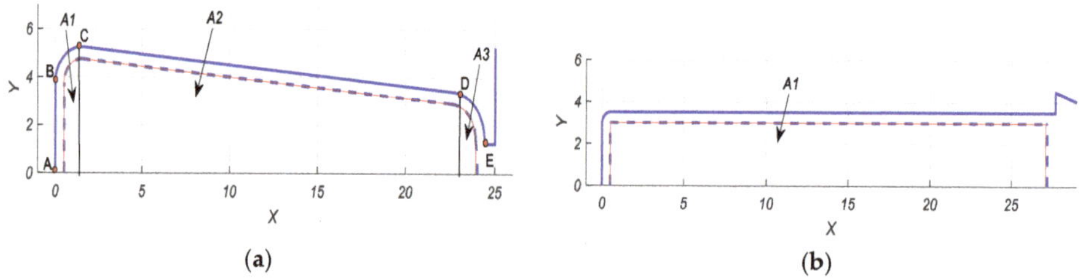

Figure 3. Half cross-section of the standard slot (**a**) and non-standard slot (**b**) divided into sections.

3.2. Magnet Wire Data

In this study, investigations into round, rectangular and hexagonal wires have been carried out. The geometrical data of the round and rectangular wires used in this paper are obtained from [25,26]. By these standards, the nameplate data, the insulation degree, tolerances and maximum dimensions allowed for each wire shape are reported. For rectangular shapes, the standard defines the shape of the wire with rounding corners. However, there is not an international standard for hexagonal wires. The use of the latter could be an innovative idea, as they can substantially improve the filling factors. Improvements must be compared with the production costs which, compared to the past, have been reduced thanks to the technological evolution of wire manufacturing processes [13,17]. Therefore, the geometric data of the hexagonal wires will be hypothesized. The geometrical reference quantities for each wire shape are defined below for:

1. Round wire geometrical data:

    - d_{cu} diameter wire without insulation;
    - d_{max} maximum diameter of wire with insulation;
    - x_c and y_c coordinates of wire center.

2. Rectangular wire geometrical data:

    - L_{1cu} width of the rectangular wire without insulation;
    - L_{2cu} height of the rectangular wire without insulation;
    - L_{1max} maximum width of the rectangular wire with insulation;
    - L_{2max} maximum height of the rectangular wire with insulation;
    - r_{corner} corner radius;
    - x_c and y_c coordinates of the wire center.

3. Hexagonal wire geometrical data:

 - r_{cu} radius of circumference circumscribed to the hexagon without insulation;
 - r_{max} maximum radius of circumference circumscribed to the hexagon with insulation;
 - x_c and y_c coordinates of the wire center.
 - *phi* (ϕ) rotation angle of the hexagon.

Figure 4 shows the wire cross-sections with the related contour data.

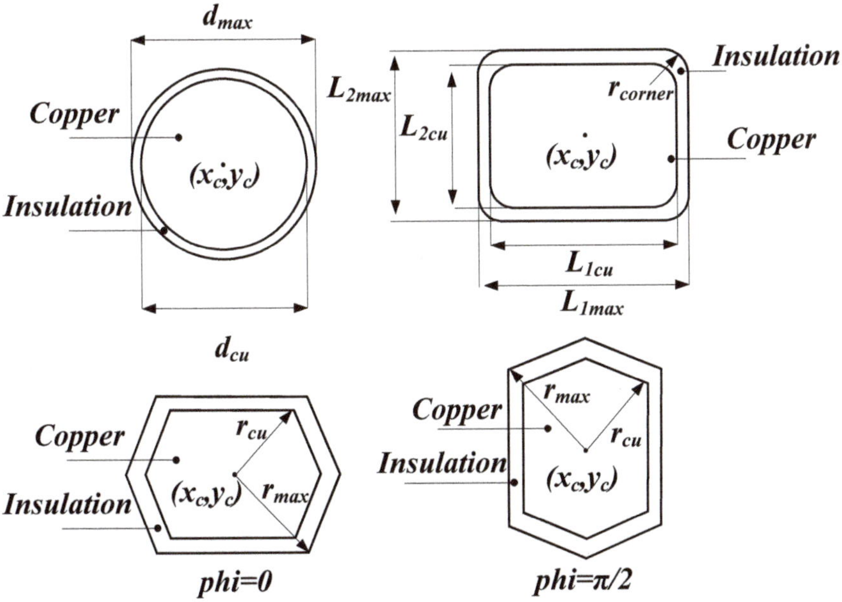

Figure 4. Wire cross-sections with contour data.

3.3. Constraints for Wire Distribution and Placement

In the algorithm, the slot models have been implemented in a xy plane with the bottom or the ground of each slot parallel to the y-axis. Each slot profile has been divided into several reference sections Figure 3a,b. These sections are referred to a one-half slot. The distribution and placement conditions of wires are defined for each section and applied in a specular way for each half of the slot. In order to avoid overlapping between the wire and the slot profile and between adjacent wires, several constraints are taken into account. Therefore, for each wire that must be positioned within the slot profile, the following gaps are considered and investigated:

- distance between the wire and a part (arc or line) of the slot insulation sheet profile;
- distance between the new wire and an existing wire.

In Figure 5, the distances between the slot profile and the round and rectangular wires are shown. In each region of the slot profile, the minimum value of the various distances is evaluated. In particular, for round wires, the distances are defined with respect to the center of the wire and it is imposed that

$$min\{|d_1|, |d_2|, \ldots, |d_n|\} \geq r_{max}, \tag{5}$$

whereas, in the case of rectangular and hexagonal wires, these distances are evaluated with respect to the vertices of each wire shape (Figure 6) and it is imposed that

$$min\{|d_1|, |d_2|, \ldots, |d_n|\} \geq 0. \tag{6}$$

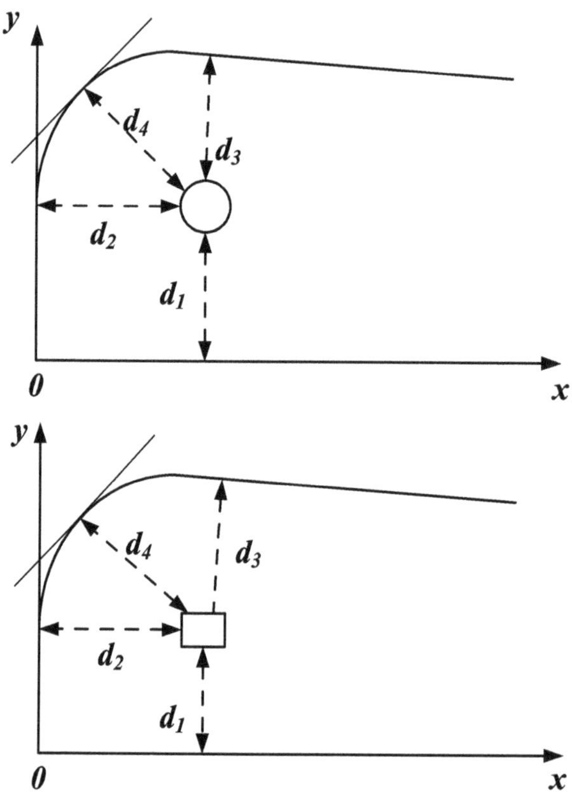

Figure 5. Distances between the wires and the slot profile.

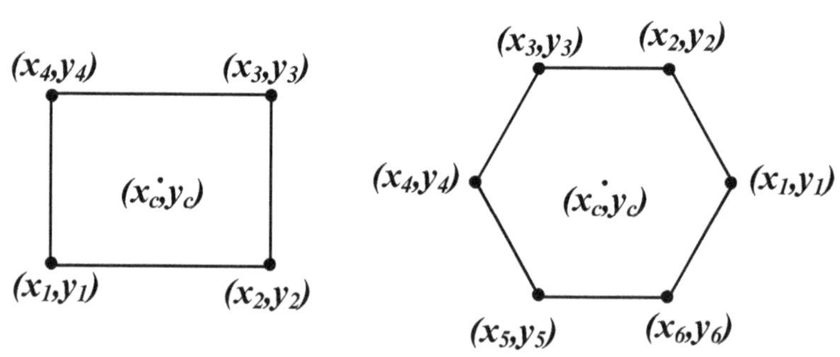

Figure 6. Characteristic points of the rectangular and hexagonal wires.

Regarding the distance between wires, it is possible to add an additional safety distance t_w. The winding positioning pattern adopted in this work is the orthocyclic one that can be obtained in the case of round and hexagonal wires. However, an ordered arrangement was taken into consideration for rectangular shaped wires. Moreover, it is possible to vary the angle formed between the line joining the centers of the wires and the horizontal one of the xy plane (Figure 7). In this case, the new arrangement is obtained by applying the following coordinate transformation to the center of wires for round wires, and also to the vertices for rectangular and hexagonal wires:

$$\begin{pmatrix} x_{inew} \\ y_{inew} \end{pmatrix} = \begin{pmatrix} \cos\beta & -\sin\beta \\ \sin\beta & \cos\beta \end{pmatrix} \begin{pmatrix} x_i \\ y_i \end{pmatrix}. \tag{7}$$

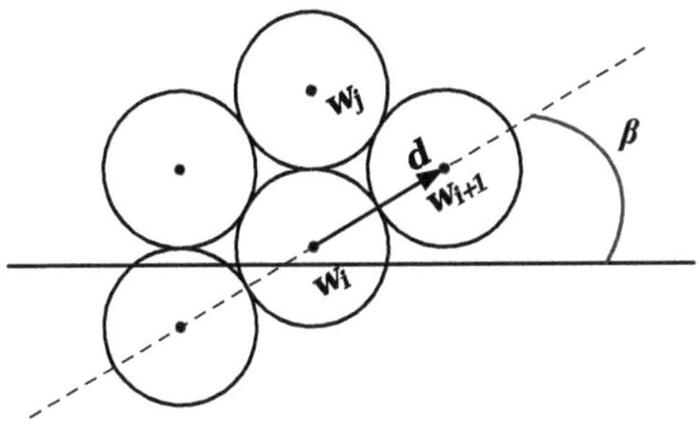

Figure 7. Rotation of the orthocyclic distribution.

In this sense, it is possible to evaluate the value of the slot filling factors in the case where the wires are arranged parallel to the flank or the bottom of the slot. This study has been performed with particular attention paid to round and rectangular wires since. in the hexagonal wire case, as described above, it is possible to set the rotation angle.

3.4. The Algorithm Procedure

As described above, the algorithm has been designed to automatically define several positions for the first wire. In this procedure, the position of the first wire is defined by its center (x_{w1}, y_{w1}), which it is selected in such a way that it will be located in the region of space near the lower wedge of the slot (i.e., in the lower rounded corner):

$$x_{w1} = x_{init} + d_{ins} \tag{8}$$

$$y_{w1} = y_{init} + d_{ins} - \frac{w}{2} \tag{9}$$

where d_{ins}, x_{int}, y_{int} are the slot insulation foil thickness and the initial wire coordinates, respectively, which are chosen in such a way as to avoid overlaps with the slot profile. In order to evaluate the filling factors, several pairs of values of x_{int}, y_{int} are assigned. For this purpose, it is necessary to evaluate the extremes of the range, within which the initial coordinates must vary. These extremes depend on the slot geometry and the wire geometrical features. In more detail, since the slot geometry is specular with respect to the x-axis (see Figure 3) and the first wire is positioned in the region of space near the lower wedge of the slot, the lower extremes are chosen in such way that they are tangent with the lower slot wedge profile, whereas the upper extremes of the range are chosen so that an additional increase in the initial coordinates causes the loss of significant useful wires. Therefore, in the case of circular, hexagonal and rectangular shapes located in the STSL model, the extremes are respectively equal to:

$$\begin{aligned} x_{init\_circ/hex_{min}} &= d_{max}/2 & x_{init\_circ/hex_{max}} &= \tfrac{3}{2}d_{max} \\ y_{init\_circ/hex_{min}} &= d_{max}/2 & y_{init\_circ/hex_{max}} &= -y_B + \tfrac{3}{2}d_{max} - d_{ins} + \tfrac{w}{2} \\ x_{init\_rect_{min}} &= L_{1max}/2 & x_{init\_rect_{max}} &= \tfrac{3}{2}L_{1max} \\ y_{init\_rect_{min}} &= L_{2max}/2 & y_{init\_rect_{max}} &= -y_B + \tfrac{3}{2}L_{2max} \end{aligned}$$

where y_B is the y-coordinate of the characteristic point B (Figure 3a). On the other hand, for NSTSL, the extremes are equal to, respectively:

$$\begin{aligned} x_{init\_circ/hex_{min}} &= d_{max}/2 & x_{init\_circ/hex_{max}} &= \tfrac{3}{2}d_{max} \\ y_{init\_circ/hex_{min}} &= d_{max}/2 & y_{init\_circ/hex_{max}} &= \tfrac{3}{2}d_{max} \\ x_{init\_rect_{min}} &= L_{1max}/2 & x_{init\_rect_{max}} &= \tfrac{3}{2}L_{1max} \\ y_{init\_rect_{min}} &= L_{2max}/2 & y_{init\_rect_{max}} &= \tfrac{3}{2}L_{2max} \end{aligned}$$

The evaluation of the filling factors, when the first wire center position (x_{w1}, y_{w1}) varies, is carried out by using two loops with indices k and z, with maximum values k_{max} and z_{max}. They are calculated by the following relationships:

$$k_{max} = ceil\left(\frac{x_{init_{max}} - x_{init_{min}}}{\delta_t}\right) \tag{18}$$

$$z_{max} = ceil\left(\frac{y_{init_{max}} - y_{init_{min}}}{\delta_t}\right) \tag{19}$$

where *ceil*(x) is a function that rounds x to the nearest integer greater or equal to x and δ_t is the variation step of the initial coordinates. The last quantity starting value δ_1 is chosen at least 1/50 of the maximum dimension of the considered wire (d_{max} for the circular case, L_{2max} for the rectangular case and $2r_{max}$ for the hexagonal case). In this way, the first wire center position (x_{w1}, y_{w1}) varies according the following relationships:

$$x_{w1}(1,k) = x_{init_{min}} + d_{ins} + \delta_t \cdot (k-1) \tag{20}$$

$$y_{w1}(z,1) = y_{init_{min}} + d_{ins} - \frac{w}{2} + \delta_t \cdot (z-1) \tag{21}$$

Furthermore, in order to investigate the possibility of finding a better wires arrangement when δ_t varies, several iterations with a progressive halving of δ_t are performed until to the maximum value of the number of wires positioned n_{max} presents the same value for three consecutive iterations. Obviously, this choice determines the increasing of computational time, but a more accurate research of the maximum filling factors.

In order to place the next wires, the cross-section of the slot is divided into a grid formed by i columns and j rows, where both the first column and the first row are defined as starting from the first wire position. The maximum number of columns i_{max} and rows j_{max} is calculated according to the slot and wire dimensions. For round, rectangular and hexagonal wires these values are calculated by means of the following relationships, respectively:

$$i_{max} = round\left(\frac{h}{\frac{\sqrt{3}}{2}d_{max}}\right) \tag{22}$$

$$j_{max} = round\left(\frac{w}{d_{max}}\right) \tag{23}$$

$$i_{max} = ceil\left(\frac{h}{L_{1max}}\right) \tag{24}$$

$$j_{max} = ceil\left(\frac{w}{L_{2max}}\right) \tag{25}$$

$$i_{max} = ceil\left(\frac{h}{\frac{\sqrt{3}}{2}r_{max}}\right) \tag{26}$$

$$j_{max} = ceil\left(\frac{w}{\frac{3}{2}r_{max}}\right) \tag{27}$$

where $round(x)$ is a function that rounds each element of x to the nearest integer. The distribution of the wires occurs through two loops that change the indices i and j in order to position the wires along with the whole slot profile. In the cases of the round and hexagonal wires, in order to obtain an orthocyclic winding pattern, the coordinates of the successive wire are calculated as previously described in Section 2. The rectangular case is widely described in [13]. The positioning of the wires is considered valid only if it meets the overlapping conditions described above. The algorithm calculates, for each possible pair of coordinates $(x_{w1}(1,k)\ y_{w1}(z,1))$, the number of positioned wires n, the electrical and mechanical slot filling factor values f_{cu} and f_{me} and the coordinates of the positioned wires (x_{wire}, y_{wire}). In this way, each of the quantities of interest is represented by an array, in which each element is associated to the coordinates of the first positioned wire. The search for the best wire arrangement for a fixed wire shape, its geometric dimension, type of positioning and slot profile can be achieved by identifying the maximum value of number of wires positioned n_{max} within the array. This procedure, as mentioned above, is repeated until the quantity, $n_{max}(\delta_t)$, presents the same value for three consecutive δ_t values. Therefore, once the coordinates of the first wire (x_{w1}, y_{w1}) associated to the best wire arrangement is determined, the algorithm provides a graphical representation of the slot containing the wires. Furthermore, the coordinates of the positioned wires (x_{wire}, y_{wire}) are available and the manufacturer has the ability to know and choose a reference wire for the production phase according to the winding technological process adopted. In short, the algorithm works via the following steps:

1. The algorithm asks for, as input data, the wire shape, the type of positioning and the slot profile to be used;
2. The geometric dimensions of the wire and the slot and the value of the safety distance are defined in an input file;
3. The algorithm proceeds by plotting the slot profile;
4. Subsequently, the algorithm proceeds to position the wires; for fixed coordinates of the first wire, for each of them, it checks the overlapping conditions and calculates the number of positioned wires and the value of the electrical and mechanical filling factors;
5. The algorithm repeats the evaluation procedure for each possible pair of the initial wire coordinates and generates an array for each quantity of interest;
6. It searches the best wire arrangement case (characterized by n_{max} wires per slot, $f_{cu\text{-}max}$ and $f_{me\text{-}max}$) and their corresponding coordinates to the first wire;
7. The algorithm repeats the procedure until to the maximum value of the number of wires positioned n_{max} presents the same value for three consecutive δ_t values;
8. Finally, from the knowledge of the first wire position, the algorithm proceeds by plotting the slot profile containing the wires and provides, in an output, the maximum number of positioned wires n_{max}, the best values of the electrical and mechanical slot filling factors $f_{cu\text{-}max}$ and $f_{me\text{-}max}$ and the coordinates of all positioned wires (x_{wire}, y_{wire}).

In addition, for an accurate description and understanding, the algorithm flowchart is reported in Figure 8.

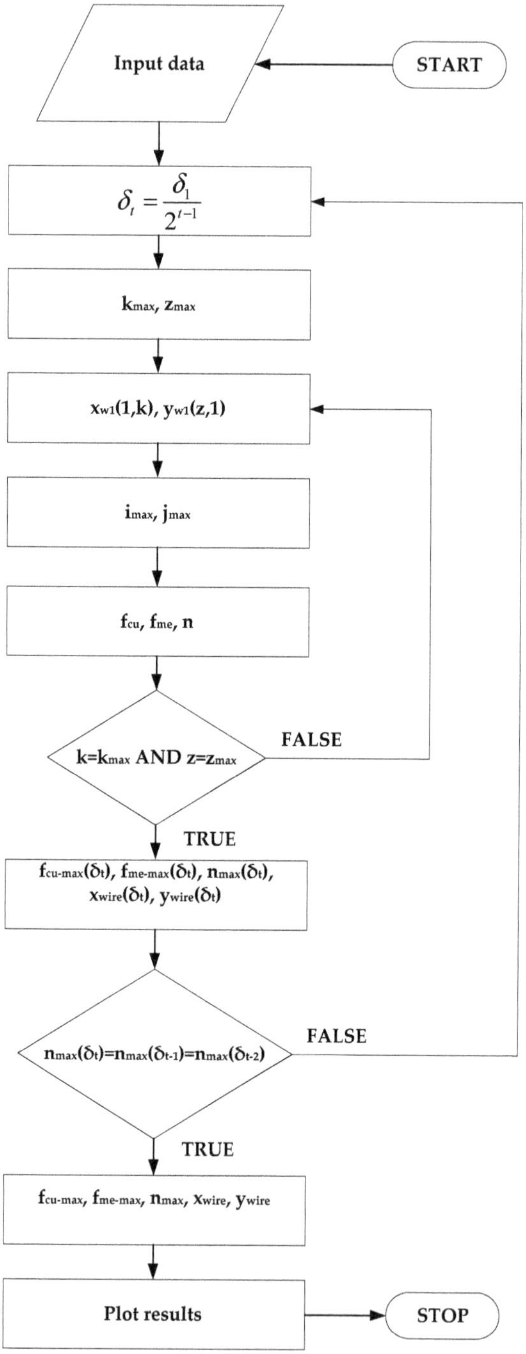

Figure 8. Algorithm flowchart.

4. Cases of Study

In order to validate the proposed algorithm, several cases of study have been performed. This study has been carried out both for the STSL and for the NSTSL and for each wire shape. From international standards [25,26], it is possible to deduce the nominal dimensions, the insulation grades, the tolerances and the outer maximum dimensions for each typology of wire. In this work, the largest allowed outer dimensions are adopted to consider the worst case for the evaluation of the slot electrical filling factor. An insulation grade (According to [25,26] the grade is defined as "the range of thickness of the insulation wire".) 3 and a grade 2 for the round-shaped wire and for the rectangular-shaped one have been taken into consideration, respectively. The dimensions of the hexagonal wire have been deduced, assuming the cross-section is equal to that of the round wire. In this way, it is possible to compare the slot filling factors and evaluate the benefits derived from the use of hexagonal wires. In this study, six different dimensions have been chosen for each wire shape, respectively. For each dimension, three additional interspaces t_w, respectively equal to 0, 0.05 and 0.1 mm, were taken into consideration. Furthermore, regarding the standard slot, a study was carried out on the possibility of positioning the wires parallel to the ground or the bottom (PG) and to the flank (PF) of the slot. The dimensions of the wires, used for this study and expressed in mm, are reported in Tables 3 and 4.

Table 3. Geometrical features of round and hexagonal wires.

s_{cu} (mm^2)	d_{cu} (mm)	d_{max} (mm)	r_{cu} (mm)	r_{max} (mm)
0.636	0.90	1.018	0.495	0.560
0.785	1.00	1.124	0.550	0618
0.985	1.12	1.248	0.616	0.686
1.227	1.25	1.381	0.687	0.759
1.539	1.40	1.535	0.770	0.854
2.010	1.60	1.740	0.880	0.957

Table 4. Geometrical features of rectangular wires.

s_{cu} (mm^2)	L_{1cu} (mm)	L_{2cu} (mm)	L_{1max} (mm)	L_{2max}(mm)	r_{corner} (mm)
1.626	2.00	0.90	2.17	1.07	0.45
2.025	2.24	1.00	2.41	1.17	0.50
2.920	2.50	1.25	2.67	1.42	0.50
3.705	2.80	1.40	2.97	1.57	0.50
4.825	3.15	1.60	3.32	1.77	0.50
5.465	3.55	1.60	3.72	1.77	0.50

Figures 9–12, it is possible to notice how the algorithm returns the slot profile with the desired distribution of the wires and also shows the number of inserted wires. In particular, in the case of round wires, the winding pattern presents an orthocyclic structure where the positioning is parallel to the bottom in one case and parallel to the flank of the slot in the other case. In the case of rectangular wire, the winding pattern is ordered. In the case of the hexagonal wire, an orthocyclic winding pattern is obtained, both with *phi* equal to zero and with *phi* equal to $\pi/2$. Furthermore, it is possible to notice how the structure remains ordered with the use of an additional interspace t_w.

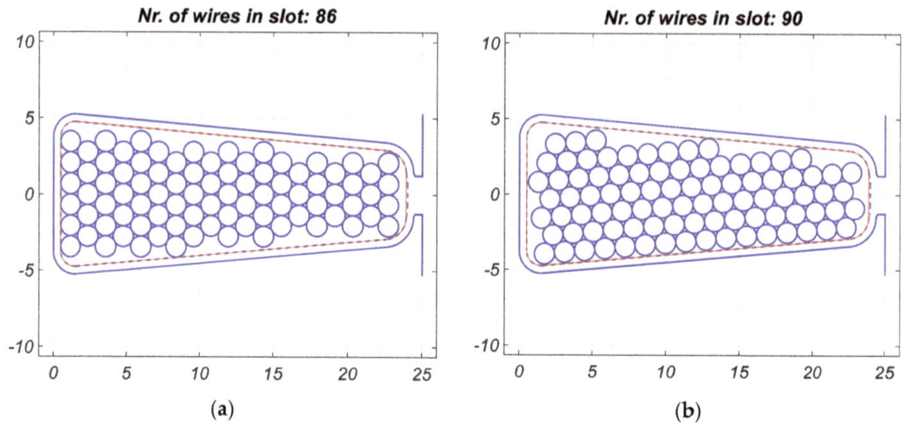

Figure 9. Orthocyclic distribution of the round wires with the parallel ground disposition (**a**) and with the parallel flank disposition in the standard slot (**b**) (t_w = 0 mm, d_{max} = 1.381 mm).

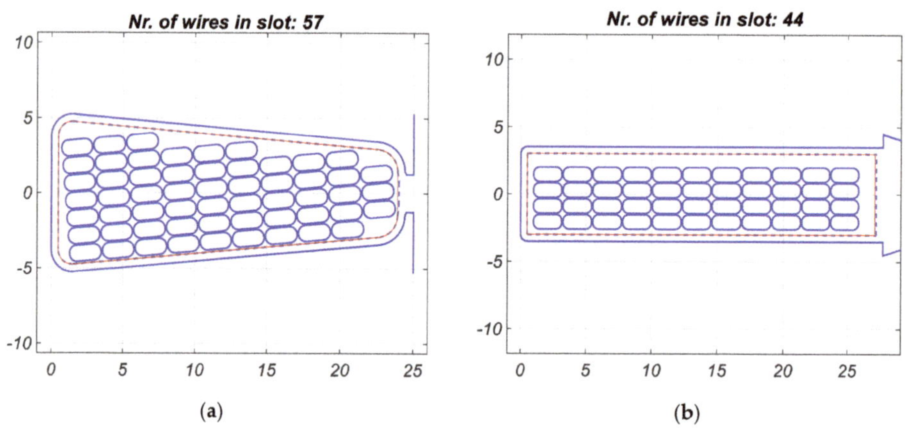

Figure 10. Ordered distribution of the rectangular wires with parallel disposition to the flank in standard slot (**a**) and in non-standard slot (**b**) (t_w = *0.1* mm, L_{1max} = 2.17 mm, L_{1max} = 1.07 mm).

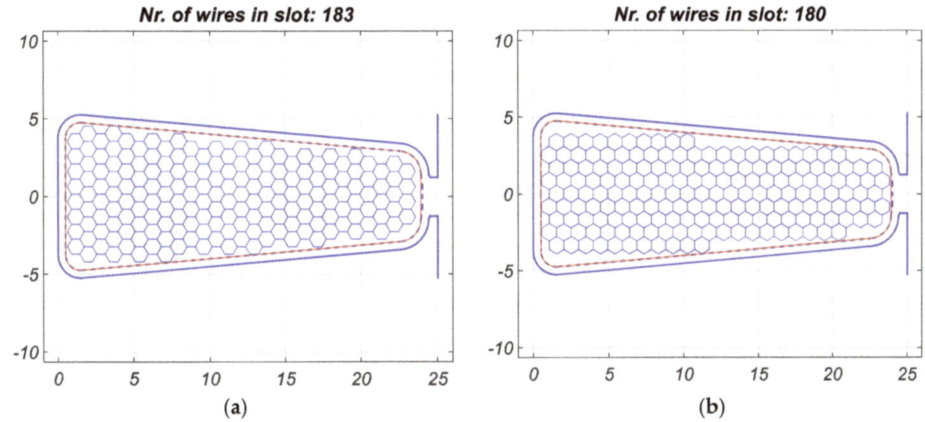

Figure 11. Orthocyclic distribution of the hexagonal wires in standard slot (*phi* = 0, t_w = 0 mm, r_{max} = 0.313 mm) (**a**) and (*phi* = $\pi/2$, t_w = 0 mm, r_{max} = 0.313 mm) (**b**).

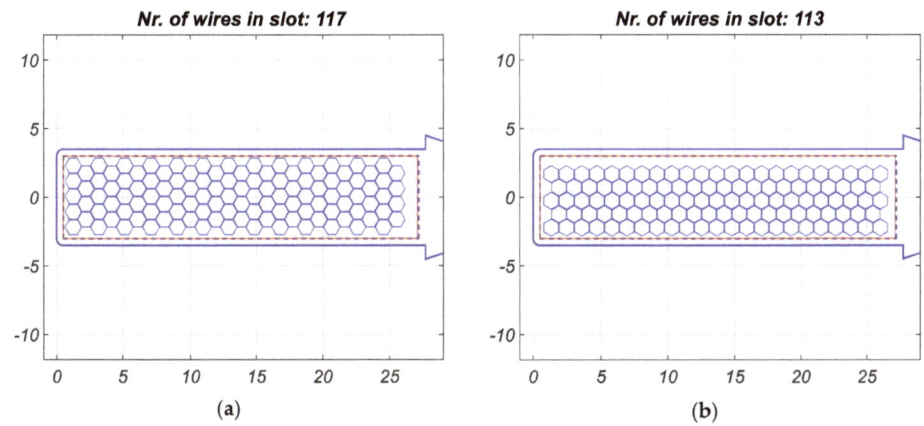

Figure 12. Orthocyclic distribution of the hexagonal wires in non-standard slot (*phi* = 0, t_w = 0 mm, r_{max} = 0.618 mm) (**a**) and (*phi* = $\pi/2$, t_w = 0.05 mm, r_{max} = 0.618 mm) (**b**).

In Figure 13, the value of slot electrical filling factors obtained for the case of the round wire, with a parallel bottom disposition, both for the STSL and the NSTSL, are reported. A comparison, between the slot electrical filling factors obtained with a parallel bottom disposition and the one obtained with the parallel flank disposition of the standard slot, are reported in Figure 14. This comparison shows that the slot electrical filling factor is higher for a parallel flank disposition for each additional interspace t_w. A similar study was performed for rectangular wires. Figure 15 shows the trend of slot electrical filling factors as a function of the cross-section area. On average, the filling factors obtained in the rectangular case are higher than those of the round case for the dimensions chosen. Additionally, in the rectangular case, the slot electrical filling factor is higher in a parallel flank disposition (Figure 16). In this work, the comparison between the slot electrical filling factors obtained in the round and rectangular wire cases, with equal cross-section, has not been reported because it has been widely discussed in [18] and in [19]. Furthermore, the rectangular wires are used for medium and high-power applications, whereas the round ones are mostly used for low-power applications. Figures 17 and 18 show the comparison between the value of electrical slot filling factors obtained for hexagonal wires with *phi* (ϕ) equal, respectively, to 0 and $\pi/2$. This study is performed both for the standard slot and for the non-standard slot, for each value of the additional interspace t_w. In the standard slot, the slot electrical filling factors are higher when $phi = 0$ whereas, in the non-standard slot, in some cases, higher electrical slot filling factors are obtained with $phi = \pi/2$. Particularly interesting is the comparison between the slot filling factors obtained in the case of the round wire and the hexagonal one, with the same cross-section. In Figure 19, the comparison between the slot electrical filling factors of round wires with the parallel flank disposition and the slot electrical filling factors of hexagonal wires with $phi = 0$, is reported. From this comparison, it can be seen that the filling factors obtained in the case of hexagonal wires are always higher than those obtained in the case of round wires. This difference decreases as the additional interspace increases. Therefore, the use of hexagonal wires can provide innovation in the field of electrical machine windings. Obviously, this improvement must be contextualized with any additional costs and the current state-of-the-art of specially shaped wire manufacturing process. Furthermore, with the choice of δ_1, equal to 1/50 of the maximum dimension of the considered wire, the worst-case scenario of the study has presented a computation time equal to about 10 min, corresponding to about 8000 different first wire positions. In order to validate the choice of δ_1 equal to 1/50, further investigations have been carried out, both increasing and decreasing the value of δ_1.

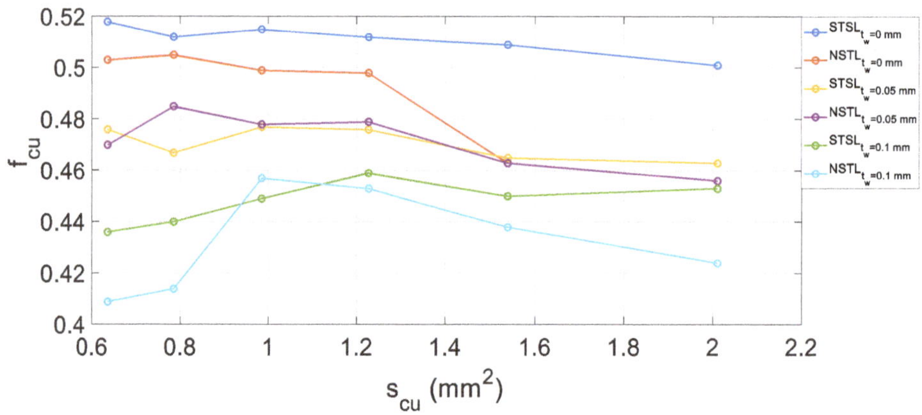

Figure 13. Slot electrical filling factor as a function of wire copper cross-section area for round wires.

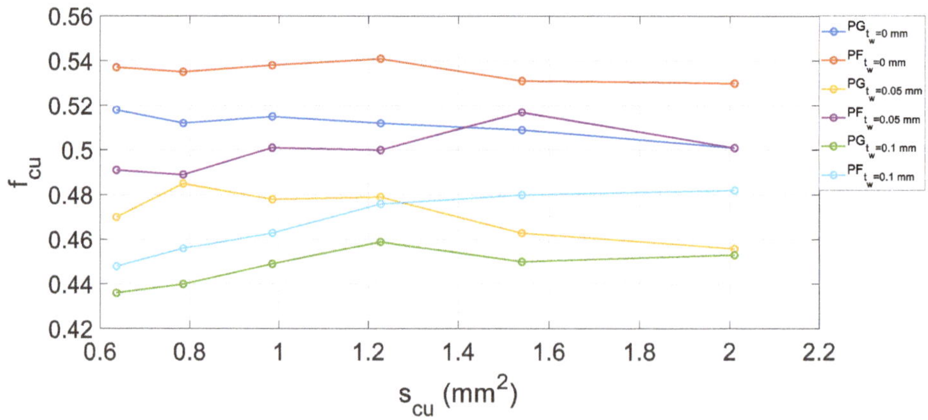

Figure 14. Comparison between the slot electrical filling factors obtained with round wires distribution parallel to the ground/bottom (PG) and to the flank (PF) of the standard slot.

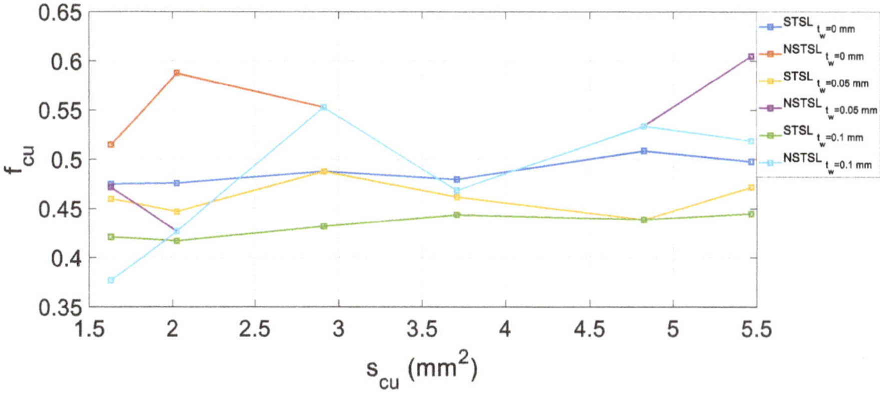

Figure 15. Slot electrical filling factor as a function of wire copper cross-section area for rectangular wires.

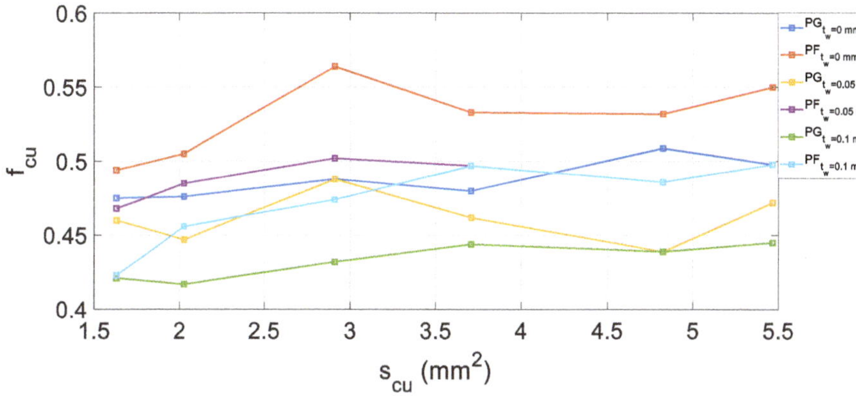

Figure 16. Comparison between the slot electrical filling factors obtained with rectangular wires distribution parallel to the ground/bottom (PG) and to the flank (PF) of the standard slot.

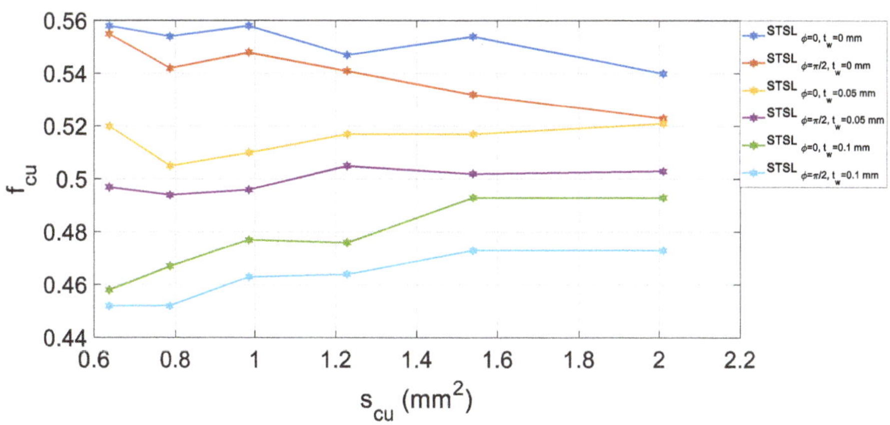

Figure 17. Slot electrical filling factor as a function of wire copper cross-section area for hexagonal wires in a standard slot (STSL).

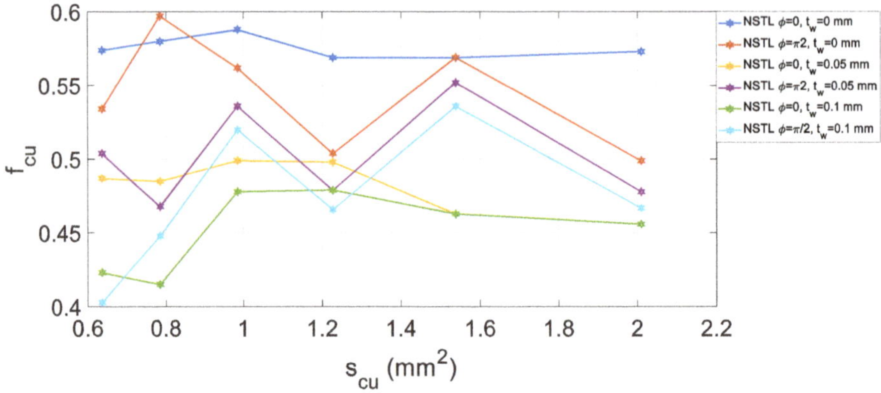

Figure 18. Slot electrical filling factor as a function of wire copper cross-section for hexagonal wires in a non-standard slot (NSTSL).

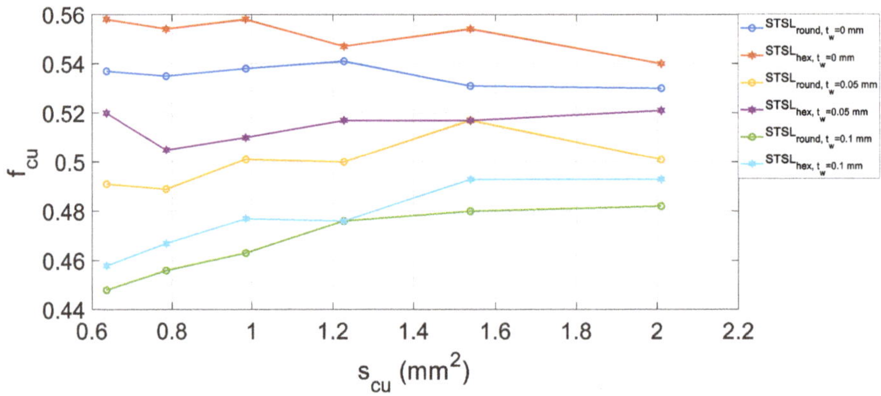

Figure 19. Comparison between the slot electrical filling factors obtained with round wires distributed parallel to the flank of the slot and the one obtained with hexagonal wires as a function of wire copper cross-section area.

In more detail, values of δ_1 smaller than 1/50 only provide longer computational times with the identification of the same best wire arrangements. On the contrary, in almost all analyzed cases, the values of δ_1 in the range from 1/50 to 1/25 allow us to identify the same best wire arrangements with shorter computational times compared to those obtained with δ_1 being equal to 1/50. Values of δ_1 lower than 1/25 generate longer computational times due the higher number of δ_i iterative variations. Therefore, the choice of δ_1 equal to 1/50 of the maximum dimension of the wire under study provides a good compromise between computation time and identification of the optimal wire arrangements.

The case studies shown here demonstrate how the algorithm allows us to accurately estimate the best value of the filling factors and reveal that it is a useful tool for the design of electrical machine windings. The results, like electrical filling factors, are calculated under the assumption of the ideal geometric shape of each wire and slot filling paper. For practical considerations, tolerances of shape and diameter must be taken into account; also, the winding process itself does not give an ideal orthocyclic winding distribution in the slot. This deviations from ideal geometry and winding process can be considered using additional geometric tolerances. With the given algorithm, the influence of non-ideal conditions on the filling factor and maximal turns can be calculated within a few minutes and motor design engineers and process engineers can make decisions based on a reliable calculation concept.

5. Conclusions

High filling factors make it possible to improve electrical machines' performances, meeting the design requirements of several application fields. This improvement requires an optimal arrangement of the wires inside the slots and a careful evaluation of the maximum slot filling factor obtainable in the design phase. In this paper, an algorithm approach is proposed to perform the calculation of slot filling factors in electrical machines. The algorithm requires, as input data, the geometrical data of the slot, the insulation thickness, the shape of the wire, the dimensions of the wire and the type of disposition inside the slot. From this data, the algorithm determines the maximum slot filling factors, the number of wires positioned, a graphic distribution of the wires inside the slot and the coordinates of the wires positioned. The algorithm has a high degree of flexibility and requires a reasonable computation time (about 10 min in the worst case). The conducted study proves that the algorithm is very simple and can give useful results in the designing processes of winding layouts. Furthermore, the algorithm can be used as investigation tool because it allows us to compare electrical filling factor values when different wire shapes with the same cross-section are employed. From the investigations here presented, it has been shown that the use of hexagonal wires provides higher filling factors than those obtained with the use of round wires. Although the comparison between the slot filling factors obtainable with rectangular wires and hexagonal wires with the same wire cross-section is not presented in this paper, the use of hexagonal wires allows us to obtain higher filling factors than those obtained with the use of rectangular wires. This result is due to the orthocyclic arrangement of the hexagonal wires that optimally occupy the slot area, unlike the ordered arrangement of the rectangular wires. Non-ideal geometry of the magnetic wires and the tolerances of the winding process can be easily considered using additional geometric factors and practical problems can be addressed. These results may be of considerable interest in the optimization of electrical machine windings. Future developments will concern the extension of this study to a greater number of cases in terms of slots and an in-depth study on the producibility of hexagonal wires. In particular, the evaluation of the effects of real cross-section deviation with respect to ideal deviation should be modelled and implemented in future research. Furthermore, experimental investigations will be conducted in order to validate the algorithm results.

Author Contributions: Authors contributed equally to this work. Authors of this manuscript jointly conceived the theoretical developments, revised the state-of-the-art algorithms and provided suggestions to obtain a flexible algorithm for the optimization of slot filling factors in the design phase of electrical machines. All authors have read and agreed to the published version of the manuscript.

Funding: This research received no external funding.

Acknowledgments: This work was financially supported by MIUR—Ministero dell'Istruzione dell'Università e della Ricerca (Italian Ministry of Education, University and Research) and by SDESLab—Sustainable Development and Energy Saving Laboratory of the University of Palermo.

Conflicts of Interest: The authors declare no conflict of interest.

References

1. You, Y.-M. Optimal Design of PMSM Based on Automated Finite Element Analysis and Metamodeling. *Energies* **2019**, *12*, 4673. [CrossRef]
2. Caruso, M.; di Tommaso, A.O.; Lombardo, M.; Miceli, R.; Nevoloso, C.; Spataro, C. Maximum Torque Per Ampere control algorithm for low saliency ratio interior permanent magnet synchronous motors. In Proceedings of the 2017 IEEE 6th International Conference on Renewable Energy Research and Applications (ICRERA), San Diego, CA, USA, 5–8 November 2017; pp. 1186–1191.
3. Caruso, M.; di Tommaso, A.O.; Miceli, R.; Nevoloso, C.; Spataro, C.; Trapanese, M. Maximum Torque per Ampere Control Strategy for Low-Saliency Ratio IPMSMs. *Int. J. Renew. Energy Res.* **2019**, *9*, 374–383.
4. L-Refaie, A.M.E. Fractional-slot concentrated-windings synchronous permanent magnet machines: Opportunities and challenges. *IEEE Trans. Ind. Electron.* **2010**, *57*, 107–121. [CrossRef]

5. Schiefer, M.; Doppelbauer, M. Indirect slot cooling for high-power-density machines with concentrated winding. In Proceedings of the 2015 IEEE International Electric Machines Drives Conference (IEMDC), Coeur d'Alene, ID, USA, 10–13 May 2015; pp. 1820–1825.
6. Hofmann, B.; Bickel, B.; Bräuer, P.; Leder, M.; Franke, J. Theoretical benefits of powder-coating based insulation layers regarding copper fill factor in electric drives. In Proceedings of the 2016 6th International Electric Drives Production Conference (EDPC), Nuremberg, Germany, 30 November–1 December 2016; pp. 172–176.
7. Kulan, M.C.; Baker, N.J.; Widmer, J.D. Design of a high fill factor permanent magnet integrated starter generator with compressed stator windings. In Proceedings of the 2016 XXII International Conference on Electrical Machines (ICEM), Lausanne, Switzerland, 4–7 September 2016; pp. 1513–1519.
8. Fyhr, P.; Domingues, G.; Reinap, A.; Andersson, M.; Alaküla, M. Performance and manufacturability tradeoffs of different electrical machine designs. In Proceedings of the 2017 IEEE International Electric Machines and Drives Conference (IEMDC), Miami, FL, USA, 21–24 May 2017; pp. 1–7.
9. Torreggiani, A.; Bianchini, C.; Davoli, M.; Bellini, A. Design for Reliability: The Case of Fractional-Slot Surface Permanent-Magnet Machines. *Energies* **2019**, *12*, 1691. [CrossRef]
10. Cheng, L.; Sui, Y.; Zheng, P.; Yin, Z.; Wang, C. Influence of Stator MMF Harmonics on the Utilization of Reluctance Torque in Six-Phase PMA-SynRM with FSCW. *Energies* **2018**, *11*, 108. [CrossRef]
11. Caruso, M.; Di Tommaso, A.O.; Marignetti, F.; Miceli, R.; Ricco Galluzzo, G. A General Mathematical Formulation for Winding Layout Arrangement of Electrical Machines. *Energies* **2018**, *11*, 446. [CrossRef]
12. Stenzel, P.; Dollinger, P.; Richnow, J.; Franke, J. Innovative needle winding method using curved wire guide in order to significantly increase the copper fill factor. In Proceedings of the 2014 17th International Conference on Electrical Machines and Systems (ICEMS), Hangzhou, China, 22–25 October 2014; pp. 3047–3053.
13. Hagrn, J.; Blanc, F.S.-L.; Fleischer, J. *Handbook of Coil Winding, Technologies for Efficient Electrical Wound Products and Their Automated Production*; Springer: Berlin, Germany, 2017.
14. Gerngroß, M.; Herrmann, P.; Westermaier, C.; Endisch, C. Highly flexible needle winding kinematics for traction stators based on a standard industrial robot. In Proceedings of the 2017 7th International Electric Drives Production Conference (EDPC), Würzburg, Germany, 5–6 December 2017; pp. 1–7.
15. Raabe, N. An algorithm for the filling factor calculation of electrical machines standard slots. In Proceedings of the 2014 International Conference on Electrical Machines (ICEM), Berlin, Germany, 2–5 September 2014; pp. 981–986.
16. Richter, R.; Brüderlink, R. *Elektrische Maschinen*; Springer: Berlin/Heidelberg, Germany, 1954; Volume 4.
17. Jaksic, D. Getting rid of the air, or how to maximize winding fill factor (ID 81). In Proceedings of the 2011 1st International Electric Drives Production Conference, Nuremberg, Germany, 28–29 September 2011; pp. 84–87.
18. di Tommaso, A.O.; Genduso, F.; Miceli, R.; Nevoloso, C. Fast procedure for the calculation of maximum slot filling factors in electrical machines. In Proceedings of the 2017 Twelfth International Conference on Ecological Vehicles and Renewable Energies (EVER), Monte Carlo, Monaco, 11–13 April 2017; pp. 1–8.
19. Caruso, M.; di Tommaso, A.O.; Miceli, R.; Nevoloso, C. Algorithmic Approach for Slot Filling Factors Determination in Electrical Machines. In Proceedings of the 2018 7th International Conference on Renewable Energy Research and Applications (ICRERA), Paris, France, 14–17 October 2018; pp. 1489–1494.
20. Herrmann, P.; Stenzel, P.; Vögele, U.; Endisch, C. Optimization algorithms for maximizing the slot filling factor of technically feasible slot geometries and winding layouts. In Proceedings of the 2016 6th International Electric Drives Production Conference (EDPC), Nuremberg, Germany, 30 November–1 December 2016; pp. 149–155.
21. Lubachevsky, B.D. How to simulate billiards and similar systems. *J. Comput. Phys.* **1991**, *94*, 255–283. [CrossRef]
22. Graham, R.; Lubachevsky, B.; Nurmela, K.; Östergård, P. Dense packings of congruent circles in a circle. *Discret. Math.* **1998**, *181*, 139–154. [CrossRef]
23. Graham, R.L.; Lubachevsky, B.D. Dense packings of equal disks in an equilateral triangle: From 22 to 34 and beyond. *Electr. J. Comb.* **1995**, *2*, 1–39.
24. Lubachevsky, B.D.; Graham, R.L. Curved hexagonal packings of equal disks in a circle. *Discret. Comput. Geom.* **1997**, *18*, 179–194. [CrossRef]

25. The National Standards Authority of Ireland. *IEC 60317-0-1 Specifications for Particular Types of Winding Wires Part 0-1: General Requirements- Enamelled Round Copper Wire*; NSAI: North Wood, Ireland, 2014.
26. The National Standards Authority of Ireland. *IEC 60317-0-2 Specifications for Particular Types of Winding Wires Part 0-2: General Requirements- Enamelled Rectangular Copper Wire*; NSAI: North Wood, Ireland, 2014.

energies

MDPI

Article

Non-Salient Brushless Synchronous Generator Main Exciter Design for More Electric Aircraft

Filip Kutt *,†, **Michał Michna** † and **Grzegorz Kostro** †

Faculty of Electrical and Control Engineering, Gdańsk University of Technology, 80-233 Gdańsk, Poland; michal.michna@pg.edu.pl (M.M.); grzegorz.kostro@pg.edu.pl (G.K.)

* Correspondence: filip.kutt@pg.edu.pl; Tel.: +48-58-347-19-39

† These authors contributed equally to this work.

Received: 19 April 2020; Accepted: 26 May 2020; Published: 27 May 2020

Abstract: This paper presents a prototype of high speed brushless synchronous generators (BSG) design for the application in autonomous electric power generation systems (e.g., airplane power grid). Commonly used salient pole field of the main generator part of BSG was replaced with a prototype non-salient pole field. The main objective of the research is an investigation into the advantages and disadvantages of a cylindrical field of the main generator part of BSG over the original salient pole field. The design process of the prototype generator is presented with a focus on the electromagnetic and mechanical finite element method (FEM) analysis. The measurements of prototype and commercial BSG were conducted for the nominal speed of 8 krpm. The advantages and disadvantages of the proposed solution were established based on measurements in load and no-load conditions.

Keywords: autonomous systems; brushless synchronous generator; electric power generation; high speed generator

1. Introduction

Modern commercial planes are designed according to the concept of more electric aircraft (MEA) [1–7]. This concept states that the electric power system in future aircraft should replace pneumatic and hydraulic systems supplied from main turbofan engines. The electrical power system should provide the ability to control the aircraft via electromechanical and electrohydraulic actuators and also provide deicing protection and control the pressurization of the cabin. The high increase in electric power demand is the main result of such an approach. High electric power demand, in turn, requires much more powerful generators to supply it. In addition, an important role of the generator is also the ability to work as a starter to accelerate a turbofan engine [7–10]. The increase in power output of the generator should not affect the volume and weight of the generators in a significant way. One of the obvious directions is the increase in the rotational speed of the machine [11–13]. However, one of the consequences of an increase in rotational speed is a higher centrifugal force acting on the rotor parts and the limitation in the rotor and shaft diameters.

To meet the requirements of the MEA concept considers various types of machines with high power densities defined as the power to weight ratio [7,11–14]. In addition to standard, wound field brushless synchronous generator (BSG), it is also proposed to use a permanent magnet (PM) [15,16], a switched reluctance (SR) [17–19] or an induction (IM) [8,20] machine as a starter-generator for aircraft. Despite the diversity of machine types proposed in research and prototype solutions, the BSG with a wound rotor is still the most often used in commercial aircraft. BSG has been used for decades, so its design is well known, tested and relatively reliable. Among its main advantages are the ability to easily control the voltage at the terminals by changing the excitation field, a simple adjustment system, the ability to easily reconfigure the system. Unfortunately, standard BSG solutions are characterized by low efficiency (<80%), the maximum speed limit is less than 28 krpm and, consequently, also low power density (about 2.5 kW/kg) [11]. Therefore, research and design works are underway on new, alternative constructions for aircraft generators, including PM, SR, and IM generators [11–14]. These machines are characterized by a simpler rotor design, which allows operation at higher rotational velocity with less maintenance required. Permanent magnet generators achieve the highest power densities in relation to weight (from 3.3 to 8 kW/kg), and high efficiency (up to 95%) [10,11,15,16]. The disadvantage of PM generators is the complex regulation of the excitation field and voltage at the terminals and also the problem with excitation during a short circuit fault. The main advantage of SR generators is a very simple construction of the rotor, and therefore high reliability. SR generators have lower efficiency and lower power density than PM machines; however, it is still better than BSG [17–19]. On the other hand, SR requires complex control systems and algorithms. IM machines are also a prudent alternative, especially in the multiphase, fault-tolerance design [8,20]. Simple and cheap design, proven control systems and algorithms guarantee reliable work required in the aviation industry.

Only proven and reliable solutions can be used in aviation. The use of new solutions and devices for a commercial aircraft requires a time-consuming and expensive design, verification and certification process [21]. Therefore, research is still being carried out to improve the BSG design both in the context of achieving a better power-to-weight ratio and working at higher speeds and frequencies. These requirements can be met when materials, cooling, and mechanical structure would be improved [11].

High requirements regarding the parameters of the aircraft generator can be met by optimizing their design. Therefore, both new construction solutions, as well as new methods of their design, project management and production, are being developed. A wide range of issues related to the design and optimization of electrical machines is of interest to both manufactures and academic centers [14,21–25]. The requirements for high power density and high speed mean that it is important to optimize the generator system as a whole, and not its individual components. The design of the electric machine ceases to be an independent problem, and the optimal solution from the component point of view does not necessarily mean an optimal solution for the entire drive system. The design of the generator system is, therefore, a multi-level, multi-task, multi-disciplinary problem and at the same time nonlinear with many independent variables [26–29].

Computer modelling and simulation methods are widely used in the design process of electrical machines. Optimal use of computer modelling methods and simulations allows one to accelerate design work, verify the importance of potential solutions and, as a result, achieve a product with better parameters. Product development uses both simple analytical methods as well as more complex finite element method (FEM). Obtaining a low mass of the generator can be achieved by increasing the speed or reducing the mass of active materials. High-speed operation requires special attention when designing the rotor. The basis for design should be the selection of appropriate materials and FEM simulations to achieve the expected electromagnetic and mechanical parameters [30]. The reduction of the mass of conductive materials can be achieved by increasing the density of currents. Conducting computational fluid dynamics (CFD) simulation based on a complete conjugate heat transfer (CHT) allows the design of an appropriate cooling system [31].

Even if the final optimization of the machine requires the use of advanced simulation tools (FEM, CFD, etc.), the basis of the project is an analytical method. One of the methods widely used in the pre-design of electric machines is the sizing equation method [32–34].

The brushless synchronous generator is a complex power generation device (Figure 1). Three machines are installed on one shaft: the main generator, the exciter, and the subexciter. Both the main generator and the exciter are electromagnetically excited synchronous generators and the subexciter is a permanent magnet synchronous generator. The main generator field is supplied from the armature winding of the exciter through a 6-phase diode rectifier. The exciter armature is on the rotor along with the field of the main generator and the exciter field winding is on the stator. The exciter field winding is supplied from the subexciter via a controlled rectifier. The generator control unit (GCU) controls the voltage and the protection systems for the generator. In the variable high-speed operation of the generator, the excitation system is designed to ensure the voltage control (RMS value) requirements of the power grid.

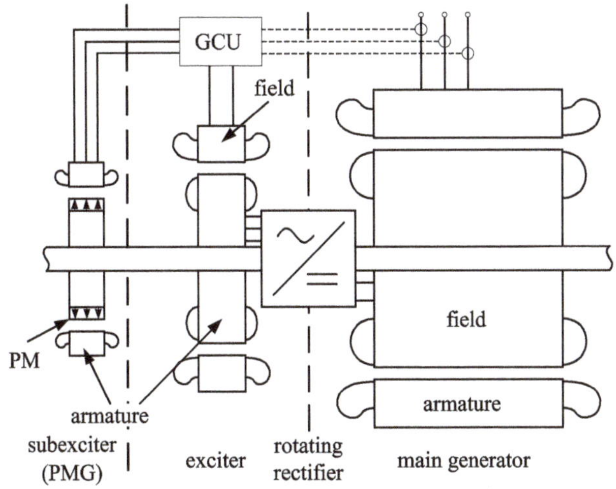

Figure 1. The brushless synchronous generator—based on a three-stage electrical machine topology: the subexciter—permanent magnet generator (PMG); the brushless exciter—synchronous machine with stationary field winding, rotating armature winding and rotating diode rectifier; the main generator—synchronous machine with rotating field winding; GCU—generator control unit.

In modern aircraft such as the Boeing 787 or the Airbus A350XWB the variable frequency generation system is used, where generators are directly driven from the turbofan engines. The system voltage is regulated at 230 V and the bus frequency varies from 350 to 800 Hz. Changing the way in MEA electric power generation requires adaptation of the distribution system and the use of more power electronic converters supplying loads [4]. This affects the power quality, which is also an important issue, as it has a direct effect on the reliability and efficiency of the power system [35]. The harmonic content of produced voltage and current of the main generator can also have a significant influence on the system power quality. In the salient pole machines, the effect of unsymmetrical pole shoe saturation during loaded conditions can increase the harmonic content in induced voltage [36].

2. Objectives and Scope

The main objective of the research is an investigation into the advantages and disadvantages of a cylindrical field of the main generator part of BSG over the original salient pole field. The main motivation for the research was the analysis of the possibility of introducing a cylindrical main filed construction without any detriment to the generated power quality in reference to the original construction. The proposed construction should allow for:

- more sinusoidal back EMF of the generated power due to more sinusoidal excitation winding magneto-motive force (MMF) spatial distribution,
- the possibility of reaching higher rotational velocities of modern MEA variable frequency power systems in future developments of BSG.

The development of a modern aircraft power system is based on MEA concept. However, only repetitively proven and reliable solutions can be used in commercial avionic applications. The MOET FP6 [37] project investigated the component level power system development. The practical result of the investigation was an evolutional change in the power generation system components and particularly the BSG. The focus of those changes was the improvement of the generator structure for higher efficiency without any detriment to its relatability. That meant the introduction of a variable frequency system with a higher rotational velocity of the generator and improved power quality. The proposal in this paper solution is aiming to investigate the possible advantages in this regard of a cylindrical structure of the main exciter of the BSG. The literature on the subject of the construction of BSG main exciter is very limited and only investigates the possibility of using claw pole design [38] for the main exciter of the machine or the possibility of using a cylindrical structure for operation with higher rotational velocity [39].

The comparative study is conducted on a commercial and redesigned generator construction. Both machines are three stage machines consisting of a permanent magnet generator (PMG) subexciter, an inverted synchronous generator exciter and the main synchronous generator. The only difference between the two machines is the construction of the main generator field (Figure 2). The commercially available machine has a salient pole main generator field construction and the proposed prototype has a cylindrical (non-salient pole) construction of the main generator field. The proposed prototype was designed as a cylindrical rotor machine in order to obtain sinusoidal distributed excitation winding MMF with the possibility of achieving adequate mechanical strength when working at a higher rotational speed. The reverse engineering approach was used on the commercial salient pole generator to determine the electromagnetic parameters of the machine. The generator performance in the steady state was measured and analyzed. The main factor for the analysis was the minimization of the high-order harmonic content of currents and voltages waveforms during the no-load and load operation.

The design of the prototype generator is based on the salient pole brushless synchronous generator type GT40PCz8 (S_n = 40 kVA, U_n = 208 V, $p.f.$ = 0.8, n_n = 8000 rpm) used in the Russian MI-28 helicopter and is a typical construction for avionic application. The commercial generator has a salient pole field of the main generator part and the proposed and constructed prototype uses a non-salient pole field (Figure 2).

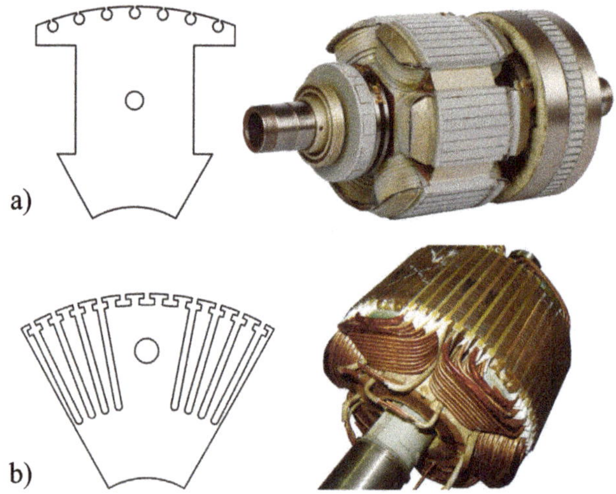

Figure 2. Two types of exciter of the main generator: (**a**) salient pole (original commercial structure, exciter, and subexciter on one shaft); (**b**) non-salient pole (prototype structure, exciter, and subexciter temporally removed from the shaft).

The commercial GT40PCz8 generator was the subject of the reverse engineering process using analytical calculations and FEM simulations. This approach was used to determine the value of electromagnetic parameters such as current densities in the armature and field windings and magnetic flux density in various parts of the machine. Based on those values, the new main generator field was designed and developed.

The main contribution of the research is the design of the prototype of the main generator part cylindrical exciter that provides better power quality in no-load and under load conditions. This will allow for future developments for variable frequency power system higher maximum frequency.

3. Solution

The development and the implementation of a new design of the electrical machine requires a comprehensive approach to the physical phenomena taking part in the electromechanical energy conversion process in the drive system.

The first step of the design and analysis process is an electric machine design based on analytical calculations (Figure 3). The dimensions of a machine are calculated based on the sizing equations approach [22,34]. The sizing equation describes the relationship between the output power of the machine (P) and its main dimensions design features, material parameters, and rotational speed:

$$P = \frac{\pi}{2} K_I K_P K_E \frac{f_s}{p} (A_s B_m)(D_s^2 l_s) \tag{1}$$

where: f_s—the stator voltage frequency, p—the number of pole pairs; the machine main dimensions: D_s—the inside stator diameter, l_s—the length of stator core; the machine design features: K_I—the current waveform factor, K_P—the electrical power waveform factor, K_E—the EMF (electromotive force or induced voltage) factor which incorporates the winding distribution factor and the air gap magnetic flux distribution, and the material parameters: A_s—the stator electric loading, and B_m—the air gap flux magnitude. The values of the design feature factors depend on the type of the machine, the type of the power supply (shape of the current waveform), the air gap flux distribution and the field and armature winding distributions [22].

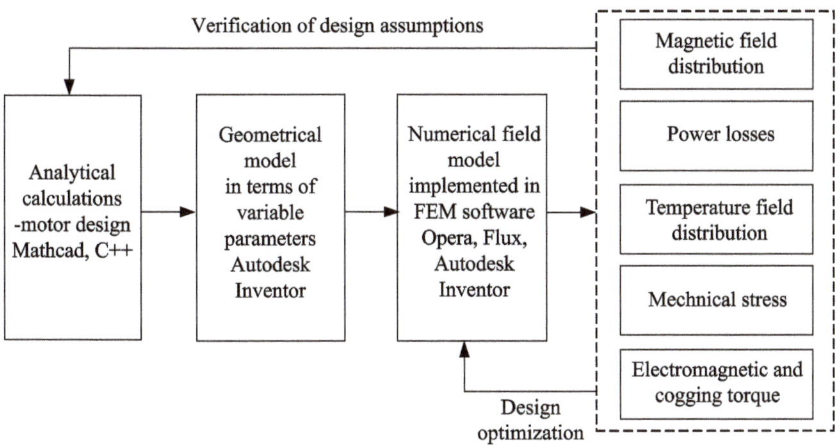

Figure 3. Design process and analysis of the electrical machine—analytical calculation, numerical field model implemented in FEM software.

The main generator stage designed process was conducted based on the sizing Equation (1) and the parameters presented in Table 1. The parameters for the design process were derived from the reverse engineering of the original generator. The design process assumed that the new exciter should allow for the higher rotational speed of the BSG. In addition, the excitation winding should generate a sinusoidal distribution of magnetomotive force (MMF) to generate a sinusoidal distribution of the air gap flux density. Part of the solution for the proposed assumption was to use a high number of rotor slots. The outer diameter of the original and new field structure of the main generator is $D_r = 173.8$ mm, the airgap length is $\delta = 0.6$ mm and the core length is $l = 73$ mm. The most important difference is that in a salient pole solution the air gap length varies, and, in a non-salient solution, it is constant. An additional advantage of the uniform air gap length is the decrease in non-uniform pole shoe saturation of the field during the under load operation of the BSG.

Table 1. Assumptions for the main generator stage design process.

Parameter	Value	Description
S_n	40 kVA	nominal power
U_s	208 V	nominal voltage
I_s	111 A	nominal current
f_n	400 Hz	nominal frequency
PF_n	0.8	nominal load power factor
I_fn	45 A	nominal field winding current
p	3	number of pole pairs
m	3	number of phases
Q_s	81	number of stator slots
Q_r	66	number of rotor slots
N_s	27	number of armature winding turns
δ	0.6 mm	air gap length
D_s	175 mm	inner diameter of the stator
l	73 mm	machine length
b_1	2 mm	stator and rotor slot opening

The prototype field dimensions are partly based on the original generator construction and partly on the selected manufacturing technology, in particular, the technology of field winding construction and manufacturing. In the proposed manufacturing process, a square profile wire was selected for the field winding to maximize the fill factor. The minimum tooth thickness was calculated and then verified using FEM models for the machine to sustain the centrifugal forces and also due to the maximum value of the flux density in the field core. Figure 4 shows the dimensions of the main generator stator and both the commercial and the prototype generator rotors. The original salient pole excitation winding has 42 turns per pole and the design cylindrical prototype has 40 turns per pole. The cross section of the original and prototype excitation winding wire is the same and is equal to 4.5 mm^2.

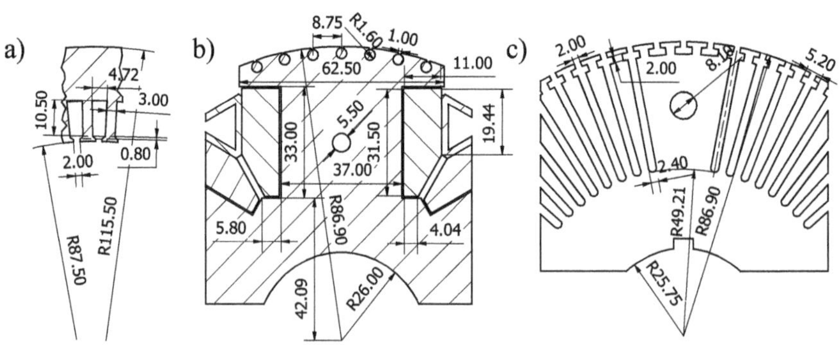

Figure 4. Dimensions of the: (**a**) main generator stator; (**b**) original salient pole rotor; (**c**) prototype non-salient pole rotor.

The proposed design prototype apart from allowing higher rotational velocity of the machine should also decrease the high-order harmonic content of the induced EMF. Both the air gap flux density spatial distribution and the field and armature winding spatial distribution influence the harmonic content of the induced EMF. To analyze this influence, the air gap length distribution functions and field and armature winding distribution functions have to be defined. The air gap length distribution function is defined as:

$$\delta\left(\phi_s-\theta_r\right)=\frac{1}{\alpha_\delta}\frac{1}{\alpha_0-\sum\limits_{k=1}^{k_\delta}\alpha_{2k-1}\cos\left(2\left(2k-1\right)\left(\phi_s-\theta_r\right)\right)} \tag{2}$$

where α_δ defines the average length of the air gap, α_0 and α_{2k-1} are the relative components of the air gap length distribution function ($\alpha_0 = 1$), ϕ_s is the angular position along the stator and θ_r is the displacement angle of the rotor axis in reference to the stator axis. In case of the prototype machine with cylindrical stator and rotor, this function is defined as:

$$\delta=\frac{1}{\alpha_\delta} \tag{3}$$

The armature winding distribution is defined based on the winding MMF distribution as:

$$N_{xs}\left(\phi_s\right)=\frac{1}{i_{xs}}\frac{\partial\, MMF_{xs}\left(i_{xs},\phi_s\right)}{\partial\,\phi_s} \tag{4}$$

where x refers to phase a, b or c in three phase machine, i_{xs} is the x'th phase armature current and the $MMF_{xs}\left(i_{xs},\phi_s\right)$ is the x'th phase magnetomotive force distribution. This force is defined as:

$$MMF_{xs}\left(i_{xs},\phi_s\right)=\frac{N_s}{2}i_{xs}\sum_{k=1}^{k_{MMF_s}}A_{s,2k-1}\cos\left(\left(2k-1\right)\left(\phi_s+\theta_{xs}\right)\right) \tag{5}$$

where N_s represents the number of turns of equivalent sinusoidally distributed armature winding, $A_{s,2k-1}$ are the relative amplitudes of armature winding MMF distribution ($A_{s,1} = 1$), θ_{xs} is the angular displacement of x'th phase in reference to phase a ($\theta_{as} = 0$) and k_{MMF_s} is the number of odd harmonics used to approximate the armature MMF distribution. The same winding distribution and magnetomotive force distribution functions can be defined for field winding:

$$N_{fd}\left(\phi_r\right)=\frac{1}{i_{fd}}\frac{\partial\, MMF_{fd}\left(i_{fd},\phi_r\right)}{\partial\,\phi_r} \tag{6}$$

$$MMF_{fd}\left(i_{fd},\phi_r\right)=\frac{N_{fd}}{2}i_{fd}\sum_{k=1}^{k_{MMF_{fd}}}A_{fd,2k-1}\sin\left(\left(2k-1\right)\left(\phi_r\right)\right) \tag{7}$$

where i_{fd} is the field current, ϕ_r is the angular position along the rotor, N_{fd} represents the number of turns of the equivalent sinusoidally distributed field winding, $A_{fd,2k-1}$ are the relative amplitudes of field winding MMF distribution ($A_{fd,1} = 1$) and $k_{MMF_{fd}}$ is the number of odd harmonics used to approximate the field MMF distribution. Based on the winding distribution the MMF distribution and the air gap length distribution, one can calculate the mutual inductance between field winding and phase winding:

$$L_{xsfd}(\theta_r)=\frac{1}{i_{fd}}\int\limits_{\pi}^{2\pi}\left(N_{xs}\left(\phi_s\right)\int\limits_{\phi_s}^{\phi_s+\pi}\frac{MMF_{xs}\left(i_{xs},\phi_s\right)}{\delta(\zeta-\theta_r)}d\zeta\right)d\phi_s \tag{8}$$

If Equations (3), (4) and (7) are applied to relation (8), the function for mutual inductance is defined as:

$$L_{xsfd}(\theta_r) = L_{sfd} \sum_{k=1}^{k_{L_{sfd}}} \Big((2k-1)\, \alpha_0 A_{s,2k-1} A_{fd,2k-1} \sin\left((2k-1)\left(\theta_r + \theta_{xs}\right)\right) \Big) \tag{9}$$

where:

$$L_{sfd} = \frac{N_s N_{fd}}{4} \mu_0 l \pi r \alpha_d \tag{10}$$

where μ_0 is the magnetic permeability of vacuum, l is the machine length and r is the distance from the axis of machine to the middle of the air gap. As can be noticed, the mutual inductance harmonic components depend on the air gap length and the armature and field winding distribution high-order harmonics. In the designed prototype, the armature winding was not changed. However, both the air gap length distribution and the field winding distribution were changed. The proposed solution for more sinusoidally distributed excitation winding MMF was to use a high number of rotor slots. Because the prototype has a cylindrical construction and more sinusoidally distributed winding, then the commercial generator the resulting EMF is much more sinusoidal. Figure 5 shows the MMF distribution for both the commercial and the prototype main generator field windings.

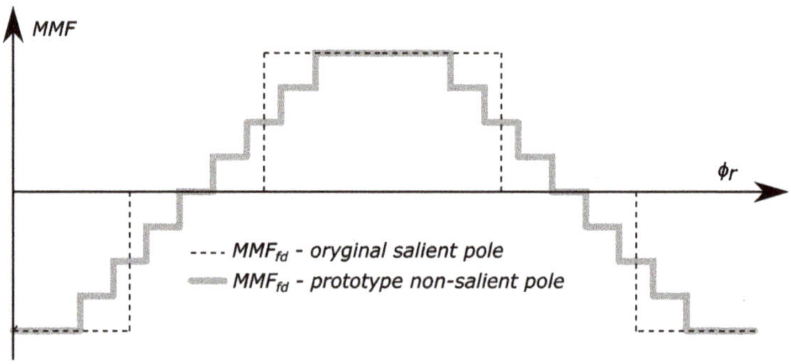

Figure 5. MMF distribution for the commercial and the prototype machine main generator field windings.

Simplified mechanical calculations have been performed to approximate forces and pressures acting on the prototype main generator field core. The force acting on the weakest part of the rotor was calculated as a sum of two forces:

- The centrifugal force acting on rotor tooth—F_{tooth},
- The centrifugal force acting on the winding in the rotor slot F_{Cu}.

The centrifugal force acting on the rotor tooth was calculated as:

$$F_{tooth} = \omega^2 \int_0^c \rho_{Fe} \cdot l \left(a + \left(x \frac{b-a}{c} \right) \right) (r_1 + x)\, dx \tag{11}$$

where ω is the mechanical rotational velocity of the rotor, ρ_{Fe} is the density of rotor core, l is the length of the main generator part of the machine, a is tooth thickness at the bottom of the tooth, b is the tooth thickness at the top of the tooth, c is the height of the tooth and r_1 is the bottom radius of the rotor tooth. The force acting on the copper in the slot was calculated as:

$$F_{Cu} = \omega^2 \int_0^c \rho_{Cu} \cdot l \cdot d\,(r_1 + x)\,dx \tag{12}$$

where d is the width of the copper profile wire in the rotor slot. The resulting force acting on the rotor tooth from both the tooth itself and the copper in the slot is about 23 kN. The resulting pressure in the thinnest part of the rotor tooth is about 126 MPa.

Because the resulting machine dimensions and the resistance of excitation winding did not change for the prototype and original commercial generator, the thermal analysis of the prototype was not conducted. However, if this solution was to be implemented in a commercial power system, such analysis would have to be conducted.

Three-dimensional (3D) geometric models of the machine are then developed as the parametric model using CAD software. This approach enables easy modification of the geometric models and studying of the influence of the geometrical parameters and material properties on the magnetic field distribution, power losses, integral parameters (inductances, torque), mechanical stresses, etc. (Figure 3).

4. Results

4.1. FEM Simulation

The FEM simulations (using Cedrat/FLUX2D software (version 9.3, Cedrat, Grenoble, France)) have been carried out to verify the designed BSG. Due to a lack of information about the materials used in the commercial generator, the FEM simulations were verified using measurement data. As a result of reverse engineering processes for the given dimensions and measured behavior of the commercial machine, the material parameters of the stator and rotor magnetic core were established. Those parameters were calculated using the iteration process by minimizing the error between measurements of no-load state and FEM simulation results. The initial relative permeability $\mu_r = 8000$ and the magnetic polarization at saturation of 1.6 T parameters were set for the commercial machine stator and rotor core. For the prototype field core of the main generator, M530-50A steel sheets have been used. Unfortunately, the manufacturing process and relatively poor stacking factor caused the necessity for additional varnish coating insulation between steel sheets. During the laser cutting process of the prototype core steel sheets, a slag caused by laser cutter needed to be removed. This process also damaged the original varnish insulation coating on the surface of the steel. Because of this decrease in the stacking factor and the damage done to the cut edge of the electrical steel, the material magnetic characteristic has changed. The field core parameters for the simulation were set to $\mu_r = 500$ (initial relative permeability) and the magnetic polarization at saturation of 1.55 T.

The model was implemented in the Cedrat/FLUX2D software using a dedicated Python script. For the discretization of the FEM model, the air gap was divided into three sections: the stator static mesh section, the rotor moving mesh region and the air gap re-meshing region for the mesh generation during the rotor movement. The resulting mesh and geometry of both the commercial and prototype generator are shown in Figure 6.

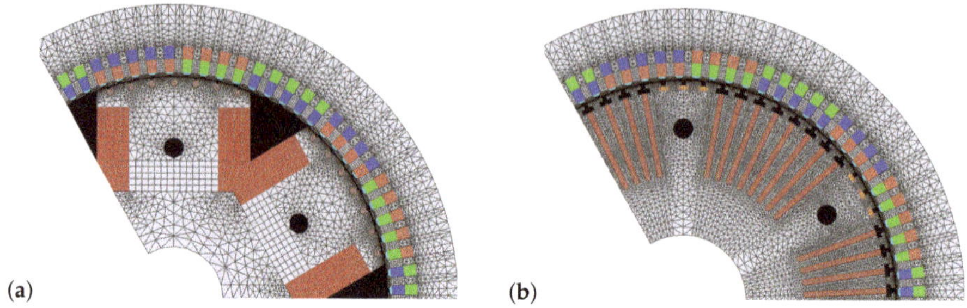

Figure 6. FEM model mesh of the: (**a**) commercial salient pole field BSG; (**b**) prototype cylindrical field BSG.

The resulting mesh generation process led to a total number of about 36 thousand nodes for the original salient pole main generator of BSG and about 53 thousand for the prototype. The simulation process was conducted with set maximum variation between the iterations of integral calculation of less than 0.1%. Table 2 shows the maximum values of magnetic field densities in the crucial parts of the commercial and designed machines.

Table 2. Maximum values of flux density of the main generator part of BSG.

Part of the Magnetic Circuit	Flux Density (T) in Commercial BSG	Flux Density (T) in Prototype BSG
Airgap flux density	0.8	0.8
Stator yoke flux density	1.45	1.46
Stator tooth flux density	1.58	1.6
Rotor yoke flux density	1.05	1.22
Rotor slot/pole flux density	1.66	1.66

FEM simulation results at no-load conditions for the commercial salient-pole and prototype cylindrical generators are shown on Figure 7.

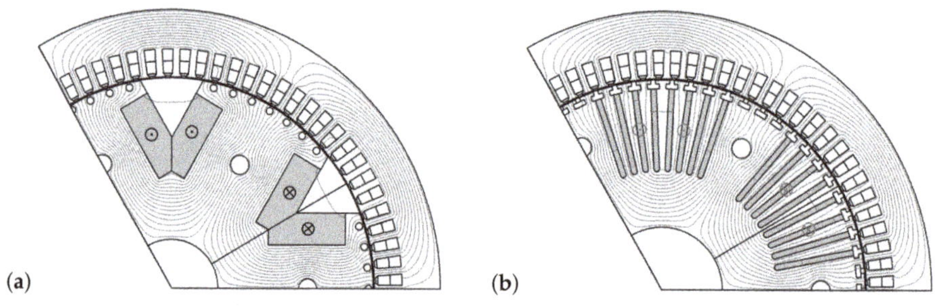

Figure 7. FEM simulation results in no-load conditions: (**a**) commercial salient pole field BSG; (**b**) prototype cylindrical field BSG (marked direction of current in field windings).

The magnetic equipotential lines are shown for both the commercial and the prototype BSG main generator part. In this simulation, the main generator field winding has been supplied with nominal no-load current and FEM computations in magnetostatic conditions have been conducted. The amplitude of the normal component of the air gap flux density of both types of the main generator field has the same maximum value of 0.8 T (Figure 8). The prototype cylindrical generator has got more sinusoidal distribution of air gap flux density (Figure 8).

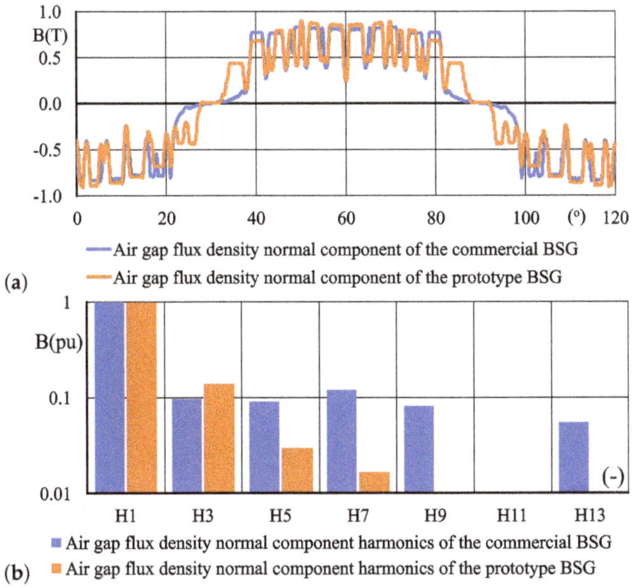

Figure 8. FEM simulation results at no-load conditions—air gap flux density normal component distribution (0.8T—amplitude of the fundamental component): (**a**) commercial salient pole and prototype air gap flux distribution BSG; (**b**) harmonic components of air gap flux distribution.

In addition to electromagnetic calculations, the mechanical calculations concerning the centrifugal forces have been conducted. The simulation of this force acting on one of the rotor teeth was analyzed. The stress analysis module of Autodesk Inventor software (version 2014, Autodesk, San Rafael, CA, USA) was used to perform the mechanical FEM simulation. The rotation speed of 16 krpm has been applied to the rotor. The materials used for stator (steel sheets), windings (copper), and slot wedges (bras) were defined. The stress analysis results are shown in Figure 9. In the simulation, only the copper inside of the slots is considered as the copper in the end windings will not be held by the rotor core itself.

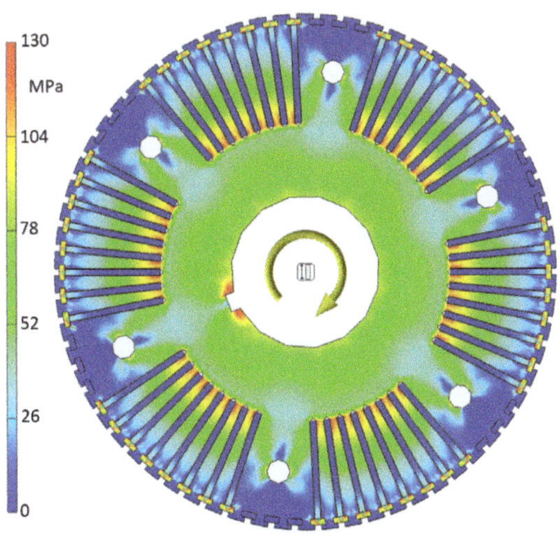

Figure 9. FEM simulation stress analysis results of centrifugal force acting on the rotor.

The electromagnetic and mechanical FEM simulations verified the assumptions and results of the analytical calculations. The proposed solution delivers more sinusoidal distribution of the excitation winding MMF and the maximum tensile stress due to centrifugal force does not exceed maximum tensile stress allowed for M530-50A steel.

4.2. Measurements

Measurements have been carried out for the no-load (open circuit) and load conditions of the prototype and commercial BSG. Measurements of the no-load back EMF in the function of field current for constant speed are shown in Figure 10. This shows that the prototype's main generator field requires more than two times the no-load nominal excitation current. The main factor in this is the smaterial and manufacturing technology used for the core of the main generator field. The difference in number of turns between the prototype and the original machine is only 5% and the air gap length is the same. This also means that the armature reaction of the prototype generator is diminished.

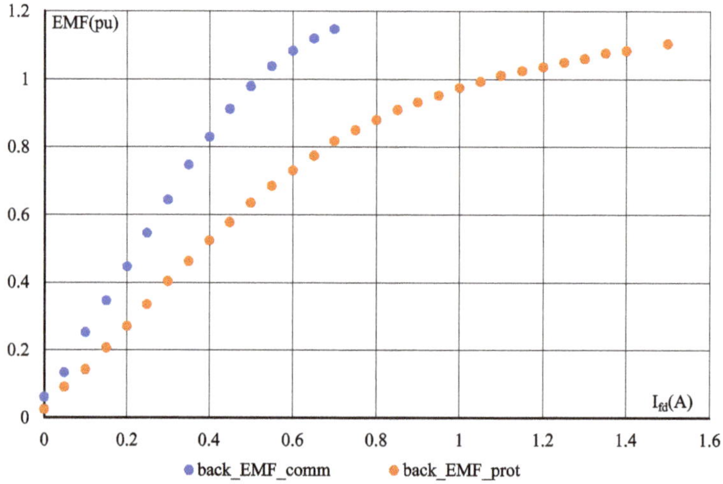

Figure 10. No-load EMF measurements (line to line value) in the function of the exciter field current for commercial generator (back_EMF_comm) and prototype generator (back_EMF_prot).

The no-load back EMF waveform of the prototype has less high-order harmonic components than the commercial one (Figure 11). In both machines, the stator winding is the same which means that the back EMF waveform is dependent on the air gap flux density spatial distribution which in case of a prototype generator is more sinusoidal.

For the load conditions measurements, a load impedance consisting of 0.8 Ω active part and 0.15 Ω reactive inductive part (@400 Hz) has been used. The measurement results of the voltage and current harmonic components in load conditions are shown in Figure 12. As can be observed, the designed prototype with non-salient pole field of the main generator produces less high-order harmonics in the back EMF waveform.

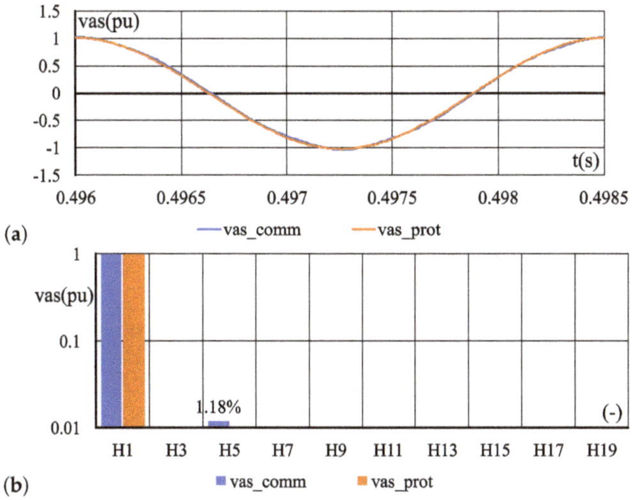

(a)

(b)

Figure 11. No-load voltage for prototype and commercial BSG: (**a**) back EMF waveform, (**b**) harmonic content of back EMF (vas_prot—measured for prototype BSG, vas_comm—measured for commercial BSG).

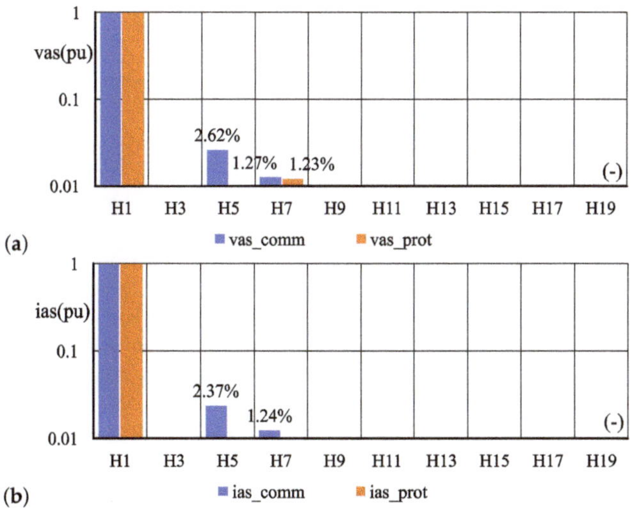

(a)

(b)

Figure 12. Measurement of armature voltage and current of the salient and non-salient BSG in load conditions (balanced star connected load $R = 0.8\ \Omega$, $X = 0.15\ \Omega$ (@400 Hz), $p.f. = 0.98$), (**a**) harmonic content of voltage waveforms, (**b**) harmonic content of current waveforms (vas_comm, ias_comm—results for salient pole generator, vas_prot, ias_prot—results for non-salient pole generator).

In addition to no-load and under load steady state test, a transient state of the symmetric short circuit was performed (Figure 13). This test was performed by shorting the armature terminals when the machine was operating at no-load condition with the nominal voltage at the terminals. As can be observed, the excitation current is higher in the prototype generator than in the commercial one and equal in steady state to the nominal no-load excitation current.

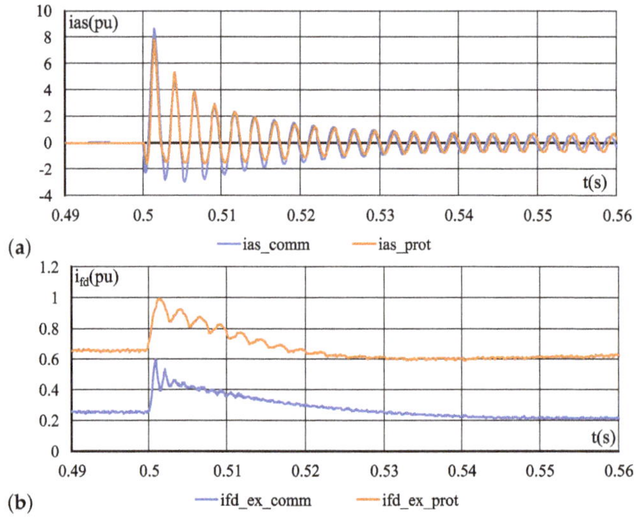

Figure 13. Measurement of armature and exciter currents during transient short circuit test: (**a**) armature currents; (**b**) exciter currents (ias_comm, ifd_ex_comm—results for salient pole generator, ias_prot, ifd_ex_prot—results for non-salient pole generator).

5. Conclusions

The comparative study of two types of BSG main generator exciter has been conducted. The salient pole and cylindrical exciter construction have been compered. The exciter and subexciter of BSG were not changed. The design process was composed of analytical calculation, electromagnetic FEM calculations, and mechanical FEM calculations. The mechanical calculations were conducted to analyze the centrifugal force impact on the designed field of the main generator at higher than nominal (16 krpm while nominal is 8 krpm) rotational velocity, whereas the electromagnetic FEM simulations have been conducted in order to verify the analytically calculated dimensions and parameters of the main generator exciter.

The BSG with the prototype non-salient field has a higher volume and mass of the magnetic core (cylindrical core—9.3 kg, salient pole core—8.1 kg). The volume and mass of the copper used for the field coils are slightly smaller in design and build a prototype (cylindrical field windings copper—2.7 kg, salient pole field windings copper—2.9 kg).

The measurements were conducted at a nominal rotational speed of 8 krpm due to the same exciter and subexciter in both generators. The results of the measurements show significant advantages of the designed solution. Because of more sinusoidal MMF distribution of the field winding, the no-load phase voltage contains less high-order harmonics than the voltage of the silent pole commercial BSG (GT40PCz8). The possibility of operation at higher rotational speeds and better quality of produced electrical energy are a good indicator for the development of the BSG for the MEA variable frequency power system.

The designed generator also has its flaws. It is slightly heavier than the original, and the cooling of the field winding will be different (lack of axial vents in the designed rotor). The selected technique of the manufacturing leaves much to be desired and will have to be modified to archive better magnetic permeability of the field core.

Author Contributions: Conceptualization, F.K., M.M., and G.K.; methodology, F.K. and M.M.; software, F.K. and G.K.; validation, F.K., M.M., and G.K.; formal analysis, M.M.; investigation, G.K.; resources, F.K. and M.M.; writing—original draft preparation, F.K.; writing—review and editing, M.M. and G.K.; visualization, F.K.; supervision, M.M. and G.K.; project administration, F.K.; funding acquisition, F.K. All authors have read and agreed to the published version of the manuscript.

Funding: This research was funded by the Polish Ministry of Science and Higher Education Grant No. N N510 328937.

Conflicts of Interest: The authors declare no conflict of interest.

Abbreviations

The following abbreviations are used in this manuscript:

BSG	Brushless Synchronous Generator
EMF	Electromotive Force
FEM	Finite Element Method
GCU	Generator Control Unit
IM	Induction Machine
MEA	More Electric Aircraft
MMF	Magnetomotive Force
PM	Permanent Magnet
PMG	Permanent Magnet Generator
SR	Switched Reluctance

References

1. Ni, K.; Liu, Y.; Mei, Z.; Wu, T.; Hu, Y.; Wen, H.; Wang, Y. Electrical and Electronic Technologies in More-Electric Aircraft: A Review. *IEEE Access* **2019**, *7*, 76145–76166. [CrossRef]
2. Emadi, K.; Ehsani, M. Aircraft power systems: Technology, state of the art, and future trends. *IEEE Aerosp. Electron. Syst. Mag.* **2000**, *15*, 28–32. [CrossRef]
3. Barzegar, A.; Su, R.; Wen, C.; Rajabpour, L.; Zhang, Y.; Gupta, A.; Gajanayake, C.; Lee, M.Y. Intelligent power allocation and load management of more electric aircraft. In Proceedings of the 2015 IEEE 11th International Conference on Power Electronics and Drive Systems, Sydney, Australia, 9–12 June 2015; pp. 533–538. [CrossRef]
4. Sarlioglu, B.; Morris, C.T. More Electric Aircraft: Review, Challenges, and Opportunities for Commercial Transport Aircraft. *IEEE Trans. Transp. Electrif.* **2015**, *1*, 54–64. [CrossRef]
5. Kim, M.; Lee, S.; Bae, S. Decentralized Power Management for Electrical Power Systems in More Electric Aircrafts. *Electronics* **2018**, *7*, 187. [CrossRef]
6. Chang, J.; Wang, A. New VF-power system architecture and evaluation for future aircraft. *IEEE Trans. Aerosp. Electron. Syst.* **2006**, *42*, 527–539. [CrossRef]
7. Henke, M.; Narjes, G.; Hoffmann, J.; Wohlers, C.; Urbanek, S.; Heister, C.; Steinbrink, J.; Canders, W.R.; Ponick, B. Challenges and Opportunities of Very Light High-Performance Electric Drives for Aviation. *Energies* **2018**, *11*, 344. [CrossRef]
8. Bu, F.; Liu, H.; Huang, W.; Hu, Y.; Degano, M.; Gerada, C.; Rajashekara, K. Induction-Machine-Based Starter/Generator Systems: Techniques, Developments, and Advances. *IEEE Ind. Electron. Mag.* **2020**, *14*, 4–19. [CrossRef]
9. Nøland, J.K.; Leandro, M.; Suul, J.A.; Molinas, M.; Nilssen, R. Electrical Machines and Power Electronics For Starter-Generators in More Electric Aircrafts: A Technology Review. In Proceedings of the IECON 2019—45th Annual Conference of the IEEE Industrial Electronics Society, Lisbon, Portugal, 14–17 October 2019; Volume 1, pp. 6994–7001. [CrossRef]
10. Zhang, Z.; Liu, Y.; Li, J. A HESM-Based Variable Frequency AC Starter-Generator System for Aircraft Applications. *IEEE Trans. Energy Convers.* **2018**, *33*, 1998–2006. [CrossRef]
11. Wang, Y.; Nuzzo, S.; Zhang, H.; Zhao, W.; Gerada, C.; Galea, M. Challenges and Opportunities for Wound Field Synchronous Generators in Future More Electric Aircraft. *IEEE Trans. Transp. Electrif.* **2020**. [CrossRef]
12. Madonna, V.; Giangrande, P.; Galea, M. Electrical Power Generation in Aircraft: Review, Challenges, and Opportunities. *IEEE Trans. Transp. Electrif.* **2018**, *4*, 646–659. [CrossRef]
13. Ganev, E. Selecting the Best Electric Machines for Electrical Power-Generation Systems: High-performance solutions for aerospace More electric architectures. *IEEE Electrif. Mag.* **2014**, *2*, 13–22. [CrossRef]
14. Zhang, Z.; Li, J.; Liu, Y.; Xu, Y.; Yan, Y. Overview and development of variable frequency AC generators for more electric aircraft generation system. *Chin. J. Electr. Eng.* **2017**, *3*, 32–40. [CrossRef]

15. Chen, Y.; Liu, B. Design and Analysis of a Five-Phase Fault-Tolerant Permanent Magnet Synchronous Motor for Aerospace Starter-Generator System. *IEEE Access* **2019**, *7*, 135040–135049. [CrossRef]
16. Zhang, Z.; Huang, J.; Jiang, Y.; Geng, W.; Xu, Y. Overview and analysis of PM starter/generator for aircraft electrical power systems. *CES Trans. Electr. Mach. Syst.* **2017**. [CrossRef]
17. Borg Bartolo, J.; Gerada, C. Design and Modeling of a 45kW, Switched Reluctance Starter-Generator for a Regional Jet Application. In Proceedings of the SAE 2014 Aerospace Systems and Technology Conference, Cincinnati, UK, 23–25 September 2014. [CrossRef]
18. Shoujun, S.; Weiguo, L.; Peitsch, D.; Schaefer, U. Detailed Design of a High Speed Switched Reluctance Starter/Generator for More/All Electric Aircraft. *Chin. J. Aeronaut.* **2010**, *23*, 216–226. [CrossRef]
19. Ferreira, C.A.; Richter, E. Detailed Design of a 250-kW Switched Reluctance Starter/Generator for an Aircraft Engine. *SAE Trans.* **1993**, *102*, 289–300.
20. Liu, H.; Bu, F.; Huang, W.; Liu, L.; Hu, Y.; Degano, M.; Gerada, C. Control Strategy for Five-Phase Dual-Stator Winding Induction Starter/Generator System. *IEEE Trans. Ind. Electron.* **2020**, *67*, 2607–2617. [CrossRef]
21. Galea, M.; Giangrande, P.; Madonna, V.; Buticchi, G. Reliability-Oriented Design of Electrical Machines: The Design Process for Machines' Insulation Systems MUST Evolve. *IEEE Ind. Electron. Mag.* **2020**, *14*, 20–28. [CrossRef]
22. Pyrhonen, J.; Jokinen, T.; Hrabovcová, V. *Design of Rotating Electrical Machines*, 2nd ed.; Wiley: Chichester, UK, 2014.
23. Asfirane, S.; Hlioui, S.; Amara, Y.; Gabsi, M. Study of a Hybrid Excitation Synchronous Machine: Modeling and Experimental Validation. *Math. Comput. Appl.* **2019**, *24*, 34. [CrossRef]
24. Sun, L.; Zhang, Z.; Yu, L.; Gu, X. Development and Analysis of a New Hybrid Excitation Brushless DC Generator With Flux Modulation Effect. *IEEE Trans. Ind. Electron.* **2019**, *66*, 4189–4198. [CrossRef]
25. Gieras, J.F. *Advancements in Electric Machines*; Power Systems; OCLC: 637451957; Springer: Dordrecht, The Netherlands, 2008.
26. Lei, G.; Zhu, J.; Guo, Y. *Multidisciplinary Design Optimization Methods for Electrical Machines and Drive Systems*; Power Systems; Springer: Berlin/Heidelberg, Germany, 2016. [CrossRef]
27. Ounis, H.; Sareni, B.; Roboam, X.; De Andrade, A. Multi-level integrated optimal design for power systems of more electric aircraft. *Math. Comput. Simul.* **2016**, *130*, 223–235. [CrossRef]
28. Amrhein, M.; O'Connell, T.C.; Wells, J.R. An integrated design process for optimized high-performance electrical machines. In Proceedings of the 2013 International Electric Machines Drives Conference, Chicago, IL, USA, 12–15 May 2013; pp. 847–854.
29. Buticchi, G.; Gerada, D.; Alberti, L.; Galea, M.; Wheeler, P.; Bozhko, S.; Peresada, S.; Zhang, H.; Zhang, C.; Gerada, C. Challenges of the Optimization of a High-Speed Induction Machine for Naval Applications. *Energies* **2019**, *12*, 2431. [CrossRef]
30. Shin, K.H.; Bang, T.K.; Cho, H.W.; Choi, J.Y. Design and Analysis of High-Speed Permanent Magnet Synchronous Generator With Rotor Structure Considering Electromechanical Characteristics. *IEEE Trans. Appl. Supercond.* **2020**, *30*, 1–5. [CrossRef]
31. Tosetti, M.; Maggiore, P.; Cavagnino, A.; Vaschetto, S. Conjugate Heat Transfer Analysis of Integrated Brushless Generators for More Electric Engines. *IEEE Trans. Ind. Appl.* **2014**, *50*, 2467–2475. [CrossRef]
32. Huang, S.; Luo, J.; Leonardi, F.; Lipo, T. A general approach to sizing and power density equations for comparison of electrical machines. *IEEE Trans. Ind. Appl.* **1998**, *34*, 92–97. [CrossRef]
33. Moreno, Y.; Almandoz, G.; Egea, A.; Madina, P.; Escalada, A.J. Multi-Physics Tool for Electrical Machine Sizing. *Energies* **2020**, *13*, 1651. [CrossRef]
34. Honsinger, V. Sizing Equations for Electrical Machinery. *IEEE Trans. Energy Convers.* **1987**, *EC-2*, 116–121. [CrossRef]
35. Kutt, F.; Michna, M.; Kostro, G.; Ronkowski, M. Modelling of steady state and transient performance of the synchronous generator considering harmonic distortions caused by non-uniform saturation of the pole shoe. *Electr. Power Syst. Res.* **2017**, *143*, 409–414. [CrossRef]
36. Kutt, F.; Michna, M.; Kostro, G. Multiple Reference Frame Theory in the Synchronous Generator Model Considering Harmonic Distortions Caused by Nonuniform Pole Shoe Saturation. *IEEE Trans. Energy Convers.* **2020**, *35*, 166–173. [CrossRef]

37. More Open Electrical Technologies—TRIMIS—European Commission. 2013. Library Catalog: Trimis.ec.europa.eu. Available online: https://trimis.ec.europa.eu/sites/default/files/project/documents/20121218_094726_85827_MOET_Public_Technical_report.pdf (accessed on 10 May 2020).
38. Arumugam, D.; Logamani, P.; Karuppiah, S. Electromagnetic & thermal analysis of synchronous generator with different rotor structures for aircraft application. *Alex. Eng. J.* **2018**, *57*, 1447–1457. [CrossRef]
39. Biais, F.; Delhasse, F.; Thalin, P. Electrical Power Generation & Start Solutions for the Falcon 5X Program. In Proceedings of the 2015 More Electric Aircraft Conference, Toulouse, France, 3–5 February 2015. Available online: https://www.see.asso.fr/bitcache/91abc2ab86a3f5a69f5b8207260be67011b46135?vid=32132 (accessed on 26 May 2020).

energies

MDPI

Article

Smart-Sensors to Estimate Insulation Health in Induction Motors via Analysis of Stray Flux

Israel Zamudio-Ramirez [1], Roque Alfredo Osornio-Rios [1], Miguel Trejo-Hernandez [1], Rene de Jesus Romero-Troncoso [1] and Jose Alfonso Antonino-Daviu [2,*]

1 Engineering Faculty, San Juan del Río Campus, Universidad Autónoma de Querétaro, Av. Río Moctezuma 249, C.P. 76808 San Juan del Río, Querétaro, México; isra.zam.ram@hotmail.com (I.Z.-R.); raosornio@hspdigital.org (R.A.O.-R.); miguel8010@yahoo.com.mx (M.T.-H.); troncoso@hspdigital.org (R.d.J.R.-T.)

2 Instituto Tecnológico de la Energía, Universitat Politècnica de València (UPV), Camino de Vera s/n, 46022 Valencia, Spain

* Correspondence: joanda@die.upv.es; Tel.: +34-963-877-592

Received: 6 April 2019; Accepted: 25 April 2019; Published: 1 May 2019

Abstract: Induction motors (IMs) are essential components in industrial applications. These motors have to perform numerous tasks under a wide variety of conditions, which affects performance and reliability and gradually brings faults and efficiency losses over time. Nowadays, the industrial sector demands the necessary integration of smart-sensors to effectively diagnose faults in these kinds of motors before faults can occur. One of the most frequent causes of failure in IMs is the degradation of turn insulation in windings. If this anomaly is present, an electric motor can keep working with apparent normality, but factors such as the efficiency of energy consumption and mechanical reliability may be reduced considerably. Furthermore, if not detected at an early stage, this degradation could lead to the breakdown of the insulation system, which could in turn cause catastrophic and irreversible failure to the electrical machine. This paper proposes a novel methodology and its application in a smart-sensor to detect and estimate the healthiness of the winding insulation in IMs. This methodology relies on the analysis of the external magnetic field captured by a coil sensor by applying suitable time-frequency decomposition (TFD) tools. The discrete wavelet transform (DWT) is used to decompose the signal into different approximation and detail coefficients as a pre-processing stage to isolate the studied fault. Then, due to the importance of diagnosing stator winding insulation faults during motor operation at an early stage, this proposal introduces an indicator based on wavelet entropy (WE), a single parameter capable of performing an efficient diagnosis. A smart-sensor is able to estimate winding insulation degradation in IMs using two inexpensive, reliable, and noninvasive primary sensors: a coil sensor and an E-type thermocouple sensor. The utility of these sensors is demonstrated through the results obtained from analyzing six similar IMs with differently induced severity faults.

Keywords: induction motor; smart-sensor; stray flux; time-frequency transforms; wavelet entropy

1. Introduction

In the companies, electric motors have gained great importance, and have been widely used as electromechanical devices for the conversion of energy, consuming more than 60% of all the energy of any industrial nation [1]. Current quality requirements consider the use of monitoring systems and the development of incipient failure detection techniques increasingly necessary in order to enhance the reliability of these industrial systems so that production is not interrupted. Machines operating under faulty conditions consume and spend more energy, causing additional economical losses. Furthermore, some failures can remain unnoticed in motors that work continuously with apparent normality;

nevertheless, if not detected in time, incipient faults can result in catastrophic and irreversible damage to the machine, and if the fault progresses it can cause collateral damages to others systems coupled to the induction motor (IM). Therefore, it is of paramount importance to study the main faults in induction motors, and there is a clear necessity to develop emergent techniques that can detect faults in the early stages, and to integrate new technologies. In this regard, some authors have adopted the concept of a smart-sensor, in which one or more primary sensors are combined with a processing unit in order to gather certain functionalities like processing, communication and integration. Smart-sensors have found an application in different research fields, including the monitoring and diagnosis of faults in distinct industrial applications [2,3], real-time high-resolution frequency measurement [4], identification of broken bars and unbalance in induction motors [5,6], among others.

Surveys on motor reliability have determined that the distribution of failures in IMs can essentially be classified into four classes: bearing faults, stator related faults, rotor related faults, and other faults (cooling, connection, terminal boxes) [7]. Some investigations have shown that most failures of electric motors can be attributed to bearings and windings [8]. Depending on the type and size of the machine, problems related to stator windings correspond to a range between 16%–36% of total reported faults [9,10], which is the second largest type of fault for IMs, just after bearings defects. Stator winding insulation failures have recently received special attention. This is mainly because the worst stator faults start from undetectable insulation degradation problems between drastically adjacent turns [11–14] that lead to the appearance of an inter-turn fault, where two or more turns become short circuited. If undetected at early stages or after its appearance, this type of fault can develop into more severe problems very quickly. Many techniques found in the literature have been proposed to detect winding faults, and focus on two main approaches: offline methods and online methods [15]. Common offline methods that are typically used in industry include insulation resistance measurement, polarization index/dielectric absorption measurement, offline partial discharge tests, and evaluation of the dissipation factor [16–18]. A disadvantage of offline tests is the necessity to remove the machine from service, a drawback that can lead to false indications caused by unrealistic operations [19]. On the other hand, online monitoring methods are desirable due to their capability to diagnose faults when a motor is in service. To this end, several techniques have been proposed to perform online diagnosis, and many physical magnitudes have been highlighted as potential sources of information, with each one having its own advantages and disadvantages, as discussed below. Vibration [20], thermographic [21,22], and partial discharge [23] are some of the online methods used to detect insulation inter-turn faults; however, most of these techniques are not yet proven to detect faults during early stages, before reaching a severe phase, and in the case of the thermographic approach, diagnosis is difficult to perform under real working environmental conditions, since optimal conditions must be met to get confident results. Other classical approaches are focused on the use of current and voltage signals: spectral analysis of the steady-state current using the Fourier transform (motor current signature analysis, MCSA) [24], analysis of the zero and negative sequence currents [25], and analysis of the zero-sequence voltage [26]. The main disadvantage of analyzing the zero-sequence voltage is that the final diagnosis can be affected by the influence of other parameters, such as voltage unbalances, measurement errors, and inherent asymmetries during the manufacturing process, which can cause false diagnoses. Although analysis of the negative sequence current overcomes these problems [27], it is required to measure three-line currents, a condition that is not always available.

Due to the need for a system that is able to automatically diagnose in an online mode and monitor the health of winding insulation in induction motors (before an irreversible fault occurs), this work introduces a smart-sensor composed of two primary sensors (a coil sensor and an E-type thermocouple sensor) and a hardware signal processing unit (HSP unit) in order to accomplish this task. The wavelet entropy (WE) of the coil sensor was used as an auxiliary parameter in the final diagnosis, since it is able to characterize the dynamism and the order/disorder of a signal using a single value [28–31]. The coil sensor was used to capture the stray flux signal, and the E-type thermocouple sensor acquired the temperature of the motor, since both of them have a non-invasive nature. In order to constantly monitor and diagnose winding insulation degradation, the smart-sensor applied a signal processing stage composed of the computation of the discrete wavelet transform (DWT) followed by the calculation of the WE. Furthermore, with the purpose of automating the full process, a trained artificial neural network (ANN) performed a regression estimation by using the wavelet entropy and the induction motor temperature as input signals. All computations were performed by a field-programmable gate array (FPGA) HSP digital unit by developing proprietary hardware cores focused on the above-mentioned tools, as described below.

2. Materials and Methods

In this section, we detail the mathematical tools and methodologies that constitute the main core of the smart-sensor. The DWT is used to obtain a representation of the frequency content for the different bands that make up the input electromotive force (*emf*) signal. Wavelet entropy is used as the main parameter that will serve to subtract relevant information about the healthiness of the winding insulation, since it is a tool capable of describing the dynamic behavior of a signal, in addition to indicating the amount of order/disorder of that signal. Furthermore, wavelet entropy shows a clear relation to the healthiness of the winding insulation, as will be shown below. Finally, a final diagnosis through an artificial neural network, whose inputs are indeed wavelet entropy and IM temperature, will indicate the healthiness of the winding insulation using an automatic process.

2.1. Discrete Wavelet Transform (DWT)

As is well known, DWT is a time-frequency analysis transform that provides significant features for the analysis of a time-variant signal, since this technique is very suitable for decomposing a signal into well-defined "wavelet signals" that cover specific frequency ranges that are known to be directly dependent on the sampling frequency used to capture the analyzed signals [32]. The DWT of a signal can be defined as

$$W(i,k) = \sum x(k)\psi_{i,k}(t) \tag{1}$$

where i is the decomposition level, k is the number of the sample, and $\psi_{i,k}(t)$ is the discrete wavelet mother function.

To compute the DWT of a signal, a Mallat's algorithm facilitates its application and improves its performance, processing time, and the computational burden that its application entails. The DWT of a signal $x[n]$ of length N is calculated by applying a mathematical convolution defined by Equation (2) with a bank of high-pass filters (HPF) with impulse response *g[n]* to analyze the high frequencies, and simultaneously with a bank of low-pass filters (LPF) with impulse response *h[n]* to analyze the low frequencies.

$$y[n] = (x * h)[n] = \sum_{k=0}^{N} x[k]h[n-k] \tag{2}$$

The DWT decomposes the time-domain signal in several levels, which are limited by the sample size N. The frequency content of every decomposition level for both aC_i and dC_i is estimated by

$$aC_i \rightarrow \left[0, \frac{f_s}{2^{i+1}}\right], dC_i \rightarrow \left[\frac{f_s}{2^{i+1}}, \frac{f_s}{2^1}\right] \tag{3}$$

where f_s is the sampling frequency and i is the desired decomposition level.

The coefficients of the HPF and LPF are determined by the selection of a mother wavelet according to the application.

In this regard, some investigations have been developed to evaluate the performance of DWT in extracting features from the current signals of induction motors. This serves the purpose of detecting eccentricities [33], rotor-asymmetries [34], broken rotor bars [35], and other factors using the Daubechies (db), Symlet (sym), Morlet, and Meyer wavelet families, and varying the orders. These works have shown that a Daubechies family of higher order is well suited to extracting the information required for the detection of motor failures. Furthermore, studies have shown that higher order filters behave as more-ideal filters, allowing less overlap between adjacent frequency bands.

2.2. Wavelet Entropy

Due to the inherent constraints of some time-frequency transforms, there can be problems when a specific window is applied to a series of data. Such is the case of the uncertainty problem given in the DWT—if the window is too narrow, the resolution of the frequency will be poor, whereas if the window is too wide, the location during the time of the signal will be less precise. This limitation is of great importance when it comes to the analysis of signals with transient components located in time, which is the case for the great majority of signals with real physical magnitudes.

To minimize the effects of this limitation, a parameter based on the entropy of a signal has been defined from a time-frequency representation of the signal provided by the wavelet transform [36]. In this regard, the entropy based on the wavelet transform (wavelet entropy) reflects the degree of order/disorder in the signal, so it can provide additional information about the underlying dynamic processes associated with the signal [29]. This is achieved by combining the information of all the wavelet bands, since data between adjacent wavelet signals is taken and combined into one index in order to avoid focusing on just one wavelet band having its own time-frequency resolution.

The total wavelet entropy (S_{WT}) is defined according to [37].

$$S_{WT} \equiv S_{WT}(p) = -\sum_{i=m}^{n} p_i \ln p_i \tag{4}$$

where m and n are the first- and last-considered decomposition levels for analysis, respectively, and p_i represent the relative wavelet energy normalized values, which can be computed as

$$p_i = \frac{E_i}{E_{tot}} \tag{5}$$

where E_i (Equation (6)) and E_{tot} (Equation (7)) represent the energy of wavelet level decomposition i and the total energy of all wavelet level decompositions, respectively.

$$E_i = \sum_k |C_i(k)|^2 \tag{6}$$

$$E_{tot} = \sum_i E_i \tag{7}$$

2.3. Stray Flux Analysis

Effective analysis of the stray flux by applying suitable signal processing techniques to detect several failures in induction motors, such as broken rotor bars, static and dynamic rotor eccentricity, bearing faults, and shorted turns in stator winding, have been proven and validated in a number of works [38–41].

The external magnetic field can be analyzed by its axial and radial components [42]. The axial radial field is generated by currents in the stator end windings or rotor cage end ring. The radial field is related to air gap flux density, which is attenuated by the stator magnetic circuit and by the external machine frame.

The coil sensor can be installed in the vicinity of the motor frame in convenient positions in order to measure the electromotive forces that are indicative of the axial and radial flux components, depending on its placement. Thus, Figure 1 shows the positions A, B, and C in which the sensor can be installed to measure both fields.

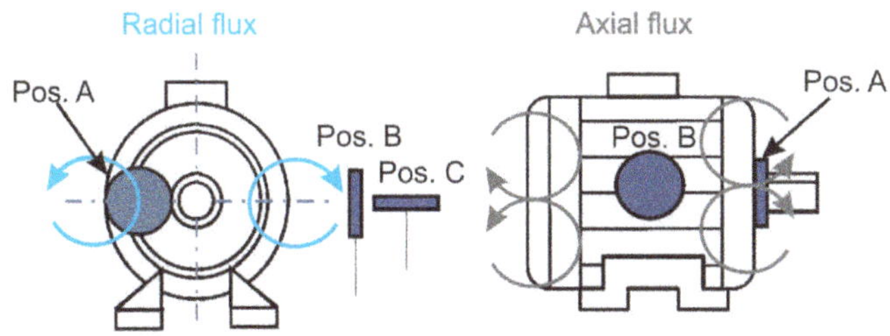

Figure 1. Coil sensor positions in the tested induction motor.

In position A, the sensor's placement enables measurement of the axial flux; on the other hand, if the sensor is placed in position B, the result of the measurements will correspond to components of the axial flux and part of the radial flux simultaneously. In position C, the coil sensor is parallel to the longitudinal cross-section of the machine, which makes the axial field null and the radial flux essentially present.

2.4. *Artificial Neural Network*

Artificial neural networks (ANNs) are computational models that simulate the neurological structure of the human brain and its capability to learn and solve problems through pattern recognition [43]. As is well known, this method has exceptional characteristics enabling it to process and extract relevant information from large amounts of data. Among the most popular ANN architectures, feed-forward neural networks (FFNNs) are widely used, since they are simple, practical, and very good at approximating real-valued functions and at classifying data. Furthermore, the operation of this kind of ANN demands a very low computational burden, which makes it appropriate for implementation in digital systems. FFNNs are composed of a layered architecture possessing essentially one input layer, one or more hidden layers, and one output layer, as shown in Figure 2a. Each layer has one or more elementary processing units called neurons (see Figure 2b), whose processing capability is stored in the connections of synaptic weights, and whose adaptation depends on learning [44]. The mathematical model describing the functionality of each neuron is given by

$$y = f\left(\sum_{i=1}^{n} w_i x_i + b\right) \tag{8}$$

where y, w_i, x_i, b, $f(\cdot)$, and n are the output, synaptic weights, inputs, bias, activation function, and the total number of inputs, respectively. To define the network weights, a training process is carried out where pairs of input–output data are presented, then a training rule is defined for adjusting these weights.

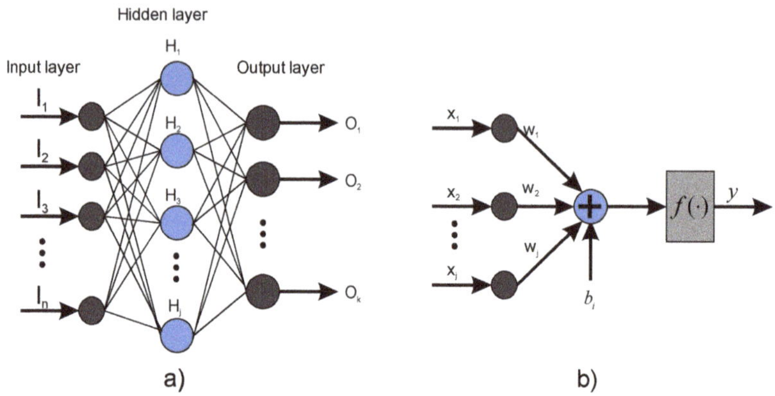

Figure 2. Artificial neural network (ANN): (**a**) feed-forward neural network (FFNN) architecture, (**b**) functional structure of a neuron.

2.5. Smart-Sensor

The smart-sensor proposed here is based on a low-cost system on a chip (SoC) field-programmable gate array (FPGA) to estimate the insulation health of an induction motor. Figure 3 shows the scheme of the structure of the proposed smart-sensor. The system uses a coil sensor and a thermocouple sensor as primary sensors that can be installed on the frame of the analyzed IM to capture the stray flux and temperature of the IM, respectively. The information coming from the primary sensors is acquired in the data acquisition system (DAS) module, then the signal processing is performed in the FPGA-based HSP unit by applying suitable time-frequency decomposition (TFD) tools and by extracting an efficient indicator based on the wavelet entropy. Finally, the estimated health of the insulation is supplied to the final user using an liquid crystal display (LCD).

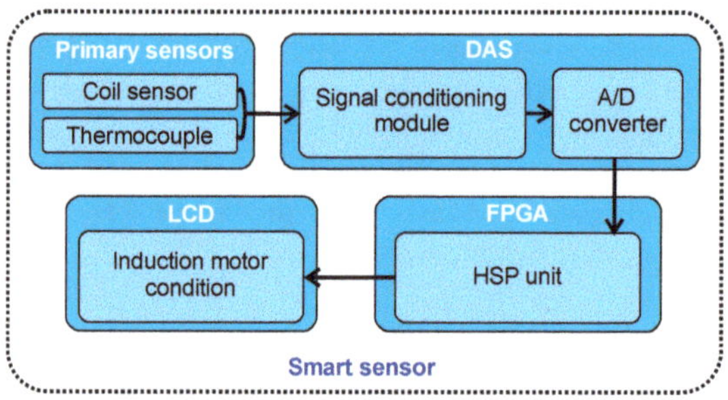

Figure 3. Block diagram of the proposed smart-sensor: primary sensor, data acquisition system (DAS), field-programmable gate array (FPGA) and liquid crystal display (LCD).

2.5.1. Primary Sensors

Two sensors are used as primary sensors—one flux sensor, and one thermocouple sensor. The flux sensor is generated by 1000 turns of a coil. Its dimensions are specified in Figure 4b. The main purpose of the coil is to detect the largest amount of stray flux possible through the induced electromotive force (*emf*) in that coil. The coil is protected with a material that is able to isolate the greater amount of electromagnetic noise coming from the outside using a special meshed cable for the transmission of the induced voltage towards the DAS. On the other hand, to capture the temperature of the analyzed IM, an E-type thermocouple sensor (Figure 4a) is used, since it is a non-magnetic sensor and has a wide temperature ranging from −50 °C to 740 °C.

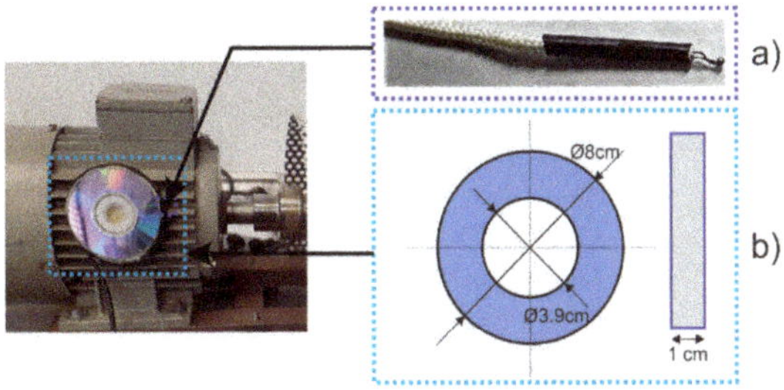

Figure 4. Primary sensors: (**a**) thermocouple sensor, (**b**) coil sensor dimensions.

2.5.2. DAS and LCD

As secondary elements, the DAS and LCD modules enable the interaction between the final user and the primary sensors. On the one hand, the DAS module is constituted by a signal-conditioning submodule and an ADS7841 analog-to-digital converter. The signal-conditioning module is composed of one operational amplifier with two processing stages, the first of which is configured to sum a constant voltage to the input flux signal, while the second amplifies it by a factor of 10 in order to standardize the input-voltage range to the analog-to-digital converter used. On the other hand, the LCD is used to display the estimated winding insulation degradation to the final user.

2.6. HSP Unit

The FPGA-based HSP unit is created by processing the DWT, feature-extraction wavelet entropy, and regression FFNN, and by mapping the min–max function to normalize inputs and outputs of the FFNN, as shown in Figure 5. First, the input *emf* signal (ϕ) is decomposed by the DWT in multiple "wavelet signals" in order to obtain the time-frequency representation of the input signal in well-known frequency bands. Then, the feature extraction is performed by applying Equation (3). Note that the S_{WT} value of a signal is a normalized parameter ranging from zero to 1, where a minimum value indicates a light disorder in the analyzed signal (that is, the signal is mainly represented by one wavelet signal having the highest amplitude). On the other hand, if the value is near 1, the analyzed signal is considered to have a high disorder, since it penetrates several wavelet signals, each one having high relative amplitudes. Next, the min–max function map normalizes the IM temperature signal (T) and the S_{WT}, in order to perform the mathematical operations inside the FFNN in a defined closed range. Finally, the FFNN unit performs the regression diagnosis by using the normalized values of the extracted S_{WT} and the temperature of the induction motor as inputs.

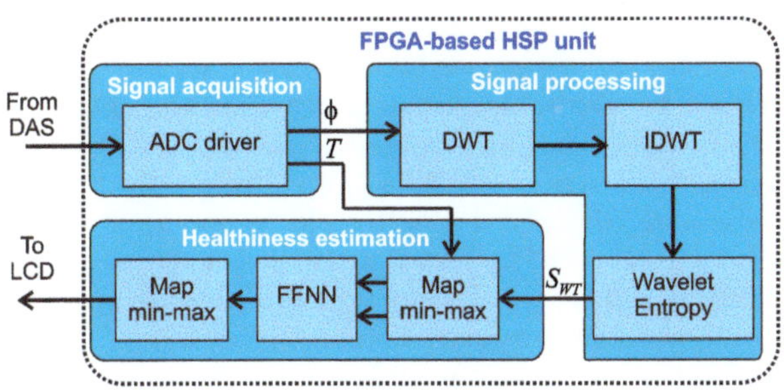

Figure 5. FPGA-based Hardware Signal Processing (HSP) unit: Analogical-Digital Converter (ADC), Discrete Wavelet Transform (DWT), Inverse Discrete Wavelet Transform (IDWT), Wavelet Entropy and Healthiness estimation module.

Figure 6 shows the *emf* and induction motor temperature signals processing the flow to the FFNN evaluation for the purpose of obtaining a health estimation of the winding insulation. The processing flow starts with the *emf* signal acquisition; then, it is computed by the DWT to a level defined by the final user. The next step is to obtain the S_{WT} parameter from the detail decomposition signals after performing the DWT by applying Equation (4). Finally, wavelet entropy and induction machine temperature are used as inputs to the FFNN, so that the information of both parameters can be combined to offer an automated estimation of the health of the winding insulation. Note that the FFNN is composed of one input layer with two input neurons (the wavelet entropy and the induction machine temperature); two hidden layers with four and two neurons, respectively; and one output neuron (the estimated health of the winding insulation, a parameter shown to the final user via the LCD). To specify the health of the winding insulation, results are shown in a continuous scale ranging from 10% (indicating a severe degradation) up to 95% (implying a healthy winding insulation system).

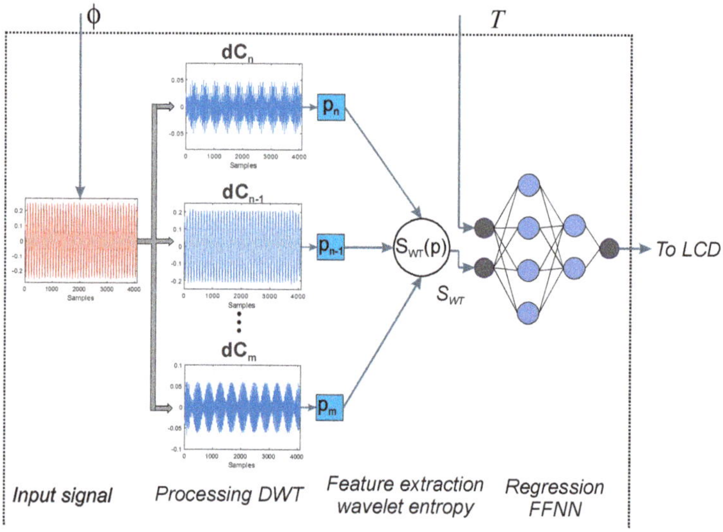

Figure 6. Proposed methodology flow.

2.6.1. DWT–IDWT Digital Structure

In Figure 7, the counters *n* and *rwdir* indicate the index of the sample $x[n]$ to read and the read/write direction of the approximation and detail coefficients, respectively. Note that it is necessary to fill the $RAMemf[n]$ with the number of samples from the coil sensor specified by the user prior to testing. When asserting *strDWT*, the module starts to compute the convolution operation defined by Equation (2), which is essentially composed of a multiply–accumulate (MAC) process. The MAC operation requires the coefficients of a filter obtained from four Read-only Memory (ROM)previously filled with the corresponding coefficients (*ROM Lo-D* and *ROM Hi-D* for low-pass and high-pass decomposition filters, respectively; and *ROM Lo-R* and *ROM Hi-R* for low-pass and high-pass reconstruction filters, respectively), as well as the $emf[n\text{-}k]$ signal, which is obtained by passing $emf[n]$ through a *k*-level pipeline register. Finally, when the computation process is finished, signal *rdyDWT* is set to high. The approximation and detail coefficients will be given by the output signals aC_{ik} and dC_{ik}, respectively.

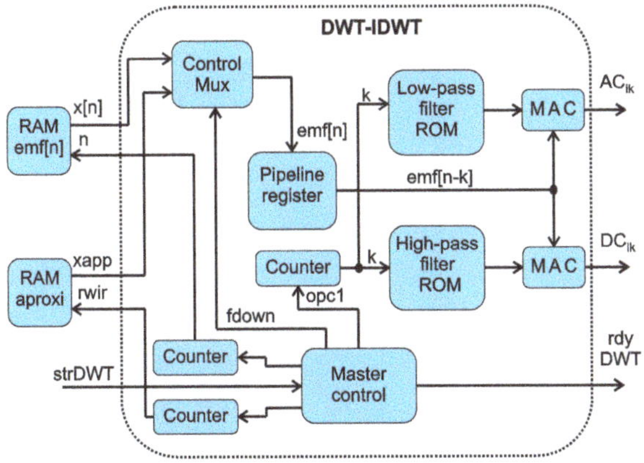

Figure 7. Block diagram of the DWT–IDWT unit: including Random Access Memory (RAM), Multiplier Accumulator (MAC), Read-only Memory (ROM), Alternating current AC and Direct current (DC) modules.

2.6.2. Wavelet Entropy Digital Structure

Figure 8 shows a block diagram of the proposed digital architecture used to obtain the wavelet entropy of a signal by applying Equation (3) where the MAC process is the main operation. Note that it is essential to fill the RAM and *RAM-pi* with the corresponding relative wavelet energies p_i of the corresponding "wavelet signals" prior to testing. To start the S_{WT} computation, the signal StrWE is asserted. After that, the counter *Rddir* selects the signal p_i to be processed, and the $\log_2(p_i{}^2)$ is computed by applying the algorithm proposed in [45], since it offers an easy implementation in hardware. Next, to obtain the required $\log_e(p_i{}^2)$ value, a simple multiplication factor defined by Equation (9) is applied to $\log_2(p_i{}^2)$.

$$\log_e(x) = \frac{\log_2(x)}{\log_2(e)} \approx 0,693147 * \log_2(x) \tag{9}$$

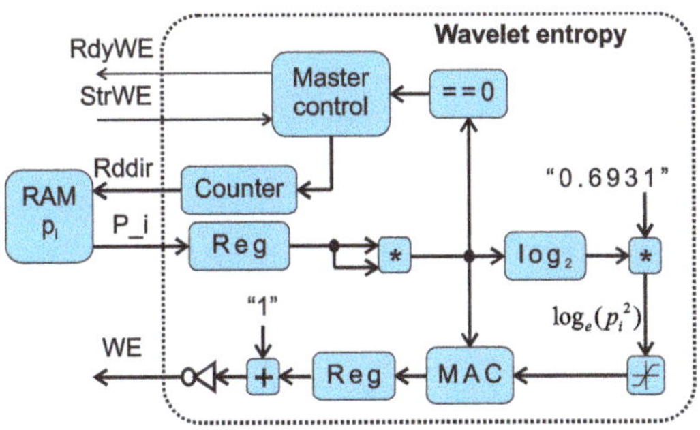

Figure 8. Block diagram of the wavelet entropy unit.

2.6.3. FFNN Digital Structure

Figure 9 shows the block diagram of the proposed digital architecture to compute a regression FFNN. When the signal *strANN* is asserted, the input signals S_{WT} and T are stored in in the first two memory elements of the *Ni* submodule. Note that the submodule *Ni* works as a storage memory for the total number of neurons that constitute the FFNN architecture used here (that is, neurons on the input layer, neurons on the hidden layers and neurons on the output layer). Memory ROMs *ROMidx-rd*,

ROMwi, *ROM-layer*, and *ROMb[i]* store the indices of each neuron to read/write, depending on the actual layer (*layer*), synaptic weights, and biases. This design is based on a MAC operation in order to save element resources and use only one multiplier. The inputs for the MAC operation are the synaptic weights (w_i) and the corresponding neuron outputs (x_i). Finally, when the MAC process is finished, its output is summed by the corresponding bias (b_i) in order to compute the activation function tansig(x) using the piece-wise linear function defined in [46].

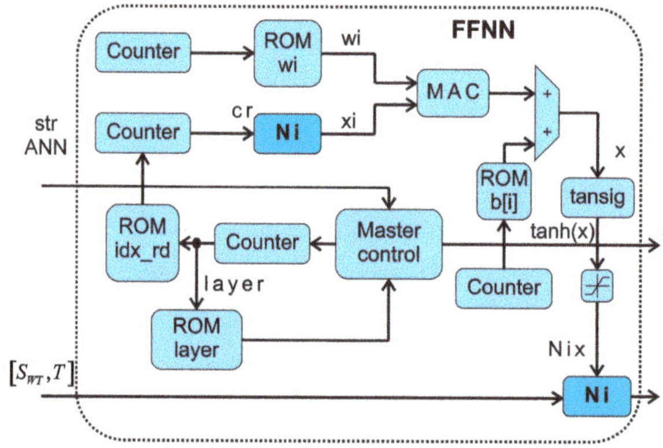

Figure 9. Block diagram of the FFNN digital unit.

To train the FFNN, the information of the wavelet entropy and motor frame temperature are used. In addition, to establish a frame of reference between a motor with healthy winding insulation and a motor with a degraded insulation system, the index of dielectric absorption of the motor is extracted by means of a megger device during the degradation tests. These data (wavelet entropy, motor frame temperature, and dielectric absorption index) are obtained during the process of induced degradation to the winding insulation system on a three-phase induction motor whose characteristics are specified in the results section. The wavelet entropy and the temperature of the motor frame are used as the input data set for the training of the neural network and as the desired output, then the interpolation between the dielectric absorption rates obtained for each test are carried out and limits are established for the purposes of this work (that is, 95% for a healthy motor, and 10% for motor with serious winding insulation degradation).

3. Results

The diagnostic procedure and functionality of the smart-sensor proposed in this paper was validated in the laboratory on six IMs with the following same characteristics: 1.1 kW, 400 V, Y-connected, 50 Hz, 4 poles, where several experiments were carried out for healthy IMs and IMs with induced winding insulation degradation, as explained below.

Experimental Set-Up

An experimental test bench was designed to develop and implement the diagnostic technique proposed here and simulate a load using a three-phase squirrel-cage induction motor connected to a Direct Current (DC) generator, as shown in Figure 10b,c, respectively. The coil sensor and the thermocouple sensor (see Figure 10a), which was connected to the encased proprietary FPGA-based HSP unit, were attached to the frame of the motor. The laboratory room where the experiments were carried out was a closed space where the ambient temperature was maintained at an approximate value of 26 °C. Other elements that could potentially interfere with the experiments were removed (inverters, other test benches, etc.) to ensure that no other factor might influence the results.

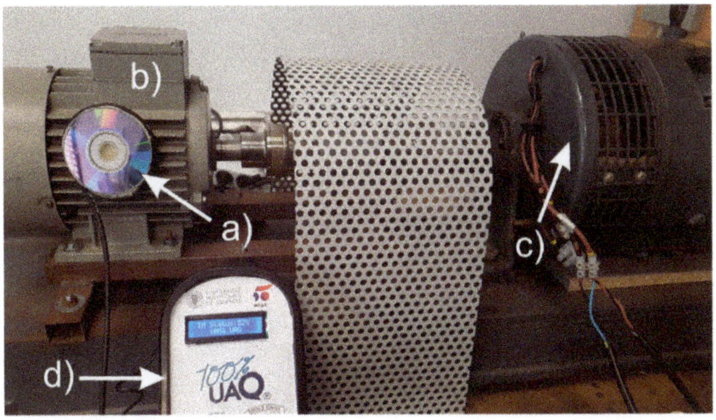

Figure 10. Laboratory test bench: (**a**) coil sensor and thermocouple sensor, (**b**) induction motor, (**c**) DC generator acting as the load, (**d**) proprietary FPGA-based HSP unit.

Two different experiments were carried out in order to probe the functionality and effectiveness of the proposed smart-sensor. In the first experiment, five IMs with the same characteristics but different health statuses where diagnosed. The smart-sensor was placed on the frame of a healthy motor, then on an IM with one or two broken bars (one of the most common failures in this type of motor), but with a healthy winding insulation. Finally, smart-sensors were added to one IM with light winding insulation degradation, and one with severe winding insulation degradation. All experiments in this first stage were developed maintaining the same operating temperature in order to keep the winding insulation temperature in a controlled range of approximately 26 °C.

In the second experiment, a winding insulation degradation was progressively induced to one IM, in order to fully diagnose several levels of deterioration. With the purpose of establishing a reference between a healthy motor and a faulty motor in the insulation system, the first IM used was in a healthy condition at the beginning of the tests. Afterwards, an overheating of the winding insulation was artificially created by connecting and disconnecting one of the motor supply phases in successive cycles. In that way, while one phase was disconnected, the other two were overloaded, leading to abrupt thermal increments that produced higher temperatures than those defined by the thermal class of the insulation (class F). The connection–disconnection cycles of one supply phase were repeated a large number of times, a fact that led to an accelerated degradation of the insulation due to thermal effects. It is worth noting that this experimental setup tries, for the first time, to study the thermal degradation that the insulation system of an induction motor suffers when it is in service (that is, before a short circuit occurs between turns). A total of 100 tests were carried out on the same induction motor, thus generating a premature and irreversible wear to the insulation of the winding, since the temperatures reached in the machine frame exceeded 150 °C. This level implies that much higher temperatures were present inside the motor that clearly exceeded the limit for class insulation (155 °C at the hottest point).

4. Discussion

In this section, the results obtained from testing the smart-sensor by installing it on six similar induction motors with different induced failures are shown.

Firstly, to probe the effectiveness of the smart-sensor and diagnose the winding insulation degradation over different faults, the smart-sensor was installed on five IMs at an ambient temperature (26 °C), all with the same constructive characteristics, but with each one possessing a special failure case, namely: minimal insulation degradation, light insulation degradation, severe insulation degradation, and an IM with one and two broken bars, but with a healthy winding insulation.

It can be clearly seen in Figure 11 how the S_{WT} parameter amplitudes were highly related to the winding insulation degradation, and it is also evident that the combined failures negligibly affected negligibly the results (that is, other failures like broken bars—one of the most frequent failures in IMs—did not affect the proposed methodology).

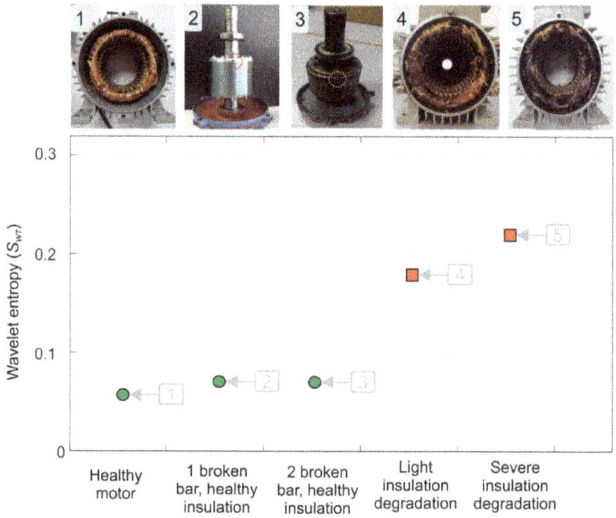

Figure 11. Wavelet entropy for different Induction Motors (IM) winding insulation degradations.

Figure 12 shows the results obtained when using the smart-sensor to compute the wavelet entropy for different prematurely induced degradation stages on the winding insulation of an IM. At the top of the same figure, the temperatures reached on the IM frame for different test points are shown, in order to contrast the results of when the temperature changes drastically. Similarly, five results displayed by the LCD of the smart-sensor proposed here are included, and shown at the top of Figure 12. These results correspond to tests labeled as A, B, C, and D for different winding insulation health states. For this purpose, over 100 tests were run. In each test, the winding insulation was degraded continuously. Evidently, the more severe the winding insulation degradation, the higher the amplitude of the wavelet entropy. Furthermore, note how the S_{WT} parameter is also dependent on the temperature of the motor, since numerous tests showed an increase of S_{WT} amplitudes with higher temperatures (especially in frame temperatures above 130 °C). Additionally at the top of Figure 12, the diagnosis offered by the smart-sensor proposed here is shown. The final results ranged from 10% to 95%, indicating the healthiness of the winding insulation (where 10% indicates a severe degradation, and 95% indicates very low or null degradation).

Considering a wavelet entropy value over 0.18 at ambient temperature (26 °C), a threshold value of 35% or below could be set as the criterion for discriminating between healthy and severe winding insulation system (requiring immediate maintenance) conditions (see Figure 12).

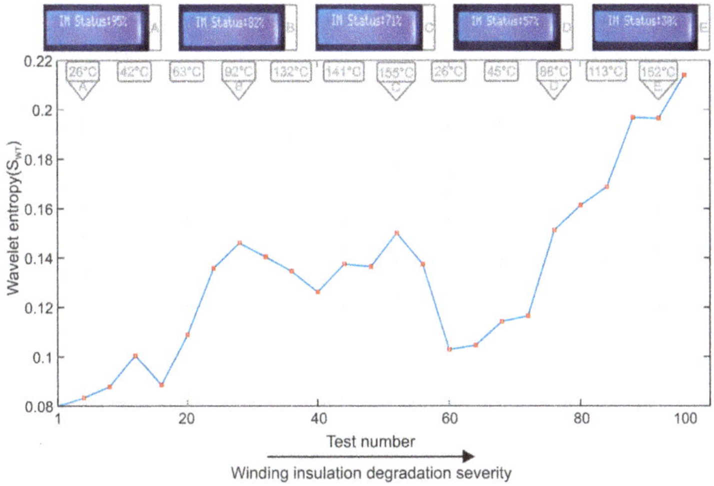

Figure 12. Wavelet entropy for different IM winding insulation degradation.

5. Conclusions

This work has introduced a new approach to performing an online estimation of the status of winding insulation degradation in IMs (a very common failure in this type of motors). The methodology is implemented in an FPGA in order to generate a smart-sensor, which is achieved by developing the digital cores needed to compute the DWT, the S_{WT} index, and the regression FFNN. These tools provide the smart-sensor with the capability to automatically diagnose the health of the winding insulation, specifically before incipient faults progress into irreversible damage to the motor, making the smart-sensor an excellent device for the online diagnosis of winding insulation degradation.

What makes the smart-sensor proposed here even more attractive, is that the signal processing tools rely on a stray-flux analysis combined with the temperature of the analyzed IM, where the signals are obtained from a coil sensor and an E-type thermocouple sensor, both of which are installed externally on the motor frame. The coil sensor complies with several characteristics that make it an excellent alternative as a source of information, including its simple design, small size, low cost, installation flexibility, and non-invasive nature.

Furthermore, the proposed methodology relies on the study of the WE parameter obtained from the stray flux captured by a coil sensor. The wavelet entropy provides a very useful and practical index to gather information related to the analyzed signal, since it can characterize and combine the dynamism and order/disorder of this signal in a single value.

In addition, it can be deduced from the results obtained here that the S_{WT} was sensitive to temperature variations in the analyzed IM, so it is very important to take this fact into account when diagnosing the severity degradation of the winding insulation, which was situation-controlled by the FFNN in this proposal.

For future work, the authors propose to further research in this open field, and suggest testing different IMs in order to deepen knowledge of the relationship discovered between the IM temperature and winding insulation degradation.

Author Contributions: J.A.A.-D. and R.A.O.-R. conceived and designed the experiments; I.Z.-R. performed the experiments and processed data; J.A.A.-D., R.A.O.-R., and R.d.J.R.-T. analyzed the data; I.Z.-R. and M.T.-H. wrote the paper.

Funding: We would like to thank Consejo Nacional de Ciencia y Tecnología (CONACYT) for providing economic support in this work (scholarship). Finally, thanks to the next projects: SEP-CONACYT 222453-2013, and FOFIUAQ-FIN201812. This wok was also funded by Spanish 'Ministerio de Ciencia Innovación y Universidades' and FEDER program in the framework of the 'Proyectos de I+D de Generación de Conocimiento del Programa Estatal de Generación de Conocimiento y Fortalecimiento Científico y Tecnológico del Sistema de I+D+i, Subprograma Estatal de Generación de Conocimiento' (ref: PGC2018-095747-B-I00).

Conflicts of Interest: The authors declare no conflict of interest.

References

1. Al Badawi, F.S.; AlMuhaini, M. Reliability modelling and assessment of electric motor driven systems in hydrocarbon industries. *IET Electr. Power Appl.* **2015**, *9*, 605–611. [CrossRef]
2. Trejo-Hernandez, M.; Osornio-Rios, R.A.; Romero-Troncoso, R.D.; Rodriguez-Donate, C.; Dominguez-Gonzalez, A.; Herrera-Ruiz, G. FPGA-based fused smart-sensor for tool-wear area quantitative estimation in CNC machine inserts. *Sensors* **2010**, *10*, 3373–3388. [CrossRef] [PubMed]
3. Cabal-Yepez, E.; Garcia-Ramirez, A.G.; Romero-Troncoso, R.J.; Garcia-Perez, A.; Osornio-Rios, R.A. Reconfigurable monitoring system for time-frequency analysis on industrial equipment through STFT and DWT. *IEEE Trans. Ind. Inform.* **2013**, *9*, 760–771. [CrossRef]
4. Granados-Lieberman, D.; Romero-Troncoso, R.; Cabal-Yepez, E.; Osornio-Rios, R.; Franco-Gasca, L. A real-time smart sensor for high-resolution frequency estimation in power systems. *Sensors* **2009**, *9*, 7412–7429. [CrossRef]
5. Cabal-Yepez, E.; Fernandez-Jaramillo, A.A.; Romero-Troncoso, R.J.; Garcia-Perez, A.; Osornio-Rios, R.A. Smart sensor for electrical machine monitoring through statistical analysis. In Proceedings of the 2012 XXth International Conference on Electrical Machines, 2–5 September 2012.
6. Garcia-Ramirez, A.G.; Osornio-Rios, R.A.; Granados-Lieberman, D.; Garcia-Perez, A.; Romero-Troncoso, R.J. Smart sensor for online detection of multiple-combined faults in VSD-fed induction motors. *Sensors* **2012**, *12*, 11989–12005. [CrossRef]
7. Zhang, P.; Du, Y.; Habetler, T.G.; Lu, B. A survey of condition monitoring and protection methods for medium-voltage induction motors. *IEEE Trans. Ind. Appl.* **2011**, *1*, 34–46. [CrossRef]
8. Bonnett, A.H.; Yung, C. Increased efficiency versus increased reliability. *IEEE Ind. Appl. Mag.* **2008**, *14*, 29–36. [CrossRef]
9. Motor Reliability Working Group. Report of large motor reliability survey of industrial and commercial installations, Part I. *IEEE Trans. Ind. Appl.* **1985**, *1*, 865–872.
10. Thorsen, O.V.; Dalva, M. A survey of faults on induction motors in offshore oil industry, petrochemical industry, gas terminals, and oil refineries. *IEEE Trans. Ind. Appl.* **1995**, *31*, 1186–1196. [CrossRef]
11. Capolino, G.A.; Antonino-Daviu, J.A.; Riera-Guasp, M. Modern diagnostics techniques for electrical machines, power electronics, and drives. *IEEE Trans. Ind. Electron.* **2015**, *62*, 1738–1745. [CrossRef]
12. Wang, L.; Li, Y.; Li, J. Diagnosis of inter-turn short circuit of synchronous generator rotor winding based on volterra kernel identification. *Energies* **2018**, *11*, 2524. [CrossRef]
13. Malekpour, M.; Phung, B.T.; Ambikairajah, E. Online technique for insulation assessment of induction motor stator windings under different load conditions. *IEEE Trans. Dielectr. Electr. Insul.* **2017**, *24*, 349–358. [CrossRef]
14. Cabanas, M.F.; Norniella, J.G.; Melero, M.G.; Rojas, C.H.; Cano, J.M.; Pedrayes, F.; Orcajo, G.A. Detection of stator winding insulation failures: Online and offline tests. In Proceedings of the 2013 IEEE Workshop on Electrical Machines Design, Control and Diagnosis (WEMDCD), Paris, France, 11–12 March 2013.
15. Siddique, A.; Yadava, G.S.; Singh, B. A review of stator fault monitoring techniques of induction motors. *IEEE Trans. Energy Conver.* **2005**, *20*, 106–114. [CrossRef]
16. Brown, A.; David, E.; Essalihi, M. Insulation resistance measurements for machine insulation. In Proceedings of the 2011 Electrical Insulation Conference (EIC), Annapolis, MD, USA, 5–8 June 2011.
17. Stone, G.C. Recent important changes in IEEE motor and generator winding insulation diagnostic testing standards. *IEEE Trans. Ind. Appl.* **2005**, *41*, 91–100. [CrossRef]
18. Yang, J.; Lee, S.B.; Yoo, J.; Lee, S.; Oh, Y.; Choi, C.A. Stator winding insulation condition monitoring technique for inverter-fed machines. *IEEE Trans. Power Electron.* **2007**, *22*, 2026–2033. [CrossRef]
19. Lee, S.B.; Yang, J.; Younsi, K.; Bharadwaj, R.M. An online ground wall and phase-to-phase insulation quality assessment technique for AC-machine stator windings. *IEEE Trans. Ind. Appl.* **2006**, *42*, 946–957.
20. Seshadrinath, J.; Singh, B.; Panigrahi, B.K. Vibration analysis based interturn fault diagnosis in induction machines. *IEEE Trans. Ind. Inf.* **2014**, *10*, 340–350. [CrossRef]

21. Singh, G.; Kumar, T.C.; Naikan, V.N. Induction motor inter turn fault detection using infrared thermographic analysis. *Infrared Phys. Tech.* **2016**, *77*, 277–282. [CrossRef]
22. Glowacz, A.; Glowacz, Z. Diagnosis of the three-phase induction motor using thermal imaging. *Infrared Phys. Tech.* **2017**, *81*, 7–16. [CrossRef]
23. Stone, G.C. A perspective on online partial discharge monitoring for assessment of the condition of rotating machine stator winding insulation. *IEEE Electr. Insul. Mag.* **2012**, *28*, 8–13. [CrossRef]
24. Wolkiewicz, M.; Tarchała, G.; Orłowska-Kowalska, T.; Kowalski, C.T. Online stator interturn short circuits monitoring in the DFOC induction-motor drive. *IEEE Trans. Ind. Electron.* **2016**, *63*, 2517–2528. [CrossRef]
25. Gyftakis, K.N.; Kappatou, J.C. The zero-sequence current as a generalized diagnostic mean in Δ-connected three-phase induction motors. *IEEE Trans. Energy Conver.* **2014**, *29*, 138–148. [CrossRef]
26. Cash, M.A.; Habetler, T.G.; Kliman, G.B. Insulation failure prediction in AC machines using line-neutral voltages. *IEEE Trans. Ind. Appl.* **1998**, *34*, 1234–1239. [CrossRef]
27. Lee, S.B.; Tallam, R.M.; Habetler, T.G. A robust, online turn-fault detection technique for induction machines based on monitoring the sequence component impedance matrix. *IEEE Trans. Power Electron.* **2003**, *18*, 865–872.
28. Dasgupta, A.; Nath, S.; Das, A. Transmission line fault classification and location using wavelet entropy and neural network. *Electric Power Compon. Syst.* **2012**, *40*, 1676–1689. [CrossRef]
29. Rosso, O.A.; Blanco, S.; Yordanova, J.; Kolev, V.; Figliola, A.; Schürmann, M.; Başar, E. Wavelet entropy: A new tool for analysis of short duration brain electrical signals. *J. Neurosci. Methods.* **2001**, *105*, 65–75. [CrossRef]
30. İşler, Y.; Kuntalp, M. Combining classical HRV indices with wavelet entropy measures improves to performance in diagnosing congestive heart failure. *Comput. Biol. Med.* **2007**, *37*, 1502–1510. [CrossRef]
31. Guojun, L.; Yong, W.; Le, L.; Shaofeng, G.; Haiqing, N.; Xuemei, W. Suppressing white noise in PD signal based on wavelet entropy and improved threshold function. In Proceedings of the 2017 IEEE 11th International Symposium on Diagnostics for Electrical Machines, Power Electronics and Drives (SDEMPED), Tinos, Greece, 29 August–1 September 2017.
32. Antonino-Daviu, J.A.; Riera-Guasp, M.; Folch, J.R.; Palomares, M.P. Validation of a new method for the diagnosis of rotor bar failures via wavelet transform in industrial induction machines. *IEEE Trans. Ind. Appl.* **2006**, *42*, 990–996. [CrossRef]
33. Antonino-Daviu, J.A.; Jover, P.; Riera, M.; Arkkio, A.; Roger-Folch, J. DWT analysis of numerical and experimental data for the diagnosis of dynamic eccentricities in induction motors. *Mech. Syst. Signal Proc.* **2007**, *21*, 2575–2589. [CrossRef]
34. Riera-Guasp, M.; Antonino-Daviu, J.A.; Pineda-Sanchez, M.; Puche-Panadero, R.; Pérez-Cruz, J.A. General approach for the transient detection of slip-dependent fault components based on the discrete wavelet transform. *IEEE Transa. Ind. Electron.* **2008**, *55*, 4167–4180. [CrossRef]
35. Antonino-Daviu, J.A.; Riera-Guasp, M.; Pineda-Sánchez, M.; Pons-Llinares, J.; Puche-Panadero, R.; Pérez-Cruz, J. Feature extraction for the prognosis of electromechanical faults in electrical machines through the DWT. *Int. J. Comput. Intell. Syst.* **2009**, *2*, 158–167. [CrossRef]
36. Quiroga, R.Q.; Rosso, O.A.; Başar, E.; Schürmann, M. Wavelet entropy in event-related potentials: A new method shows ordering of EEG oscillations. *Biol. Cybern.* **2001**, *84*, 291–299. [CrossRef] [PubMed]
37. Blanco, S.; Figliola, A.; Quiroga, R.Q.; Rosso, O.A.; Serrano, E. Time-frequency analysis of electroencephalogram series. Wavelet packets and information cost function. *Phys. Rev. E.* **1998**, *57*. [CrossRef]
38. Ramirez-Nunez, J.A.; Antonino-Daviu, J.A.; Climente-Alarcón, V.; Quijano-López, A.; Razik, H.; Osornio-Rios, R.A.; Romero-Troncoso, R.D. Evaluation of the detectability of electromechanical faults in induction motors via transient analysis of the stray flux. *Trans. Ind. Appl.* **2018**, *54*, 4324–4332. [CrossRef]
39. Iglesias-Martínez, M.E.; Antonino-Daviu, J.A.; Fernández de Córdoba, P.; Conejero, J.A. Rotor Fault Detection in Induction Motors Based on Time-Frequency Analysis Using the Bispectrum and the Autocovariance of Stray Flux Signals. *Energies* **2019**, *12*, 597. [CrossRef]
40. Frosini, L.; Harlişca, C.; Szabó, L. Induction machine bearing fault detection by means of statistical processing of the stray flux measurement. *IEEE Trans. Ind. Electron.* **2015**, *62*, 1846–1854. [CrossRef]

41. Dehghan, H.; Haghjoo, F.; Cruz, S.M. A Flux-Based Differential Technique for Turn-to-Turn Fault Detection and Defective Region Identification in Line-Connected and Inverter-Fed Induction Motors. *IEEE Trans. Energy Conver.* **2018**, *33*, 1876–1885. [CrossRef]
42. Romary, R.; Roger, D.; Brudny, J.F. Analytical computation of an AC machine external magnetic field. *Euro. Phys. J.-Appl. Phys.* **2009**, *47*. [CrossRef]
43. Camarena-Martinez, D.; Valtierra-Rodriguez, M.; Garcia-Perez, A.; Osornio-Rios, R.A.; Romero-Troncoso, R.D. Empirical mode decomposition and neural networks on FPGA for fault diagnosis in induction motors. *Sci. World J.* **2014**, *2014*. [CrossRef]
44. Rairán-Antolines, J.D. Reconstruction of periodic signals using neural networks. *Tecnura* **2014**, *18*, 34–46.
45. Mitchell, J.N. Computer multiplication and division using binary logarithms. *IRE Trans. Electron. Comput.* **1962**, *4*, 512–517. [CrossRef]
46. Tlelo-Cuautle, E.; de la Fraga, L.G.; Rangel-Magdaleno, J. *Engineering applications of FPGAs*; Springer: New York, NY, USA, 2016.

energies

MDPI

Article

Characteristics Analysis and Measurement of Inverter-Fed Induction Motors for Stator and Rotor Fault Detection

Jing Tang [1,*], Yongheng Yang [2], Jie Chen [1], Ruichang Qiu [1] and Zhigang Liu [1,3]

1 School of Electrical Engineering, Beijing Jiaotong University, Beijing 100044, China; jiechen@bjtu.edu.cn (J.C.); rchqiu@bjtu.edu.cn (R.Q.); zhgliu@bjtu.edu.cn (Z.L.)

2 Department of Energy Technology, Aalborg University, 9220 Aalborg, Denmark; yoy@et.aau.dk

3 Beijing Engineering Research Center for Electrical Rail Transit, Beijing 100044, China

* Correspondence: 15117396@bjtu.edu.cn; Tel.: +86-131-6129-4848

Received: 25 November 2019; Accepted: 20 December 2019; Published: 24 December 2019

Abstract: Inverter-fed induction motors (IMs) contain a serious of current harmonics, which become severer under stator and rotor faults. The resultant fault components in the currents affect the monitoring of the motor status. With this background, the fault components in the electromagnetic torque under stator faults considering harmonics are derived in this paper, and the fault components in current harmonics under rotor faults are analyzed. More importantly, the monitoring based on the fault characteristics (both in the torque and current) is proposed to provide reliable stator and rotor fault diagnosis. Specifically, the fault components induced by stator faults in the electromagnetic torque are discussed in this paper, and then, fault components are characterized in the torque spectrum to identify stator faults. To achieve so, a full-order flux observer is adopted to calculate the torque. On the other hand, under rotor faults, the sidebands caused by time and space harmonics in the current are analyzed and exploited to recognize rotor faults, being the motor current signature analysis (MCSA). Experimental tests are performed on an inverter-fed 2.2 kW/380 V/50 Hz IM, which verifies the analysis and the effectiveness of the proposed fault diagnosis methods of inverter-fed IMs.

Keywords: characteristics analysis; fault detection; stator fault; rotor fault; torque estimation; induction motor

1. Introduction

In recent years, researches on the condition monitoring and health prognosis of electrical equipment are drawing more and more concerns. The reliabilities of converters and machines are directly related to the system, even human life, in applications such as photovoltaic (PV), electrical vehicle, etc. Fault-detection and tolerant operation of modular multilevel converters (MMC) are investigated in [1,2], where the latter deals with the insulated gate bipolar transistor (IGBT) open circuit fault. Fault diagnosis techniques related to machines are more diverse, where synchronous generators (SG) [3], permanent magnet (PM) machines [4–6], multi-phase machines [5,7,8], and induction motors (IM) [9–11] are involved, and winding short circuit fault [4,5], rotor broken bar fault [12,13], and eccentric and bearing faults [13] are discussed.

Induction motors (IMs) are widely applied in industry [11] and usually in harsh environments. This causes early motor faults, which may grow to irreparable failures if not properly treated. Motor faults can mainly be divided into two categories: Mechanical and electrical faults, as shown in Figure 1, where mechanical faults include bearing and eccentricity faults, and electrical faults include stator and rotor faults. According to [12], stator and rotor faults account for 37% and 10%, respectively, in all motor failures. Thus, many fault diagnosis techniques are presented. These fault diagnosis methods are divided into: Signal-based techniques, artificial intelligence (AI)-based techniques, and model-based techniques.

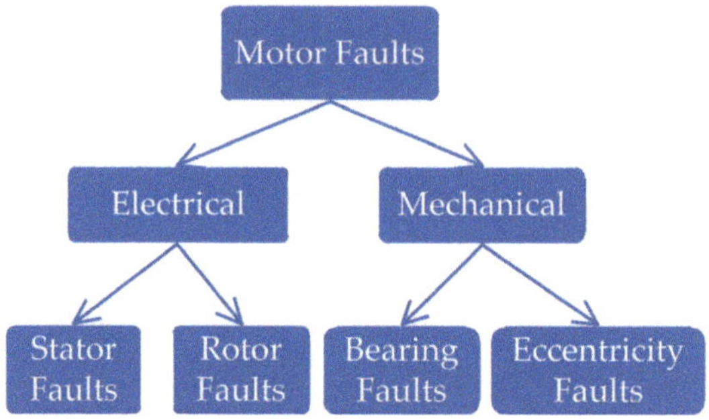

Figure 1. Motor faults classification.

The fault diagnosis methods based on signal processing [13,14] are very popular in industry. A novel Park's vector approach is investigated in [15,16], and the extended Park's [17] and 3-D-ellipse [18] methods are then developed for stator fault diagnosis. The winding asymmetry of stator faults causes un-balanced three-phase currents. Thus, the methods based on negative current and impedance are applied to stator monitoring [19]. Many signal spectrum analysis techniques [16,20–22] based on time domain, frequency domain, and time–frequency domain of motor current signature analysis (MCSA) are proposed to analyze the $(1 - 2s)f_1$ characterization for rotor fault detection, where s is the slip and f_1 represents the base frequency. The fast Fourier transform (FFT) is used for the MCSA. However, the fault frequency $(1 - 2s)f_1$ is easily covered by the fundamental frequency in the current spectrum, as s is usually small for inverter-fed IMs, where the rotor speed is very close to the stator frequency; thus, the time-windowing methods are proposed to weaken the influence of the spectrum leakage, and methods by removing the base frequency are considered. Furthermore, the high-resolution spectrum techniques such as the estimation of signal parameters via rotational invariance techniques (ESPRITs) [23,24], multiple signal classification (MUSIC) [25,26], root-MUSIC [25], and zoom-MUSIC (ZMUSIC) [26] are investigated in rotor faults detection. Finally, the time–frequency analysis methods of short time Fourier transform (STFT) [27,28], wavelet [29–31], and Wigner–Ville transform [32] are also applied to the MCSA.

The AI-based methods are investigated in the fault diagnosis. The diagnosis procedure usually includes the process of the signature extraction, signal processing, fault classification, and fault decision. For each fault, the signatures can be current, voltage, power, magnetic flux, torque, and vibration. The fault classification technique is crucial for the AI-based fault diagnosis. There are many artificial techniques that are applied to faults classification. Among them, the neural network (NN) is very popular. In [33], an early stator fault was detected through the NN. In [34], a cascaded NN was developed to classify faults. In [35], an artificial neural network (ANN) is used to train and classify different faults. In addition, [36] uses a hybrid fuzzy min-max NN and classification and regression tree (FMM-CART) to undertake data classification and rule extraction problems. Furthermore, the fault diagnosis method based on the support vector machine (SVM) is employed in [37,38] for motor faults. Moreover, classifiers based on C4.5, K-nearest neighbors (k-NNs), and multilayer perceptron (MLP) are discussed in [39,40] to recognize faults.

Model-based or parameter-based fault diagnosis methods utilize the deviation of critical parameters under fault conditions. For example, in [41], stator and rotor faults were modeled firstly, and then, the fault severity is estimated, and position is located by considering the prior information. As presented in [42], the Kalman filter can be adopted to identify parameters online to detect faults. Additionally, broken rotor faults will cause rotor resistance R_r to increase, which can be used for fault detection.

When the pulse width modulation (PWM) voltage is supplied to a motor, current harmonics can be used to characterize faults in IMs under stator and rotor faults. Therefore, this paper is devoted to the study of stator and rotor faults monitoring for IMs with voltage source inverters. The rest of this paper is organized as follows: Section 2 analyzes the fault components under stator faults in torque, and furthermore, the stator fault detection methods based on torque spectral analysis by using a full-order flux observer are proposed. In Section 3, rotor fault components caused by current harmonics and space magnetomotive force (MMF) are considered and analyzed in detail, and then the rotor fault diagnosis based on the MCSA is presented. In Section 4, the proposed fault diagnosis methods are verified through experimental results. Finally, the conclusion is presented in Section 5.

2. Stator Fault Diagnosis

2.1. Motor Voltages

The inverter consists of switching devices of IGBTs and diodes as shown in Figure 2, where T_1-T_6 are the switching elements of the three-phase inverter. Considering a balanced voltage condition, the three-phase voltage of the induction motor can be expressed as:

$$\begin{cases} u_{sa}(t) = \sum\limits_{k=1,5,7,\ldots}^{\infty} U_k \cos k\omega_1 t \\ u_{sb}(t) = \sum\limits_{k=1,5,7,\ldots}^{\infty} U_k \cos k\left(\omega_1 t - \frac{2\pi}{3}\right) \\ u_{sc}(t) = \sum\limits_{k=1,5,7,\ldots}^{\infty} U_k \cos k\left(\omega_1 t + \frac{2\pi}{3}\right) \end{cases} \tag{1}$$

where u_{sa}, u_{sb}, and u_{sc} are the voltage of phase A, B, and C, respectively; U_k (k = 1, 5, 7, ...) is the magnitude of the k-th harmonic; and ω_1 is the angular frequency ($\omega_1 = 2\pi f_1$) with f_1 being the fundamental frequency. As observed in Equation (1), the motor voltage is a PWM voltage that contains a series of harmonics at $k = 6i \pm 1$ (i= 0, 1, 2, 3 ...), whose magnitude decreases with the increasing harmonic order.

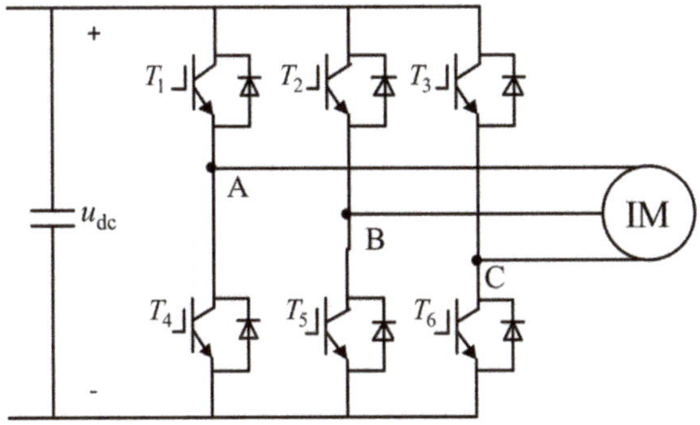

Figure 2. A typical three-phase voltage source inverter for induction motors (IMs), where u_{dc} is the DC voltage.

2.2. Currents of IM Under Stator Faults

When motors have stator faults, the three-phase stator windings become asymmetry. The stator current will be un-balanced, and thus, the negative-sequence current will be introduced. In this case, the motor currents are described by:

$$\begin{cases} i_{\text{sa,SF}}(t) = \sum\limits_{k=1,5,7,\ldots}^{\infty} I_{\text{kp}} \cos\left(k\omega_1 t + \varphi_{\text{kp}}\right) + \sum\limits_{k=1,5,7,\ldots}^{\infty} I_{\text{kn}} \cos(k\omega_1 t + \varphi_{\text{kn}}) \\ i_{\text{sb,SF}}(t) = \sum\limits_{k=1,5,7,\ldots}^{\infty} I_{\text{kp}} \cos\left[k\left(\omega_1 t - \frac{2\pi}{3}\right) + \varphi_{\text{kp}}\right] + \sum\limits_{k=1,5,7,\ldots}^{\infty} I_{\text{kn}} \cos\left[k\left(\omega_1 t + \frac{2\pi}{3}\right) + \varphi_{\text{kn}}\right] \\ i_{\text{sc,SF}}(t) = \sum\limits_{k=1,5,7,\ldots}^{\infty} I_{\text{kp}} \cos\left[k\left(\omega_1 t + \frac{2\pi}{3}\right) + \varphi_{\text{kp}}\right] + \sum\limits_{k=1,5,7,\ldots}^{\infty} I_{\text{kn}} \cos\left[k\left(\omega_1 t - \frac{2\pi}{3}\right) + \varphi_{\text{kn}}\right] \end{cases} \tag{2}$$

in which $i_{\text{sa,SF}}$, $i_{\text{sb,SF}}$, and $i_{\text{sc,SF}}$ are the stator fault currents of phase A, B, and C, respectively; I_{kp} and I_{kn} are the magnitudes of the kth positive and negative harmonic currents; φ_{kp} and φ_{kn} are the phase angle of the kth positive and negative harmonic currents.

2.3. Stator Fault Components in the Torque Spectrum

According to the voltage in Equation (1) and the stator current in Equation (2), the motor voltage and current in the stationary α-β reference frame are given as:

$$u_{s\alpha} = \sum_{k=1,5,7,\ldots}^{\infty} U_{\text{k}} \cos(k\omega_1 t) \tag{3}$$

$$u_{s\beta} = \frac{2}{\sqrt{3}} \sum_{k=1,5,7,\ldots}^{\infty} U_{\text{k}} \sin(k\frac{2\pi}{3}) \sin(k\omega_1 t) \tag{4}$$

$$i_{s\alpha} = \sum_{k=1,5,7,\ldots}^{\infty} \left[I_{\text{kp}} \cos\left(k\omega_1 t + \varphi_{\text{kp}}\right) + I_{\text{kn}} \cos(k\omega_1 t + \varphi_{\text{kn}})\right] \tag{5}$$

$$i_{s\beta} = \frac{2}{\sqrt{3}} \sum_{k=1,5,7,\ldots}^{\infty} \sin(k\frac{2\pi}{3})\left[I_{\text{kp}} \sin\left(k\omega_1 t + \varphi_{\text{kp}}\right) - I_{\text{kn}} \sin(k\omega_1 t + \varphi_{\text{kn}})\right] \tag{6}$$

where $u_{s\alpha}$ and $u_{s\beta}$ are the stator voltages in the α-β reference frame; $i_{s\alpha}$ and $i_{s\beta}$ are the stator currents in the α-β reference frame.

To derive the stator flux under stator faults and simplify the calculation, it is assumed that the voltage-drop caused by the stator resistance is ignored. Then, the stator flux in the stationary α-β reference frame can be expressed through the voltage integral, and then, simplified as:

$$\lambda_{s\alpha} = \omega_1^{-1} \sum_{k=1,5,7,\ldots}^{\infty} k^{-1} U_{\text{k}} \sin(k\omega_1 t) \tag{7}$$

$$\lambda_{s\beta} = -\omega_1^{-1} \frac{2}{\sqrt{3}} \sum_{k=1,5,7,\ldots}^{\infty} k^{-1} U_{\text{k}} \sin(k\frac{2\pi}{3}) \cos(k\omega_1 t) \tag{8}$$

in which $\lambda_{s\alpha}$ and $\lambda_{s\beta}$ are the stator flux in the α-β reference frame. According to [43,44], when a stator fault is present in the motor, the electromagnetic torque is calculated as:

$$T_{\text{e}} = \frac{3}{2}\frac{P}{2} L_{\text{m}}\left(i_{r\alpha} i_{s\beta} - i_{r\beta} i_{s\alpha}\right) - \frac{P}{2} \mu L_{\text{m}} i_f i_{r\beta} \tag{9}$$

where P is the number of poles, μ is the fault severity factor, i_{f} is the circulating fault current in the short-circuit path. In practice, it is unavailable to measure μ and i_{f} directly, and thus, the last item in (9) caused by the stator fault is neglected here. Nevertheless, the rest of Equation (9) contains fault components induced by the unbalance of three-phase currents, which are further analyzed to find fault indicators for motor condition monitoring. Accordingly, the electromagnetic torque can be approximately expressed as:

$$T_e = \frac{3}{2}\frac{P}{2}\left(\lambda_{s\alpha} i_{s\beta} - \lambda_{s\beta} i_{s\alpha}\right) \tag{10}$$

Substituting Equations (5)–(8) into Equation (10), gives:

$$T_e = \left\{\begin{array}{l}
\underbrace{\frac{3}{2}n_p\omega_1^{-1}\sum_{\substack{m=6i+1\\ i=0,1,2\ldots}}^{\infty}\sum_{\substack{k=6j+1\\ j=0,1,2\ldots}}^{\infty} m^{-1}U_m I_{kp}\cos\left[(m-k)\omega_1 t-\varphi_{kp}\right]}_{healthypart}+\underbrace{\frac{3}{2}n_p\omega_1^{-1}\sum_{\substack{m=6i+1\\ i=0,1,2\ldots}}^{\infty}\sum_{\substack{k=6j+1\\ j=0,1,2\ldots}}^{\infty} m^{-1}U_m I_{kn}\cos\left[(m+k)\omega_1 t+\varphi_{kn}\right]}_{faultypart}\\
\underbrace{-\frac{3}{2}n_p\omega_1^{-1}\sum_{\substack{m=6i+1\\ i=0,1,2\ldots}}^{\infty}\sum_{\substack{k=6j-1\\ j=1,2\ldots}}^{\infty} m^{-1}U_m I_{kp}\cos\left[(m+k)\omega_1 t+\varphi_{kp}\right]}_{healthypart}\underbrace{-\frac{3}{2}n_p\omega_1^{-1}\sum_{\substack{m=6i+1\\ i=0,1,2\ldots}}^{\infty}\sum_{\substack{k=6j-1\\ j=1,2\ldots}}^{\infty} m^{-1}U_m I_{kn}\cos\left[(m-k)\omega_1 t-\varphi_{kn}\right]}_{faultypart}\\
\underbrace{+\frac{3}{2}n_p\omega_1^{-1}\sum_{\substack{m=6i-1\\ i=1,2\ldots}}^{\infty}\sum_{\substack{k=6j+1\\ j=0,1,2\ldots}}^{\infty} m^{-1}U_m I_{kp}\cos\left[(m+k)\omega_1 t+\varphi_{kp}\right]}_{healthypart}+\underbrace{\frac{3}{2}n_p\omega_1^{-1}\sum_{\substack{m=6i-1\\ i=1,2\ldots}}^{\infty}\sum_{\substack{k=6j+1\\ j=0,1,2\ldots}}^{\infty} m^{-1}U_m I_{kn}\cos\left[(m-k)\omega_1 t-\varphi_{kn}\right]}_{faultypart}\\
\underbrace{-\frac{3}{2}n_p\omega_1^{-1}\sum_{\substack{m=6i-1\\ i=1,2\ldots}}^{\infty}\sum_{\substack{k=6j-1\\ j=1,2\ldots}}^{\infty} m^{-1}U_m I_{kp}\cos\left[(m-k)\omega_1 t-\varphi_{kp}\right]}_{healthypart}\underbrace{-\frac{3}{2}n_p\omega_1^{-1}\sum_{\substack{m=6i-1\\ i=1,2\ldots}}^{\infty}\sum_{\substack{k=6j-1\\ j=1,2\ldots}}^{\infty} m^{-1}U_m I_{kn}\cos\left[(m+k)\omega_1 t+\varphi_{kn}\right]}_{faultypart}
\end{array}\right\} \tag{11}$$

According to the above formula, it can be found that when $m = 6i + 1$ ($i = 0, 1, 2, \ldots$) and $k = 6j + 1$ ($j = 0, 1, 2, \ldots$), the healthy part produces a dc component ($m = k$) and harmonic components ($m \neq k$) with the frequencies of $6h\omega_1$ ($h = 1, 2, 3, \ldots$); the fault part produces harmonic components with the frequencies of $(6h + 2)\omega_1$. Similarly, for other conditions, the $(6h + 2)\,\omega_1$ or $(6h - 2)\omega_1$ is generated according to the fault parts in Equation (11). In all, for a healthy motor, the negative current I_{kn} ($k = 1, 5, 7, \ldots$) is equal to zero, and there are only a dc component and harmonics whose frequencies are $6h\omega_1$ in the torque spectrum. When the motor has stator faults, it is observed from Equation (11) that additional fault frequencies of $(6h \pm 2)\omega_1$ will be produced, which can be used as indicators of the motor status.

The frequency components induced by the interaction of the fundamental, the fifth and seventh harmonics under stator faults in the torque spectrum are listed in Table 1. It can be seen from Table 1 that the fault frequencies of $2f_1$, $4f_1$, $8f_1$, $10f_1$, and $14f_1$ are produced in the torque spectrum. It is also observed in Table 1 that the components with frequencies of $2f_1$, $4f_1$, and $8f_1$ have higher magnitudes in the torque spectrum, as these frequencies components are the results of the interaction of the fundamental and harmonics. Consequently, interactive components with the frequencies of $2f_1$, $4f_1$, and $8f_1$ are expected to detect stator faults.

Table 1. Stator Fault Components Induced by the Fundamental, Fifth, and Seventh Harmonics in Torque.

Harmonics		Torque Spectrum	
		Health Components	Fault Components
f_1 $(m = 1)$	f_1 $(k = 1)$	dc	$2f_1$
	$5f_1$ $(k = 5)$	$6f_1$	$4f_1$
	$7f_1$ $(k = 7)$	$6f_1$	$8f_1$
$5f_1$ $(m = 5)$	f_1 $(k = 1)$	$6f_1$	$4f_1$
	$5f_1$ $(k = 5)$	dc	$10f_1$
	$7f_1$ $(k = 7)$	$12f_1$	$2f_1$
$7f_1$ $(m = 7)$	f_1 $(k = 1)$	$6f_1$	$8f_1$
	$5f_1$ $(k = 5)$	$12f_1$	$2f_1$
	$7f_1$ $(k = 7)$	dc	$14f_1$

2.4. Torque Measurement

The above stator fault diagnosis method depends on the electromagnetic torque. However, the motor is usually not equipped with a torque sensor in practice. Therefore, the acquisition of the flux is crucial for the torque calculation according to (10). A full-order flux observer is very popular and employed for rotor flux estimation because of its high robustness and fast convergence. Even the IM has parameter variations, the observed current from the closed-loop observer still can track its actual value. The novel full-order flux observer is expressed as:

$$\begin{cases} \dot{\hat{\mathbf{x}}}(t) = \mathrm{A}\hat{\mathbf{x}}(t) + \mathbf{Bu}(t) + \mathrm{G}[\hat{\mathbf{y}}(t) - \mathbf{y}(t)] \\ \hat{\mathbf{y}}(t) = \mathrm{C}\hat{\mathbf{x}}(t) \end{cases} \tag{12}$$

with

$$\hat{\mathbf{x}}(t) = \begin{bmatrix} \hat{i}_{s\alpha} & \hat{i}_{s\beta} & \hat{\lambda}_{r\alpha} & \hat{\lambda}_{r\beta} \end{bmatrix}, \hat{\mathbf{y}}(t) = \begin{bmatrix} \hat{i}_{s\alpha} & \hat{i}_{s\beta} \end{bmatrix}, \mathbf{u}(t) = \begin{bmatrix} u_{s\alpha} & u_{s\beta} \end{bmatrix}^T$$

$$A = \begin{bmatrix} -\left(\frac{R_r L_m^2}{\sigma L_s L_r^2} + \frac{R_s}{\sigma L_s}\right) & 0 & \frac{R_r L_m}{\sigma L_s L_r^2} & \frac{L_m \omega_r}{\sigma L_s L_r} \\ 0 & -\left(\frac{R_r L_m^2}{\sigma L_s L_r^2} + \frac{R_s}{\sigma L_s}\right) & -\frac{L_m \omega_r}{\sigma L_s L_r} & \frac{R_r L_m}{\sigma L_s L_r^2} \\ \frac{R_r L_m}{L_r} & 0 & -\frac{R_r}{L_r} & -\omega_r \\ 0 & \frac{R_r L_m}{L_r} & \omega_r & -\frac{R_r}{L_r} \end{bmatrix}$$

$$B = \begin{bmatrix} \frac{1}{\sigma L_s} & 0 & 0 & 0 \\ 0 & \frac{1}{\sigma L_s} & 0 & 0 \end{bmatrix}^T, C = \begin{bmatrix} 1 & 0 & 0 & 0 \\ 0 & 1 & 0 & 0 \end{bmatrix}, G = \begin{bmatrix} g_1 & g_2 & g_3 & g_4 \\ -g_2 & g_1 & -g_4 & g_3 \end{bmatrix}^T$$

in which $\hat{\mathbf{x}}$ is the observed state variables; $\hat{\mathbf{y}}$ is the output variables; A, B, and C are the state, input, and output matrices; G is the feedback gain matrix; $\hat{i}_{s\alpha}$, $\hat{i}_{s\beta}$, $\hat{\lambda}_{s\alpha}$, and $\hat{\lambda}_{s\beta}$ are the estimated stator current and rotor flux in the α-β reference frame; R_s, R_r, L_s, and L_r are motor resistances and inductances; σ is the total leakage coefficient; ω_r is the rotor angular frequency. The elements g_1 to g_4 in G are defined as

$$\begin{aligned} g_1 &= (\rho - 1)\left[-\left(\frac{R_r L_m^2}{\sigma L_s L_r^2} + \frac{R_s}{\sigma L_s}\right) - \frac{R_r}{L_r}\right] \\ g_2 &= (\rho - 1)\omega_r \\ g_3 &= \left(\rho^2 - 1\right)\left[-\frac{\sigma L_s L_r}{L_m}\left(\frac{R_r L_m^2}{\sigma L_s L_r^2} + \frac{R_s}{\sigma L_s}\right) + \frac{R_r L_m}{L_r}\right] - (\rho - 1)\frac{\sigma L_s L_r}{L_m}\left[-\left(\frac{R_r L_m^2}{\sigma L_s L_r^2} + \frac{R_s}{\sigma L_s}\right) - \frac{R_r}{L_r}\right] \\ g_4 &= -(\rho - 1)\omega_r \sigma L_s L_r / L_m \end{aligned}$$

where ρ $(\rho > 1)$ determines the poles of the observer. Accordingly, the lower the pole ρ is, the slower the convergence is, whereas the more stable the observer is; on the contrary, the higher the pole ρ is, the faster convergence is, whereas the less stable the observer is. The value of ρ balances the convergence speed and stability.

The stator flux is estimated according to the observer in Equation (12), and subsequently, the torque is calculated as

$$T_e = \frac{3}{2}\frac{P}{2}\frac{L_m}{L_r}\left(\hat{\lambda}_{r\alpha} i_{s\beta} - \hat{\lambda}_{r\beta} i_{s\alpha}\right) \tag{13}$$

Then, the torque spectrum is obtained through the FFT method. Depending on the magnitude of the fault frequencies $2f_1$, $4f_1$, and $8f_1$, the monitoring of stator faults can be realized. This stator fault diagnosis procedure is illustrated in Figure 3, where PI represents a proportional-integral controller; VSI is the voltage source inverter.

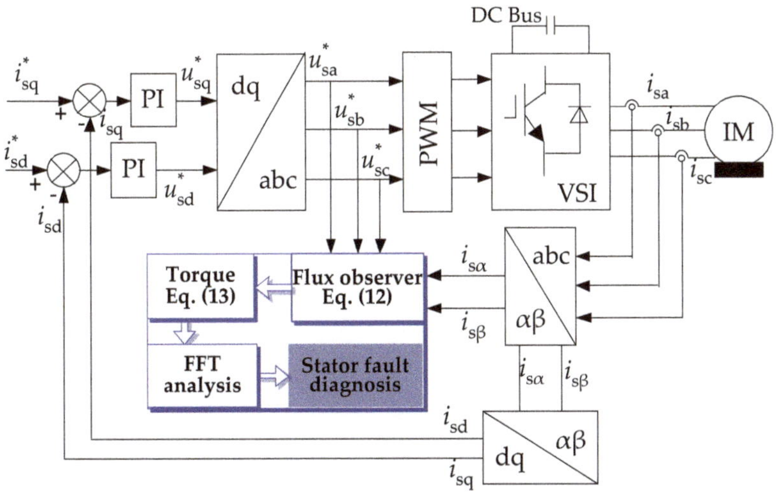

Figure 3. Proposed stator fault diagnosis procedure implemented in the entire control of the IM.

3. Rotor Fault Diagnosis

3.1. Fundamental Sidebands of Rotor Faults

Rotor faults lead to the rotor asymmetry, which produces fault components with a frequency $(1-2s)f_1$ in the stator current. Then, the fault components will induce motor torque and speed ripples, and furthermore, the frequency component of $(1+2s)f_1$ in the current is introduced. Eventually, sideband components with the frequency shown in Equation (14) can be used for fault detection. It is clear that the component with a higher harmonic order k has a lower current magnitude until the magnitude is close to zero, as shown in Equation (15).

$$f_{b,b} = (1 \pm 2ks)f_1, \qquad (k = 1,2,3,\ldots) \tag{14}$$

$$|1 \pm 2s| > |1 \pm 4s| > |1 \pm 6s| \ldots \tag{15}$$

3.2. Space Harmonics Sidebands of Rotor Faults

The non-sinusoidal distribution of the stator winding produces a series of space harmonic MMFs with the order being $\mu = 6i \pm 1$ (i = 1, 2, 3, …). If a rotor fault occurs, these space harmonics will produce fault components in the stator current. According to [45,46], the induced fault frequencies can be described as

$$f_{b,\mu} = (\mu(1-s) \pm s)f_1, \qquad (\mu = 5,7,\ldots) \tag{16}$$

where the frequencies produced by the fifth- and seventh-order space harmonics, i.e., $(5-4s)f_1$, $(5-6s)f_1$, $(7-6s)f_1$, $(7-8s)f_1$, are the largest among all the space harmonic components.

3.3. Time Harmonics Sidebands of Rotor Faults

The time-domain current harmonics of the inverter-connected healthy motor are given as

$$\begin{cases} i_{\text{sa,h}}(t) = I_1 \cos(\omega_1 t + \varphi_1) + \sum\limits_{\upsilon=5,7\ldots}^{\infty} I_\upsilon \cos(\upsilon\omega_1 t + \varphi_\upsilon) \\ i_{\text{sb,h}}(t) = I_1 \cos(\omega_1 t - \frac{2\pi}{3} + \varphi_1) + \sum\limits_{\upsilon=5,7\ldots}^{\infty} I_\upsilon \cos[\upsilon(\omega_1 t - \frac{2\pi}{3}) + \varphi_\upsilon] \\ i_{\text{sc,h}}(t) = I_1 \cos(\omega_1 t + \frac{2\pi}{3} + \varphi_1) + \sum\limits_{\upsilon=5,7\ldots}^{\infty} I_\upsilon \cos[\upsilon(\omega_1 t + \frac{2\pi}{3}) + \varphi_\upsilon] \end{cases} \quad (17)$$

where $i_{\text{sa,h}}$, $i_{\text{sb,h}}$, and $i_{\text{sc,h}}$ are the currents of phase A, B, and C; I_1 represents the fundamental current magnitude and I_υ is the magnitude of the υth harmonic current; φ_1 and φ_υ represent the phase angles of the fundamental and the υth harmonic current, respectively.

If a rotor fault occurs in the motor, the above time-domain harmonics will produce fault sidebands in the stator current. It is clear that the current contains harmonics with the order being $\upsilon = 6i \pm 1$ ($i = 1, 2, 3, \ldots$). The harmonic signals rotate in the positive direction ($\upsilon = 7, 13, \ldots$) and in the negative direction ($\upsilon = 5, 11, \ldots$), as described by

$$\upsilon = \begin{cases} 6i + 1, \text{positive} \quad \text{direction}(\text{"}+\text{"}) \\ 6i - 1, \text{negative} \quad \text{direction}(\text{"}-\text{"}) \end{cases} \quad (i = 1, 2, \ldots) \quad (18)$$

Therefore, the υth time harmonics rotate at the speed of f_υ that can be expressed by

$$f_\upsilon = \begin{cases} \upsilon f_1 (\upsilon = 7, 13, \ldots) \\ -\upsilon f_1 (\upsilon = 5, 11, \ldots) \end{cases} \quad (19)$$

$$f_\upsilon = \upsilon(\upsilon - 6i) f_1 \qquad (\upsilon = 5, 7, 11, \ldots) \quad (20)$$

where the sign of ($\upsilon - 6i$) represents the rotation direction. Then, the rotation speed of harmonics with respect to the rotor can be represented by

$$f_{\upsilon,r} = f_\upsilon - (1 - s) f_1 \quad (21)$$

which is the difference between f_υ and the rotor speed $f_r = (1 - s) f_1$.

As a consequence, the rotor faults will produce negative sequence currents at frequencies of $-f_{\upsilon,r}$ in the rotor current, which induces $f_{b,\upsilon} = -f_{\upsilon,r} + f_r$ fault components in the stator current. Therefore, the υth time harmonic produces the fault component at $(\upsilon \pm 2s) f_1$ frequency in the stator current. Similarly, a series of time harmonic sidebands appear as

$$f_{b,\upsilon} = (\upsilon \pm 2ks) f_1, \qquad (\upsilon = 5, 7, \ldots; k = 1, 2, \ldots) \quad (22)$$

According to the aforementioned fundamental sidebands, space, and time harmonic sidebands, the rotor fault components are summarized, as shown in Figure 4, where the frequencies $f_{b,\mu}$, $f_{b,b}$, $f_{b,\upsilon}$ in the stator current spectrum are the fault features that ensure reliable monitoring of rotor faults.

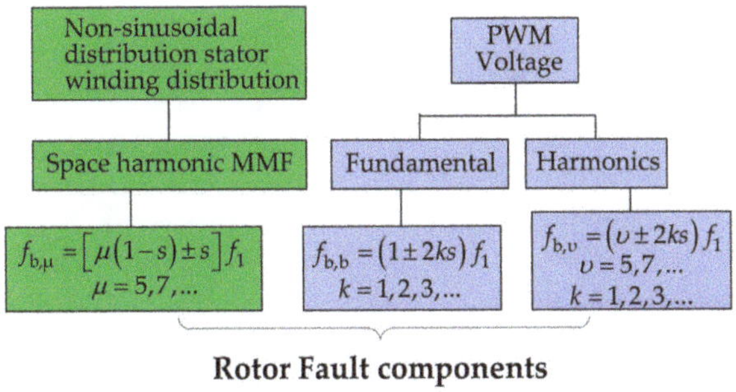

Figure 4. Rotor Fault Components in the Stator Current for Fault Monitoring.

3.4. Rotor Faults Diagnosis Based on MCSA

It can be observed in Equation (1) that the fifth and seventh harmonics have larger magnitudes, and thus, the rotor fault monitoring based on the detection of fault components produced by the fundamental, the fifth- and seventh-time and space harmonics with the frequencies of $(1 \pm 2s)f_1$, $(5 \pm 2s)f_1$, $(5 - 4s)f_1$, $(5 - 6s)f_1$, $(7 \pm 2s)f_1$, $(7 - 6s)f_1$, and $(7 - 8s)f_1$ is significantly advantageous. Therefore, the rotor fault is detected through the comparison of the magnitudes of these rotor fault signatures (RFS) under health and fault conditions. It should be noted that the slip s can be directly calculated by

$$s = (f_1 - f_r)/f_1 \tag{23}$$

where f_r is the rotor speed that can be obtained through the speed sensor. This rotor fault diagnosis procedure is illustrated as Figure 5.

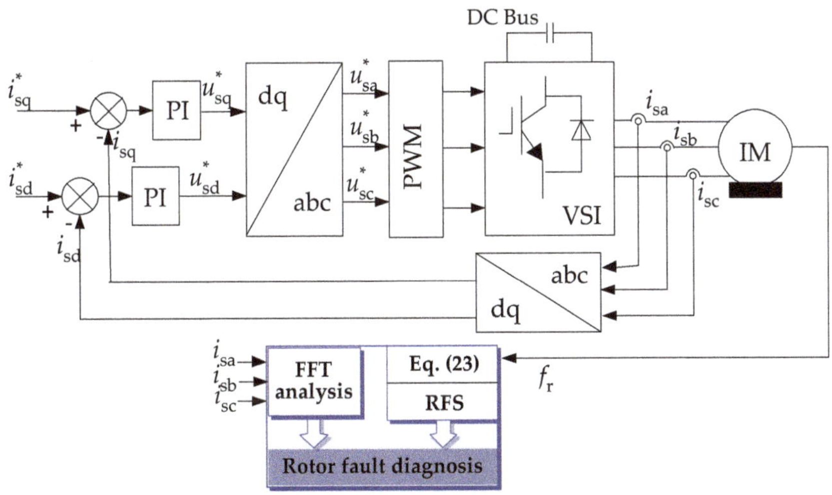

Figure 5. Proposed Rotor Fault Diagnosis Procedure Based on the MCSA.

4. Experimental Results

4.1. Experimental Setup

The basic schematic of the experimental setup is shown in Figure 6, where the three-phase voltage (380 V/50 Hz) is adjusted through an auto-transformer and a rectifier. Then, the dc voltage is fed to the IM through the variable-voltage and variable-frequency VSI. The load is controlled by adjusting the field voltage of a 1.5 kW dc generator coupled to the motor. The photo of entire experimental rig is shown in Figure 7. The control frequency of the IM is 2 kHz, and the switching frequency of the inverter is 1 kHz. Motor currents are measured and analyzed with a 10 kHz sampling frequency.

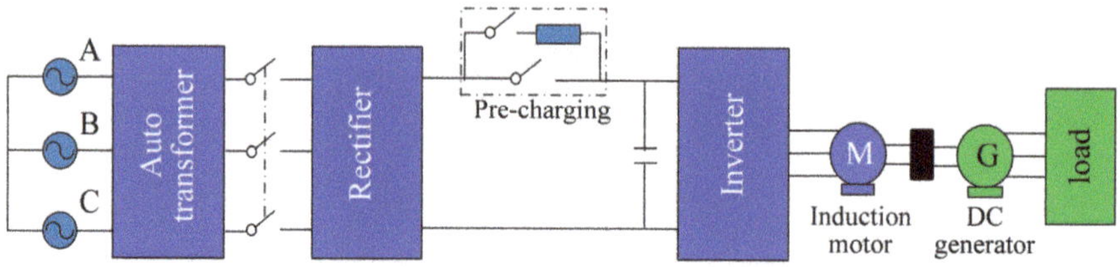

Figure 6. Schematic of the Experimental System.

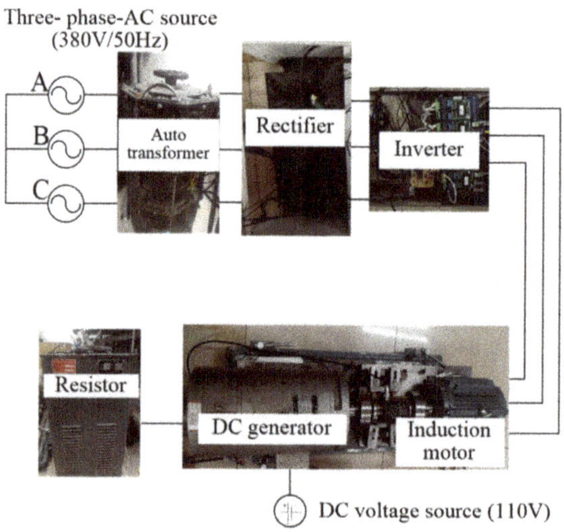

Figure 7. Experimental Test-Rig.

To verify the aforementioned, experimental tests are performed on the 2.2-kW/380-V/50-Hz three-phase IM system shown in Figure 6; Figure 7. The motor has 36 stator slots and 28 rotor slots, and the number of turns in each phase is 252, where there are 6 coils of windings for a stator phase, and each coil has 42 turns, as shown in Figure 8. One stator with turn-to-turn insulation failures and three rotors with broken bars are arbitrarily designed. For the fault stator, the stator winding was modified through the addition of taps connected to the stator winding turns, and the ends of these external wires are placed in the motor terminal box, as shown in Figure 9. When it needs to perform the stator inter-turn short fault tests, any two additions of the taps are collected to short-circuit several turns. For the rotor fault, the bar breakage is created by drilling holes in the contiguous aluminum bars, and then, the dynamic balance of the damaged rotors is tested and adjusted by the machine manufacturer. Photos of damaged rotors are shown in Figure 10.

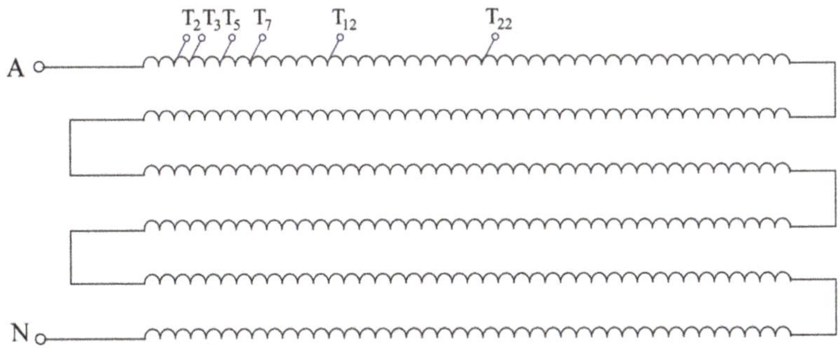

Figure 8. Windings of a Phase for the Stator Turn-to-Turn Short.

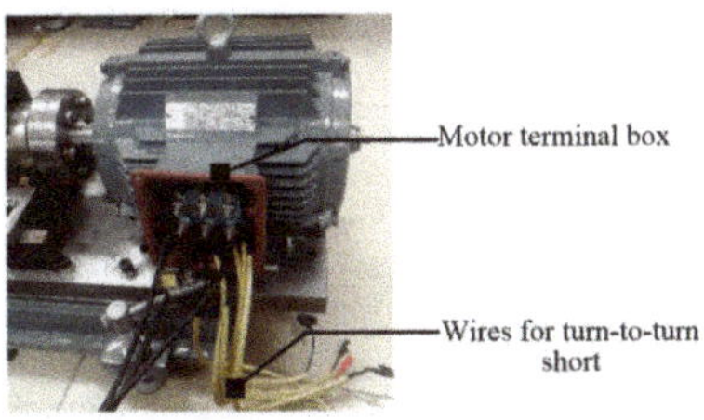

Figure 9. Photos of the Motor with a Stator Turn-to-Turn Short-Circuit Fault.

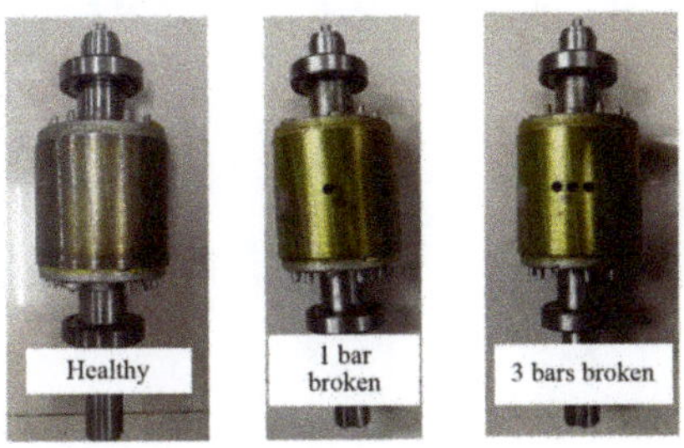

Figure 10. Photos of the Rotors with Holes.

4.2. Experimental Results

4.2.1. Stator Faults

To evaluate the performance of the proposed stator fault diagnosis method, the flux should be observed for further torque calculation. The three-phase currents were measured, and the full-order flux observer was performed to obtain the flux. The results are shown in Figure 11.

Then, the torque was calculated according to Equation (13), and the torque spectrum was analyzed. When 10 turns are short-circuited, the torque spectra of the healthy and fault stators are shown in Figure 12, which compares the magnitudes of the fault components ($2f_1$, $4f_1$, $8f_1$, and $10f_1$) under healthy and fault conditions. The magnitudes of these frequencies are normalized, as illustrated in Figure 12. It can be seen from Figure 12 that the measured ($2f_1$, $4f_1$, $8f_1$, and $10f_1$) components of the stator fault in the torque spectra have higher magnitudes than those of the healthy stator.

To further observe the torque spectrum, the fault frequencies ($2f_1$, $4f_1$, $8f_1$, and $10f_1$) in the torque spectrum were zoomed in and shown in Figure 13. By contrast, $2f_1$ and $4f_1$ components in the torque spectrum have higher magnitudes compared to the healthy case. Thus, those can be used as the fault indicator of the stator fault. In all, the results show that the fault components with the frequencies of $2f_1$ and $4f_1$ can provide an effective stator fault detection.

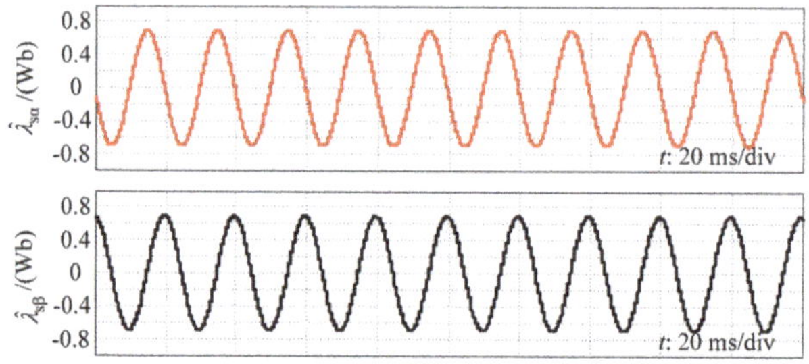

Figure 11. Observed Flux in the Experiments through the Measured Three-Phase Currents for Further Torque Calculation.

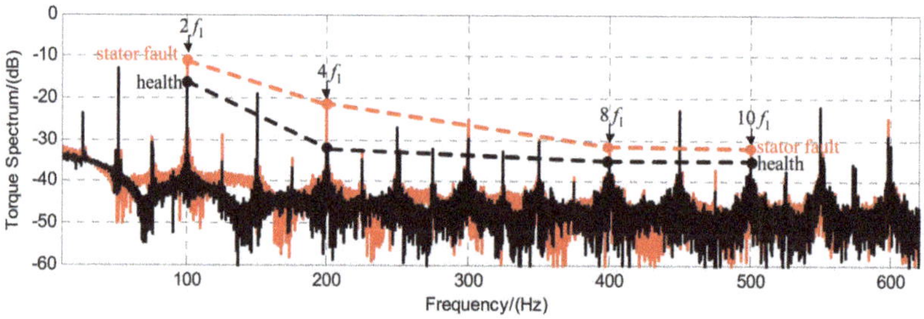

Figure 12. Torque Spectrum Showing from 0 to 600 Hz for the Healthy and Fault Stator.

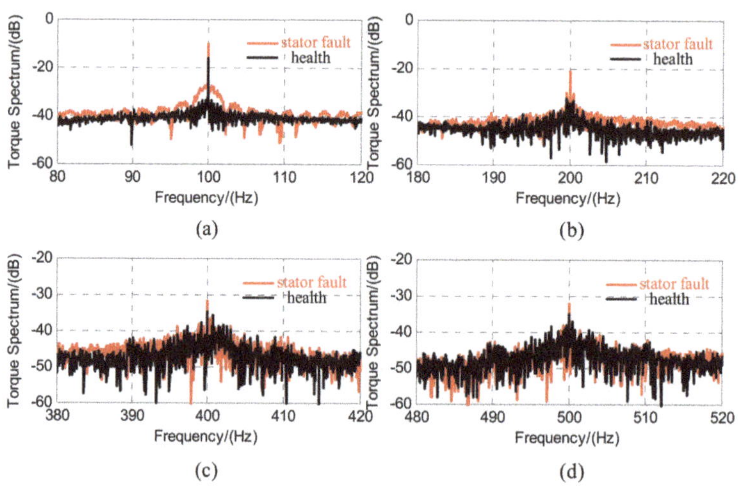

Figure 13. Zoom-in individual stator fault components: (**a**) Torque spectrum of $2f_1$, (**b**) torque spectrum of $4f_1$, (**c**) torque spectrum of $8f_1$, and (**d**) torque spectrum of $10f_1$.

4.2.2. Rotor Faults

To evaluate the performance of the rotor with and without broken bars, the currents were measured firstly. Then, current spectrums are obtained and normalized. The fundamental sidebands of the currents are shown in Figure 14. It can be observed from the extra components under broken bar fault conditions in Figure 14 that the magnitudes of $(1 \pm 2s)f_1$ and $(1 \pm 4s)f_1$ components increase with the higher number of the broken bars. The results clearly show that $(1 \pm 2s)f_1$ and $(1 \pm 4s)f_1$ signatures can not only identify the healthy and fault rotors, but also can be used as an indicator of the severity of the broken bar fault.

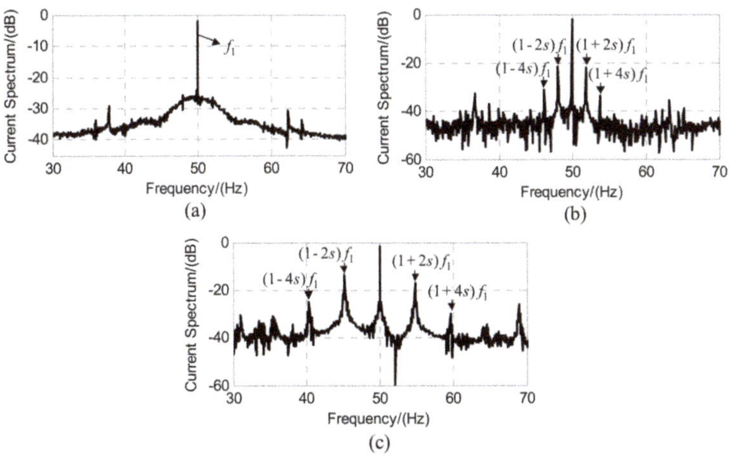

Figure 14. Fundamental Current Sidebands Without and With Broken Bars: (**a**) Healthy, (**b**) one broken bar, and (**c**) three broken bars.

The current sidebands of the fifth and seventh harmonics include the fault components induced by space and time harmonics when the motor has rotor faults. The current spectra from 220 to 380 Hz are shown in Figure 15, where the space and time fault harmonic characterized frequencies are marked. For the healthy rotor, the magnitudes of these components are significantly smaller than those under rotor faults. As observed in Figure 15b,c, the left and right sidebands of the fifth and seventh current harmonics are significant.

It can be seen from Figure 15b that additional $(5 \pm 2s)f_1$ components are produced, and then $(5 \pm 4s)f_1$ components are induced, where the magnitude of $(5 + 4s)f_1$ is small. This feature is not obvious in the spectrum. In theory, the $(5 \pm 4s)f_1$-components are induced by the $(5 \pm 2s)f_1$-components. Hence, the current magnitudes have a relationship of $|(5 \pm 4s)f_1| < |(5 \pm 2s)f_1|$. Nevertheless, it is found that the $(5 - 4s)f_1$-current has a larger magnitude than the $(5 - 2s)f_1$ component. This is because the $(5 - 4s)f_1$ frequency is not only produced by the fifth time harmonic but also by the fifth space fault harmonic. It is the superposition of the fifth time and the fifth space harmonics in the current under rotor faults. In addition, the space MMF also introduces the $(5 - 6s)f_1$ component, and then the $(5 - 8s)f_1$-component is induced. For the seventh harmonic current sidebands, the $(7 \pm 2s)f_1$-fault components of the time harmonics are observed. The $(7 - 2s)f_1$-component induces the $(7 - 4s)f_1$-component. However, it can be seen that the magnitude of $(7 - 4s)f_1$ is larger than the $(7 - 2s)f_1$-component, which is similar to the phenomenon for the $(5 - 4s)f_1$-component. Here, the $(7 - 4s)f_1$-component is produced by the $(7 - 2s)f_1$-component induced by the seventh time harmonic and the $(7 - 6s)f_1$-component induced by the seventh space harmonic under rotor faults. Furthermore, it also can be observed the $(7 - 8s)f_1$-component from the current spectra is in accordance with the above analysis. The results clearly show that the derived space and time fault components can be used for rotor monitoring.

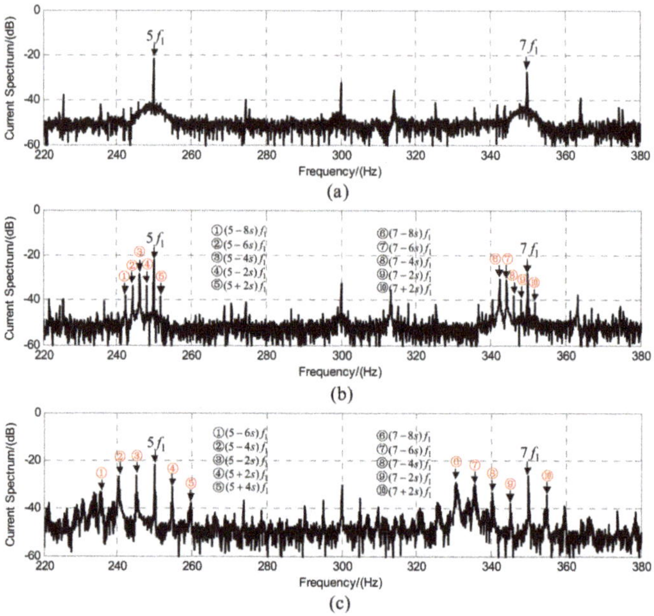

Figure 15. Fifth and Seventh Current Sidebands Without and With Broken Bars: (**a**) Healthy, (**b**) one broken bar, and (**c**) three broken bars.

Figure 15c shows that the additional fault components with the frequencies of $(5 \pm 2s)f_1$ and $(5 \pm 4s)f_1$ of the time harmonic current, and fault components of $(5 - 6s)f_1$ and $(5 - 4s)f_1$ of the space harmonic current are recognized. The generation of $(5 - 4s)f_1$-component is similar to the behavior of the case of one bar broken. In addition, the $(7 - 8s)f_1$, $(7 - 6s)f_1$, $(7 - 4s)f_1$, and $(7 \pm 2s)f_1$ fault components are observed in Figure 15c, which can provide the detection information of the rotor fault. When comparing the current spectrums described in Figure 15b,c, it can be observed that the magnitude of the individual fault component increases with a high number of broken bars. The results show that the derived sidebands have a significant increase in the current spectrum when a rotor fault is presented, which can provide an effective monitoring of the rotor fault.

5. Conclusions

Reliable fault diagnosis is important for high-performance inverter-fed IMs. This paper thus proposed novel fault diagnosis methods for IMs under stator and rotor faults, considering the harmonics in the currents. The diagnosis is achieved based on the fault characteristics. That is, for stator faults, the corresponding fault components are obtained in the torque spectrum; for rotor faults, the time and space harmonic sidebands are characterized in the currents. Experimental tests on an inverter-fed 2.2 kW/380 V/50 Hz IM have been provided. The test results have verified the effectiveness of the proposed methods and the analysis. In a word, the characterized fault components under the rotor and stator faults in this paper can be used for the reliable detection of stator and rotor faults for IMs.

Author Contributions: All authors contributed equally to this paper. All authors have read and agreed to the published version of the manuscript.

Funding: This research was funded by the Fundamental Research Funds for the Central Universities (2018JBZ004).

Acknowledgments: The author would like to thank the editor and the reviewers who provided many helpful comments and thereby contributed to the final manuscript.

Conflicts of Interest: The authors declare no conflict of interest.

References

1. Wu, W.; Wu, X.; Yin, J.; Jing, L.; Wang, S.; Li, J. Characteristic Analysis and Fault-Tolerant Control of Circulating Current for Modular Multilevel Converters under Sub-Module Faults. *Energies* **2017**, *10*, 1827. [CrossRef]
2. Li, W.; Li, G.; Zeng, R.; Ni, K.; Hu, Y.; Wen, H. The Fault Detection, Localization, and Tolerant Operation of Modular Multilevel Converters with an Insulated Gate Bipolar Transistor (IGBT) Open Circuit Fault. *Energies* **2018**, *11*, 837. [CrossRef]
3. Salomon, C.P.; Ferreira, C.; Sant'Ana, W.C.; Lambert-Torres, G.; Borges da Silva, L.E.; Bonaldi, E.L.; de Oliveira, L.E.L.; Torres, B.S. A Study of Fault Diagnosis Based on Electrical Signature Analysis for Synchronous Generators Predictive Maintenance in Bulk Electric Systems. *Energies* **2019**, *12*, 1506. [CrossRef]
4. Ullah, Z.; Hur, J. A Comprehensive Review of Winding Short Circuit Fault and Irreversible Demagnetization Fault Detection in PM Type Machines. *Energies* **2018**, *11*, 3309. [CrossRef]
5. Candelo-Zuluaga, C.; Riba, J.-R.; López-Torres, C.; Garcia, A. Detection of Inter-Turn Faults in Multi-Phase Ferrite-PM Assisted Synchronous Reluctance Machines. *Energies* **2019**, *12*, 2733. [CrossRef]
6. Gao, C.; Nie, Y.; Si, J.; Fu, Z.; Feng, H. Mode Recognition and Fault Positioning of Permanent Magnet Demagnetization for PMSM. *Energies* **2019**, *12*, 1644. [CrossRef]
7. Gao, H.; Zhang, W.; Wang, Y.; Chen, Z. Fault-Tolerant Control Strategy for 12-Phase Permanent Magnet Synchronous Motor. *Energies* **2019**, *12*, 3462. [CrossRef]
8. Gonçalves, P.; Cruz, S.; Mendes, A. Finite Control Set Model Predictive Control of Six-Phase Asymmetrical Machines—An Overview. *Energies* **2019**, *12*, 4693. [CrossRef]
9. Siddique, A.; Yadava, G.S.; Singh, B. A review of stator fault monitoring techniques of induction motors. *IEEE Trans. Energy Convers.* **2005**, *20*, 106–114. [CrossRef]
10. Motor Reliability Working Group. Report of large motor reliability survey of industrial and commercial installations, part I. *IEEE Trans. Ind. Appl.* **1985**, *IA-21*, 853–864. [CrossRef]
11. Benbouzid, M.E.H. A review of induction motors signature analysis as a medium for faults detection. In Proceedings of the 24th Annual Conference of the IEEE Industrial Electronics Society (Cat. No.98CH36200), Aachen, Germany, 31 August–4 September 1998; pp. 1950–1955. [CrossRef]
12. Sun, L.; Xu, B. An Improved Method for Discerning Broken Rotor Bar Fault and Load Oscillation in Induction Motors. *Energies* **2018**, *11*, 3130. [CrossRef]
13. Nemec, M.; Ambrožič, V.; Fišer, R.; Nedeljković, D.; Drobnič, K. Induction Motor Broken Rotor Bar Detection Based on Rotor Flux Angle Monitoring. *Energies* **2019**, *12*, 794. [CrossRef]
14. Riera-Guasp, M.; Antonino-Daviu, J.A.; Capolino, G. Advances in Electrical Machine, Power Electronic, and Drive Condition Monitoring and Fault Detection: State of the Art. *IEEE Trans. Ind. Electron.* **2015**, *62*, 1746–1759. [CrossRef]
15. Marques Cardoso, A.J.; Cru, S.M.A.Z.; Fonseca, D.S.B. Inter-turn stator winding fault diagnosis in three-phase induction motors, by Park's vector approach. *IEEE Trans. Energy Convers.* **1999**, *14*, 595–598. [CrossRef]
16. Benbouzid, M.E.H.; Kliman, G.B. What stator current processing-based technique to use for induction motor rotor faults diagnosis? *IEEE Trans. Energy Convers.* **2003**, *18*, 238–244. [CrossRef]
17. Cruz, S.M.; Cardoso, A.A.J.M. Stator winding fault diagnosis in three-phase synchronous and asynchronous motors, by the extended Park's vector approach. *IEEE Trans. Ind. Appl.* **2001**, *37*, 1227–1233. [CrossRef]
18. Eftekhari, M.; Moallem, M.; Sadri, S.; Hsieh, M. Online Detection of Induction Motor's Stator Winding Short-Circuit Faults. *IEEE Syst. J.* **2014**, *8*, 1272–1282. [CrossRef]
19. Cheng, S.; Zhang, P.; Habetler, T.G. An Impedance Identification Approach to Sensitive Detection and Location of Stator Turn-to-Turn Faults in a Closed-Loop Multiple-Motor Drive. *IEEE Trans. Ind. Electron.* **2011**, *58*, 1545–1554. [CrossRef]
20. Puche-Panadero, R.; Pineda-Sanchez, M.; Riera-Guasp, M.; Roger-Folch, J.; Hurtado-Perez, E.; Perez-Cruz, J. Improved Resolution of the MCSA Method Via Hilbert Transform, Enabling the Diagnosis of Rotor Asymmetries at Very Low Slip. *IEEE Trans. Energy Convers.* **2009**, *24*, 52–59. [CrossRef]

21. Pezzani, C.; Donolo, P.; Bossio, G.; Donolo, M.; Guzmán, A.; Zocholl, S.E. Detecting Broken Rotor Bars With Zero-Setting Protection. *IEEE Trans. Ind. Appl.* **2014**, *50*, 1373–1384. [CrossRef]
22. Douglas, H.; Pillay, P.; Ziarani, A.K. Broken Rotor Bar Detection in Induction Machines With Transient Operating Speeds. *IEEE Trans. Energy Convers.* **2005**, *20*, 135–141. [CrossRef]
23. Xu, B.; Sun, L.; Xu, L.; Xu, G. An ESPRIT-SAA-Based Detection Method for Broken Rotor Bar Fault in Induction Motors. *IEEE Trans. Energy Convers.* **2012**, *27*, 654–660. [CrossRef]
24. Xu, B.; Sun, L.; Xu, L.; Xu, G. Improvement of the Hilbert Method via ESPRIT for Detecting Rotor Fault in Induction Motors at Low Slip. *IEEE Trans. Energy Convers.* **2012**, *27*, 654–660. [CrossRef]
25. Benbouzid, M.E.H.; Vieira, M.; Theys, C. Induction motors faults detection and localization using stator current advanced signal processing techniques. *IEEE Trans. Power Electron.* **1999**, *14*, 14–22. [CrossRef]
26. Kia, S.H.; Henao, H.; Capolino, G. A High-Resolution Frequency Estimation Method for Three-Phase Induction Machine Fault Detection. *IEEE Trans. Ind. Electron.* **2007**, *54*, 2305–2314. [CrossRef]
27. Jerkan, D.G.; Reljić, D.D.; Marčetić, D.P. Broken Rotor Bar Fault Detection of IM Based on the Counter-Current Braking Method. *IEEE Trans. Energy Convers.* **2017**, *32*, 1356–1366. [CrossRef]
28. Antonino-Daviu, J.A.; Quijano-López, A.; Rubbiolo, M.; Climente-Alarcon, V. Advanced analysis of motor currents for the diagnosis of the rotor condition in electric motors operating in mining facilities. *IEEE Trans. Ind. Appl.* **2018**, *54*, 3934–3942. [CrossRef]
29. CusidÓCusido, J.; Romeral, L.; Ortega, J.A.; Rosero, J.A.; GarcÍaGarcia Espinosa, A. Fault Detection in Induction Machines Using Power Spectral Density in Wavelet Decomposition. *IEEE Trans. Ind. Electron.* **2008**, *55*, 633–643. [CrossRef]
30. Pons-Llinares, J.; Antonino-Daviu, J.A.; Riera-Guasp, M.; Bin Lee, S.; Kang, T.; Yang, C. Advanced Induction Motor Rotor Fault Diagnosis Via Continuous and Discrete Time–Frequency Tools. *IEEE Trans. Ind. Electron.* **2015**, *62*, 1791–1802. [CrossRef]
31. Antonino-Daviu, J.A.; Pons-Llinares, J.; Lee, S.B. Advanced Rotor Fault Diagnosis for Medium-Voltage Induction Motors Via Continuous Transforms. *IEEE Trans. Ind. Appl.* **2016**, *52*, 4503–4509. [CrossRef]
32. Climente-Alarcon, V.; Antonino-Daviu, J.A.; Riera-Guasp, M.; Vlcek, M. Induction Motor Diagnosis by Advanced Notch FIR Filters and the Wigner–Ville Distribution. *IEEE Trans. Ind. Electron.* **2014**, *61*, 4217–4227. [CrossRef]
33. Chow, M.-Y.; Mangum, P.M.; Yee, S.O. A neural network approach to real-time condition monitoring of induction motors. *IEEE Trans. Ind. Electron.* **1991**, *38*, 448–453. [CrossRef]
34. Ghate, V.N.; Dudul, S.V. Cascade Neural-Network-Based Fault Classifier for Three-Phase Induction Motor. *IEEE Trans. Ind. Electron.* **2011**, *58*, 1555–1563. [CrossRef]
35. Sadeghian, A.; Ye, Z.; Wu, B. Online Detection of Broken Rotor Bars in Induction Motors by Wavelet Packet Decomposition and Artificial Neural Networks. *IEEE Trans. Instrum. Meas.* **2009**, *58*, 2253–2263. [CrossRef]
36. Seera, M.; Lim, C.P.; Ishak, D.; Singh, H. Fault Detection and Diagnosis of Induction Motors Using Motor Current Signature Analysis and a Hybrid FMM–CART Model. *IEEE Trans. Neural Netw. Learn. Syst.* **2012**, *23*, 97–108. [CrossRef]
37. Yagami, Y.; Araki, C.; Mizuno, Y.; Nakamura, H. Turn-to-turn insulation failure diagnosis of stator winding of low voltage induction motor with the aid of support vector machine. *IEEE Trans. Dielectr. Electr. Insul.* **2015**, *22*, 3099–3106. [CrossRef]
38. Keskes, H.; Braham, A. Recursive Undecimated Wavelet Packet Transform and DAG SVM for Induction Motor Diagnosis. *IEEE Trans. Ind. Informat.* **2015**, *11*, 1059–1066. [CrossRef]
39. Godoy, W.F.; da Silva, I.N.; Goedtel, A.; Cunha Palácios, R.H.; Lopes, T.D. Application of intelligent tools to detect and classify broken rotor bars in three-phase induction motors fed by an inverter. *IET Electr. Power App.* **2016**, *10*, 430–439. [CrossRef]
40. Martin-Diaz, I.; Morinigo-Sotelo, D.; Duque-Perez, O.; Romero-Troncoso, R.J. An Experimental Comparative Evaluation of Machine Learning Techniques for Motor Fault Diagnosis Under Various Operating Conditions. *IEEE Trans. Ind. Appl.* **2018**, *54*, 2215–2224. [CrossRef]
41. Bachir, S.; Tnani, S.; Trigeassou, J.-C.; Champenois, G. Diagnosis by parameter estimation of stator and rotor faults occuring in induction machines. *IEEE Trans. Ind. Electron.* **2006**, *53*, 963–973. [CrossRef]

42. Bazine, I.B.A.; Tnani, S.; Poinot, T.; Champenois, G.; Jelassi, K. On-line detection of stator and rotor faults occurring in induction machine diagnosis by parameters estimation. In Proceedings of the 8th IEEE Symposium on Diagnostics for Electrical Machines, Power Electronics & Drives, Bologna, Italy, 5–8 September 2011; pp. 105–112. [CrossRef]
43. Tallam, R.M.; Habetler, T.G.; Harley, R.G. Transient model for induction machines with stator winding turn faults. *IEEE Trans. Ind. Appl.* **2002**, *38*, 632–637. [CrossRef]
44. Berzoy, A.; Mohammed, O.A.; Restrepo, J. Analysis of the Impact of Stator Interturn Short-Circuit Faults on Induction Machines Driven by Direct Torque Control. *IEEE Trans. Energy Convers.* **2018**, *33*, 1463–1474. [CrossRef]
45. Kim, J.; Shin, S.; Lee, S.B.; Gyftakis, K.N.; Drif, M.; Cardoso, A.J.M. Power Spectrum-Based Detection of Induction Motor Rotor Faults for Immunity to False Alarms. *IEEE Trans. Energy Convers.* **2015**, *30*, 1123–1132. [CrossRef]
46. Kim, H.; Lee, S.B.; Park, S.; Kia, S.H.; Capolino, G. Reliable Detection of Rotor Faults Under the Influence of Low-Frequency Load Torque Oscillations for Applications With Speed Reduction Couplings. *IEEE Trans. Ind. Appl.* **2016**, *52*, 1460–1468. [CrossRef]

energies

MDPI

Review

High Technology Readiness Level Techniques for Brushless Direct Current Motors Failures Detection: A Systematic Review

Vito Mario Fico [1,*], María Ángeles Martín Prats [2] and Carmelina Ierardi [3]

1 Skylife Engineering, 41092 Seville, Spain
2 Escuela Técnica Superior de Ingeniería, Electronics Engineering Department, Universidad de Sevilla, 41092 Seville, Spain; mmprats@us.es
3 Engineering Department, Universidad Loyola Andalucía, 41704 Seville, Spain; cierardi@uloyola.es
* Correspondence: vito.fico@skylife-eng.com

Received: 3 February 2020; Accepted: 18 March 2020; Published: 1 April 2020

Abstract: Many papers related to this topic can be found in the bibliography; however, just a modest percentage of the introduced techniques are developed to a Technology Readiness Level (TRL) sufficiently high to be implementable in industrial applications. This paper is focused precisely on the review of this specific topic. The investigation on the state of the art has been carried out as a systematic review, a very rigorous and reliable standardised scientific methodology, and tries to collect the articles which are closer to a possible implementation. This selection has been carefully done with the definition of a series of rules, drawn to represent the adequate level of readiness of fault detection techniques which the various articles propose.

Keywords: failure; PMSM; detection; diagnosis; BLDC; brushless; systematic review

---

1. Introduction

The topic of the present review is subjected to a growing interest, both from the academic and the industrial worlds, due to the parallel increase of the usage of electric machines for high reliability tasks as motoring of electric vehicles and actuation of flight surface for the future More Electric Aircrafts.

Above all, when aerospace applications are involved, reliability becomes of vital importance; indeed, the performance of flight actuators on a damaged aircraft is not as important as ensuring that the remaining actuators continue operation until the aircraft can land safely. In most cases, an adequate level of reliability can be reached only by using diagnostic tools [1].

The availability of an accurate and efficient mean of condition monitoring and machine fault diagnosis can be of paramount importance, as it improves the reliability and stability of the plant and at the same time it reduces costs, ideally leading to a system without programmed maintenance. Statistical studies [2] show that expected reliability can be improved up to 5–6 percentage points with the use of monitoring.

In this context arises the need to precisely know the evolution and the current state of the literature about the fault detection and diagnosis techniques for Brushless DC (BLDC) and in particular which techniques are closer to a possible implementation, i.e., which techniques have the highest Technology Readiness Level (TRL).

In the area of engineering, and especially for aerospace engineering, the narrative revision is usually preferred. In this kind of review, the authors decide which papers include in the survey, based on their wide knowledge and experience and offering a personal point of view and interpretation of the chosen theme.

The SR is a rigorous standardised scientific methodology, used to produce reliable literature reviews, mainly recognised by its objectivity. It is employed with exceptional results in many areas, including bio-science [3,4], computer science [5] and in recent years particularly in software engineering [6,7]. In particular, these last papers and the work proposed by [8] have been used as guides to undertake the systematic review in this work.

In fact, the first objective of this work is precisely to adapt the guidelines mentioned above to our field, while the second objective is to apply the systematic review to a specific topic: high TRL techniques for BLDC motors failures detection.

To be more precise, the authors are interested in those detection techniques that are not restricted to a particular machine or with special set-ups, configurations, loads or motor manoeuvres. Furthermore the techniques shall have been tested at various operation point and the algorithm shall have tested with success for at least one of some cases later described in the inclusion criteria. Concluding, in order to be accepted in the systematic review, the paper shall demonstrate that the proposed algorithm is capable to discern between healthy and faulty motor. These constitute, mainly, the inclusion criteria for the studies appearing in the review.

The fundamental scope of this review is indeed to detect which techniques are presently being profitably used for motor fault detection and diagnosis and to provide the industry with some high readiness level and tested techniques. In this prospect, most of the inclusion and exclusion criteria have been defined to focus the investigation on those techniques with demonstrated fault detection performances at various operation points and easily automatable or already automated.

An additional question considered in this SR has been the possibility to embed in the motor body, the hardware needed for fault detection. Although most articles are focused on the detection by utilising commonly measured variables (mainly speed, current, voltage) some authors have elaborated fault detection techniques based on the analysis of images from external cameras or sensitive accelerometers. Those techniques are appropriate to be implemented only in particular applications [1] and have been discarded from the scope of the present work.

After the screening of more than 3000 possible papers, only 44 primary studies have been found to satisfy the aforementioned criteria. The authors have carefully revised those papers and have collected the following data: the type of fault detected, the technique that was used for the detection and the sensors used, the inclusion of experiments or simulations, if the technique has been tested at different operation points (diverse speed or loads or in the best case a combination of both), the working condition (stationary or not) and some other limitations/advantages.

This information is then gathered in a feature table, which is an useful strategy to get a complete, rigorous and objective view of the chosen topic. At the end of the process, the research questions initially formulated are answered, providing a full perspective of the topic [8].

The SR consists of three sequential phases, each of which is subdivided in turn into sub-phases, as detailed below [7]:

1. Planning the review
    - identification of the need
    - research questions
    - review protocol
    - evaluating protocol
2. Conducting the review
    - selection of primary studies
    - study quality assessment
    - extraction and synthesis of data
3. Reporting the review
    - specifying dissemination mechanisms
    - formatting the main report

- evaluating the report

The rest of the paper is organised as follows.. Section 2 presents the description and the adaptation to our case of planning phase of the SR. Conducting and reporting of the systematic review are given in Sections 3 and 4 respectively. Finally, the conclusions are drawn in Section 5. Additionally, in Appendix A is given a definition of the main terminology used along the document.

2. Planning

The first step of the systematic review consists in planning, which is the foundation of the entire revision. It is at this stage that the main tools are developed, such as the Boolean function, the inclusion and exclusion criteria, the choice of the different databases in which to carry out the research and above all the development and evaluation of a protocol that regulates all the phases.

The need to undertake a systematic review, arises first of all because the research topic is very wide and a rigorous method was needed to correctly extract the needed information. As said, in the engineering field this type of methodology is not usual because, even by being scientific and rigorous, it is difficult and complicated to carry out. Currently there is no systematic review on the fault detection techniques for brushless DC motors, and in reality there is not even a traditional revision so detailed on the chosen theme (Should be clarified that this sentence is referred to the reviews taking into account the readiness level of the technique.).

2.1. Research Questions

Once the concrete topic has been identified, there are some criteria that help to clearly formulate research questions. Among the most used criteria in other sectors there are the criteria called PICOC (Population, Intervention, Comparison, Outcome, Context). In this work we have considered those presented in [8] and adapted to our case.

In this case, just some of these criteria have been used to formulate and process the questions that this SR is trying to answer. In the final section of the review, defined as reporting, there is a sub-section called Discussion (Section 4.1) where the relative answers are discussed and analysed.

The questions formulated for the present work are listed below:

RQ.1: Which are the most common faults of BLDC motors?
RQ.2: Which parameters are used for fault detection in BLDC motors?
RQ.3: Which type of failure can be detected by each technique?
RQ.4: Which technique requires less computational power?
RQ.5: Which technique requires less sensors?
RQ.6: Which technique gives the best results for each type of failure?

2.2. Review Protocol

The revision protocol is nothing more than a set of rules and criteria to be followed during all the stages, in order to reduce the bias and make the SR as objective as possible. In the bioscience field, the protocol is sometimes recorded in a prospective register, such as PROSPERO (https://www.crd.york.ac.uk/prospero/). Unfortunately, these type of registers do not exist in the aerospace field.

A very important aspect to be considered for the SR is the clarity wherewith the protocol is exposed and elaborated, as at least two persons are involved in the review drafting. A common, but very time consuming, approach consists in the implementation of the SR by two independent persons, who carry out the part of the conducting and reporting separately and then compare and discuss the obtained results. Another method, that is the one used in this work, is that a person performs all the phases individually and a second person randomly checks some data, as for example, some of the rows of the features tables (Tables 1–5).

Table 1. Features table (a).

Cite	Year	Fault Type	Technique Used	Sensors Used	Experiments or Simulations	Various Speed/Loads	Working Condition	Limitations/ Advantages
[9]	2019	Armature faults	Parameters Estimation	Voltage, Current and Position sensors	Both	Both	Stationary conditions	It propose indicators deduced from symmetrical component of phase currents in the reference frame. The method has been validate at various speed, loads and short circuit resistance magnitude for ITSC and at constant speed, load, resistance magnitude for PPSC. The algorithm computational load is nos specified.
[10]	2019	Permanent Magnetic faults	Model, AI and neural-network-based techniques	Voltage and Current sensors	Both	Various Speed	Stationary conditions	Experimentally tested with 5 motor conditions (1 healthy, 4 faulty) with good detection performances. Proposes two failure extraction methods and compares them. Training time and computational load not specified.
[11]	2019	Mechanical faults	MCSA	Voltage and Current sensors	Both	Both	Stationary conditions	Uses wavelet decomposition of the current signal and an adaptive filter to estimate and remove the fundamental component. Tested using two case studies, i.e., broken magnet and eccentricity fault, automatic fault classification with SVM and average accuracy of 96%. Training time and computational load not specified.
[12]	2019	Permanent Magnetic faults	Other (Hall Effect Sensors flux analysis)	Hall Effect Sensors	Experiments	Various Speed	Non-Stationary conditions	Method capable of detecting bearing and permanent magnets faults by analysing respectively the cascade DWT-CWT transform of the speed signal and the kurtosis index of the duty cycle signal of the hall sensor output. Electrically independent from the motor. Influence of load not specified. Computational load not specified.
[1]	2019	Permanent Magnetic faults	Other (Signals Similarity Analysis)	Voltage and Current sensors	Both	Various Speed	Stationary Conditions	The method has been tested with FEM simulations and experimentally with good results. The test effect of torque on the method has not benn evaluated. The method can be used only for multipole motors.
[13]	2019	Armature faults	Electromagnetic field monitoring	Tunneling Magnetoresistive sensors	Both	Both	Stationary Conditions	The method is capable of detecting both location and severity of inter-turn short-circuit by sensing the stray magnetic field outside the stator yoke. It needs the installation of TM sensors around the motor body. Computational load not specified.
[14]	2018	Mechanical faults	Other (Angular Resample)	Voltage and Current sensors	Experiments	Both	Non-Stationary conditions	Method based on the angular resample of speed obtained with a sensorless observer. Computational load not specified. Torque variation not specified.
[15]	2018	Permanent Magnetic faults	Other (Voltage Angle)	Voltage and Current sensors	Both	Various Speed	Stationary Conditions	The method takes advantage from the variations of the voltage angle observed during demagnetisation and inter-turns short faults to identify their presence. The method is temperature-dependant. A clear detection threshold is not defined. Computational load not specified.

Table 2. Features table (b).

Cite	Year	Fault Type	Technique Used	Sensors used	Experiments or Simulations	Various Speed/Loads	Working Condition	Limitations/ Advantages
[16]	2018	Armature faults	Parameters Estimation	Current and Voltage sensors	Both	Both	Stationary condition	The method is affected by the magnitude of the stator current, should be used in constant torque conditions.
[17]	2017	Mechanical faults	Electromagnetic field monitoring, search coils, coils wound around motor shafts	Search coil	Simulations	Various Speeds	Stationary Conditions	The method is independent from motor variables, but needs the search coil to be installed on the stator.
[18]	2017	Armature faults	Model, AI, and neural-network -based techniques, Parameters Estimation	Current, Voltage and Speed sensors	Experiments	Both	Stationary Conditions	This method has been tested experimentally on an aeronautical motor, but the experiment set-up has not been presented. The algorithm is very fast ($\approx$ 15 msec), but the computation time for the features is not taken into account. Also it needs a large database for training the algorithm.
[19]	2017	Permanent Magnetic and Mechanical faults	Model, AI, and neural -network-based techniques	Current sensor	Experiments	Various Loads	Stationary Conditions	Two failures introduced on an experimental platform and a 10-fold validation of the algorithm is executed. The algorithm is fast (30 msec), but the training time is not specified.
[20]	2017	Armature faults	Parameters Estimation	Current and Voltage sensors	Both	Both	Non-Stationary Conditions	The proposed method is just slightly dependent from the load and speed. The authors also demonstrated robustness against parameters variation and inaccuracies by introducing a threshold, but the allowed tolerance is not specified and this quantity can also depend on the motor.
[21]	2017	Armature faults	MCSA	Current and Voltage sensors	Both	Both	Stationary Conditions	The proposed method is particularised for intermittent faults. The test set-up is not presented. Tests at various loads and speeds have been executed, but their influence on the method is not specified.
[22]	2017	Mechanical faults	Other (Hall Effect Sensors flux analysis)	Analogue or Digital Hall sensors	Both	Various Loads	Stationary Conditions	The method is independent from speed and demonstrates only a slight dependence from loads. The best accuracy is obtained with analogue Hall effect sensor which are not common, even if the authors provide an alternative based on digital Hall effect sensors. This method can be used only if the Hall sensors are placed in the radial direction.
[23]	2017	Permanent Magnetic faults	Other (Torque Ripple Analysis)	Torque sensor	Both	Various Loads	Stationary Conditions	No information is given about how the motor speed affects the proposed method. A torque transducer is needed to apply the algorithm.

Table 3. Features table (c).

Cite	Year	Fault Type	Technique Used	Sensors Used	Experiment or Simulations	Various Speed/Loads	Working Condition	Limitations/ Advantages
[24]	2016	Armature faults	Parameters estimation	Current, Voltage and Speed sensor	Both	Various Speeds	Non-Stationary Conditions	The method is based on a modified motor model. It is based on computations in the rotating frame (dq). No information about how the load affects the detection method is given.
[25]	2016	Armature faults	Parameters Estimation	Current and Speed sensors	Experiments	Various speeds	Stationary Conditions	The proposed fault index has a very reduced dependence from motor speed.
[26]	2016	Mechanical faults	MCSA, Model, AI and NN-based techniques	Current Sensor	Experiments	Various Speeds	Stationary Conditions	The implementation is very close to a real scenario, but in some studied conditions the fail rate of the classifier is relatively high.
[27]	2016	Permanent Magnetic Faults	Parameters Estimation	Current, Speed and Angle Sensors	Both	Both	Non-Stationary Conditions	The method needs the knowledge of various motor parameters and their variation (or incorrectness) can result in poor diagnosis performances. The authors demonstrated good performances with various demagnetisation levels and working conditions.
[28]	2016	Armature faults	Other (PWM Ripple Current Measurements)	Current and Voltage sensors	Both	Both	Stationary Conditions	The method needs an electric model of the motor valid for high frequencies. The author demonstrated good sensitivity also at low speed.
[29]	2015	Mechanical faults	Noise and Vibration Monitoring, Model, AI, NN based techniques, MCSA	Current sensor and Accelerometer	Both	Both	Non-Stationary Conditions	The method is capable to detect and distinguish different bearing failures. Need an accelerometer to be placed close to the bearing.
[30]	2015	Mechanical faults	Model, AI, NN-based techniques	Current sensor	Experiments	Both	Stationary Conditions	The authors performed an extensive experiment campaign with good results. From the images in the article, the damages reproduced on the bearing appear to be considerable.
[31]	2015	Armature faults	Model, AI and NN-based techniques, MCSA	Current sensor	Experiments	Both	Stationary Conditions	High detection ratio. The frequency analysis is did with the FFT, this means that during the 12 s of signal acquisition the motor speed and load shall be constant.
[32]	2015	Armature faults	Electromagnetic field monitoring, Search Coils, Coils wound around motor shaft	Search coil	Both	Various Speeds	Stationary Conditions	The presented method is invasive for an already built motor. The detection time is very short (3–5) ms but it seems be dependent on motor speed; furthermore the hardware used for computations is not presented.
[33]	2015	Armature faults	Model, AI and NN-based techniques, MCSA	Current sensors	Both	Both	Stationary Conditions	The motor used for experiments has an inherent anomaly, but the algorithm is capable to discern it from the short-circuit. The presented method does not imply any previous knowledge on the motor. They uses FFT, it implies that during the signal acquisition the conditions need to be stationary.

Table 4. Features table (d).

Cite	Year	Fault Type	Technique Used	Sensors Used	Experiment or Simulations	Various Speed/Loads	Working Condition	Limitations/ Advantages
[34]	2013	Permanent Magnets faults	Model, AI and NN-based techniques, Parameters Estimation	Current, Voltage and Speed sensors	Experiments	Various loads	Stationary Conditions	Proposes a method for demagnetisation. Comparison of the proposed method with various other established methods.
[35]	2013	Armature, Permanent Magnets and Mechanical Faults	Model, AI, NN-based techniques, MCSA	Current sensors	Simulations	Both	Stationary Conditions	The paper presents a good variety of motors, faults and working conditions. The detection accuracy obtained is very high, but it can be due to the use of clean signals from simulations.
[36]	2013	Armature faults	Model, AI and NN-based techniques	Current and Voltage Sensors	Both	Both	Stationary Conditions	The training of the AI has been executed with data from both experiments and simulations. The method uses current measurements in time domain with no need of frequency domain transformation. The method is capable of detecting fault severity and location.
[37]	2013	Armature faults	Parameters estimation	Current, Voltage and Speed sensors	Both	Various loads	Stationary Conditions	The method compares an estimated back-EMF with a reference one for fault detection. The reference is obtained from a FEM model or from an healthy machine. This method can be very sensible to motor parameters change. The influence of the load is not discussed.
[38]	2013	Armature faults	MCSA	Current, Voltage and Speed sensors	Both	Both	Non-Stationary Conditions	The proposed method has a low computational burden, but needs access to the motor neutral point to be applied.
[39]	2011	Mechanical faults	Model, AI and NN-based techniques	Current, Speed and Torque sensors,	Simulations	Both	Non-Stationary Conditions	The method tries to detect mechanical faults by estimating the bearing health status. A torque sensor is used, which is not usually mounted in motors and the validation is carried out by simulation without added noise.
[40]	2011	Armature faults	Parameters estimation	Current, Voltage and Speed sensors	Both	Both	Non-Stationary Conditions	The back-EMF is estimated when the machine is healthy and then frozen, which causes dependence on motor parameters changes. There is a model for compensate the inverter losses compensation.
[41]	2011	Armature faults	Other (High Frequency Injection)	Current and Angular Position sensors	Both	Various loads	Stationary Conditions	The method has a very good resolution, but the detection is based on a look-up table. This makes the algorithm ignore all the failures (if any) present before the table creation.
[42]	2011	Armature faults	MCSA	Current sensors	Experiments	Both	Stationary Conditions	Capable of detecting two failures. Use the FFT for the frequency analysis, but detect periods of stationarity of the motor. Use linear interpolation to define the healthy comparison term.

Table 5. Features table (e).

Cite	Year	Fault Type	Technique Used	Sensors Used	Experiment or Simulations	Various Speed/Loads	Working Condition	Limitations/ Advantages
[43]	2011	Armature and Mechanical faults	Model, AI NN-based techniques	Current and Voltage senors	Experiments	Various Loads	Stationary Conditions	Perform an interesting multi-class classification based on seven parameters. The algorithm does not seem capable to classify failures not present the training set.
[44]	2010	Mechanical faults	Model, AI, NN-based techniques, MCSA	Current sensors	Both	Various loads (Load independence demonstrated analytically)	Stationary Conditions	Study on the impact of SNR.
[45]	2010	Armature faults	Model, AI, NN-based techniques, MCSA	Current sensors	Both	Various loads (Load independence demonstrated analytically)	Stationary Conditions	Study on the impact of SNR. Two failures studied, with fault severity estimation.
[46]	2008	Armature faults	Model, AI and NN-based techniques	Current, Voltage and Speed sensors	Experiments	Various Loads	Non-Stationary Conditions	Training of a neural-network to predict current and include initial asymmetries. The predicted value of the current is used as a reference to detect failures under load variations.
[47]	2007	Mechanical faults	MCSA	Current and Voltage sensors	Both	Both	Non-Stationary Conditions	The experimental set-up is not described. There is a comparison between three technique for time-frequency analysis and relative fault detection.
[48]	2007	Mechanical faults	MCSA	Current and Voltage sensors	Both	Both	Non-Stationary Conditions	The experimental set-up is not described.
[49]	2007	Mechanical faults	MCSA	Current and Voltage sensors	Both	Various Speeds	Non-Stationary Conditions	The influence of load is not taken into account. The experimental set-up is not described.
[50]	2006	Armature, Permanent Magnets and Mechanical faults	MCSA	Current and Speed sensors	Experiments	Both	Non-Stationary Conditions	Method for tracking the fault frequencies during variable speed operations. The test set-up is not described. Various faults have been implemented.
[51]	2006	Mechanical faults	MCSA	Current and Speed sensors	Both	Both	Non-Stationary Conditions	Two methods presented based on different frequency tracking algorithms. Real-time implementation with processor execution time is also included.

One of the basic steps of the protocol is the creation of a Boolean function that comprehensively includes all the terms related to the chosen theme, including all the synonyms and terms that may be related to the words of interest for the topic. To carry out this research based on keywords, it is appropriate to deeply read about the theme to detect which words are most frequently used by the authors.

The articulate Boolean function created for this work is as follows:

```
((("brushless DC" OR "permanent magnet electrical") AND (motor OR machine)) OR BLDC OR PMSM)
AND
(((condition OR health) AND monitoring) OR ((diagnosis OR detection) AND (fault OR failure)))
```

The first part of the Boolean function defines the type of motor, while the second one defines the detection of the defect.

A difficulty encountered during the research is that the different bibliographic databases are not prepared for this kind of revision, as they do not allow certain researches or to search in certain fields of the papers. Indeed, the Boolean function based research was carried out in the title, abstract and keywords of the papers.

Due to the research restrictions of the databases, as specified in [52], and thanks to the good coverage of the editorials obtained shown in Table 6, the following databases have been used:

- IEEE Xplore Digital Library
- Scopus
- ACM Digital Library
- Science Direct
- Web of Science

Table 6. Databases coverage with respect to the content of the publishers: IE = IEEE, IT = IET, PE = Pegamon-Elsevier, ES = Elsevier Science, WB = Wiley Blackwell, TF = Taylor & Francis, SP = Springer, SI = SIAM Publications, OX = Oxford University Press, KO = Korean Inst. Electrical Eng., SA = Sage Publications, AS = ASME, MP = Microtome Publications [52].

	IE	IT	PE	ES	WB	TF	SP	SI	OX	KO	SA	AS	MP
IEEEX	✓	✓			✓								
ACM	✓		✓	✓		✓	✓	✓	✓		✓		✓
Scopus	✓		✓	✓	✓	✓	✓	✓	✓	✓	✓	✓	
WoS	✓		✓	✓	✓	✓	✓	✓	✓		✓		
SD			✓	✓									

Once the research questions have been identified and the relative Boolean function created, it must be introduced in the different bibliographic databases, adapting it according to the search language of each database. In this work, the research has been carried out by searching only in the abstract, title and keywords of the papers, obtaining a total of 3167 items until of November 2019, as detailed in Table 7.

Table 7. Studies obtained by the chosen database.

Database	Studies
IEEE Xplore	842
Web of Science	590
Scopus	697
ScienceDirect	600
ACM Digital Library	17
	3167

A further, fundamental step of the Planning phase is the formulation of the inclusion and exclusion criteria. This criteria are very important, as they are used as objective rules for the selection of the studies that can become part of the review.

For a paper to become part of the review it must respect all the inclusion criteria, presented in Table 8 and must not contain any of the exclusion criteria presented in Table 9.

Table 8. Inclusion Criteria [1].

Num.	Description
1	The technique should not be restricted to a particular machine (number of phases, etc)
2	Shall propose at least one detection technique (a parameter or an index that clearly and uniquely identifies the failure or an automatic detection algorithm)
3	The technique shall have been tested at various operation points (different speed or loads or a combination of both)
4	The technique shall not need special set-ups, configurations, loads or motor manoeuvres
5	Characteristics of the detection and diagnosis algorithm: a The paper shall demonstrate that the algorithm is capable to discern between healthy and faulty. b The algorithm shall have been tested with success for at least one of the following cases: • Different levels of the same failure, demonstrating coherence (simulation or test on real motor) • On different real motors, demonstrating coherence • Capability do discern between different failures • On simulation and then real motor, demonstrating coherence

Table 9. Exclusion Criteria [1].

Num.	Description
1	Grey literature and secondary studies
2	Non English written papers
3	Duplicated studies
4	Full paper not available
5	Lack of tests or simulations
6	Use of big sensors, not embeddable in the motor
7	Not focused on the selected topic

The inclusion criteria are all related to the research theme itself, i.e., they explain exactly what is expected to be found in an article in order to be accepted. Among the exclusion criteria, however, there are some points that are proper to the systematic review, such as the criterion that excludes all secondary studies or grey literature (books, book chapters, PhD thesis, reviews, etc.). In this way it is ensured that the selected articles have already passed through a peer-review process.

In addition to all these tools, needed to reduce the 3167 papers coming from the Boolean function search (see Table 7), it is needed to define some guidelines on how to carry out the selection of the studies and finally how to extract the data considered important from every paper.

The authors decided to extrapolate the following features from each article:

- Data of the article (e.g., author, year of publication, etc.)
- The type of failure
- The type of technique used
- The type of sensors used
- If it presents experiments or simulations
- If the algorithm has been tested at different speeds or loads
- If the algorithm has been tested in dynamic or stationary conditions

3. Conducting

The second phase of the SR is the conducting and is mainly divided into three parts: the selection of primary studies, a study quality assessment and finally the extraction and synthesis of data for each paper.

The main objectives of this phase are the following:

- to significantly reduce the large number of studies that have been obtained, by using the inclusion and exclusion criteria,
- to extract a features table in which the main characteristics of each article are highlighted.

3.1. Selection of Primary Studies

To guide the selection process of primary studies, it is very common to use the PRISMA [53] (Preferred Reporting Items for Systematic Reviews and Meta-Analyses) method. The method consists of 27 points and a flow chart built with the aim of making the whole process simpler and more ordered.

Referring to Figure 1, of the 3167 papers initially selected, the first screening takes place eliminating the duplicates, which in this case turn out to be 1217; this is due to the fact that many databases share some publishers, as shown in Table 6. It is advisable to use a software (Mendeley, Zotero, JabRef, etc.) to automatically detect the duplicates. Once the duplicates have been removed, it is necessary to take care of the grey literature as for some databases was not possible to exclude it during the research.

The following step has been to read the title, abstracts and keywords of the remaining 1950 papers, and to apply the inclusion and exclusion criteria to them.

Frequently, only by screening these particular fields, it is not possible to verify whether all the requirements have been met and, as a consequence, it is necessary to read the whole paper. This difficulty is principally due to the fact that generally, in the field of engineering, there is no normalisation of rules to create these fields, as they exist in other sectors such as medicine or psychology.

After this step, from the total amount of 1950 articles, 365 of them have been accepted for full paper review and finally, only 44 papers have met the selected criteria for being included in the revision. They will also be used to try to answer to all the questions previously formulated.

In conclusion, the papers that definitively became part of the review are 44 (see Figure 1), a much smaller amount if compared to the initial 1950, as shown in Figure 2.

The Conducting phase is certainly the most difficult phase and where perhaps more time has been spent compared to the whole systematic review, since 1950 papers have been carefully analysed.

For many of these, the full-text review was necessary to verify and ascertain whether all the criteria had actually been met or not.

At this stage it's also where another reviewer comes in. As said there were two possibilities: either to revise all 1950 papers and then compare the results, or randomly choose some paper and review them to see if the SR has been carried out correctly.

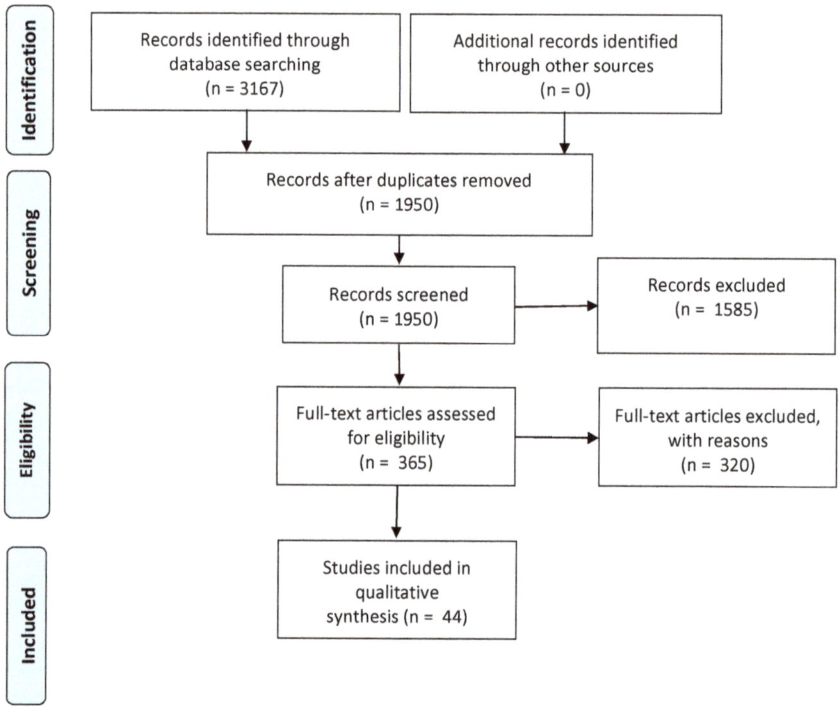

Figure 1. Preferred Reporting Items for Systematic Reviews and Meta-Analyses (PRISMA) flowchart [53].

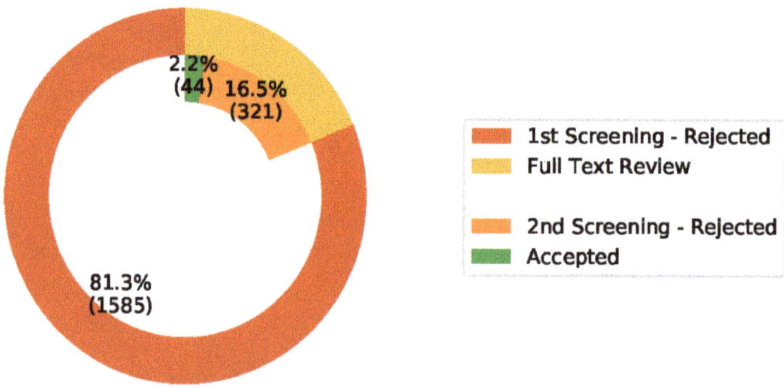

Figure 2. Paper inclusion statistics

3.2. Study Quality Assessment

For the current systematic review 44 papers have been chosen.

A quality study has been carried out on them, i.e., all the points presented in Table 10 have been analysed. The results of this quality assessment are shown in Table 11.

Table 10. Checklist for quality assessment

	Question	Score
Q_1	Is the problem presented clearly?	Yes/Partly/No
Q_2	Is the methodology used presented clearly?	Yes/Partly/No
Q_3	Is there a discussion of the results?	Yes/Partly/No
Q_4	Does it answer to the presented problem(s)?	Yes/Partly/No
Q_5	Number of cites	Cites
Q_6	Where was it published?	Journal / Conference

Table 11. Checklist for quality assessment for the selected papers

	Q1	Q2	Q3	Q4	Q5	Q6
[9]	Y	Y	P	Y	3	Journal
[10]	Y	P	Y	Y	12	Journal
[11]	Y	Y	Y	Y	5	Journal
[12]	Y	P	Y	P	1	Journal
[1]	Y	Y	Y	Y	3	Journal
[13]	Y	Y	P	P	1	Journal
[14]	Y	Y	Y	Y	2	Conference
[15]	P	Y	Y	P	1	Conference
[16]	Y	Y	Y	Y	1	Journal
[17]	Y	Y	P	P	0	Conference
[18]	Y	P	P	P	0	Conference
[19]	Y	P	Y	Y	1	Conference
[20]	Y	Y	Y	Y	13	Journal
[21]	Y	N	N	P	2	Journal
[22]	Y	Y	P	Y	1	Conference
[23]	Y	P	Y	Y	7	Journal
[24]	P	Y	P	Y	7	Journal
[25]	Y	P	P	Y	5	Conference
[26]	Y	Y	Y	Y	21	Journal
[27]	Y	P	Y	Y	18	Journal
[28]	Y	Y	Y	Y	15	Journal
[29]	Y	Y	Y	Y	5	Journal
[30]	Y	Y	P	Y	4	Journal
[31]	Y	Y	P	Y	5	Conference
[32]	Y	Y	P	P	3	Conference
[33]	Y	P	Y	P	30	Journal
[34]	P	N	Y	Y	20	Journal
[35]	Y	P	N	Y	19	Conference
[36]	Y	P	Y	Y	24	Journal
[37]	Y	P	P	P	70	Journal
[38]	Y	P	Y	Y	68	Journal
[39]	Y	P	P	Y	5	Conference
[40]	N	Y	Y	Y	38	Journal
[41]	Y	Y	Y	Y	45	Journal
[42]	Y	Y	Y	Y	35	Journal

Table 11. *Cont.*

	Q1	Q2	Q3	Q4	Q5	Q6
[43]	Y	P	P	Y	6	Conference
[44]	Y	P	Y	Y	45	Journal
[45]	Y	P	Y	Y	84	Journal
[46]	Y	P	P	Y	14	Conference
[47]	P	P	P	P	3	Conference
[48]	P	P	P	P	13	Conference
[49]	P	P	P	P	10	Conference
[50]	P	Y	P	P	18	Conference
[51]	Y	Y	Y	Y	108	Journal

Table 11 contains only qualitative information related to the selected articles. Some answers may be more or less objective, as for example if the problem or methodology have been presented clearly, while others are completely objective, as for example the cites of each article with respect to year of publication or if it has been published in a journal or in a conference.

3.3. Extraction and Synthesis Of Data

In order to extract and synthesise the data of the selected papers, a summary table (Tables 1–4) has been created. It has to be pointed out that the data presented in this work is updated to November 2019. In this tables are listed the more important characteristics of the papers related to the research topic.

- Year of publication;
- The category of the detected fault. This field can be one or more of the following items: armature faults, mechanical faults or permanent magnet faults;
- The class of the technique used for the detection. The main categories have been selected to be:
  - Radio-frequency emissions monitoring,
  - Electromagnetic field monitoring,
  - Infrared recognition,
  - Noise and vibration monitoring,
  - Model, AI, and neural-network-based techniques,
  - Temperature measurements,
  - Chemical analysis,
  - Acoustic noise measurements,
  - Parameters estimation,
  - MCSA,
  - Other (specify).

  The last item has been left if some technique cannot be classified into the previously listed categories;
- The sensor(s) needed for the failure detection technique implementation;
- If the demonstration of the effectiveness of the proposed detection techniques has been carried out with simulations and/or experiments;
- If the simulations and/or experiments have been carried out under different working conditions, such as different speeds and/or different loads;
- If the proposed technique is capable of working during changes in load and/or speed or the working conditions should be kept stationary;
- Other limitations and/or advantages. This feature is the only one that is subjective, but it contains useful information that cannot be collected in the other features.

The papers are presented in chronological order. This is usually done when performing a systematic review to put in evidence the major number of articles on the topic published during the last years. It is also useful to show if there is some clear change in research trend about one of the considered parameters.

4. Reporting

The SR is completed here by carrying out the reporting, i.e., a discussion about the obtained results. Additionally, a detailed answers to each of the research questions previously formulated and general considerations about the techniques are provided.

4.1. Discussion

This section will present some statistical data about the study and the present review and, more in general, about the chosen topic. Figure 3 shows this trend by representing the number of papers per year published on this thematic (results from 1990 to 2018) emerged from the research after that the first screening has been passed (year 2019 was not included in Figure 3, because it was not considered the whole year). This is a clear sign that research community is highly interested in the topic and this interest is growing very fast during the last years.

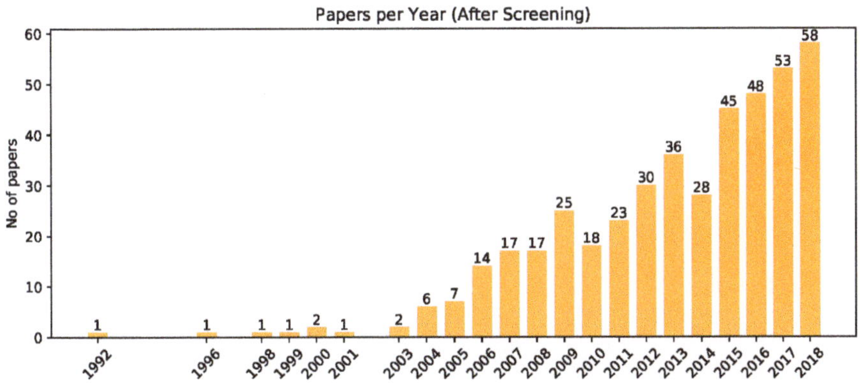

Figure 3. Papers on the topic, after the first screening of title and abstract and keywords.

This tendency is perhaps due to the current impulse in developing high reliability aircraft and electric vehicles, and similarly to the recent availability of new techniques and more powerful processors which promoted innovative applications [1]. In this view, it is useful to analyse how the number of articles based on certain techniques are distributed over the year of publication (Figure 4).

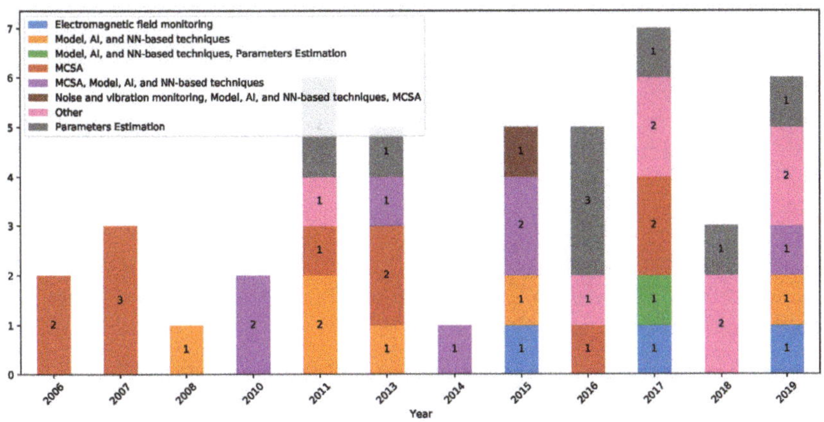

Figure 4. Distribution of the used techniques over the publication year.

Figure 4 is not representative of the whole literature, but it is possible to use it to investigate the progresses on the topic. During the last years, the techniques based on artificial intelligence, parameters estimation and models, are being utilised with increasing frequency, often as classifiers, in conjunction with established methods like the MCSA. On the other hand, the number of articles presenting detection techniques based on MCSA has drastically reduced, probably because these techniques have been intensively studied in the past years and there is less space left for innovations. Starting from 2016 techniques tagged with Other, i.e., the techniques not classifiable in the previously defined categories, have steadily increased in number. This indicates that previously unexplored phenomena are being used for BLDC fault detection and outlines that the research on the chosen topic is in turmoil.

Figure 5 shows the overall distribution of the papers according to the used technique. The following techniques have been omitted from the graph because they have not been found:

- radio-frequency emissions monitoring,
- temperature measurements,
- infrared recognition,
- chemical analysis.

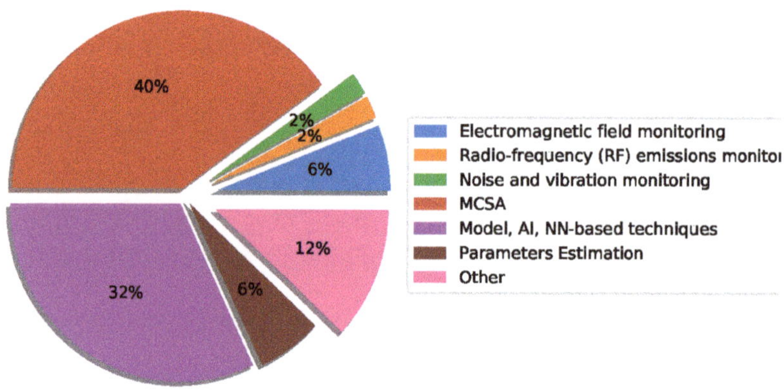

Figure 5. Distribution articles according to the used techniques

The MCSA is the most used technique, followed by the AI algorithms. It is important to point out that frequently the techniques based on Artificial Intelligence are used as classifiers of results obtained with other, already established, methods for failure detection. This association demonstrated to be have a great impact in improving the detection rate and in extending the use of the technique for a wider range of both speed and load.

One other key aspect in the graph, is the presence of a good amount of papers using techniques which were not previously classified (grouped under the tag other). Between them it is possible to find innovative techniques based on High Frequency Injection [41], hall effect sensors measurements [22] or innovative motor signals analysis [1,14,15].

An interesting alternative for the techniques based on Electromagnetic field monitoring is represented by [13]. The authors use external sensor to sense the stray magnetic fields outside the stator to detect armature failures severity and location, solving one of the biggest drawback of this powerful techniques category, i.e., the invasive procedure of placing additional windings inside the stator core.

The next paragraphs are dedicated to answer to the previously formulated research questions by using the selected papers.

4.1.1. Rq.1: Most Common Failures of Bldc Motors

Figure 6 shows the distribution of the papers in relation to the type of failure discussed. The results are in accordance with the failure distribution presented in various papers [29,31,36,54], and, in turn, this means that the research efforts are consistent with the failures occurrence.

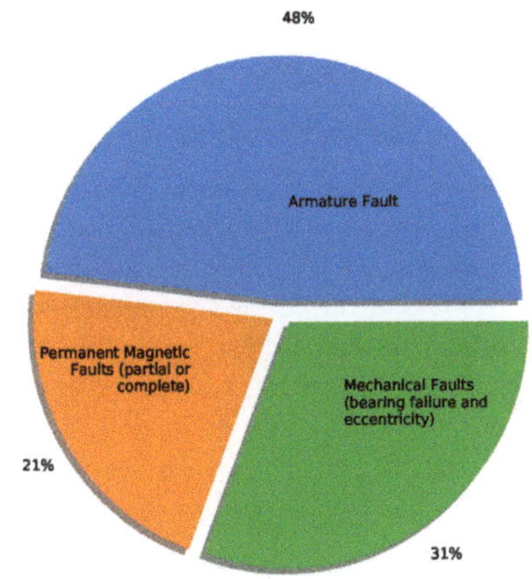

Figure 6. Distribution of the papers according the type of failure

4.1.2. Rq.2: Parameters Used for Failure Detection in Bldc Motors

Due to the intense research in this field, many of the motor parameters have been used for fault detection purposes. In the following, the variables used will be listed, dividing them between those directly measurable and those estimated.

Directly measurable quantities

The quantities listed below are directly measurable by using specific sensors.

Output torque Torque-meters shall be used to measure this variable and it can provide very useful information. The problem resides in the fact that this type of sensors are often big and expensive.

Current The current is always already measured by the motor controller and there are an immense quantity of failure detection algorithms based on this variable.

Voltage The voltage is also commonly measured by the motor controller.

Vibrations By placing accelerometers on the motor, it is possible to measure its vibration level. The algorithms based on vibration analysis could present problems when used in moving systems, like aircraft, due to the coupling of external and unpredictable vibrations.

Magnetic flux The magnetic flux gives a deep insight on how the motor is working. In order to measure it, it is usually necessary to include in the motor winding so called search coils, i.e., some additional windings not connected to the phases. The inclusion of these additional coils is not common and, although being a simple procedure, it need to unmount the motor, rewound it and to extract from the interior as many pairs of wires as many search coils as are inserted. An alternative to this procedure is to place external magnetic sensors on the stator to sense the stray magnetic fields.

Estimated quantities

The procedures based on parameters estimation can identify failures by evaluating the changes within the measured motor parameters as well as evaluating factors which are not straightforwardly quantifiable, such as:

- Back-EMF,
- Magnetic flux,
- Winding resistance,
- Winding inductance.

Estimation could be a effective instrument which allows the use of variables straightforwardly related to the fault and something else not measurable. The drawback is that it depends on models which can be constrained to particular working points and affected by the shift of some parameter.

4.1.3. Rq.3: Type of Failures Detectable by Each Technique

Figure 7 represents the distribution of the different papers according to the various techniques proposed for detecting different types of faults. This allows to evaluate which techniques are most suitable for detecting and distinguishing between different types of failure or if some techniques are more suitable for detecting specific faults or can be used as a broad spectrum analysis tool.

The techniques are widely distributed among the types detection methods, with the exception of vibration monitoring which appears limited to the detection of mechanical faults; however, this specific item can be biased due to the presence of only one single paper in the review for this category.

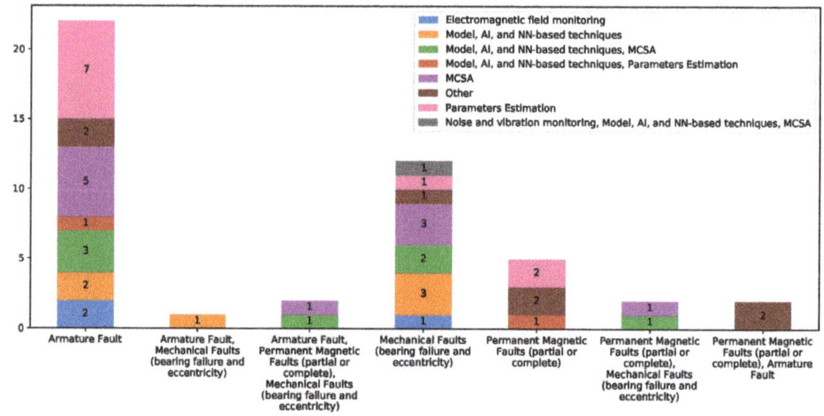

Figure 7. Distribution of the papers according the type of failure and the used technique.

4.1.4. Rq.4: Computational Power Needed for Each Technique

Only a few papers ([18,19,55]) offered a clear quantification of the computational power needed to implement the proposed technique, and therefore this question can only be answered in a qualitative manner. By what has emerged it can be seen that by a theoretical point of view the most costly strategies are those focused on models. That is because of the necessity of running the computer model parallel to the machine itself when comparing the outputs.

In addition, the complexity increases with the level of detail of the model, the parameters involved, etc. The following techniques in terms of computational cost are those based on the estimation of the parameter and then those that use the NNs. In any case, it mostly depends on the way the algorithms are implemented. The least expensive techniques are the MCSA and other techniques which directly analyse sensor data.

4.1.5. Rq.5: Sensors Needed for Each Technique

Generally the techniques that require less sensors are based on current or voltage analysis, such as MCSA. The near-totality of the reviewed methods must at least measure current consumption and motor voltage, although this is not an issue as these quantities are already available in most drivers.

Many methods also employ the motor speed to diagnose the fault. It can be speed can be estimated, from back-EMF measurements, or obtained directly from Hall effect frequency sensors or from a resolver [1].

4.1.6. Rq.6: Best Detection Results

Only some selected papers ([10,11,18,19,26,29–31,35,43,45]) provide statistics on the rate of error detection and are mostly based on the use of AI. It is very complex to compare all these results, since the test conditions are not uniform.

This is considered a weak point in this topic, which, although being very rich in ideas and proposed techniques, lacks validation and verification of the same. A possible solution to this problem, would be to propose a minimum standard set of tests to be performed in order to validate a fault detection algorithm and generate a set of minimum comparable outcomes [1].

4.2. General Considerations About The Techniques

In Table 12 the main characteristics of the detection methods have been grouped. As previously mentioned, the techniques based on AI are very effective. These can be used either as stand-alone fault detectors or in combination with other techniques to significantly improve their performance in detection. It must be noted, in any case, that their success relies on an intense process of learning and they take considerable time before working properly [1].

Referring to the statistics listed above, in particular Figure 4, it is possible to notice how fault detection techniques based on parameter estimation have also seen an increase in number.

Such techniques can provide continuous access to otherwise unobservable variables such as back-EMF or magnetic flux, facilitate the task of fault detection or identify more explicitly observable fault indicators. Noteworthy is their characteristic of being able of working while the motor is running in non-steady-state conditions of speed and/or torque. On the other hand, the potential problem with these techniques is that they are based on assumptions, models and measurements of motor variables whose limited validity and inaccuracy could hinder fault detection. According to the results of this research, about one third of the selected articles propose a technique able to work in non-stationary conditions and most of them are based on AI, NN and parameters estimation.

The need to operate the motor under steady-state conditions can be a significant limitation, above all if this is necessary to measure signals over a long period of time. Such condition may be achieved with large industrial machines working at constant load but rarely in aircraft actuators [1].

Table 12. Techniques summary [1].

	Noise and Vibration Monitoring	Electromagnetic Field Monitoring	Motor Current Signature Analysis	Model and AI based techniques	Parameters Estimation
Advantages	Most suitable method for detecting mechanical faults, as the accelerometers can be placed close to the vibration source	Can directly measure the electromagnetic field inside the motor, does not need complicated algorithm to detect failures, can virtually detect all the motor failures	Does not need additional sensors, can detect a large variety of failures, is the most used technique	Can be used during non-stationary motor operation, can be used in conjunction with other techniques	Can be used during non-stationary motor operation, can virtually monitor every motor parameter
Disadvantages	Need to install accelerometers on the motor, measurements can be corrupted by environmental vibrations, difficult to use in non-stationary motor operation	Need to rewind the stator and to extract as many additional cables as many coils inserted	Need to transform the signal in the frequency domain, the motor current depend on the load, cannot be used during non-stationary motor operation	Need extensive training	The method depends on the knowledge of various motor parameters and on the accuracy of the model, their variation (or incorrectness) can result in poor diagnosis performance

5. Conclusions

This paper presents a systematic review of high TRL techniques for BLDC motors failures detection, that have been published in the period of time from the early 1990s to November 2019. In addition, the article itself can be considered as a proof of concept of applying the SR to a particular study case in the aerospace field, in order to demonstrate its feasibility.

The studies presented in this work, have been analysed to respond to the research questions posed, that is, what are the techniques applied for fault detection, the sensors used, the working condition, what are their advantages and limitations. These results have been included in multiple tables to illustrate the findings and ease the consultation.

The greatest difficulty encountered during this study has been the impossibility of comparing the the different proposed algorithms in terms of performance, due to the lack of uniformity in tests, features measurement and estimation and presentation of the results. The authors would suggest, as a possible solution to this issue, a study to introduce a standardised benchmark and a set of parameters to be presented in order to harmonise the evaluation of the fault detection algorithms.

Author Contributions: Conceptualisation, V.M.F.; Funding acquisition, M.Á.M.P.; Investigation, V.M.F. and C.I.; Methodology, V.M.F. and C.I.; Supervision, M.Á.M.P.; Validation, C.I.; Writing—original draft, V.M.F.; Writing—review & editing, V.M.F. and C.I. All authors have read and agreed to the published version of the manuscript.

Funding: The research leading to these results has been partly funded by the European Research Council under the European Union's Seventh Framework Programme (FP7/2007-2013)/ ERC grant agreement n. 785332

Conflicts of Interest: The authors declare no conflict of interest.

Abbreviations

The following abbreviations are used in this manuscript:

AI	Artificial Intelligence
BLDC	Brushless Direct Current
FEM	Finite Element Method
HF	High Frequency
MCSA	Motor Current Signature Analysis
NN	Neural Network
SR	Systematic Review
SVM	Support Vector Machine
TRL	Technology Readiness Level

Appendix A Nomenclature

By going through the literature, the terminology in this field appears non-uniform. This is due to the fact that fault detection and diagnosis is usually distributed over many different disciplines. The definition of the following terms is specified in the glossary section and is based on [1,56]. This terminology will be used along the entire document.

Fault : Unpermitted deviation of at least one feature (characteristic property) of the system out of the acceptable standard condition threshold. The fault is a state of the system and can be of various types (manufacturing, assembly, maintenance, software, operators, wrong operation). It may not affect the correct functioning of the overall system

Failure : Permanent interruption of a system's ability to perform a required function under determined operating conditions.

Malfunction : Intermittent irregularity in the fulfilment of a system's function. It can arise from one or more faults.

From the description it is possible to draw the relationship between faults, failures and malfunctions (Figure A1).

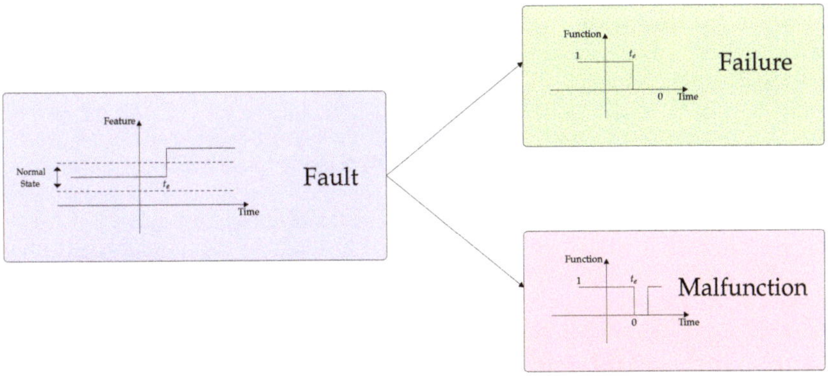

Figure A1. Scheme of the relation between faults, failures and malfunctions [1].

References

1. Fico, V.M.; Vázquez, A.L.R.; Prats, M.Á.M.; Bernelli-Zazzera, F. Failure detection by signal similarity measurement of brushless DC motors. *Energies* **2019**, *12*. [CrossRef]
2. Basak, D.; Tiwari, A.; Das, S.P. Fault diagnosis and condition monitoring of electrical machines-A review. In Proceedings of the IEEE International Conference on Industrial Technology, Mumbai, India, 15–17 December 2006; pp. 3061–3066. [CrossRef]
3. Buchwald, H.; Avidor, Y.; Braunwald, E.; Jensen, M.D.; Pories, W.; Fahrbach, K.; Schoelles, K. Bariatric surgery: A systematic review and meta-analysis. *JAMA* **2004**, *292*, 1724–1737. [CrossRef] [PubMed]
4. Renehan, A.G.; Tyson, M.; Egger, M.; Heller, R.F.; Zwahlen, M. Body-mass index and incidence of cancer: A systematic review and meta-analysis of prospective observational studies. *Lancet* **2008**, *371*, 569–578. [CrossRef]
5. Kitchenham, B.; Charters, S. Guidelines for performing systematic literature reviews in software engineering. In *EBSE Technical Report*; Elsevier: Amsterdam, The Netherlands, 2007; pp. 1–65.
6. Hall, T.; Beecham, S.; Bowes, D.; Gray, D.; Counsell, S. A systematic literature review on fault prediction performance in software engineering. *IEEE Trans. Softw. Eng.* **2011**, *38*, 1276–1304. [CrossRef]
7. Kitchenham, B.; Brereton, O.P.; Budgen, D.; Turner, M.; Bailey, J.; Linkman, S. Systematic literature reviews in software engineering—A systematic literature review. *Inf. Softw. Technol.* **2009**, *51*, 7–15. [CrossRef]
8. Ierardi, C.; Orihuela, L.; Jurado, I. Distributed Estimation Techniques for Cyber-Physical Systems: A Systematic Review. *Sensors* **2019**, *19*, 4720. [CrossRef]
9. Ge, Y.; Song, B.; Pei, Y.; Mollet, Y.A.; Gyselinck, J.J. Analytical Expressions of Isolation Indicators for Permanent-Magnet Synchronous Machines Under Stator Short-Circuit Faults. *IEEE Trans. Energy Convers.* **2019**, *34*, 984–992. [CrossRef]
10. Kao, I.H.; Wang, W.J.; Lai, Y.H.; Perng, J.W. Analysis of Permanent Magnet Synchronous Motor Fault Diagnosis Based on Learning. *IEEE Trans. Instrum. Meas.* **2019**, *68*, 310–324. [CrossRef]
11. Heydarzadeh, M.; Zafarani, M.; Nourani, M.; Akin, B. A Wavelet-Based Fault Diagnosis Approach for Permanent Magnet Synchronous Motors. *IEEE Trans. Energy Convers.* **2019**, *34*, 761–772. [CrossRef]
12. Zandi, O.; Poshtan, J. Fault Diagnosis of Brushless DC Motors Using Built-In Hall Sensors. *IEEE Sensors J.* **2019**, *19*, 8183–8190. [CrossRef]
13. Liu, X.; Miao, W.; Xu, Q.; Cao, L.; Liu, C.; Pong, P.W.T. Inter-Turn Short-Circuit Fault Detection Approach for Permanent Magnet Synchronous Machines Through Stray Magnetic Field Sensing. *IEEE Sens. J.* **2019**, 1. [CrossRef]
14. Ye, M.; Huang, J. Bearing Fault Diagnosis under Time-Varying Speed and Load Conditions via Speed Sensorless Algorithm and Angular Resample. In Proceedings of the 2018 23rd International Conference on Electrical Machines (ICEM 2018), Alexandroupoli, Greece, 3–6 September 2018; pp. 1775–1781. [CrossRef]

15. Ullah, Z.; Lee, S.; Hur, J. A Novel Fault Diagnosis Technique for IPMSM Using Voltage Angle. In Proceedings of the 2018 IEEE Energy Conversion Congress and Exposition (ECCE), Portland, OR, USA, 23–27 September 2018; pp. 3236–3243. [CrossRef]
16. Moon, S.; Jeong, H.; Lee, H.; Kim, S.W. Interturn Short Fault Diagnosis in a PMSM by Voltage and Current Residual Analysis With the Faulty Winding Model. *IEEE Trans. Energy Convers.* **2018**, *33*, 190–198. [CrossRef]
17. Ahsanullah, K.; Jeyasankar, E.; Vignesh, A.; Panda, S.; Shanmukha, R.; Nadarajan, S. Eccentricity fault analysis in PMSM based marine propulsion motors. In Proceedings of the 20th International Conference on Electrical Machines and Systems, Sydney, NSW, Australia, 11–14 August 2017; pp. 1–6.
18. Breuneval, R.; Clerc, G.; Nahid-Mobarakeh, B.; Mansouri, B. Hybrid diagnosis of intern-turn short-circuit for aircraft applications using SVM-MBF. In Proceedings of the 2017 IEEE International Conference on Fuzzy Systems (FUZZ-IEEE), Cincinnati, OH, USA, 1–5 October 2017; pp. 1–6.
19. Heydarzadeh, M.; Zafarani, M.; Ugur, E.; Akin, B.; Nourani, M. A model-based signal processing method for fault diagnosis in PMSM machine. In Proceedings of the Energy Conversion Congress and Exposition (ECCE), Naples, Italy, 9–12 July 2017; pp. 3160–3164.
20. Mazzoletti, M.A.; Bossio, G.R.; De Angelo, C.H.; Espinoza-Trejo, D.R. A model-based strategy for interturn short-circuit fault diagnosis in PMSM. *IEEE Trans. Ind. Electron.* **2017**, *64*, 7218–7228. [CrossRef]
21. Obeid, N.H.; Battiston, A.; Boileau, T.; Nahid-Mobarakeh, B. Early Intermittent Interturn Fault Detection and Localization for a Permanent Magnet Synchronous Motor of Electrical Vehicles Using Wavelet Transform. *IEEE Trans. Transp. Electrif.* **2017**, *3*, 694–702. [CrossRef]
22. Park, Y.; Fernandez, D.; Lee, S.B.; Hyun, D.; Jeong, M.; Kommuri, S.K.; Cho, C.; Reigosa, D.; Briz, F. On-line detection of rotor eccentricity for PMSMs based on hall-effect field sensor measurements. In Proceedings of the Energy Conversion Congress and Exposition (ECCE), Cincinnati, OH, USA, 1–5 October 2017; pp. 4678–4685.
23. Zhu, M.; Hu, W.; Kar, N.C. Torque-ripple-based interior permanent-magnet synchronous machine rotor demagnetization fault detection and current regulation. *IEEE Trans. Ind. Appl.* **2017**, *53*, 2795–2804. [CrossRef]
24. Gu, B.G. Study of IPMSM interturn faults part II: Online fault parameter estimation. *IEEE Trans. Power Electron.* **2016**, *31*, 7214–7223. [CrossRef]
25. Jeong, H.; Moon, S.; Lee, J.; Kim, S.W. Inter-turn short fault diagnosis of permanent magnet synchronous machines using negative sequence components. In Proceedings of the 2016 IEEE International Conference on Industrial Technology (ICIT), Taipei, Taiwan, 14–17 March 2016; pp. 170–174.
26. Mbo'o, C.P.; Hameyer, K. Fault diagnosis of bearing damage by means of the linear discriminant analysis of stator current features from the frequency selection. *IEEE Trans. Ind. Appl.* **2016**, *52*, 3861–3868. [CrossRef]
27. Moon, S.; Lee, J.; Jeong, H.; Kim, S.W. Demagnetization fault diagnosis of a PMSM based on structure analysis of motor inductance. *IEEE Trans. Ind. Electron.* **2016**, *63*, 3795–3803. [CrossRef]
28. Sen, B.; Wang, J. Stator interturn fault detection in permanent-magnet machines using PWM ripple current measurement. *IEEE Trans. Ind. Electron.* **2016**, *63*, 3148–3157. [CrossRef]
29. Abed, W.; Sharma, S.; Sutton, R.; Motwani, A. A robust bearing fault detection and diagnosis technique for brushless DC motors under non-stationary operating conditions. *Control. Autom. Electr. Syst.* **2015**, *26*, 241–254. [CrossRef]
30. Akar, M.; Hekim, M.; Orhan, U. Mechanical fault detection in permanent magnet synchronous motors using equal width discretization-based probability distribution and a neural network model. *Electr. Eng. Comput. Sci.* **2015**, *23*, 813–823. [CrossRef]
31. Çira, F.; Arkan, M.; Gümüş, B. A new approach to detect stator fault in permanent magnet synchronous motors. In Proceedings of the 2015 IEEE 10th International Symposium on Diagnostics for Electrical Machines, Power Electronics and Drives (SDEMPED), Guarda, Portugal, 1–4 September 2015; pp. 316–321.
32. Lee, S.T.; Kim, K.T.; Hur, J. Diagnosis technique for stator winding inter-turn fault in BLDC motor using detection coil. In Proceedings of the 2015 9th International Conference on Power Electronics and ECCE Asia (ICPE-ECCE Asia), Seoul, Korea, 1–5 June 2015; pp. 2925–2931.
33. Moosavi, S.; Djerdir, A.; Ait-Amirat, Y.; Khaburi, D. ANN based fault diagnosis of permanent magnet synchronous motor under stator winding shorted turn. *Electr. Power Syst. Res.* **2015**, *125*, 67–82. [CrossRef]
34. Chakraborty, S.; Keller, E.; Ray, A.; Mayer, J. Detection and estimation of demagnetization faults in permanent magnet synchronous motors. *Electr. Power Syst. Res.* **2013**, *96*, 225–236. [CrossRef]

35. Haddad, R.Z.; Strangas, E.G. Fault detection and classification in permanent magnet synchronous machines using Fast Fourier Transform and Linear Discriminant Analysis. In Proceedings of the 2013 9th IEEE International Symposium on Diagnostics for Electric Machines, Power Electronics and Drives (SDEMPED), Valencia, Spain, 27–30 August 2013; pp. 99–104.
36. Nyanteh, Y.D.; Srivastava, S.K.; Edrington, C.S.; Cartes, D.A. Application of artificial intelligence to stator winding fault diagnosis in Permanent Magnet Synchronous Machines. *Electr. Power Syst. Res.* **2013**, *103*, 201–213. [CrossRef]
37. Sarikhani, A.; Mohammed, O.A. Inter-turn fault detection in PM synchronous machines by physics-based back electromotive force estimation. *IEEE Trans. Ind. Electron.* **2013**, *60*, 3472–3484. [CrossRef]
38. Urresty, J.C.; Riba, J.R.; Romeral, L. Diagnosis of interturn faults in PMSMs operating under nonstationary conditions by applying order tracking filtering. *IEEE Trans. Power Electron.* **2013**, *28*, 507–515. [CrossRef]
39. Abu-Rub, H.; Ahmed, S.M.; Iqbal, A.; Toliyat, H.A.; Rahimian, M.M. Incipient bearing fault detection for three-phase Brushless DC motor drive using ANFIS. In Proceedings of the IEEE International Symposium on Diagnostics for Electric Machines, Power Electronics & Drives, Bologna, Italy, 5–8 September 2011; pp. 620–625.
40. Leboeuf, N.; Boileau, T.; Nahid-Mobarakeh, B.; Clerc, G.; Meibody-Tabar, F. Real-time detection of interturn faults in PM drives using back-EMF estimation and residual analysis. *IEEE Trans. Ind. Appl.* **2011**, *47*, 2402–2412. [CrossRef]
41. Arellano-Padilla, J.; Sumner, M.; Gerada, C. Winding condition monitoring scheme for a permanent magnet machine using high-frequency injection. *IET Electr. Power Appl.* **2011**, *5*, 89–99. [CrossRef]
42. Kim, K.H.; Gu, B.G.; Jung, I.S. Online fault-detecting scheme of an inverter-fed permanent magnet synchronous motor under stator winding shorted turn and inverter switch open. *IET Electr. Power Appl.* **2011**, *5*, 529–539. [CrossRef]
43. Ondel, O.; Boutleux, E.; Clerc, G. Diagnosis by pattern recognition for PMSM used in more electric aircraft. In Proceedings of the IECON 2011-37th Annual Conference on IEEE Industrial Electronics Society, Melbourne, VIC, Australia, 7–10 November 2011; pp. 3452–3458.
44. Ebrahimi, B.; Faiz, J.; Araabi, B. Pattern identification for eccentricity fault diagnosis in permanent magnet synchronous motors using stator current monitoring. *IET Electr. Power Appl.* **2010**, *4*, 418–430. [CrossRef]
45. Ebrahimi, B.M.; Faiz, J. Feature extraction for short-circuit fault detection in permanent-magnet synchronous motors using stator-current monitoring. *IEEE Trans. Power Electron.* **2010**, *25*, 2673–2682. [CrossRef]
46. Quiroga, J.; Cartes, D.; Edrington, C.; Liu, L. Neural network based fault detection of PMSM stator winding short under load fluctuation. In Proceedings of the 13th Power Electronics and Motion Control Conference, Poznan, Poland, 1–3 September 2008; pp. 793–798.
47. Rosero, J.; Cusido, J.; Garcia, A.; Romeral, L.; Ortega, J. Fault detection of eccentricity by means of joint time-frequency analysis in PMSM under dynamic conditions. In Proceedings of the International Symposium on Intelligent Signal Processing, Alcala de Henares, Spain, 3–5 October 2007; pp. 1–6.
48. Rosero, J.; Cusido, J.; Ortega, J.; Romeral, L.; Garcia, A. PMSM Bearing Fault Detection by means of Fourier and Wavelet transform. In Proceedings of the IECON 2007—33rd Annual Conference of the IEEE Industrial Electronics Society, Taipei, Taiwan, 5–8 November 2007; pp. 1163–1168.
49. Rosero, J.; Cusido, J.; Garcia, A.; Romeral, L.; Ortega, J. Detection of stator short circuits in PMSM by mean of joint time-frequency analysis. In Proceedings of the IEEE International Symposium on Diagnostics for Electric Machines, Power Electronics and Drives (SDEMPED 2007), Cracow, Poland, 6–8 September 2007; pp. 420–425.
50. Barendse, P.; Pillay, P. A new algorithm for the detection of faults in permanent magnet machines. In Proceedings of the IECON 2006 - 32nd Annual Conference on IEEE Industrial Electronics, Paris, France, 6–10 November 2006; pp. 823–828.
51. Rajagopalan, S.; Aller, J.M.; Restrepo, J.A.; Habetler, T.G.; Harley, R.G. Detection of rotor faults in brushless DC motors operating under nonstationary conditions. *IEEE Trans. Ind. Appl.* **2006**, *42*, 1464–1477. [CrossRef]
52. Ierardi, C.; Orihuela, L.; Jurado, I. Guidelines for a systematic review in systems and automatic engineering. Case study: Distributed estimation techniques for cyber-physical systems. In Proceedings of the European Control Conference, Limassol, Cyprus, 12–15 June 2018; pp. 2230–2235.
53. Moher, D.; Liberati, A.; Tetzlaff, J.; Altman, D. Reprint-preferred reporting items for systematic reviews and meta-analyses: The PRISMA statement. *Phys. Ther.* **2009**, *89*, 873–880. [CrossRef] [PubMed]

54. Nandi, S.; Toliyat, H.A.; Li, X. Condition monitoring and fault diagnosis of electrical motors-A review. *IEEE Trans. Energy Convers.* **2005**, *20*, 719–729. [CrossRef]
55. Heydarzadeh, M.; Zafarani, M.; Akin, B.; Nourani, M. Automatic fault diagnosis in PMSM using adaptive filtering and wavelet transform. In Proceedings of the 2017 IEEE International Electric Machines and Drives Conference (IEMDC), Miami, FL, USA, 21–24 May 2017; pp. 1–7.
56. Isermann, R. *Fault-Diagnosis Systems: An Introduction from Fault Detection to Fault Tolerance*; Springer: Berlin, Germany, 2006; pp. 1–475, arXiv:1011.1669v3. [CrossRef]

energies

MDPI

Article

High Resistance Fault-Detection and Fault-Tolerance for Asymmetrical Six-Phase Surface-Mounted AC Permanent Magnet Synchronous Motor Drives

Claudio Rossi [1], Yasser Gritli [1,2,*], Alessio Pilati [1], Gabriele Rizzoli [1], Angelo Tani [1] and Domenico Casadei [1]

1 DEI—Department of Electrical, Electronic and Information Engineering—"Guglielmo Marconi", University of Bologna, 40126 Bologna, Italy; claudio.rossi@unibo.it (C.R.); alessio.pilati@unibo.it (A.P.); gabriele.rizzoli@unibo.it (G.R.); angelo.tani@unibo.it (A.T.); domenico.casadei@unibo.it (D.C.)

2 DEE—Department of Electrical Engineering, LR-11-ES18, National Engineering School of Tunis, University of Tunis El Manar, Tunis, Belvedere 1002, Tunisia

* Correspondence: yasser.gritli@unibo.it

Received: 19 April 2020; Accepted: 12 June 2020; Published: 15 June 2020

Abstract: In the last decade, the interest for higher reliability in several industrial applications has boosted the research activities in multiphase permanent magnet synchronous motors realized by multiple three-phase winding sets. In this study, a mathematical model of an asymmetric surface-mounted six-phase permanent magnet synchronous motor under high resistance connections was developed. By exploiting the intrinsic properties of multiphase machines in terms of degrees of freedom, an improved field-oriented control scheme is presented that allows online fault detection and a quite undisturbed operating condition of the machine under high resistance connections. More specifically, the proposed strategies for online fault-detection and fault-tolerance are based on the use of multi-reference frame current regulators. The feasibility of the proposed approach was theoretically analyzed, then confirmed by numerical simulations. In order to validate experimentally the proposed strategies, the entire control system was implemented using TMS-320F2812 based platform.

Keywords: stator fault; high resistance connection; fault-detection; fault-tolerant control; six-phase permanent magnet synchronous machines; field-oriented control

1. Introduction

Multiphase permanent magnet synchronous machines (m-ϕ PMSMs) have gained significant attention, in variable-speed drives and generation systems, during the last decade. They have attracted much interest becoming a viable solution for a wide range of modern industry applications such as aerospace applications, naval propulsion, energy generation and transportation electrification [1–3].

The main reasons of this interest are justified by combining the well-known advantages of permanent magnet synchronous machines (PMSMs) in terms of high efficiency, high power density and high dynamic performances, with the strengths of multiphase machines, which provide lower torque ripple, lower current harmonics, fault tolerance capabilities and higher torque/power switch rating [1,4].

As mentioned in [5], it is known that stator windings faults account for 40% of the overall electric machine failures in different industrial applications. As with any rotating electrical machine in healthy conditions, PMSMs undergo mechanical and electrical stresses symmetrically distributed inside the machine. In particular, the stator windings are subjected to several stresses induced by a combination of several factors, including thermal effects, vibrations, voltage spikes caused by adjustable-speed drives and environmental conditions [5,6].

Effectively, under healthy operating conditions, the phase impedances of the stator windings are identical, leading to balanced phase currents. When a stator fault occurs, phase currents are no longer balanced determining too high peak values, which may affect the performance and reliability of the motor.

Stator winding faults can be roughly classified as open-circuit or short-circuit, both affecting the phases and/or the terminal connections. Some other anomalies are not destructive at incipient stage but can evolve and initiate serious damages to the motors. During the last decade, high resistance connection (HRC) has been clearly identified as the main initiator of the stator failures. In fact, HRC is a progressive failure mode that can affect any power connection and/or end winding and is mostly caused by a combination of excessive vibration levels, poor workmanship, metal fatigue, overheating and/or corrosion of the power contact surfaces. Comprehensive descriptions of HRC mechanism propagation, initiated by the above factors, are available in [6–8].

Although the advantageous performances of m-ϕ PMSMs over their classical 3-ϕ PMSMs counterparts, they are more subjected to stator faults owing to the higher number of stator windings. Thus, if such a fault is not properly cleared in a timely manner, it spreads and may conduct to rotor magnets demagnetization and eventual dramatic damages with serious unexpected outages [9,10].

Several techniques have been developed for the diagnosis of stator faults for three-phase machines. Classical off-line techniques such as measurement and comparison of winding resistances or related voltage drops, analysis of the temperature distribution by infrared thermography and partial discharge analysis, have been successfully applied [11–13]. Even if standard off-line techniques provide reliable results for stator asymmetry, they are limited by the necessity of full or partial motor disassembly, and/or dedicated equipment or setup.

Actually, the four main diagnostic strategies, adopted to cope with the above limitations and to provide useful fault indexes and fault-tolerant control strategies, are known as model-based fault diagnosis, knowledge-based fault diagnosis, signal-based fault diagnosis and hybrid fault diagnosis approaches [14–16]. Although the advantages of each approach, in general signal-based approach is the preferred strategy.

Stator fault diagnosis for 3-ϕ PMSMs has been extensively investigated in the literature [5]. The main focus was on inter-turn short-circuit faults [17,18] and open-phase faults [19], while investigations on HRC are relatively few, except recent studies presented in [8,20,21]. Based on high order sliding mode principle, an interesting current-control scheme designed to simultaneously detect and tolerate the existence of HRC, was investigated in [8]. The fault compensation is obtained by canceling extra current dynamics, which provides more effective d–q currents components tracking. In [20], a full online diagnosis of HRC is developed for delta-connected PMSM using zero-sequence current component. The proposed approach has shown interesting performances for detecting and quantifying the extend of the fault. Another relevant approach dedicated to detecting and estimate the HRC severity, for vector-controlled PMSM drive system, is investigated in [21]. The proposed technique is based on a signal injection in the reference signals applied to the controlled PMSM under its normal operation, leading to the appearance of DC components in the stator phase currents, used thereby for detecting and estimating the propagation degree of HRC.

With reference to m-ϕ PMSMs, much more efforts have been directed toward effective fault tolerant strategies than diagnosis approaches. Among several stator configurations of m-ϕ PMSMs, multiple three-phase winding sets are probably the most preferred for numerous industrial applications. The interest for these configurations of m-ϕ PMSMs is mainly justified by the fact that each stator winding set can be separately supplied by standard three-phase inverters, which allows crucial power flow modularity control, particularly useful under stator fault conditions.

Different control strategies for 6-ϕ PMSMs can be found in the literature, namely, constant V/f control, field-oriented control (FOC) and direct torque control (DTC). A recent comprehensive review, including theory, simulations, and experimental tests is presented in [22]. Under the context of stator fault risks, several fault tolerant control techniques have been developed for exploiting the inherent active redundancy due to the associated precious degrees of freedom, for ensuring a continuity of operation even in case of more than one phase affected by a stator fault. An open-phase fault-tolerant control for 6-ϕ PMSM is developed in [23], where the torque capability is maximized considering the overcurrent protection limits. In [24], two optimal current control modes that tolerate open-phase fault, with minimum stator losses and maximum torque output, have been analyzed. A novel optimized open-phase fault tolerant control strategy is developed in [25], where a genetic algorithm is used to maximize the average torque and minimize the torque ripple for post-fault operating condition. In [26], an intelligent complementary sliding-mode control approach was developed for effective open-phase fault tolerance. To maintain the stability of the fault-tolerant control of the 6-ϕ PMSM drive system, a Takagi–Sugeno–Kang–type fuzzy neural network with asymmetric membership function was developed to estimate unknown lumped uncertainty including parameter variations, external disturbances and nonlinear friction force online.

Recently, effective diagnostic techniques dedicated to 6-ϕ PMSMs under stator fault conditions have been presented in [26–28], for open-phase faults or inverter related-faults and in [29] for short-circuits.

In [27], both open-phase fault and open-switch fault are tolerated using a voltage compensation-based fault tolerant control. The detection process is based on real time monitoring of the current amplitude in a specific subspace, considering a predetermined threshold for alert. Diagnosis and fault tolerant approaches have been successfully developed in [28] for open-phase faults, open-switch faults and short-switch faults in T-type three-level inverter fed dual-three phase PMSM drives. After fault detection process, which is based on the amplitude variations of a specific current space vector, an effective open-phase fault compensation was achieved without changing the machine model, nor the control framework. Open-switch faults and short-switch faults are tolerated by making full use of the remaining healthy-phases after faults. Although the verified good performances in terms of detection and fault-tolerance, the use of current amplitude space vector for the proposed diagnosis technique may show some limitations when changing the operating conditions of the machine. In [29], a new magnetic equivalent circuit model for dual-three phase PMSM under winding short-circuit is proposed for accurate prediction of the fault impact.

Based on the above observations, the present contribution is aimed to present a new strategy of fault-detection and fault-tolerant control for 6-ϕ PMSMs affected by HRC. The existing papers on this type of fault are dealing with three-phase machines, but to the best of the author's knowledge, no recent papers were published investigating HRC in 6-ϕ PMSMs. The presented strategy is based on the use of multiple space vector transformations for developing a new mathematical model able to deal with stator winding affected by HRC, and on the use of multi-reference frame current regulators for implementing an improved field-oriented control (IFOC) scheme.

The proposed strategy allows online fault-detection and fault-tolerance to be achieved without the need of additional hardware, in both stationary and dynamic operating conditions, as it is based on detecting the DC component of a new variable representing the Fault Index. In this way it is possible to avoid the critical problem of detecting certain current harmonic components having variable frequency depending on the operating speed.

This study is organized as follows. Modeling of the investigated 6-ϕ PMSM under HRC, in terms of multiple space vector, is presented in Section 2. The proposed fault-detection and fault-tolerant control strategies are detailed in Section 3. Numerical simulations and experimental tests are presented and commented in Sections 4 and 5, respectively. The recommended Fault Index for quantifying the degree of HRC, as well as the corresponding simulation and experimental evaluations are presented in Section 6.

2. Motor Modeling under HRC

In this Section, the concept of multiple space vector transformations is presented for a set of six variables. Then, the mathematical modeling of an asymmetrical six-phase surface mounted permanent magnet synchronous motor, affected by HRC, is presented.

2.1. Multiple Space Vector Transformations for Six-phase Systems

Multiple space vector transformation concept is an effective approach for multiphase electrical systems representation [30]. It is particularly useful for modeling, analysis and control design for multiphase machines and drives. For a given electrical system composed by six real variables x_{a1}, x_{a2}, x_{b1}, x_{b2}, x_{c1}, x_{c2}, a new set of three complex variables $\overline{y}_1$, $\overline{y}_3$, $\overline{y}_5$, can be obtained by means of the symmetrical linear direct and inverse transformations expressed by Equations (1) and (2), respectively.

$$\begin{cases} \overline{y}_{S1} = \frac{1}{3}\left[x_{a1} + x_{a2}\,\overline{\alpha} + x_{b1}\,\overline{\alpha}^4 + x_{b2}\,\overline{\alpha}^5 + x_{c1}\,\overline{\alpha}^8 + x_{c2}\,\overline{\alpha}^9\right] \\ \overline{y}_{S3} = \frac{1}{3}\left[x_{a1} + x_{a2}\,\overline{\alpha}^3 + x_{b1} + x_{b2}\,\overline{\alpha}^3 + x_{c1} + x_{c2}\,\overline{\alpha}^3\right] \\ \overline{y}_{S5} = \frac{1}{3}\left[x_{a1} + x_{a2}\,\overline{\alpha}^5 + x_{b1}\,\overline{\alpha}^8 + x_{b2}\,\overline{\alpha} + x_{c1}\,\overline{\alpha}^4 + x_{c2}\,\overline{\alpha}^9\right] \end{cases} \tag{1}$$

$$\begin{cases} x_{a1} = \Re_e[\overline{y}_{S3}] + \overline{y}_{S1}\cdot 1 + \overline{y}^*_{S5}\cdot 1 \\ x_{b1} = \Re_e\left[\overline{y}_{S3}\right] + \overline{y}_{S1}\cdot\overline{\alpha}^4 + \overline{y}^*_{S5}\cdot\overline{\alpha}^4 \\ x_{c1} = \Re_e\left[\overline{y}_{S3}\right] + \overline{y}_{S1}\cdot\overline{\alpha}^8 + \overline{y}^*_{S5}\cdot\overline{\alpha}^8 \\ x_{a2} = \Im_m\left[\overline{y}_{S3}\right] + \overline{y}_{S1}\cdot\overline{\alpha} + \overline{y}^*_{S5}\cdot\overline{\alpha}^7 \\ x_{b2} = \Im_m\left[\overline{y}_{S3}\right] + \overline{y}_{S1}\cdot\overline{\alpha}^5 + \overline{y}^*_{S5}\cdot\overline{\alpha}^{11} \\ x_{c2} = \Im_m\left[\overline{y}_{S3}\right] + \overline{y}_{S1}\cdot\overline{\alpha}^9 + \overline{y}^*_{S5}\cdot\overline{\alpha}^3 \end{cases} \tag{2}$$

where $\overline{\alpha} = e^{j\,\pi/6}$, the symbol "·" represents the scalar product and "*" the complex conjugate.

It is worth noting that the obtained three space vectors, involved in the transformations (1) and (2), can arbitrarily move in the respective independent subspaces, namely α_1–β_1, $\alpha_3 \| \beta_3$ and α_5–β_5.

2.2. Model of the 6-ϕ PMSM under HRC

The multiple space vector transformation principle allows the modeling of six-phase AC PMSMs by means of vectors expressed in three α–β subspaces. The considered machine is a 30 asymmetrical 6-ϕ PMSM, as illustrated by Figure 1.

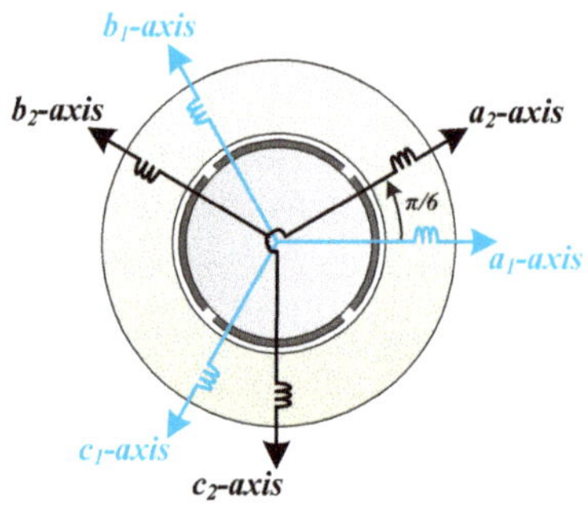

Figure 1. Six-phase surface-mounted permanent magnet synchronous motor with two sets of three phase windings and separated neutral points.

The model is developed under the conventional assumptions usually adopted for the analysis of AC rotating electrical machines and considers up to the eleventh spatial harmonic of the magnetic field in the air gap. Based on the concept of multiple space vector representation, the electrical quantities are developed in a stationary reference frame.

Assuming a set of six different stator phase resistances R_{Sa1}, R_{Sb1}, R_{Sc1}, R_{Sa2}, R_{Sb2} and R_{Sc2}, under the effect of HRC, the two sets ($k = 1, 2$) of three voltage equations can be expressed as

$$\begin{cases} v_{Sak} = R_{Sak}\, i_{Sak} + \frac{d\varphi_{Sak}}{dt}; \\ v_{Sbk} = R_{Sbk}\, i_{Sbk} + \frac{d\varphi_{Sbk}}{dt}; \\ v_{Sck} = R_{Sck}\, i_{Sck} + \frac{d\varphi_{Sck}}{dt}; \end{cases} \qquad k = 1,\ 2. \tag{3}$$

Considering the direct transformation (1), the six stator voltage expressions given by Equation (3) can be reformulated, leading to the following new set of three stator voltage space vectors

$$\bar{v}_{S1} = R_{S0}^{+}\, \bar{i}_{S1} + \overline{R}^{*}_{S10}\, \bar{i}_{S1}^{\,*} + \overline{R}_{S10}\, \bar{i}_{S3} + \overline{R}_{S4}\, \bar{i}_{S3}^{\,*} + \overline{R}^{*}_{S4}\, \bar{i}_{S5} + R_{S0}^{-}\, \bar{i}^{*}_{S5} + \frac{d\overline{\varphi}_{S1}}{dt} \tag{4}$$

$$\bar{v}_{S3} = \overline{R}^{*}_{S10}\, \bar{i}_{S1} + \overline{R}_{S4}\, \bar{i}_{S1}^{\,*} + R_{S0}^{+}\, \bar{i}_{S3} + R_{S0}^{-}\, \bar{i}_{S3}^{\,*} + \overline{R}_{S10}\, \bar{i}_{S5} + \overline{R}^{*}_{S4}\, \bar{i}^{*}_{S5} + \frac{d\overline{\varphi}_{S3}}{dt} \tag{5}$$

$$\bar{v}_{S5} = \overline{R}_{S4}\, \bar{i}_{S1} + R_{S0}^{-}\, \bar{i}_{S1}^{\,*} + \overline{R}^{*}_{S10}\, \bar{i}_{S3} + \overline{R}^{*}_{S4}\, \bar{i}_{S3}^{\,*} + R_{S0}^{+}\, \bar{i}_{S5} + \overline{R}_{S10}\, \bar{i}^{*}_{S5} + \frac{d\overline{\varphi}_{S5}}{dt} \tag{6}$$

where,

$$R_{S0}^{+} = \frac{1}{6}\left[R_{Sa1} + R_{Sa2} + R_{Sb1} + R_{Sb2} + R_{Sc1} + R_{Sc2}\right] \tag{7}$$

$$R_{S0}^{-} = \frac{1}{6}\left[R_{Sa1} - R_{Sa2} + R_{Sb1} - R_{Sb2} + R_{Sc1} - R_{Sc2}\right] \tag{8}$$

$$\overline{R}_{S4} = \frac{1}{6}\left[R_{Sa1} + R_{Sa2}\overline{\alpha}^{4} + R_{Sb1}\,\overline{\alpha}^{4} + R_{Sb2}\,\overline{\alpha}^{8} + R_{Sc1}\,\overline{\alpha}^{8} + R_{Sc2}\right] \tag{9}$$

$$\overline{R}_{S10} = \frac{1}{6}\left[R_{Sa1} + R_{Sa2}\overline{\alpha}^{10} + R_{Sb1}\,\overline{\alpha}^{4} + R_{Sb2}\,\overline{\alpha}^{2} + R_{Sc1}\,\overline{\alpha}^{8} + R_{Sc2}\,\overline{\alpha}^{6}\right] \tag{10}$$

From the previous equations it is evident that in case of balanced stator resistances the only resistance component different from zero is R_{S0}^{+}.

The stator flux space vectors can be expressed by

$$\overline{\varphi}_{S1} = L_{S1}\,\bar{i}_{S1} + 2\,\varphi_{M1}\,\cos(\gamma)\,e^{j\,\theta} + 2\,\varphi_{M11}\,\cos(11\,\gamma)\,e^{-j\,11\,\theta} \tag{11}$$

$$\overline{\varphi}_{S3} = L_{S3}\,\bar{i}_{S3} + 2\,\varphi_{M3}\,\cos(3\,\gamma)\,e^{j\,3\,\theta} + 2\,\varphi_{M9}\,\cos(9\,\gamma)\,e^{-j\,9\,\theta} \tag{12}$$

$$\overline{\varphi}_{S5} = L_{S5}\,\bar{i}_{S5} + 2\,\varphi_{M5}\,\cos(5\,\gamma)\,e^{j\,5\,\theta} + 2\,\varphi_{M7}\,\cos(7\,\gamma)\,e^{-j\,7\,\theta} \tag{13}$$

In Equations (11)–(13), $\gamma = (\pi - \beta)/2$ and the constant values φ_{M1}, φ_{M3}, φ_{M5}, φ_{M7}, φ_{M9}, φ_{M11}, are expressed as in Equations (14)–(19), respectively.

$$\varphi_{M1} = \frac{2\,\mu_0\,N_S\,L\,\tau\,H_{R,\max}\,K_{BS1}\,K_{RS1}}{\pi^2} \tag{14}$$

$$\varphi_{M3} = \frac{2\,\mu_0\,N_S\,L\,\tau\,H_{R,\max}}{\pi^2}\,\frac{K_{BS3}\,K_{RS3}}{9} \tag{15}$$

$$\varphi_{M5} = \frac{2\,\mu_0\,N_S\,L\,\tau\,H_{R,\max}}{\pi^2}\,\frac{K_{BS5}\,K_{RS5}}{25} \tag{16}$$

$$\varphi_{M7} = \frac{2\,\mu_0\,N_S\,L\,\tau\,H_{R,\max}}{\pi^2}\,\frac{K_{BS7}\,K_{RS7}}{49} \tag{17}$$

$$\varphi_{M9} = \frac{2\,\mu_0\,N_S\,L\,\tau\,H_{R,\max}}{\pi^2}\,\frac{K_{BS9}\,K_{RS9}}{81} \tag{18}$$

$$\varphi_{M11} = \frac{2\,\mu_0\,N_S\,L\,\tau\,H_{R,\max}}{\pi^2}\,\frac{K_{BS11}\,K_{RS11}}{121} \tag{19}$$

Based on the previous equations, the electromagnetic torque can be formulated as

$$\begin{aligned} T_{em} = \quad & 6\,p\,\varphi_{M1}\left[\bar{i}_{S1}\cdot j\,\cos(\gamma)\,e^{j\,\theta}\right]+18\,p\,\varphi_{M3}\left[\bar{i}_{S3}\cdot j\,\cos(3\,\gamma)\,e^{j\,3\,\theta}\right]+ \\ & +30\,p\,\varphi_{M5}\left[\bar{i}_{S5}\cdot j\,\cos(5\,\gamma)\,e^{j\,5\,\theta}\right]+42\,p\,\varphi_{M7}\left[\bar{i}^{*}_{S5}\cdot j\,\cos(7\,\gamma)\,e^{j\,7\,\theta}\right]+ \\ & +54\,p\,\varphi_{M9}\left[\bar{i}^{*}_{S3}\cdot j\,\cos(9\,\gamma)\,e^{j\,9\,\theta}\right]+66\,p\,\varphi_{M11}\left[\bar{i}^{*}_{S1}\cdot j\,\cos(11\,\gamma)\,e^{j\,11\,\theta}\right] \end{aligned} \tag{20}$$

It can be noted that the torque, besides the fundamental component, contains several additional oscillating contributions that can be compensated by using suitable machine design and current control techniques.

3. Proposed Fault-Detection and Fault-Tolerant Strategy

In this section, the proposed strategy for an online fault-detection of HRC in 6-ϕ PMSM is presented. Then, an improved field-oriented control (IFOC) scheme based on appropriate stator currents control, which provides fault-tolerance against the investigated stator fault, is presented.

In order to better understand the principle of the proposed IFOC scheme and the associated online fault detection algorithm, it is useful to analyze the fault effects in the rotating reference frames, where the needed current regulators are conventionally implemented.

Taking into account a stator windings design with isolated neutral points as shown in Figure 1, the current space vector in the α_3–β_3 subspace is equal to zero. Thus, the voltage equations expressed in the stator reference frame by Equations (4)–(6), can be limited to the 1st and 5th subspaces, corresponding to Equations (4) and (6), respectively. It is worth noting that for control purposes the voltage equations in α_1–β_1 subspace will be expressed in a reference frame (d_1–q_1) rotating at an angular speed of ω, whereas the equations in α_5–β_5 subspace will be expressed in a reference frame (d_5–q_5) rotating at an angular speed of $5\,\omega$.

The transformation of the space vectors from stator reference frame to the rotating reference frame, regarding the 1st and 5th subspaces, can be carried out by using the following relationships

$$\bar{x}^{r}_{1} = \bar{x}_{1}\,e^{-j\,\theta} \tag{21}$$

$$\bar{x}^{r}_{5} = \bar{x}_{5}\,e^{-j\,5\theta} \tag{22}$$

where $\bar{x}^{r}_{1}$ and $\bar{x}^{r}_{5}$ are the vectors in the new rotating reference frames, whereas $\bar{x}_1$ and $\bar{x}_5$ are the vectors expressed in the stationary reference frames.

Assuming isolated neutral points ($\bar{i}_{S3}=0$), substituting Equations (11) and (13) into Equations (4) and (6), respectively and taking into account Equations (21) and (22) leads to the following voltage equations written in rotating reference frames

$$\bar{v}^{r}_{S1} = R^{+}_{S0}\,\bar{i}^{r}_{S1}+\bar{v}^{r}_{S1,HR}+L_{S1}\frac{d\,\bar{i}^{r}_{S1}}{dt}+j\omega\,L_{S1}\,\bar{i}^{r}_{S1}+j\,2\,\omega\,\varphi_{m1}\cos(\gamma)-j\,22\,\omega\,\varphi_{m11}\cos(11\gamma)\;e^{-j12\theta} \tag{23}$$

$$\bar{v}^{r}_{S5} = R^{+}_{S0}\,\bar{i}^{r}_{S5}+\bar{v}^{r}_{S5,HR}+L_{S5}\frac{d\,\bar{i}^{r}_{S5}}{dt}+j\,5\,\omega\,L_{S5}\,\bar{i}^{r}_{S5}+j\,10\,\omega\,\varphi_{m5}\cos(5\gamma)-j\,14\,\omega\,\varphi_{m7}\cos(7\gamma)\;e^{-j12\theta} \tag{24}$$

where,

$$\bar{v}^{r}_{S1,HR} = \overline{R}^{*}_{S10}\,\bar{i}^{r\,*}_{S1}\,e^{-j2\theta}+\overline{R}^{*}_{S4}\,\bar{i}^{r\,*}_{S5}\;e^{j4\theta}+R^{-}_{S0}\,\bar{i}^{r\,*}_{S5}\,e^{-j6\theta} \tag{25}$$

$$\bar{v}^{r}_{S5,HR} = \overline{R}_{S4}\,\bar{i}^{r}_{S1}\,e^{-j4\theta}+R^{-}_{S0}\,\bar{i}^{r\,*}_{S1}e^{-j6\theta}+\overline{R}_{S10}\,\bar{i}^{r\,*}_{S5}\,e^{-j10\theta} \tag{26}$$

The space vectors $\bar{v}^{r}_{S1,HR}$ and $\bar{v}^{r}_{S5,HR}$ represent the voltage drops due to the stator HRC, in the d_1–q_1 and d_5–q_5 subspaces, respectively. In case of healthy conditions, $\bar{v}^{r}_{S1,HR}$ and $\bar{v}^{r}_{S5,HR}$ will be equal to zero as the only resistance component different from zero is R^{+}_{S0}, which is not present in Equations (25) and (26).

3.1. Proposed HRC Detection Approach

The effects of stator HRC can be identified by the presence of voltage space vectors $\vec{v}^{r}_{S1,HR}$ and $\vec{v}^{r}_{S5,HR}$, expressed by Equations (25) and (26), respectively. The sensitivity of these voltages with respect to the HRC is established by their dependence on the new stator resistance components R^{-}_{S0}, $\overline{R}_{S4}$ and $\overline{R}_{S10}$. Assuming that, for the proposed IFOC scheme, the current space vector $\vec{i}^{r}_{S5}$ is set to be equal to zero for reducing the torque ripple, the voltage space vectors, expressed by Equations (25) and (26), become

$$\vec{v}^{r}_{S1,HR} = \overline{R}^{*}_{S10}\, \vec{i}^{r\,*}_{S1}\, e^{-j2\theta} \tag{27}$$

$$\vec{v}^{r}_{S5,HR} = \overline{R}_{S4}\, \vec{i}^{r}_{S1}\, e^{-j4\theta} + R^{-}_{S0}\, \vec{i}^{r\,*}_{S1} e^{-j6\theta} \tag{28}$$

As can be seen, under the presence of unbalanced stator winding resistances, Equation (27) reveals the presence of a single fault harmonic component in the voltage space vector $\vec{v}^{r}_{S1,HR}$, which rotates at an angular speed of $-2\ \omega$ with respect to d_1–q_1 plane.

In addition, for the considered stator fault, Equation (28) shows the presence of two additional harmonic components in the voltage space vector $\vec{v}^{r}_{S5,HR}$, which rotate at the two angular speeds of $-4\ \omega$ and $-6\ \omega$ with respect to d_5–q_5 plane.

At this point, it is very useful to recall that under healthy conditions, i.e., the six stator phase resistances are equal, the new stator resistance components R^{-}_{S0}, $\overline{R}_{S4}$ and $\overline{R}_{S10}$ become equal to zero, as well as the voltage space vectors $\vec{v}^{r}_{S1,HR}$ and $\vec{v}^{r}_{S5,HR}$, leading to a clear identification of the healthy case.

Therefore, an effective fault-detection strategy, for 6-ϕ PMSM under HRC, can be based on tracking the contribution of the fault component at $-2\ \omega$ in the voltage space vector $\vec{v}^{r}_{S1,HR}$, as well as the fault components $-4\ \omega$ or $-6\ \omega$ in the voltage space vector $\vec{v}^{r}_{S5,HR}$.

Taking into account the above conclusions, an effective control strategy that intends to tolerate such a stator HRC for 6-ϕ PMSM, should be based on monitoring the above tracked fault components, which is the subject of the next Subsection.

3.2. Proposed Fault Tolerant Control Strategy for HRC

Field-oriented control (FOC) strategy is a widely adopted control scheme for implementing closed-loop speed control for m-ϕ PMSMs. An improved field-oriented control (IFOC) scheme, based on appropriate stator currents control, which provides fault-tolerance against the investigated stator HRC, is presented in this subsection. The block scheme of the proposed control strategy is illustrated by Figure 2.

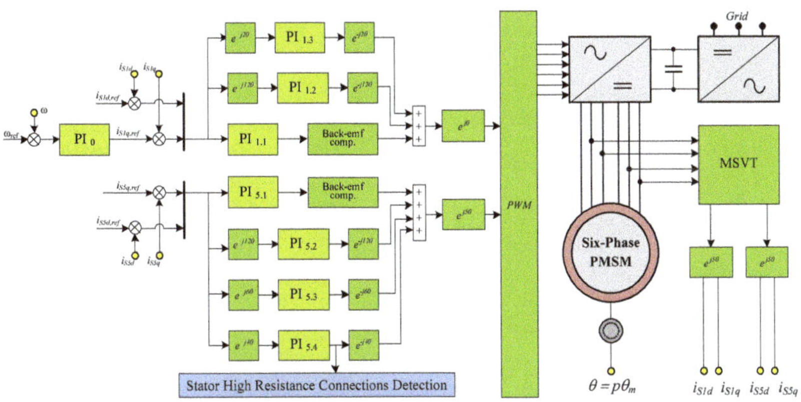

Figure 2. Block scheme of the proposed control strategy, with the capability to detect and tolerate a high resistance connection (HRC) fault.

The current regulators $PI_{1.1}$ and $PI_{5.1}$ are implemented in the synchronous reference frames d_1–q_1 and d_5–q_5, respectively, to ensure the tracking of currents references for torque control.

The reference currents $i_{S5d,ref}$ and $i_{S5q,ref}$, in the d_5–q_5 plane, are set to zero for compensating possible torque oscillations due to the 5th and 7th harmonic of the field distribution generated by permanent magnets in the air–gap.

In order to compensate the negative effects of the seventh and eleventh harmonic of the back-emf, two further current regulators $PI_{1.2}$ and $PI_{5.2}$ are necessary. They are employed in two different reference frames synchronous with the corresponding back-emf harmonics. In particular, the space vectors of eleventh and seventh back-emf harmonics are both rotating at $-12\ \omega$, as shown by Equations (23) and (24), respectively.

The four current regulators $PI_{1.1}$, $PI_{5.1}$, $PI_{1.2}$ and $PI_{5.2}$ are used in both FOC and IFOC schemes. As highlighted in the voltage Equations (23)–(26), the presence of HRC introduces disturbing voltage vector drops $\vec{v}^{r}_{S1,HR}$ and $\vec{v}^{r}_{S5,HR}$ rotating at different angular speeds.

More specifically, with the adopted mathematical model, the latter quantities are exactly equal to Equations (27) and (28). Due to the disturbances introduced by these quantities, the current regulators $PI_{1.1}$ and $PI_{5.1}$ cannot track properly the reference values $i_{S1d,ref}$, $i_{S1q,ref}$, $i_{S5d,ref}$ and $i_{S5q,ref}$.

To cope with these undesired effects, in IFOC scheme supplementary current regulators $PI_{1.3}$ and $PI_{5.3}$–$PI_{5.4}$, have been implemented in reference frames synchronized with the different angular speeds of the voltage vectors drops $\vec{v}^{r}_{S1,HR}$ and $\vec{v}^{r}_{S5,HR}$, respectively. Thus, a correct stator current reference tracking is ensured.

Finally, it is opportune to note that the generated output voltage of any supplementary regulator among $PI_{1.3}$, $PI_{5.3}$ or $PI_{5.4}$, can be used for online detection of HRC affecting the stator phases of the 6-ϕ PMSM drive. In fact, in healthy conditions these voltages are practically zero, becoming different from zero only in case of a stator phase resistance asymmetry and showing an amplitude variation proportional to the severity of the fault.

4. Simulation Results

In order to verify the effectiveness of the proposed approach, the control scheme illustrated by Figure 2 was implemented in Matlab/Simulink™ for numerical simulations. The parameters of the machine are reported in Table 1, which corresponds to the real machine used for the experimental validation. By combining the above control scheme with the 6-ϕ PMSM model, the implemented system allows a very detailed analysis of the whole drive under stator HRC, which is emulated by an additional resistance in series to Phase a1. The numerical simulations were realized at constant speed of 150 rpm, under the rated torque of 20 Nm. The reference signals $i_{s1d,ref}$, $i_{s5d,ref}$ and $i_{s5q,ref}$ were set to zero. Three different operating conditions were analyzed.

Table 1. Parameters of the 6-ϕ PMSM.

Parameter		
Phase resistance	0.36	Ohms
Pole number	4	–
Phase inductance	1.44	mH
Stator inner radius	150	mm
Stator outer radius	240	mm
Slot number	48	–
Stator winding pitch	165°	el. degrees
Magnet pole–arc	151°	el. degrees
Magnet radial thickness	5	mm

Initially, the healthy 6-ϕ PMSM is controlled by a conventional FOC scheme [31], where the six phase resistances are equal. The corresponding simulation results, in terms of stator phase currents, are reported in Figure 3a. The subsequent current space vectors, evaluated in the α_1–β_1 and α_5–β_5 planes, are reported in Figure 3b,c, respectively. As can be seen in the zoomed area of Figure 3a, the six-phase currents are balanced, leading to a circular behavior of the corresponding locus in plane α_1–β_1. The current space vector—evaluated in the α_5–β_5 plane—is equal to zero.

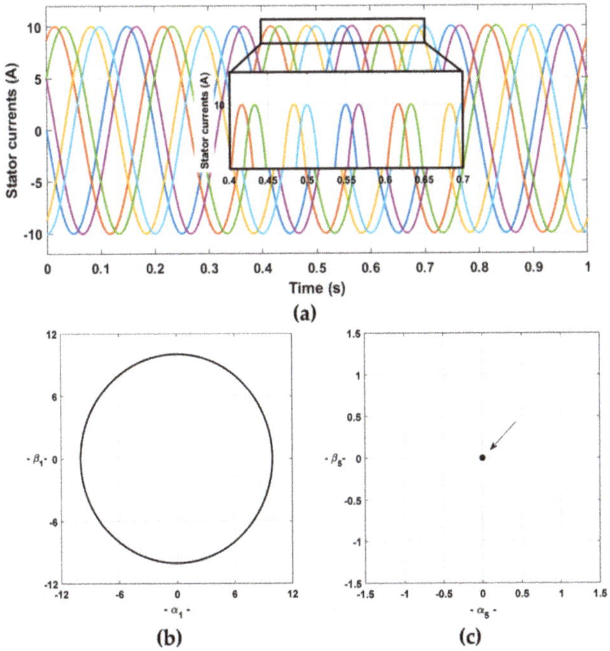

Figure 3. Simulation results: behavior of the drive using conventional field-oriented control (FOC), under healthy conditions: (**a**) stator currents waveforms, (**b**) current space vector in α_1–β_1 plane and (**c**) current space vector in α_5–β_5 plane.

Under healthy condition where the six-phase resistances are identical, the resistance components $\overline{R}_{S0}$, $\overline{R}_{S4}$ and $\overline{R}_{S10}$, expressed in Equations (8)–(10), respectively, are equal to zero. As a consequence, the voltage drops expressed by the space vectors $\overline{v}^r_{S1,HR}$ and $\overline{v}^r_{S5,HR}$ are equal to zero, leading to a healthy 6-ϕ PMSM. Thus, the circular trajectory observed in Figure 3b is clearly justified by the dominance of the fundamental harmonic and the zero value in Figure 3c is the absence proof of any resistance unbalance in the machine. From this starting point of the investigations, these results are considered as a reference for the next simulations under faulty conditions.

The second operating condition was realized with the same conventional FOC scheme, but with an additional resistance R_{add} = 250 mΩ (0.7 pu of healthy stator phase resistance Rs) in series with Phase a1, during 1.0 s of steady-state faulty conditions. The simulation results, in terms of stator phase currents and the corresponding current space vectors in planes α_1–β_1 and α_5–β_5, are reported in Figure 4a–c, respectively. Under the considered stator fault, the six stator phase resistances are no more equal, leading to the existence of voltage drops expressed by the space vectors $\overline{v}^r_{S1,HR}$ and $\overline{v}^r_{S5,HR}$. Thus, the stator symmetry of the machine is lost, which justify the current unbalance evidenced in the zoomed area of Figure 4a, when compared to the healthy case (Figure 3a).

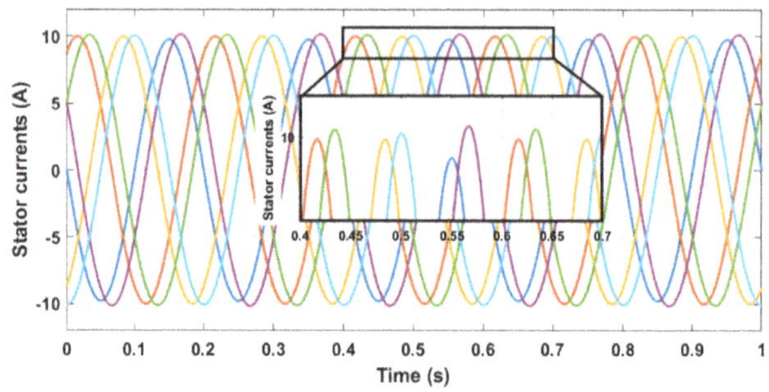

(a)

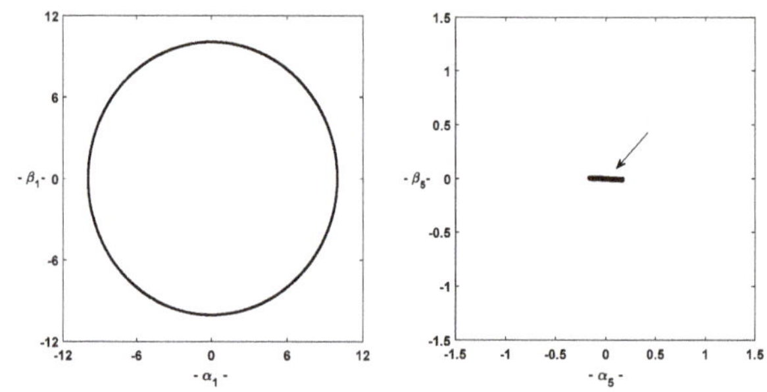

(b)

(c)

Figure 4. Simulation results: behavior of the drive using conventional FOC, under faulty phase-a1 with an additional resistance (R_{add} = 0.7 pu): (**a**) Stator currents waveforms; (**b**) current space vector in α_1–β_1 plane; (**c**) current space vector in α_5–β_5 plane.

It is worth noting that the behavior of the current space vector (Figure 4b), evaluated in the α_1–β_1 plane, is mainly determined by the dominance of the fundamental current harmonic component with respect to the small contribution of an inverse component due to the unbalanced conditions. As a result, the trajectory is practically circular. The behavior variation in the current locus evaluated in plane α_5–β_5 (Figure 4c) can be justified by the non-zero value of the space vector $\overline{v}^{r}_{S5,HR}$ under the actual stator asymmetry.

Finally, the proposed IFOC scheme illustrated by Figure 2 was implemented under a faulty Phase a1 with an additional resistance (R_{add} = 0.7 pu) during 1.0 s of steady-state faulty conditions as in the previous simulation. The corresponding simulation results are reported in Figure 5. As can be seen in the zoomed area of Figure 5a, although the presence of the stator fault, the six phase currents show a balanced behavior.

In fact, this can be explained by the compensating effect assured by the current controller $PI_{1.3}$ against the disturbance introduced by the voltage space vector $\overline{v}^{r}_{S1,HR}$ in the synchronous reference frame d_1–q_1. The disturbances introduced by the voltage space vector $\overline{v}^{r}_{S5,HR}$ are cleared by the current controllers $PI^{5.3}$ and $PI^{5.4}$, in the synchronous reference frame d_5–q_5, leading to a zero value of the current space vector in the α_5–β_5 plane as can be seen in Figure 5c.

At this level of investigation, the established theoretical analysis corroborates with numerical simulations results, confirming the effectiveness of the proposed fault-tolerant control.

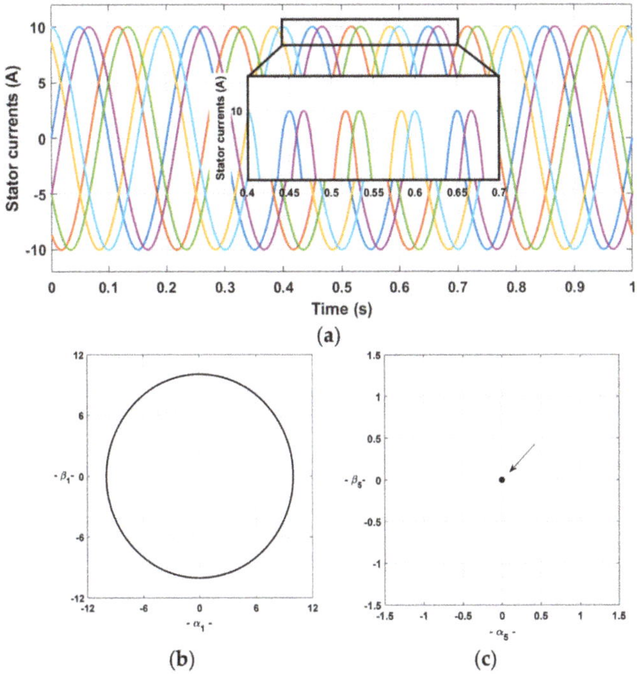

Figure 5. Simulation results: behavior of the drive using the proposed improved field-oriented control (IFOC), under faulty phase-a1 with an additional resistance (R_{add} = 0.7 pu). (**a**) Stator current waveforms; (**b**) current space vector in α_1–β_1 plane; (**c**) current space vector in α_5–β_5 plane.

5. Experimental Results

In order to experimentally validate the previous simulation results, a complete drive system was mounted in laboratory (see Figure 6) and some experimental tests were carried out to verify the effectiveness of the proposed fault-detection and fault-tolerance strategies.

Figure 6. View of the experimental setup.

The experimental setup is composed of a six-phase MOSFET inverter and a six-phase surface mounted PMSM. The 6-ϕ PMSM parameters are reported in Table 1. The controlled 6-ϕ PMSM is coupled to a three-phase induction machine used in generator mode to set the mechanical speed of the drive. During the experimental tests, the fault condition was emulated by an external resistor inserted in series with the machine Phase a1. The proposed IFOC scheme, presented in Figure 2, was implemented on a TMS-320F2812 DSP based platform. The experimental tests were realized in steady-state condition at 150 rpm, and setpoint of the torque is 20 Nm, as in simulations of Section 4. An oscilloscope with six channels was used for acquiring four phase currents and two control signals from the control platform, with a sampling rate of 100 kHz.

The behavior of the multiphase drive was assessed initially with healthy stator windings, then under a resistance increase of Phase a1 by 70%, during 1.0 s for each test. The obtained current waveforms under healthy, and faulty conditions are reported in Figure 7a,b, respectively.

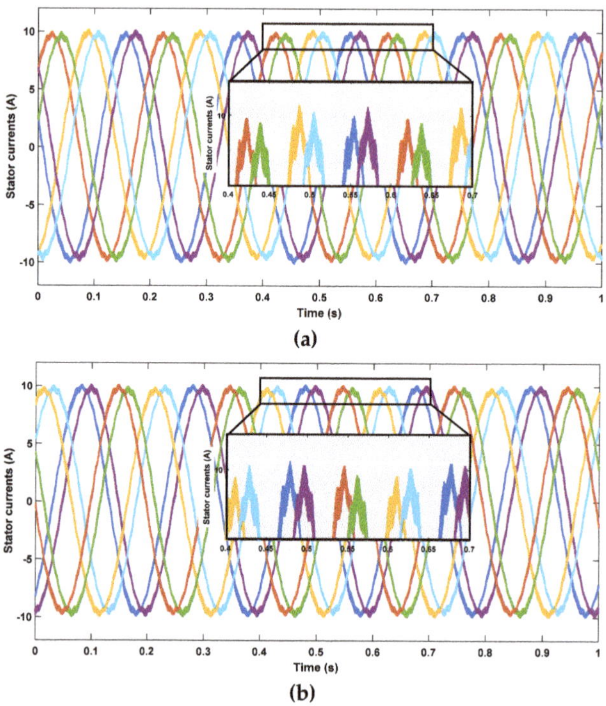

Figure 7. Experimental results: behavior of the drive using the proposed IFOC. Stator currents waveforms under (**a**) healthy and (**b**) HRC (R_{add} = 0.7 pu).

As can be seen, under healthy and faulty conditions, the six phase currents are balanced. Aside the small differences in amplitude that are probably due to measurement and acquisition chain accuracy, the six currents show practically the same waveforms.

The corresponding current space vectors in planes α_1–β_1 and α_5–β_5, under healthy and HRC, are reported in Figure 8. The current space vector in plane α_1–β_1, under healthy and HRC, shows a circular behavior (Figure 8a,b), which confirms the theoretical analysis and the simulation results presented in Figure 3b. The corresponding current loci in planes α_5–β_5, are reported in Figure 8c,d, where a quasi-zero value can be observed (note the different axis scale). In fact, the latter behavior slightly different from zero is mainly caused by non-perfectly balanced conditions among the six stator phases due to manufacturing process, switching effects and measurement noise during data acquisition.

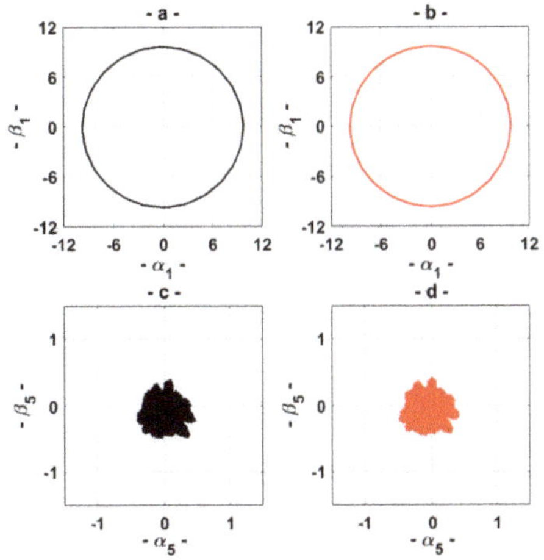

Figure 8. Experimental results: behavior of the drive using the proposed IFOC. Loci of the current space vector on plane α_1–β_1, under (**a**) healthy and (**b**) HRC (R_{add} = 0.7 pu). Loci of the current space vector on plane α_5–β_5, under (**c**) healthy and (**d**) HRC (R_{add} = 0.7 pu).

As can be seen, despite the presence of the fault, the proposed IFOC is able to ensure balanced stator phase currents. This can be clearly evidenced when comparing the obtained loci under healthy and HRC conditions (Figure 8), which are practically identical.

Finally, considering the good performances of the proposed IFOC to maintain the system of six phase current balanced even under the presence of stator asymmetry, it is obvious that the use of the classical Motor Current Signature Analysis (MCSA) cannot be adopted for detecting the presence of HRC. Thus, a more appropriate fault index is necessary for online fault detection and quantification. This problem introduces the subject of the following Section.

6. Quantitative HRC Evaluation

Using the proposed IFOC, the 6-ϕ PMSM drive can maintain good operating conditions, even under degraded mode of the stator windings affected by HRC. In this context, the presence of an online fault-detection process which ensures not only the monitoring of the stator windings state, but also quantifying the severity of the stator winding resistance unbalance, is advisable.

As already anticipated in the previous theoretical analysis, a suitable online fault-detection strategy, for 6-ϕ PMSM under HRC, can be based on tracking the contribution of the fault component at $-2\ \omega$ in the voltage space vector $\overrightarrow{v}^{r}_{S1,HR}$, as well as the fault components at $-4\ \omega$ or $-6\ \omega$ in the voltage space vector $\overrightarrow{v}^{r}_{S5,HR}$. Here, a new fault index "F_i" is proposed, that is based on the *d–q* outputs of the controller $PI_{5.4}$ defined as

$$F_i = \sqrt{v^2_{PI\ 5.4,d} + v^2_{PI\ 5.4,q}} \tag{29}$$

In order to test the sensitivity of the proposed fault index, different additional resistances (R_{add}) were tested in simulations, namely 0.10 Ω (0.28 pu), 0.25 Ω (0.70 pu), 0.75 Ω (2.09 pu) and 1.00 Ω (2.78 pu).

The spectra of the fault index, obtained from simulation results under healthy and faulty conditions, are reported in Figure 9. Observing the obtained spectra, one can see the dominance of a DC component, which has the advantage to be easily detected even during speed variations.

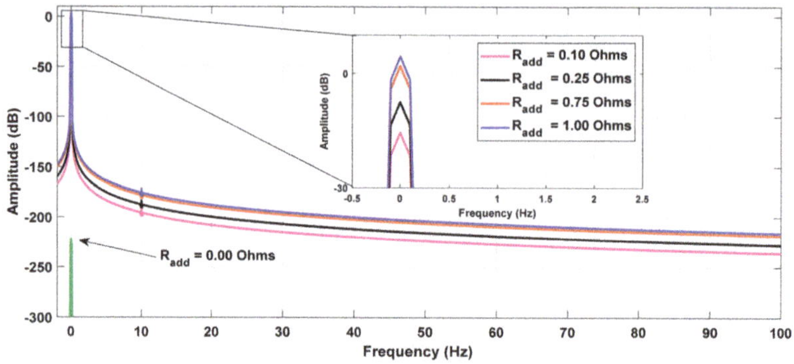

Figure 9. Simulation results: spectra of the proposed fault index under healthy and different HRC conditions affecting Phase a1.

It can be noticed that the tracked DC component shows a relevant increase in amplitude from healthy (Radd = 0.00 Ohms) to the first faulty case (R_{add} = 0.10 Ohms). It is also clearly evident that higher the severity of the fault (from 0.10 Ohms to 1.00 Ohms), higher the value of the fault index.

The corresponding spectra obtained from experimental tests for R_{add} (0.25 Ω, 0.75 Ω and 1.00 Ω) are reported in Figure 10, respectively. As can be seen, the obtained experimental spectra corroborate with those obtained by simulations. It is worth noting that the experimental signals in Figure 10 show a certain noise compared to Figure 9 owing to an unavoidable small noise present in the measurement and acquisition chain.

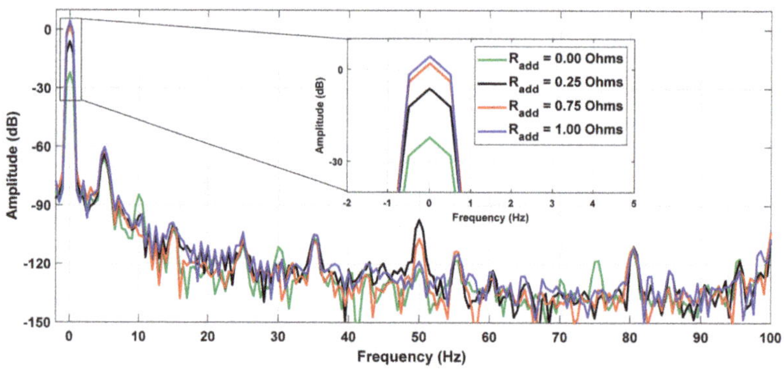

Figure 10. Experimental results: spectra of the proposed fault index under healthy and different HRC conditions affecting Phase a1.

In case of faulty conditions, the maximum difference from simulation to experimental results is lower than 2 dB. In healthy conditions, the differences appearing when comparing Figures 9 and 10 are mainly due to measurement uncertainties, noting that the fault index is practically equal to zero in both simulation and experimental tests. More specifically, the fault index shows a relevant variation in amplitude (~15 dB) from healthy to the first faulty condition (R_{add} = 0.25 Ω). The proposed fault index has shown relevant increments in agreement with the fault severity also in experimental tests, leading to an effective fault index particularly adapted not only for fault-detection, but also for fault-quantification.

Finally, in order to show the performance of the proposed fault index against the operating speed, additional numerical simulations were carried out. The results achieved using a setpoint of the torque equal to 20 Nm as in previous tests and varying the speed from 50 rpm to 250 rpm, are shown in

Figure 11. As can be noted the fault index is not dependent on the speed, allowing a further advantage of the proposed fault index to be emphasized.

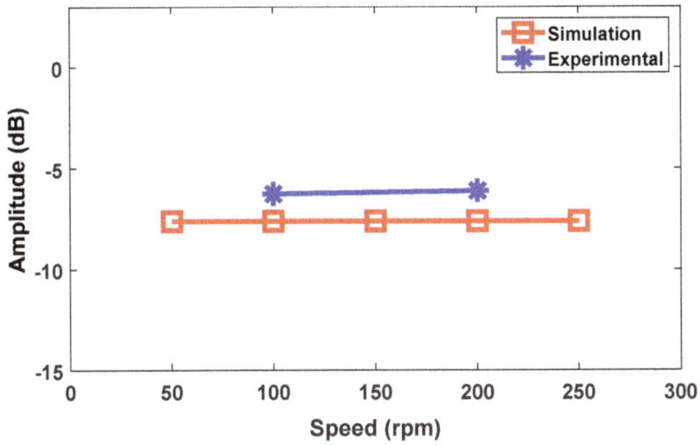

Figure 11. Behavior of the DC component of the fault index under a HRC of 0.25 Ω in Phase a1, at different speed operating conditions.

7. Conclusions

The main contribution of the study is the presentation of a new mathematical model of an asymmetric six-phase PMSM under HRC fault, which exploits the additional degrees of freedom given by the use of multi α–β planes adopted for describing the behavior of the machine. The presented mathematical model has a general validity, showing the possibility to investigate new diagnostic techniques based on monitoring the signals available in the multi α–β and d–q planes. In agreement with the developed mathematical model, an improved FOC scheme for asymmetric six-phase PMSMs based on the presence of several current regulators in the different α–β planes, was also presented. The proposed control scheme was adapted to provide not only an online HRC fault-detection and quantification, but also ensures a nearly undisturbed behavior of the drive.

The main features of the proposed HRC fault detection method are as follows:

- No need of additional hardware or measurements as it makes use of the output voltage signals of a specific current regulator already available on the control board;
- Reduced latency in detecting the fault, as it requires low computation time;
- Fault index independent of the frequency, as it is based on DC-component monitoring;
- Fault index independent of the speed of the operating conditions.

The results obtained, the theoretical analysis and the numerical simulations were compared with experimental results achieving a good agreement.

The presented results emphasize that the fault index was dependent on the severity of the fault, showing a good sensitivity. Adding a resistance of 0.25 Ω to a stator phase winding, a variation of about 15 dB was observed for the fault index magnitude.

Considering the above cited features of the proposed fault detection technique, the definition of the threshold should be adapted only to the torque demand.

Based on the results obtained the authors are encouraged to continue the experimental tests for verifying the effectiveness of the proposed diagnostic technique in dynamic conditions and to further develop the mathematical model for also achieving a reliable fault localization.

Author Contributions: Conceptualization, C.R., Y.G. and D.C.; Funding acquisition, C.R.; Investigation, Y.G. and A.P.; Methodology, Y.G., G.R. and A.T.; Project administration, C.R. and D.C.; Software, A.P. and A.T.; Writing—original draft, Y.G., G.R. and D.C. All authors have read and agreed to the published version of the manuscript.

Funding: This work was supported in part by the Italian Ministry for Education, University and Research (MIUR) under the program *"Dipartimenti di Eccellenza (2018–2022)"*.

Conflicts of Interest: The authors declare no conflicts of interest.

Nomenclature

$v_{Sa1}, v_{Sb1}, v_{Sc1}, v_{Sa2}, v_{Sb2}, v_{Sc2}$	Stator voltages, phase a/b/c, set 1/2
$i_{Sa1}, i_{Sb1}, i_{Sc1}, i_{Sa2}, i_{Sb2}, i_{Sc2}$	Stator currents, phase a/b/c, set 1/2
$\phi_{Sa1}, \phi_{Sb1}, \phi_{Sc1}, \phi_{Sa2}, \phi_{Sb2}, \phi_{Sc2}$	Stator fluxes, phase a/b/c, set 1/2
$R_{Sa1}, R_{Sb1}, R_{Sc1}, R_{Sa2}, R_{Sb2}, R_{Sc2}$	Stator resistances, phase a/b/c, set 1/2
$\overline{v}_{S1}, \overline{v}_{S3}, \overline{v}_{S5}$	Stator voltage space vectors, subspace 1/3/5
$\overline{i}_{S1}, \overline{i}_{S3}, \overline{i}_{S5}$	Stator current space vectors, subspace 1/3/5
$\overline{\varphi}_{S1}, \overline{\varphi}_{S3}, \overline{\varphi}_{S5}$	Stator flux space vectors, subspace 1/3/5
$L_{S1}, L_{S3}, L_{S5},$	Synchronous inductances, subspace 1/3/5
θ	Rotor position in electrical radians
ω	Rotor speed in electrical radians
T_{em}	Electromagnetic torque
p	Pole pairs
L	Stack length
τ	Polar pitch
N_S	Number of series-connected conductors per pole and per phase
$K_{BS\rho}$	Distribution factor of the stator winding for the ρ-th harmonic
$K_{RS\rho}$	Pitch factor of the stator winding for the ρ-th harmonic
β	Magnet pole arc in electrical radians
$H_{R,\max}$	Amplitude of the rotor magnet magnetic field

References

1. Duran, M.J.; Barrero, F. Recent advances in the design, modeling, and control of multiphase machines—Part II. *IEEE Trans. Ind. Electron.* **2016**, *63*, 459–468. [CrossRef]
2. Barrero, F.; Duran, M.J. Recent advances in the design, modeling, and control of multiphase machines—Part I. *IEEE Trans. Ind. Electron.* **2016**, *63*, 449–458. [CrossRef]
3. Salem, A.; Narimani, M. A Review on Multiphase Drives for Automotive Traction Applications. *IEEE Trans. Transp. Electrif.* **2019**, *5*, 1329–1348. [CrossRef]
4. Zhao, W.; Xu, L.; Liu, G. Overview of permanent-magnet fault-tolerant machines: Topology and design. *CES Trans. Electr. Mach. Syst.* **2018**, *2*, 51–64.
5. Choi, S.; Haque, M.S.; Tarek, T.B.; Mulpuri, V.; Duan, Y.; Das, S.; Garg, V.K.; Ionel, D.M.; Masrur, M.A.; Mirafzal, B.; et al. Fault diagnosis techniques for permanent magnet ac machine and drives—A review of current state of the art. *IEEE Trans. Transp. Electrif.* **2018**, *4*, 444–463. [CrossRef]
6. de la Barrera, P.M.; Bossio, G.R.; Solsona, J.A. High-resistance connection detection in induction motor drives using signal injection. *IEEE Trans. Ind. Electron.* **2014**, *61*, 3563–3573. [CrossRef]
7. Yun, J.; Cho, J.; Lee, S.B.; Yoo, J.Y.Y. Online detection of high-resistance connections in the incoming electrical circuit for induction motors. *IEEE Trans. Ind. Appl.* **2009**, *45*, 694–702. [CrossRef]
8. Kommuri, S.K.; Park, Y.; Lee, S.B. Higher-Order Sliding Mode Based High-Resistance Fault-Control in PMSM Drives. In Proceedings of the 2019 IEEE 28th International Symposium on Industrial Electronics (ISIE), Vancouver, BC, Canada, 12–14 June 2019; pp. 539–544.
9. Tani, A.; Gritli, Y.; Mengoni, M.; Zarri, L.; Sala, G.; Bellini, A.; Serra, G. Detection of magnet demagnetization and high-resistance connections in five-phase surface-mounted permanent magnet generators. In Proceedings of the 2015 IEEE 10th International Symposium on Diagnostics for Electrical Machines, Power Electronics and Drives (SDEMPED), Guarda, Portugal, 1–4 September 2015; pp. 487–493.
10. Gritli, Y.; Mengoni, M.; Rizzoli, G.; Rossi, C.; Tani, A.; Casadei, D. Rotor Magnet Demagnetization Diagnosis in Asymmetrical Six-Phase SurfaceMounted AC Permanent Magnet Synchronous Machine Drives. *IET Electr. Power Appl.* **2019**. [CrossRef]

11. Braunovic, M.; Myshkin, N.K.; Konchits, V.V. *Electrical Contacts: Fundamentals, Applications and Technology*; CRC Press: Boca Raton, FL, USA, 2017.
12. Singh, G.; Kumar, T.C.A.; Naikan, V.N.A. Induction motor inter turn fault detection using infrared thermographic analysis. *Infrared Phys. Technol.* **2016**, *77*, 277–282. [CrossRef]
13. Tallam, R.M.; Lee, S.B.; Stone, G.C.; Kliman, G.B.; Yoo, J.; Habetler, T.G.; Harley, R.G. A survey of methods for detection of stator-related faults in induction machines. *IEEE Trans. Ind. Appl.* **2007**, *43*, 920–933. [CrossRef]
14. Gao, Z.; Cecati, C.; Ding, S.X. A survey of fault diagnosis and fault-tolerant techniques—Part I: Fault diagnosis with model-based and signal-based approaches. *IEEE Trans. Ind. Electron.* **2015**, *62*, 3757–3767. [CrossRef]
15. Gao, Z.; Cecati, C.; Ding, S.X. A survey of fault diagnosis and fault-tolerant techniques—Part II: Fault diagnosis with knowledge-based and hybrid/active approaches. *IEEE Trans. Ind. Electron.* **2015**, *62*, 3768–3774. [CrossRef]
16. Merizalde, Y.; Hernández-Callejo, L.; Duque-Perez, O. State of the art and trends in the monitoring, detection and diagnosis of failures in electric induction motors. *Energies* **2017**, *10*, 1056. [CrossRef]
17. Qi, Y.; Bostanci, E.; Zafarani, M.; Akin, B. Severity estimation of interturn short circuit fault for PMSM. *IEEE Trans. Ind. Electron.* **2019**, *66*, 7260–7269. [CrossRef]
18. Qi, Y.; Zafarani, M.; Gurusamy, V.; Akin, B. Advanced Severity Monitoring of Interturn Short Circuit Faults in PMSMs. *IEEE Trans. Transp. Electrif.* **2019**, *5*, 395–404. [CrossRef]
19. Béthoux, O.; Labouré, E.; Remy, G.; Berthelot, E. Real-time optimal control of a 3-phase PMSM in 2-phase degraded mode. *IEEE Trans. Veh. Technol.* **2017**, *66*, 2044–2052. [CrossRef]
20. Hang, J.; Zhang, J.; Cheng, M.; Zhang, B.; Ding, S. High-resistance connection detection in permanent magnet synchronous machine using zero-sequence current component. *IEEE Trans. Power Electron.* **2015**, *31*, 4710–4719. [CrossRef]
21. Hang, J.; Tang, C.; Ding, S.; Hu, C.; Li, G.; Wang, Q. A signal injection method for fault diagnosis of high-resistance connection in vector-controlled PMSM drive system. In Proceedings of the IECON 2017-43rd Annual Conference of the IEEE Industrial Electronics Society, Beijing, China, 5–8 November 2017; pp. 5021–5026.
22. Gonçalves, P.; Cruz, S.; Mendes, A. Finite Control Set Model Predictive Control of Six-Phase Asymmetrical Machines—An Overview. *Energies* **2019**, *12*, 4693. [CrossRef]
23. Wang, W.; Zhang, J.; Cheng, M.; Li, S. Fault-Tolerant Control of Dual Three-Phase Permanent-Magnet Synchronous Machine Drives Under Open-Phase Faults. *IEEE Trans. Power Electron.* **2017**, *32*, 2052–2063. [CrossRef]
24. Zhou, C.; Tang, W.; Sun, X.D.; Zhou, Z.; Yang, G.; Su, J. Control strategy for dual three-phase PMSM based on reduced order mathematical model under fault condition due to open phases. *J. Eng.* **2018**, *2018*, 489–494.
25. Feng, G.; Lai, C.; Li, W.; Tjong, J.; Kar, N.C. Open-Phase Fault Modeling and Optimized Fault-Tolerant Control of Dual Three-Phase Permanent Magnet Synchronous Machines. *IEEE Trans. Power Electron.* **2019**, *34*, 11116–11127. [CrossRef]
26. Lin, F.J.; Hung, Y.C.; Tsai, M.T. Fault-tolerant control for six-phase PMSM drive system via intelligent complementary sliding-mode control using TSKFNN-AMF. *IEEE Trans. Ind. Electron.* **2013**, *60*, 5747–5762. [CrossRef]
27. Wang, X.; Wang, Z.; Xu, Z.; Cheng, M.; Wang, W.; Hu, Y. Comprehensive diagnosis and tolerance strategies for electrical faults and sensor faults in dual three-phase PMSM drives. *IEEE Trans. Power Electron.* **2019**, *34*, 6669–6684. [CrossRef]
28. Wang, X.; Wang, Z.; Xu, Z.; He, J.; Zhao, W. Diagnosis and Tolerance of Common Electrical Faults in T-Type Three-Level Inverters Fed Dual Three-Phase PMSM Drives. *IEEE Trans. Power Electron.* **2020**, *35*, 1753–1769. [CrossRef]
29. Forstner, G.; Kugi, A.; Kemmetmüller, W. Magnetic Equivalent Circuit Model of a Dual Three-Phase PMSM with Winding Short Circuit. In Proceedings of the IECON 2019-45th Annual Conference of the IEEE Industrial Electronics Society, Lisbon, Portugal, 14–17 October 2019; pp. 1177–1182.

30. Grandi, G.; Serra, G.; Tani, A. General analysis of multi-phase systems based on space vector approach. In Proceedings of the 2006 12th International Power Electronics and Motion Control Conference, Portoroz, Slovenia, 30 August–1 September 2006; pp. 834–840.
31. Gritli, Y.; Tani, A.; Rossi, C.; Casadei, D. Detection of Rotor Magnet Demagnetization in Asymmetrical Six-Phase Surface Mounted Permanent Magnet Synchronous Motor Drive. In Proceedings of the 2018 23rd International Conference on Electrical Machines ICEM 2018, Alexandroupoli, Greece, 3–6 September 2018; pp. 1809–1814.

energies

MDPI

Article

Comparative Analysis of Fault-Tolerant Dual-Channel BLDC and SR Motors

M. Korkosz *, P. Bogusz, J. Prokop, B. Pakla and G. Podskarbi

The Faculty of Electrical and Computer Engineering, Rzeszow University of Technology, Al. Postancow Warszawy 12, 35-959 Rzeszow, Poland

* Correspondence: mkosz@prz.edu.pl; Tel.: +48-178-651-389

Received: 20 May 2019; Accepted: 25 June 2019; Published: 28 June 2019

Abstract: This article presents the results of a comparative analysis of two electronically commutated brushless direct current machines intended for fault-tolerant drives. Two machines designed by the authors were compared: a 12/14 dual-channel brushless direct current motor (DCBLDCM) with permanent magnets and a 12/8 dual-channel switched reluctance motor (DCSRM). Information is provided here on the winding configuration, the parameters, and the power converters of both machines. We developed mathematical models of the DCBLDCM and DCSRM which accounted for the nonlinearity of their magnetization characteristics in dual-channel operation (DCO) and single-channel operation (SCO) modes. The static torque characteristics and flux characteristics of both machines were compared for operation in DCO and SCO modes. The waveforms of the current and the electromagnetic torque are presented for DCO and SCO operating conditions. For DCO mode, an analysis of the behavior of both machines under fault conditions (i.e., asymmetrical control, shorted coil, and open phase) was performed. The two designs were compared, and their strengths and weaknesses were indicated.

Keywords: multiphase machines; fault-tolerance; dual-channel; brushless direct current motor with permanent magnet (BLDCM); switched reluctance motor (SRM)

1. Introduction

In fault-tolerant electrical drives, two solutions are used: multiphase motors (with more than three phases) as the standard or three-phase motors with dual windings. Multiphase motors, regardless of the type of machine, usually require a more extensive power supply system and more complex control algorithms [1–4]. Three-phase motors with independent, three-phase dual windings require two independent power supply circuits [5–21]. As a result, two channels are obtained which, depending on the type of machine and the configuration of its windings, may be magnetically independent or partly independent. Full or significant magnetic independence between the channels makes control of such a motor much easier. This is particularly important in the case of fault-tolerant operation.

Direct current electronically commutated motors include brushless direct current motors (BLDCMs) with permanent magnets and switched reluctance motors (SRMs). BLDCMs and SRMs can be designed as standard three-phase or multiphase machines. In the case of a BLDCM, the use of more than three phases is not a typical approach [1,22,23]. In the case of switched reluctance motors, a four-phase solution is a borderline case for potential commercial applications [24]. Consequently, in fault-tolerant drives, the authors suggest using dual-channel BLDCM (DCBLDCM) or SRM (DCSRM) (i.e., ones with dual three-phase windings). In such machines, in order to achieve the same operating point, depending on the control strategy adopted, two channels or only one channel is supplied in normal operating conditions. Selection of a DCBLDCM or a DCSRM for a given drive is not self-evident.

In this study, a comparative analysis was performed of two designs of dual-channel brushless direct current electronically commutated motors with or without permanent magnets. The machines were designed for use in critical drive systems. The first machine, with permanent magnets, was designated DCBLDCM and the other, without permanent magnets, was designated DCSRM. Both machines, at the stages of their design and construction, were adapted for independent dual-channel supply, that is, for dual-channel operation (DCO) and single-channel operation (SCO).

The aim of this study was to compare the features of the DCBLDCM and the DCSRM designs in DCO and SCO, especially in the context of guaranteeing the continuation of motor operation after the occurrence of a fault condition. The rated power, rated torque and base speed of both machines are different. Therefore, the comparison of performance of both machines is difficult. For this reason, the results are presented as a ratio of value to the base values (presented in Table 1). Nevertheless, the aim of the paper is to compare the possibilities of dual-channel operation rather than the performance.

In Section 2, the DCBLDC and DCSR machine designs studied are presented and the configurations of their power supply systems, the distributions of their stator windings, and their parameters and characteristics are shown. Section 3 contains the authors' nonlinear mathematical models of DCBLDC and DCSR machines for DCO and SCO. These mathematical models take into account the nonlinearity of the magnetic circuit and the magnetic coupling between particular phases of channels A and B. The models include novel electromagnetic torque formulas of the DCBLDCM and the DCSRM for DCO and SCO proposed by the authors. It was proven that the structure of the formulas of the mathematical model of SRM was the particular simplification of the formulas of the model of the BLDC machine assuming that PM magnetization equivalent current was zero (PM flux was neglected). Section 4 presents the static characteristics of the motors studied (the electromagnetic torque and the linkage fluxes) for DCO and SCO modes. The results of the analysis of current and electromagnetic torque waveforms presented in Section 5 are divided into two parts. The first part contains the results of an analysis of DCO and SCO, assuming the electric and magnetic symmetry of both channels. The second part describes the analysis that was performed of nonstandard operating conditions in DCO mode, such as:

- asymmetry control of both channels,
- short circuits of part of the winding of one of the channels, and
- open-phase of the winding of one of the channels.

The results of the comparative analysis of the two studied machines and the conclusions are presented in Section 6.

2. Analysis of the DCBLDC and DCSR Motors

An analysis was performed of two three-phase designs of brushless direct current machines. In the case of the DCBLDC motor, this was a solution with 12 stator poles and 14 rotor poles (magnets) with an external rotor (Figure 1a). The machine was designed for a small unmanned aerial vehicle (UAV). In the case of the DCSRM, a solution with 12 stator poles and 8 rotor poles was selected (Figure 1b). The DCSRM was designed for a fan drive. Both motors have the same number of stator poles. At the design stage of both motors, provisions were made for the possibility of a dual-channel power supply. There were six stator poles per channel. The use of two independent power supply channels made it possible to achieve a dual three-phase power supply in both cases. The dual-channel power supply diagram is shown in Figure 1c for the BLDCM and in Figure 1d for the SRM.

In the case of the configuration of the windings of the DCBLDC motor, the layout shown in Figure 1a (ABABAB) was used. As demonstrated in [6,24], this is the most advantageous configuration, and not only with regard to the distribution of the magnetic pull forces. In the case of the SRM, a similar principle was adopted with regard to the positions of the windings of the different channels. The configuration used (ABABABABABAB) is shown in Figure 1b. In the case of an SRM with four poles per phase, the short-flux path solution (NSNS) is usually selected [8,9]. This configuration of the poles of a single phase may also be used in the case of a dual-channel power supply [8,9,11,12]. It ensures the most advantageous motor parameters for a dual-channel power supply. However, in the authors' opinion, in the case of a machine intended for a dual-channel power supply, the configuration with a long-flux path (NNSS) should be selected [11]. For a classic or dual-channel power supply, this configuration results in slightly poorer machine performance [11]. At the same time, it is characterized by greater magnetic independence of the different channels. The NNSS configuration of the poles of an SRM meets this condition. This is demonstrated later in this article.

Selected geometric parameters of both motors are shown in Table 1, which also contains selected electric parameters. The electric parameters were specified for the SCO power supply. The base torque obtained for the DCBLDCM was equal to $T_{bDCBLDCM}$ = 1.0 N·m, while that obtained for the DCSRM was equal to T_{bDCSRM} = 0.5 N·m. The base torque in the DCBLDC was obtained when the line current was equal to $I_{refDCBLDCM}$ = 36.5 A. In the case of the DCSRM, this value was $I_{refDCSRM}$ = 8.5 A. The above values were adopted as a reference for presentation of the waveform of the line current (Section 5) and the electromagnetic torque in the relative values (Sections 4 and 5) as a function of time.

Figure 2c,d show idealized torque–speed characteristics of the DCBLDCM (Figure 2c) and the DCSRM (Figure 2d). The torque–speed characteristics of the two motors were significantly different. In the case of the DCBLDCM in DCO and SCO modes, there were small differences in the base speeds. For DCO mode, the base speed was slightly higher (by several percent). In the case of the DCSRM, the differences between DCO and SCO modes were much more significant. In DCO operation mode, the base speed was up to two times higher than in SCO mode. In the case of the DCSRM, there was a constant power range regardless of the operation mode (DCO or SCO). In the case of the DCBLDCM, there was practically no constant power region.

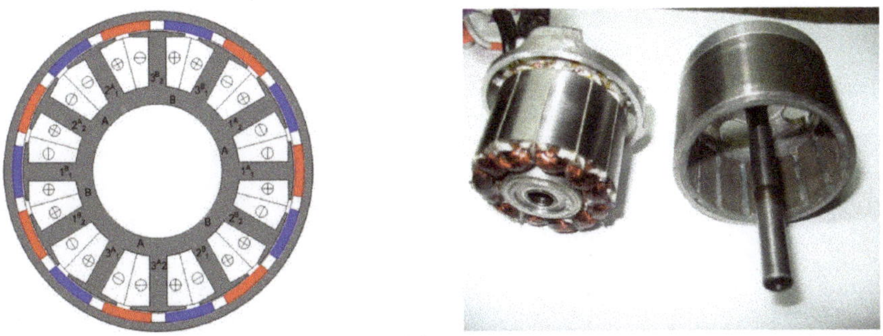

(a)

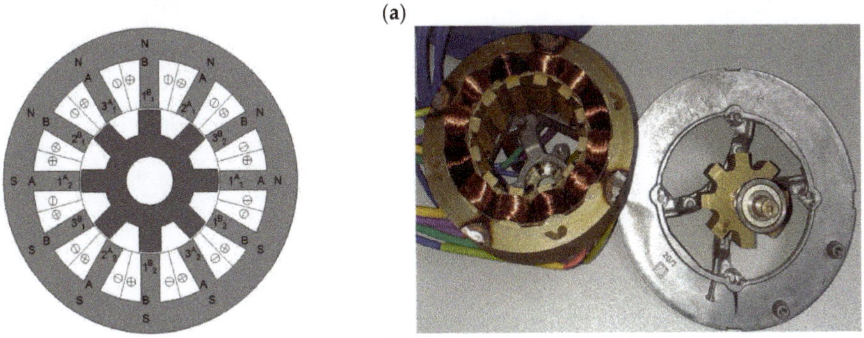

(b)

Figure 1. *Cont.*

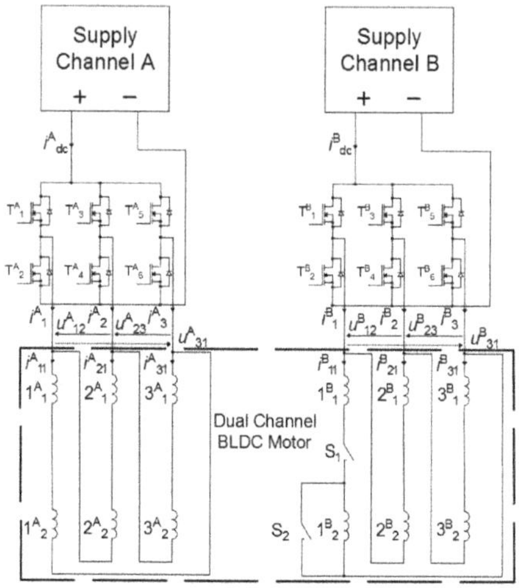

(c)

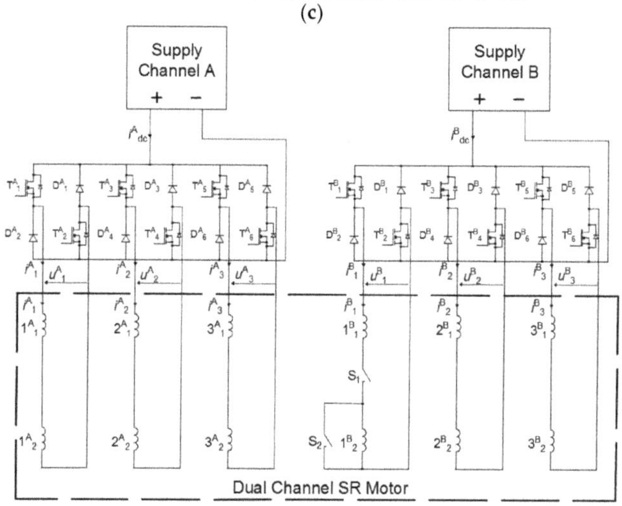

(d)

Figure 1. The geometry and prototypes of dual-channel power supplied brushless electronically commutated motors. (**a**) The geometry and prototype of a three-phase DCBLDCM. (**b**) The geometry and prototype of a DCSRM. (**c**) A scheme of a DCBLDC motor supply system in the DCO mode. (**d**) A scheme of a DCSRM supply system in the DCO mode.

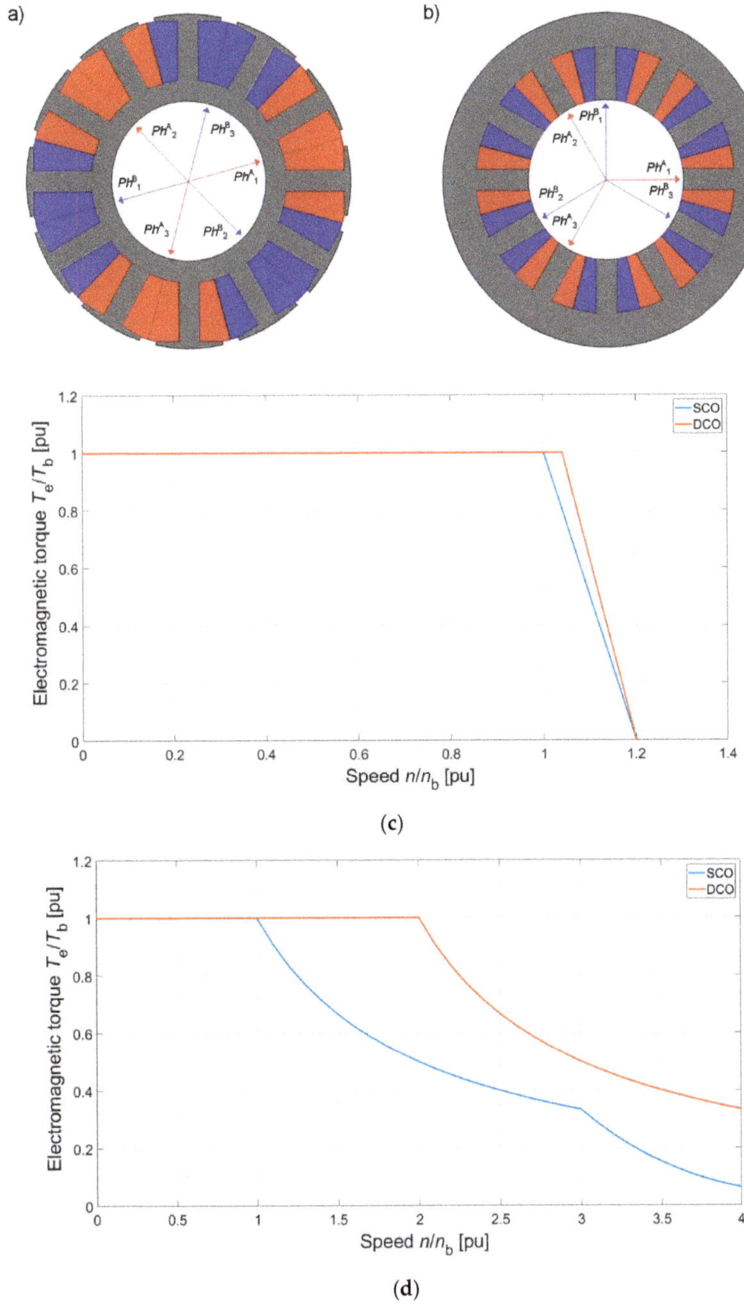

Figure 2. A scheme of winding distributions of channels A and B on a stator of the three-phase DCBLDCM and DCSRM and theoretical torque vs. speed characteristics: (**a**) DCBLDCM, (**b**) DCSRM, (**c**) torque vs. speed for DCBLDCM, and (**d**) torque vs. speed for DCSRM.

Table 1. Main parameters of the BLDC motor and the SRM for SCO modes.

Parameter	DCBLDC	DCSRM
Supply voltage	25 V	60 V
Reference current	36.5 A	8.5 A
Maximum speed on idle run	10,000 r/min	15,000 r/min
Base/rated speed	7900/8000 r/min	3000/8000 r/min
Base/rated torque	1.0/0.5 N·m	0.5/0.25 N·m
Number of phases	3	3
Number of stator slots	12	12
Number of rotor poles	14	8
Number of channels	2	2
Diameter of rotor	47 mm (inner)	45 mm (outer)
Core length	28 mm	28 mm
Winding configuration	Delta	Independent
Electrical steel	M230-23	M470-50
Permanent magnet	N42SH	-
Current density	10 A/mm^2	7 A/mm^2
Fill factor	45%	40%
Type of winding	concentrated	concentrated
Angle of stator pole	28°	15°
Angle of rotor pole	20°	18°
Total net mass	0.5 kg	1.5 kg

3. A Mathematical Model of the DCBLDC and DCSR Motors Analyzed

3.1. Main Assumptions and General Equation Structure

The subjects of the mathematical modeling were dual-channel three-phase BLDC and SR machines, for which the authors' circuit-based models, known as flux models, were proposed. The presented mathematical models of the DCBLDC and DCSR machines took into account the nonlinearity of the magnetic circuit and the magnetic couplings between particular phases within a given channel (A or B), as well as between channels (A and B). The following simplifying assumptions were made in the proposed models:

- symmetry of the magnetic circuit structure of both the stator and the rotor;
- decomposition of phase fluxes into a sum of fluxes induced by phase currents (leakage and main fluxes) and fluxes from permanent magnets in the DCBLDCM;
- omission of phenomena related to eddy currents and magnetic hysteresis;
- omission of the influence of temperature on the parameters of the machines and the fluxes generated by permanent magnets (in the DCBLDCM).

The general structure of the formulas of the circuit-based mathematical models of the three-phase dual-channel machines can be written in the following form:

$$\begin{bmatrix} \mathbf{u}^A \\ \mathbf{u}^B \end{bmatrix} = \begin{bmatrix} \mathbf{R}^A & \mathbf{0} \\ \mathbf{0} & \mathbf{R}^B \end{bmatrix} \begin{bmatrix} \mathbf{i}^A \\ \mathbf{i}^B \end{bmatrix} + \frac{d}{dt} \begin{bmatrix} \boldsymbol{\psi}^A \\ \boldsymbol{\psi}^B \end{bmatrix}, \tag{1}$$

$$J\frac{d\omega}{dt} + D\omega + T_L = T_e \tag{2}$$

$$\frac{d\theta}{dt} = \omega \tag{3}$$

where for channels A and B (k ∈ A, B), the vectors representing voltages $\mathbf{u}^k$, currents $\mathbf{i}^k$, fluxes $\boldsymbol{\psi}^k$, as well as matrixes of resistances $\mathbf{R}^k$ are defined as follows:

$$\mathbf{u}^k = \left[u_1^k, u_2^k, u_3^k\right]^T, \mathbf{i}^k = \left[i_1^k, i_2^k, i_3^k\right]^T, \boldsymbol{\psi}^k = \left[\psi_1^k, \psi_2^k, \psi_3^k\right]^T, \mathbf{R}^k = diag(R_1^k, R_2^k, R_3^k)$$

The following symbols are used in Equations (1)–(3): θ—rotor position, ω—angular velocity, J—rotor moment of inertia, D—coefficient of viscous friction, T_L—load torque, and T_e—electromagnetic torque. Electromagnetic torque T_e in Equation (2) can be calculated as a derivative of total magnetic field coenergy in the air gap with respect to rotor position.

3.2. Mathematical Models of DCBLDC Motors

3.2.1. DCBLDCM—DCO Mode

In machines with permanent magnets, the fluxes in Equation (1) depend on the rotor position θ, the winding current, and the permanent magnets' magnetization equivalent current, designated as i^{PM}. The voltage–current Equation (1) and the expression for electromagnetic torque in Equation (2) for DCO can be written in the following form:

$$\begin{bmatrix} \mathbf{u}^A \\ \mathbf{u}^B \end{bmatrix} = \begin{bmatrix} \mathbf{R}^A & \mathbf{0} \\ \mathbf{0} & \mathbf{R}^B \end{bmatrix} \begin{bmatrix} \mathbf{i}^A \\ \mathbf{i}^B \end{bmatrix} + \frac{d}{dt} \begin{bmatrix} \boldsymbol{\psi}^A(\theta, \mathbf{i}^A, \mathbf{i}^B, i^{PM}) \\ \boldsymbol{\psi}^B(\theta, \mathbf{i}^A, \mathbf{i}^B, i^{PM}) \end{bmatrix} \tag{4}$$

$$T_e = T_e(\theta, \mathbf{i}^A, \mathbf{i}^B, i^{PM}) \tag{5}$$

where flux linkages caused by windings' currents and permanent magnets for both channels A and B are defined as follows:

$$\boldsymbol{\psi}^k(\theta, \mathbf{i}^A, \mathbf{i}^B, i^{PM}) = \begin{bmatrix} \psi_1^{kPM}(\theta, i^{PM}) + L_{1\sigma}^{kk} i_1^k + \sum\limits_{j=1}^{3} \left(\sum\limits_{l=A}^{B} \psi_{1j}^{kl}(\theta, i_j^l, i^{PM}) \right) \\ \psi_2^{kPM}(\theta, i^{PM}) + L_{2\sigma}^{kk} i_2^k + \sum\limits_{j=1}^{3} \left(\sum\limits_{l=A}^{B} \psi_{2j}^{kl}(\theta, i_j^l, i^{PM}) \right) \\ \psi_3^{kPM}(\theta, i^{PM}) + L_{3\sigma}^{kk} i_3^k + \sum\limits_{j=1}^{3} \left(\sum\limits_{l=A}^{B} \psi_{3j}^{kl}(\theta, i_j^l, i^{PM}) \right) \end{bmatrix} \tag{6}$$

In Expression (6), $\psi_i^{kPM}(\theta, i^{PM})$ are the fluxes generated by permanent magnets, and $L_{i\sigma}^{kk}$ are the leakage fluxes (for k ∈ A, B and i = 1,2,3).

The expression for electromagnetic torque (5), with assumptions (6) taken into account, can be written in the following form:

$$\begin{aligned} T_e(\theta, \mathbf{i}^A, \mathbf{i}^B, i^{PM}) = & \sum_{i=1}^{3} \left(i_i^A \frac{\partial \psi_i^{APM}(\theta, i^{PM})}{\partial \theta} + i_i^B \frac{\partial \psi_i^{BPM}(\theta, i^{PM})}{\partial \theta} \right) \\ & + \sum_{i=1}^{3} \left(\int_0^{i_i^A} \frac{\partial \psi_{ii}^{AA}(\theta, \bar{i}_i^A, i^{PM})}{\partial \theta} d\bar{i}_i^A + \int_0^{i_i^B} \frac{\partial \psi_{ii}^{BB}(\theta, \bar{i}_i^B, i^{PM})}{\partial \theta} d\bar{i}_i^B \right) \\ & + \sum_{i=2}^{3} \sum_{j=1}^{i-1} \left(i_i^A \frac{\partial \psi_{ij}^{AA}(\theta, i_j^A, i^{PM})}{\partial \theta} + i_i^B \frac{\partial \psi_{ij}^{BB}(\theta, i_j^B, i^{PM})}{\partial \theta} \right) + \sum_{i=1}^{3} \sum_{j=1}^{3} \left(i_i^B \frac{\partial \psi_{ij}^{BA}(\theta, i_j^A, i^{PM})}{\partial \theta} \right) + T_{cog}(\theta, i^{PM}) \end{aligned} \tag{7}$$

Electromagnetic torque is the sum of torques from the fluxes of magnets, windings' currents, and the cogging torque T_{cog}. Equations (2) and (4) with Expressions (6) and (7) constitute the nonlinear mathematical model of the DCBLDC machine in DCO mode.

3.2.2. DCBLDCM—SCO Mode

In this particular case, when the DCBLDC machine is operating in SCO mode (e.g., only channel A or B is supplied), Equations (4) and (6), as well as the expression for torque (7), can be simplified (for $k \in A, B$):

$$\begin{bmatrix} u_1^k \\ u_2^k \\ u_3^k \end{bmatrix} = \begin{bmatrix} R_1^k & 0 & 0 \\ 0 & R_2^k & 0 \\ 0 & 0 & R_3^k \end{bmatrix} \begin{bmatrix} i_1^k \\ i_2^k \\ i_3^k \end{bmatrix} + \frac{d}{dt} \begin{bmatrix} \psi_1^{kPM}(\theta, i^{PM}) + L_{1\sigma}^{kk} i_1^k + \sum\limits_{j=1}^{3} \left(\psi_{1j}^{kk}(\theta, i_j^k, i^{PM}) \right) \\ \psi_2^{kPM}(\theta, i^{PM}) + L_{2\sigma}^{kk} i_2^k + \sum\limits_{j=1}^{3} \left(\psi_{2j}^{kk}(\theta, i_j^k, i^{PM}) \right) \\ \psi_3^{kPM}(\theta, i^{PM}) + L_{3\sigma}^{kk} i_3^k + \sum\limits_{j=1}^{3} \left(\psi_{3j}^{kk}(\theta, i_j^k, i^{PM}) \right) \end{bmatrix} \tag{8}$$

$$\begin{aligned} T_e(\theta, i_1^k, i_2^k, i_3^k, i^{PM}) = {} & \sum_{i=1}^{3} \left(i_i^k \frac{\partial \psi_i^{kPM}(\theta, i^{PM})}{\partial \theta} \right) + \sum_{i=1}^{3} \left(\int_0^{i_i^k} \frac{\partial \psi_{ii}^{kk}(\theta, \bar{i}_i^k, i^{PM})}{\partial \theta} d\bar{i}_i^k \right) + \sum_{i=2}^{3} \sum_{j=1}^{i-1} \left(i_i^k \frac{\partial \psi_{ij}^{kk}(\theta, i_j^k, i^{PM})}{\partial \theta} \right) \\ & + T_{cog}(\theta, i^{PM}) \end{aligned} \tag{9}$$

Equations (8) and (2) with Expression (9) constitute the nonlinear mathematical model of the DCBLDC machine in SCO mode.

3.3. Mathematical Models of DCSR Motors

3.3.1. DCSRM—DCO Mode

The equations of the mathematical model of DCSR machines can also be derived from the equations of DCBLDC machines by eliminating the relevant fluxes produced by permanent magnets. The voltage–current Equation (1) and the equation of motion of a circuit-based mathematical model of the DCSRM for DCO mode can be written in the following form:

$$\begin{bmatrix} \mathbf{u}^A \\ \mathbf{u}^B \end{bmatrix} = \begin{bmatrix} \mathbf{R}^A & \mathbf{0} \\ \mathbf{0} & \mathbf{R}^B \end{bmatrix} \begin{bmatrix} \mathbf{i}^A \\ \mathbf{i}^B \end{bmatrix} + \frac{d}{dt} \begin{bmatrix} \boldsymbol{\psi}^A(\theta, \mathbf{i}^A, \mathbf{i}^B) \\ \boldsymbol{\psi}^B(\theta, \mathbf{i}^A, \mathbf{i}^B) \end{bmatrix} \tag{10}$$

$$T_e = T_e(\theta, \mathbf{i}^A, \mathbf{i}^B) \tag{11}$$

where for channels $k \in A, B$, flux linkages caused only by winding currents $\boldsymbol{\psi}^k(\theta, \mathbf{i}^A, \mathbf{i}^B)$ are defined as follows:

$$\boldsymbol{\psi}^k(\theta, \mathbf{i}^A, \mathbf{i}^B) = \begin{bmatrix} L_{1\sigma}^{kk} i_1^k + \sum\limits_{j=1}^{3} \left(\sum\limits_{l=A}^{B} \psi_{1j}^{kl}(\theta, i_j^l) \right) \\ L_{2\sigma}^{kk} i_2^k + \sum\limits_{j=1}^{3} \left(\sum\limits_{l=A}^{B} \psi_{2j}^{kl}(\theta, i_j^l) \right) \\ L_{3\sigma}^{kk} i_3^k + \sum\limits_{j=1}^{3} \left(\sum\limits_{l=A}^{B} \psi_{3j}^{kl}(\theta, i_j^l) \right) \end{bmatrix} \tag{12}$$

For the DCO mode of the DCSRM, the expression for electromagnetic torque, with assumptions taken into account, can be written in the following form:

$$\begin{aligned} T_e(\theta, \mathbf{i}^A, \mathbf{i}^B) = {} & \sum_{i=1}^{3} \left(\int_0^{i_i^A} \frac{\partial \psi_{ii}^{AA}(\theta, \bar{i}_i^A)}{\partial \theta} d\bar{i}_i^A + \int_0^{i_i^B} \frac{\partial \psi_{ii}^{BB}(\theta, \bar{i}_i^B)}{\partial \theta} d\bar{i}_i^B \right) \\ & + \sum_{i=2}^{3} \sum_{j=1}^{i-1} \left(i_i^A \frac{\partial \psi_{ij}^{AA}(\theta, i_j^A)}{\partial \theta} + i_i^B \frac{\partial \psi_{ij}^{BB}(\theta, i_j^B)}{\partial \theta} \right) + \sum_{i=1}^{3} \sum_{j=1}^{3} \left(i_i^B \frac{\partial \psi_{ij}^{BA}(\theta, i_j^A)}{\partial \theta} \right) \end{aligned} \tag{13}$$

3.3.2. DCSRM—SCO Mode

In the particular case where the DCSRM is operating in SCO mode (e.g., only channel A or B is supplied), Equation (10) with the expression for torque (13) can be simplified (k ∈ A or B):

$$\begin{bmatrix} u_1^k \\ u_2^k \\ u_3^k \end{bmatrix} = \begin{bmatrix} R_1^k & 0 & 0 \\ 0 & R_2^k & 0 \\ 0 & 0 & R_3^k \end{bmatrix}\begin{bmatrix} i_1^k \\ i_2^k \\ i_3^k \end{bmatrix} + \frac{d}{dt}\begin{bmatrix} L_{1\sigma}^{kk} i_1^k + \sum\limits_{j=1}^{3}\left(\psi_{1j}^{kk}(\theta, i_j^k)\right) \\ L_{2\sigma}^{kk} i_2^k + \sum\limits_{j=1}^{3}\left(\psi_{2j}^{kk}(\theta, i_j^k)\right) \\ L_{3\sigma}^{kk} i_3^k + \sum\limits_{j=1}^{3}\left(\psi_{3j}^{kk}(\theta, i_j^k)\right) \end{bmatrix} \tag{14}$$

$$T_e(\theta, i_1^k, i_2^k, i_3^k) = \sum_{i=1}^{3}\left(\int_0^{i_i^k} \frac{\partial \psi_{ii}^{kk}(\theta, \bar{i}_i^k)}{\partial \theta} d\bar{i}_i^k\right) + \sum_{i=2}^{3}\sum_{j=1}^{i-1}\left(i_i^k \frac{\partial \psi_{ij}^{kk}(\theta, i_j^k)}{\partial \theta}\right) \tag{15}$$

Equations (14) and (2) with Expression (15) constitute the nonlinear mathematical model of the DCSRM in SCO mode.

Based on the presented equations for the BLDCM in DCO and SCO modes, it is possible to obtain models for special simplifying assumptions, for example, by omitting couplings between particular phases or channels A and B.

3.4. *Flux Characteristics for Simulation Models*

The relationships of fluxes as a function of rotor position and phase current were determined by means of 2D field methods (finite element method) and then the obtained set of relationships was used in the circuit models. For example, Figure 3 shows 3D views of flux linkage ψ_1^A of the first phase of channel A as a function of rotor position θ and current i_1^A for a DCBLDCM (Figure 3a) and a DCSRM (Figure 3b).

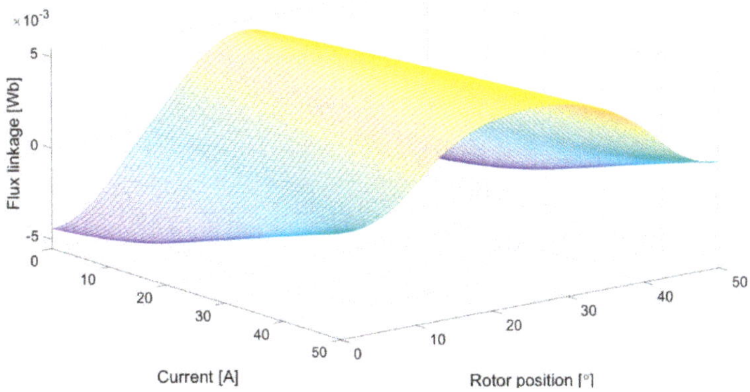

(**a**)

Figure 3. *Cont.*

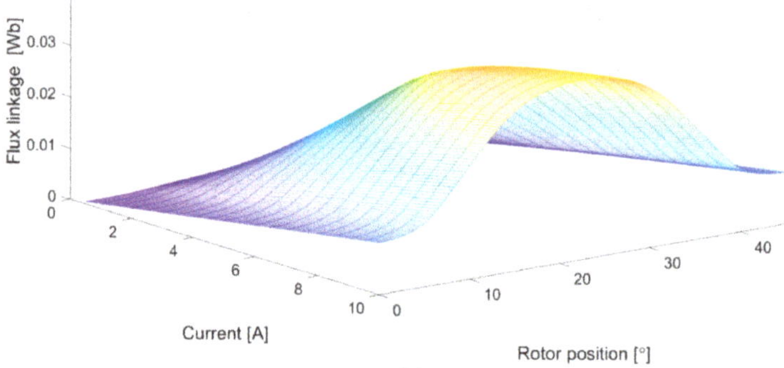

(b)

Figure 3. Flux linkage ψ_1^A vs. rotor position θ and current i_1^A: (**a**) DCBLDCM; (**b**) DCSRM.

4. Static Analysis

Static calculations were performed for the reference values of currents given in Table 1 that were required for SCO mode. For DCO mode, the reference values of currents were reduced in order to obtain a comparable shape of the static torque characteristic.

4.1. Characteristics of the DCBLDCM

Figure 4a shows the determined static characteristics of electromagnetic torque as a function of rotor position for DCO and SCO modes. In the case of DCO mode, the reference value of the current was reduced to 50% of the value specified in Table 1. The results of laboratory tests are shown in Figure 4b. Laboratory tests were performed for $I = 0.5\, I_{ref}$ (i.e., $i^A = i^B = 0.25\, I_{ref}$ for DCO mode and $i^A = 0.5\, I_{ref}$, $i^B = 0$ for SCO mode (channel A) and $i^A = 0\, I_{ref}$, $i^B = 0.5\, I_{ref}$ for SCO mode (channel B)). The fluxes linked with the different windings for DCO mode are shown in Figure 4c. The linkage fluxes for SCO mode are shown in Figure 4d.

In dual-channel mode, the brushless motor with a permanent magnet required half of the value of the reference current of SCO mode in order to obtain the required value of the base torque. This was quite beneficial from the point of view of the control algorithm of the operation of a motor supplied from two channels. The configuration used was characterized by very high magnetic independence between the two channels. The magnetic couplings between the two channels in normal operating conditions were minimal. The linkage fluxes of the windings of the channel that was not being supplied (Figure 4d) came mainly from permanent magnets. The assumption made in the mathematical model that the two channels of the presented design are magnetically independent significantly simplified the equations of the mathematical model of the DCBLDCM.

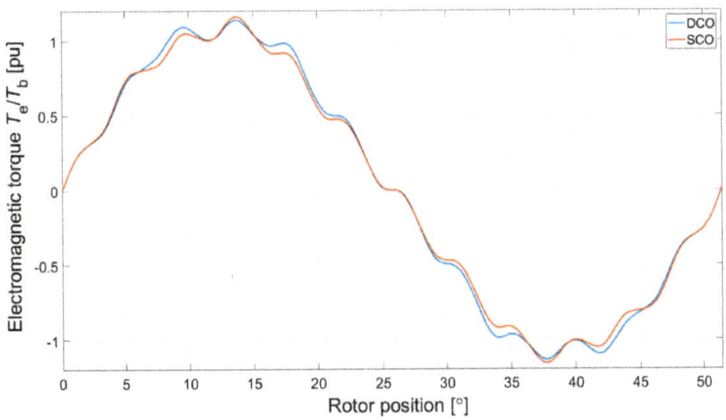

(a)

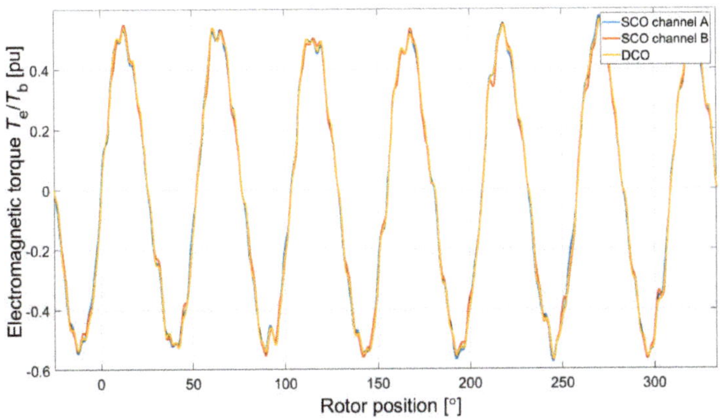

(b)

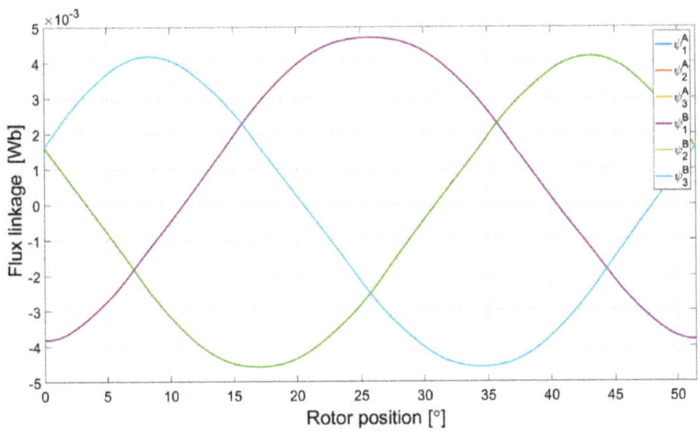

(c)

Figure 4. *Cont.*

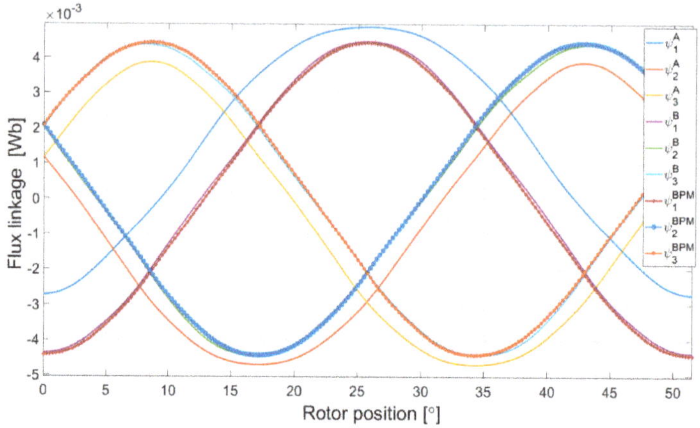

(**d**)

Figure 4. Static torque and flux linkage characteristics of the DCBLDCM. (**a**) Electromagnetic torque T_e vs. rotor position θ in the DCO mode ($i^A = i^B = 0.5\, I_{ref}$) and SCO mode ($i^A = I_{ref}$, $i^B = 0$)—simulation. (**b**) Electromagnetic torque T_e vs. rotor position θ in DCO mode ($i^A = i^B = 0.25\, I_{ref}$), channel A SCO mode ($i^A = 0.5\, I_{ref}$, $i^B = 0$), and channel B SCO mode ($i^A = 0$, $i^B = 0.5\, I_{ref}$)—laboratory test. (**c**) Flux linkage vs. rotor position θ in DCO mode ($i^A = i^B = 0.5\, I_{ref}$)—simulation. (**d**) Flux linkage vs. rotor position θ in SCO mode ($i^A = I_{ref}$, $i^B = 0$)—simulation.

4.2. Characteristics of the DCSRM

In the case of a switched reluctance motor, the relationship between DCO and SCO modes is more complex. This is largely due to this machine's principle of operation. Unfortunately, the principle that, in DCO mode, the reference value of the current should be equal to 50% of the value in SCO mode, which applies to the DCBLDCM, could not be adopted here. Due to the nonlinear relationship between the value of the generated electromagnetic torque and the current, achieving the same value of electromagnetic torque in DCO mode requires a current greater than 50% of the reference value of SCO mode. In the analyzed case, the required level was 70%. Figure 5a shows the relationship between the electromagnetic torque as a function of rotor position for DCO and SCO modes of the SRM. Examples of characteristics determined in laboratory conditions are shown in Figure 5b. Laboratory tests were performed for $I = 0.5\, I_{ref}$ (i.e., $i^A = 0.5\, I_{ref}$, $i^B = 0$ for SCO mode (channel A); $i^A = 0$, $i^B = 0.5\, I_{ref}$ for SCO mode (channel B); and $i^A = i^B = 0.35\, I_{ref}$ for DCO mode). The characteristics of the linkage flux for DCO mode are shown in Figure 5c, and those for SCO mode in Figure 5d.

In the case of the SRM, the saturation of a magnetic circuit affected the relationship between DCO and SCO modes. This can be seen in the torque characteristics (Figure 5a) and the flux characteristics (Figure 5c,d). However, the impact of magnetic couplings between the two channels was small (Figure 5d). For this configuration, it can also be assumed that both channels were characterized by a very big magnetic separation. This is why this specific configuration of pole windings was selected for the dual-channel power supply (Figure 1b).

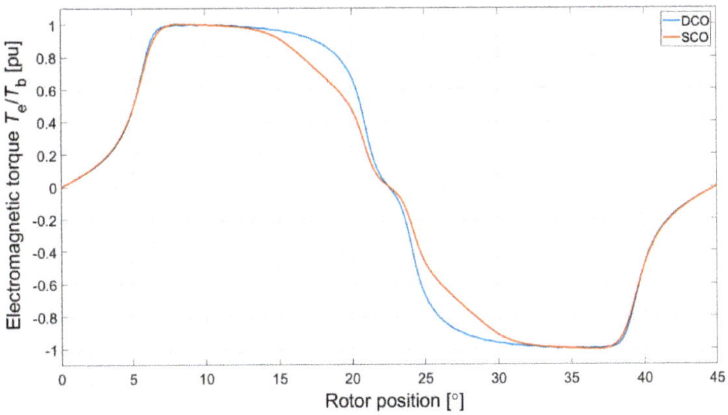

(**a**)

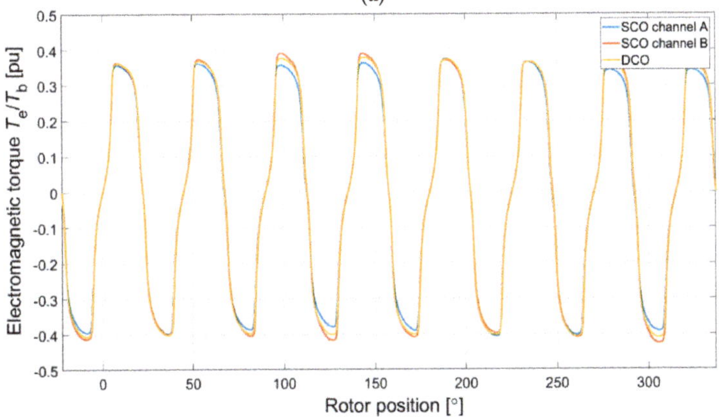

(**b**)

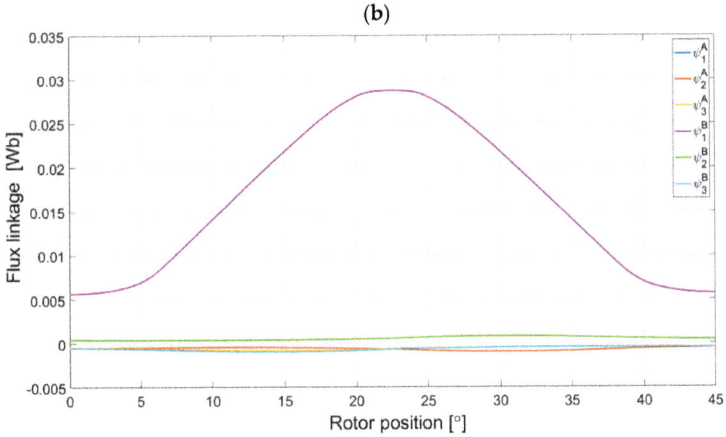

(**c**)

Figure 5. *Cont.*

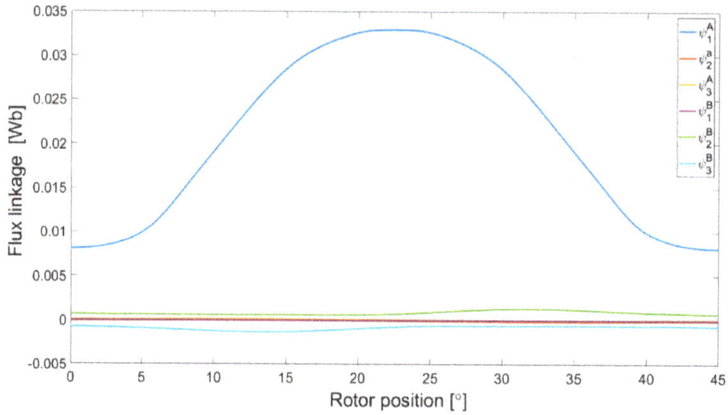

(d)

Figure 5. Static torque and flux linkage characteristics of the DCSRM. (**a**) Electromagnetic torque T_e vs. rotor position θ in DCO mode ($i^A = i^B = 0.7\ I_{ref}$) and SCO mode ($i^A = I_{ref}$, $i^B = 0$)—simulation. (**b**) Electromagnetic torque T_e vs. rotor position θ in DCO mode ($i^A = i^B = 0.35\ I_{ref}$), channel A SCO mode ($i^A = 0.5\ I_{ref}$, $i^B = 0$), and channel B SCO mode ($i^A = 0$, $i^B = 0.5\ I_{ref}$)—laboratory test. (**c**) Flux linkage vs. rotor position θ in DCO mode ($i^A = i^B = 0.7\ I_{ref}$)—simulation. (**d**) Flux linkage vs. rotor position θ in SCO mode ($i^A = I_{ref}$, $i^B = 0$)—simulation.

5. Transient Analysis

5.1. DCO and SCO

An analysis of DCO and SCO of both motors was performed for two cases: operation at low rotational speed and operation at high rotational speed. In the first operating point, both motors worked with constant torque, which required use of a current controller. The second operating point was located on the natural characteristic. Like in the previous section, the obtained waveforms of the electromagnetic torque and the line currents were compared with the base values specified in Table 1. In SCO mode, it was assumed that only channel A of the machine would be supplied.

5.1.1. Constant Torque Operation

For this operating point, the numerical calculations were performed for the speed of $n = 2000$ r/min. It was assumed that the electromagnetic torque T_e was equal to the base value given in Table 1. The obtained results are shown in Figure 6.

The comparison of the waveforms of the electromagnetic torque and the currents led to the conclusion that, in this range of operation, there were no significant differences between DCO and SCO modes for the DCBLDCM (Figure 6a,b). In DCO mode, half of the value of the reference current for SCO mode was required. The torque ripple in the case of DCBLDCM was 32% in DCO mode and 31% in SCO mode. Slightly greater differences were observed for DCO and SCO modes of the DCSRM (Figure 6d) in the waveforms of both the electromagnetic torque and the currents. In the case of SCO mode, a greater electromagnetic torque ripple (34%) was observed in comparison with DCO mode (27%). The main cause of the increase in electromagnetic torque ripple was the changed shape of the static torque characteristic (Figure 5a). In order to obtain the same value of electromagnetic torque in DCO mode, the DCSRM required a reduction of the value of the reference current (to 70%). In the case of DCO mode, the impact of magnetic couplings between the channels was unnoticeable.

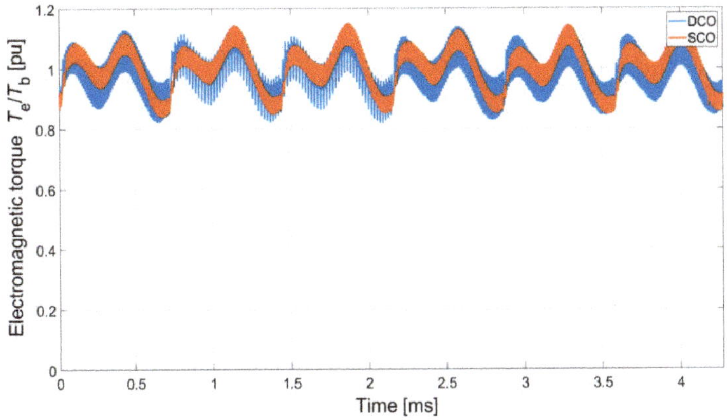

(a)

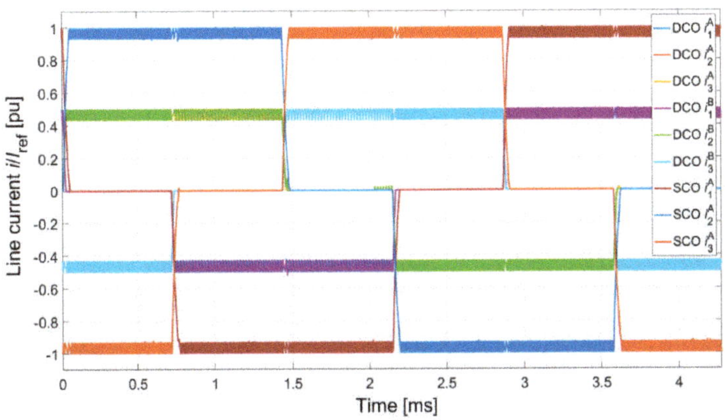

(b)

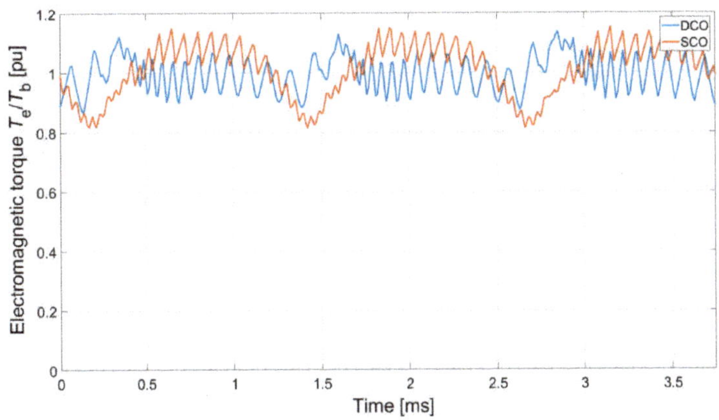

(c)

Figure 6. *Cont.*

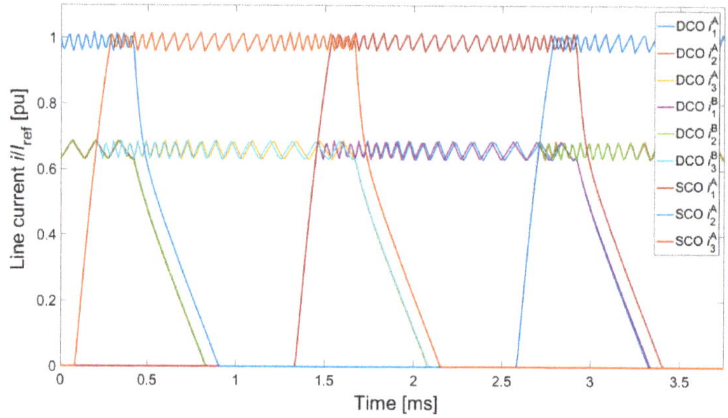

(**d**)

Figure 6. Relative electromagnetic torque and line currents in DCO and SCO for the DCBLDCM and the DCSRM at low speed. (**a**) Electromagnetic torque T_e for a DCBLDCM. (**b**) Line currents for the DCBLDCM. (**c**) Electromagnetic torque T_e for the DCSRM. (**d**) Line currents for the DCSRM.

5.1.2. Operation without Current Control

In the case of operation at the rated speed (n = 8000 r/min), there was no control of the line current. In the calculations, it was assumed that at this operating point, both motors would generate the rated torque equal to half of the value of the base torque. However, achieving the same value of electromagnetic torque in SCO mode is more complex. In the case of the DCSRM, a change of the turn-on and dwell angle provides great possibilities. In the case of the DCBLDCM, one can also change the turn-on angle in order to increase the value of the generated electromagnetic torque [25]. Another way to achieve the same value of electromagnetic torque in SCO and DCO modes in this part of the characteristic is to use PWM control. This applies to both motors. In the numerical calculations, the turn-on angle was changed in both motors. In the case of the DCBLDCM, in DCO mode, the turn-on angle was not changed. Figure 7 shows the results of the numerical calculations in the analyzed operating point.

In the case of operation at high rotational speed (without current control), the differences between DCO and SCO modes were more noticeable. This applied to both motors. In SCO mode, the electromagnetic torque ripple increased (42% (Figure 7a) and 65% (Figure 7c)). In DCO mode, the torque ripple was reduced (DCBLDCM—13% and DCSRM—39%). In the case of the DCSRM, torque ripple was always greater, regardless of the mode of operation. In the case of the DCSRM, in DCO mode, there was a noticeable small impact of magnetic couplings between the channels (Figure 7d).

Figure 8 shows examples of waveforms of line currents in DCO and SCO modes for both motors, measured in laboratory conditions. The laboratory system used in the tests is discussed in [25].

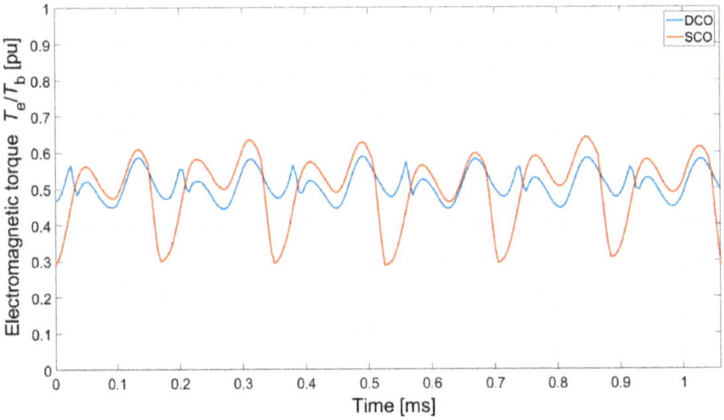

(**a**)

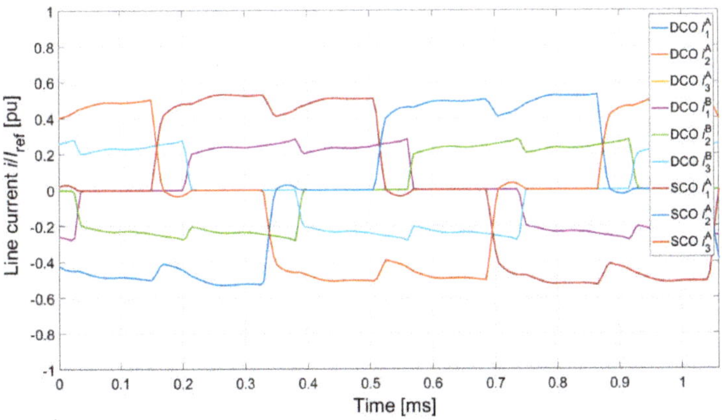

(**b**)

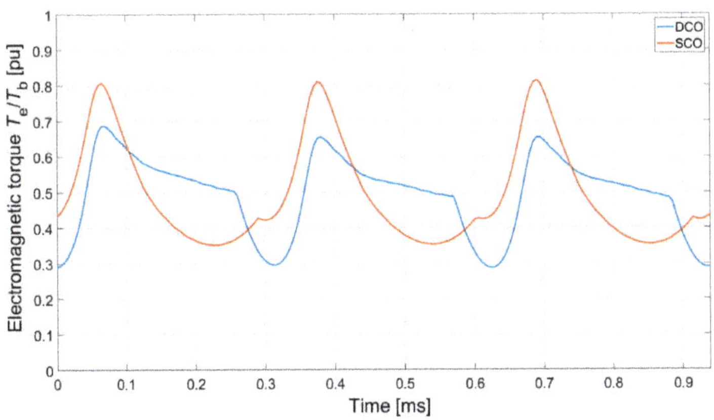

(**c**)

Figure 7. *Cont.*

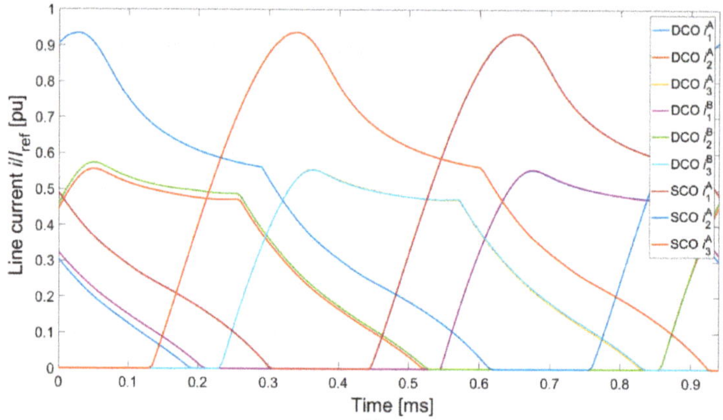

(d)

Figure 7. Relative electromagnetic torque and line currents in DCO and SCO modes of the DCBLDCM and the DCSRM at high speed. (**a**) Electromagnetic torque T_e for the DCBLDCM. (**b**) Line currents for the DCBLDCM. (**c**) Electromagnetic torque T_e for the DCSRM. (**d**) Line currents for the DCSRM.

In the real system, the differences between DCO and SCO modes were greater. This was due to the differences in the electric and magnetic parameters of the different phases of both channels. However, the waveforms of the current confirmed the results of the numerical calculations shown in Figure 7b,d.

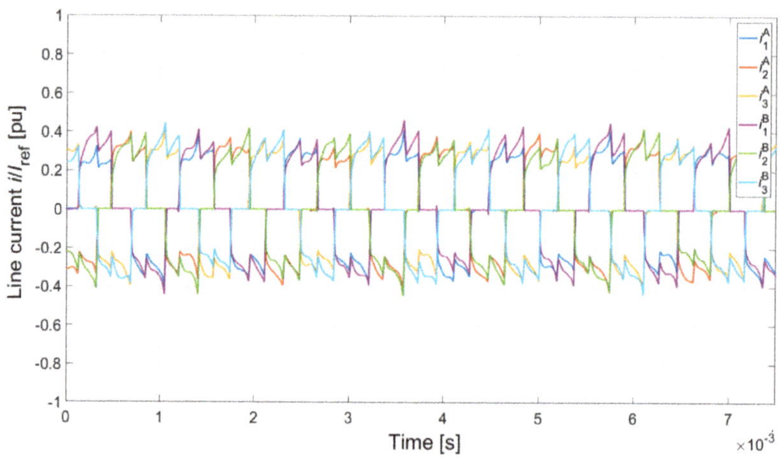

(a)

Figure 8. *Cont.*

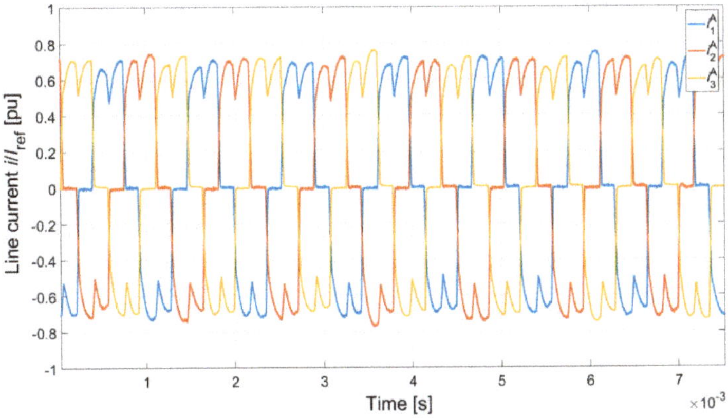

(b)

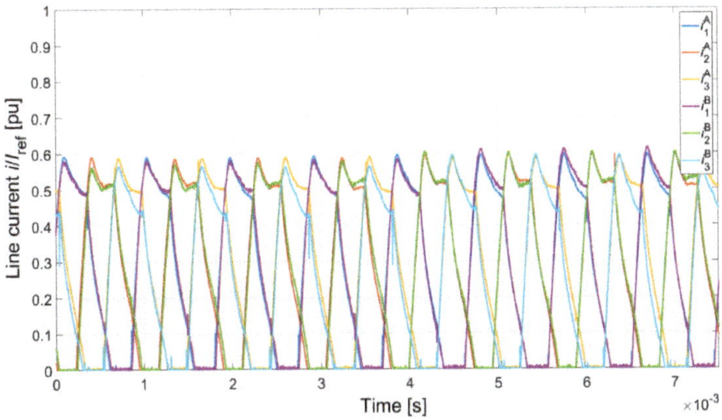

(c)

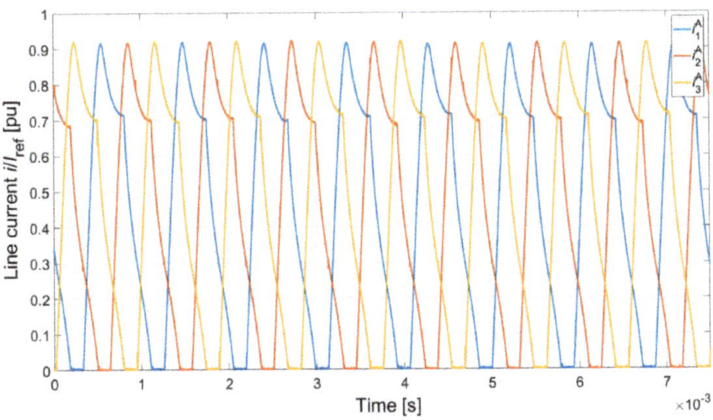

(d)

Figure 8. Relative line currents in DCO and SCO modes for the DCBLDCM and the DCSRM in a high-speed laboratory test. (**a**) Line currents for the DCBLDCM in DCO mode. (**b**) Line currents for the DCBLDCM in SCO mode. (**c**) Line currents for the DCSRM in DCO mode. (**d**) Line currents for the DCSRM in SCO mode.

5.2. Influence of Asymmetrical Control in DCO Operation

In the analysis of DCO mode described in Section 5.1, both channels were controlled symmetrically. The test of magnetic independence for both machines was an analysis of a case where they are controlled asymmetrically. Use of two independent power supply channels (Figure 1c,d) allowed independent control of each channel. An analysis of the impact of asymmetric control was performed for a high rotational speed (n = 8000 r/min). In the calculations, it was assumed that the control parameters of channel A were the same as those specified in Section 5.1. In the case of channel B, the value of the turn-on angle was changed, and in the case of the DCSRM, the dwell angle was additionally changed. The results of the calculations are shown in Figure 9. Examples of results of laboratory tests are shown in Figure 10.

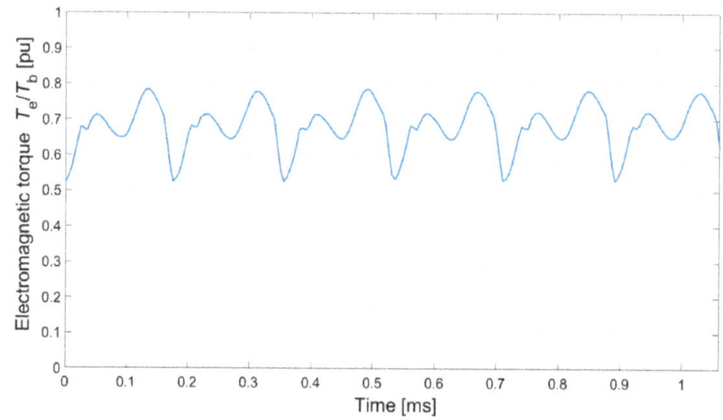

(a)

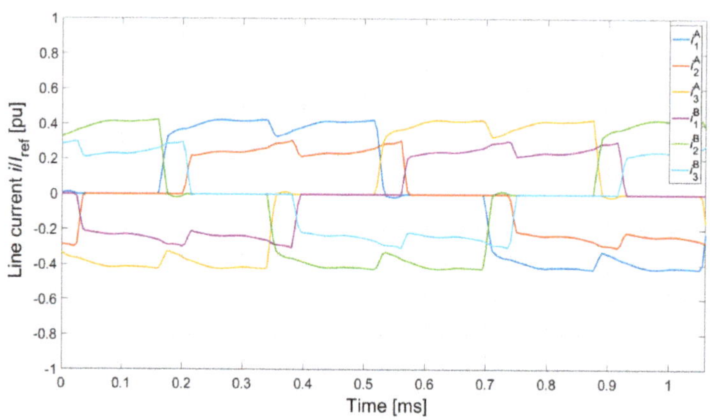

(b)

Figure 9. *Cont.*

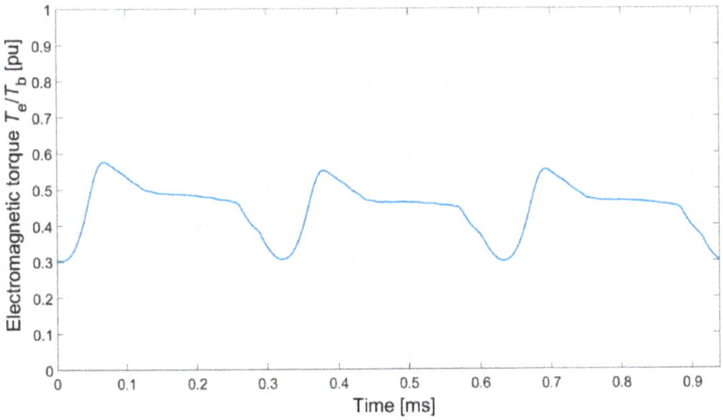

(c)

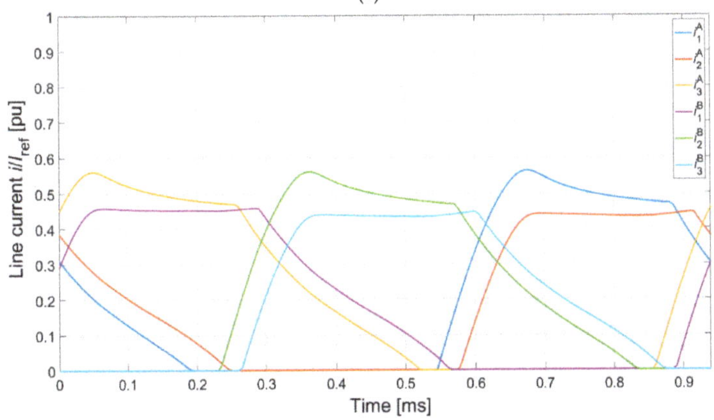

(d)

Figure 9. Relative electromagnetic torque and line currents in DCO mode for the DCBLDCM and the DCSRM at asymmetrical control. (**a**) Electromagnetic torque T_e for the DCBLDCM. (**b**) Line currents for the DCBLDCM. (**c**) Electromagnetic torque T_e for the DCSRM. (**d**) Line currents for the DCSRM.

Application of the asymmetrical control for the DCBLDCM caused an increase of the average value of electromagnetic torque and the torque ripple (23%). For the DCSRM at asymmetrical control, the trend of the average value of electromagnetic torque depends on the control angles. At the same time, it is feasible to suppress the torque ripple. In the case presented in Figure 9c, the torque ripple was 29%.

The results of the laboratory test showed that both designs were characterized by very high magnetic independence between the channels. This was confirmed by the results of laboratory tests shown in Figure 10. In the case of the DCSRM, this justified the adoption of the "NNSS" configuration. Such high magnetic independence between the channels in both solutions provides great possibilities, also in the case of generator operation.

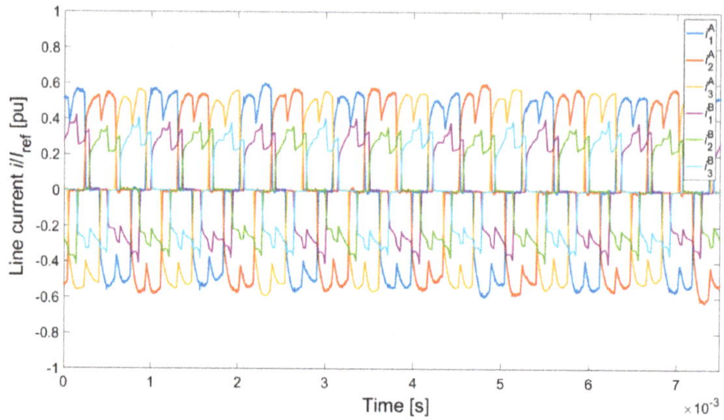

(**a**)

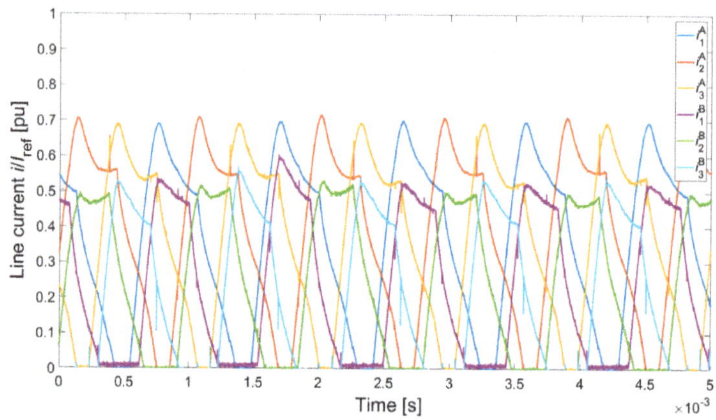

(**b**)

Figure 10. Relative line currents in DCO mode for the DCBLDCM and the DCSRM at asymmetrical control—a laboratory test. (**a**) Line currents for the DCBLDCM. (**b**) Line currents for the DCSRM.

5.3. Influence of Shorted Coil in DCO Mode

Defects inside the machine are one of the typical fault conditions. Usually, these are various short circuits inside the windings of the machine. This article includes an analysis of a short circuit in part of the winding of channel B of phase 1_2^B (the entire coil is shown in Figure 1c,d). The analysis was performed during operation at a high rotational speed (n = 8000 r/min). The shorted coil was simulated using an additional switch S_2 (Figure 1c,d). At the selected time, a short circuit in the coil occurred. In the case of the DCBLDCM, the short circuit occurred at t = 0.66 s. In the case of the DCSRM, the short circuit occurred at t = 1.0 s. Figure 11 shows the results of the numerical calculations.

Occurrence of a fault operating condition in the form of a short circuit in a part of the winding has a negative impact on the service life of machines. This applies in particular to DCBLDCM machines. In the shorted part of the winding, a very high current is present (Figure 11b), which causes thermal damage to the winding. However, this is not the only negative impact of this defect. Disconnecting the defective channel has practically no influence on the value of the short-circuit current. Consequently, further operation of a DCBLDCM machine with this defect is practically impossible. Moreover, the fault operating condition causes substantial torque ripple (497% in the presented case). Under certain conditions, it is possible to suppress it by application of appropriate drive system topology [21]. Nevertheless, it has no influence on the value of the short-circuit current.

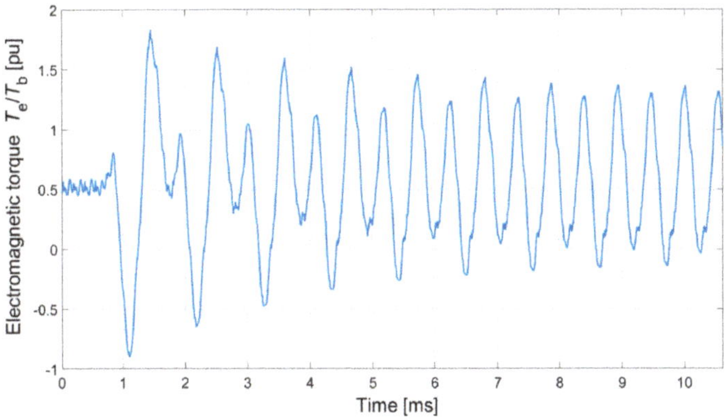

(a)

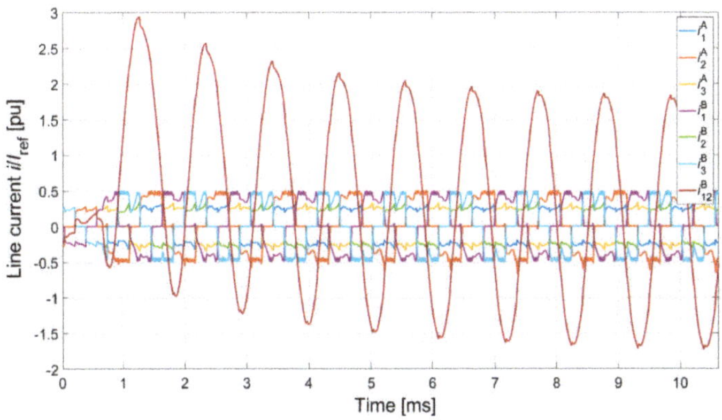

(b)

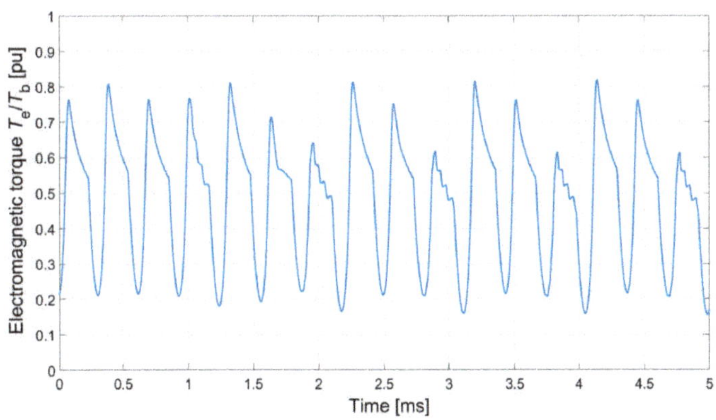

(c)

Figure 11. *Cont.*

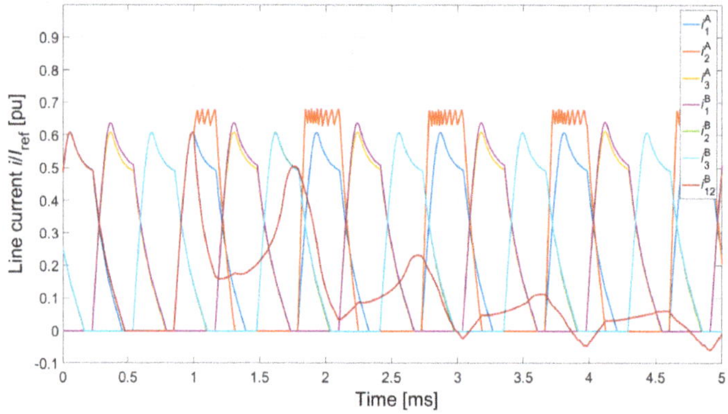

(d)

Figure 11. Relative electromagnetic torque and line currents in DCO mode for the DCBLDCM and the DCSRM at a shorted coil in channel B. (**a**) Electromagnetic torque T_e for the DCBLDCM. (**b**) Line currents for the DCBLDCM. (**c**) Electromagnetic torque T_e for the DCSRM. (**d**) Line currents for the DCSRM.

Unlike a DCBLDCM, in the case of a DCSRM, occurrence of a partial short circuit in the winding is not a critical condition. After an electromagnetic balance is reached, a small short-circuit current flows in the shorted part of the winding (Figure 11d). Much more problematic are the consequences of an unbalanced magnetic pull and the torque ripple (92%). However, the machine can continue to operate with a defective channel. Here, the experimental results of operation under the pole 1_2^B short-circuit condition are presented (Figure 12). The used motor controller could not operate in current-control mode. A small current flowed in the shorted part of the winding. In the rest of the winding, the current became two times greater. However, it had a noticeable influence on the other current waveforms. In comparison with the simulation results, this influence was more significant. The impact of the control parameters and the second channel on the short-circuit current was minor.

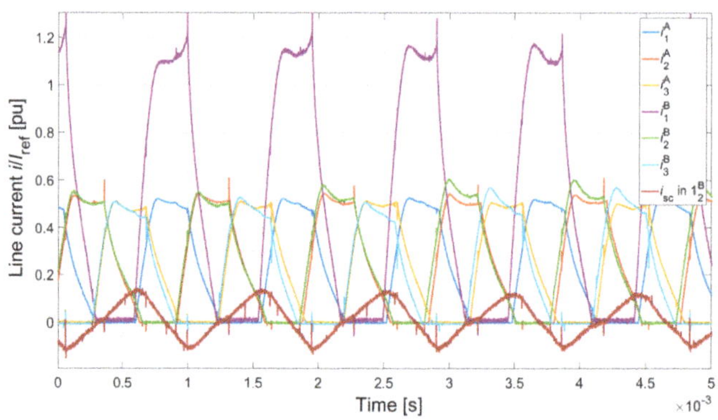

Figure 12. Line currents in DCO mode for the DCSRM at a shorted coil in channel B—laboratory test.

5.4. *Influence of an Open Phase in DCO Mode*

Another analyzed case of failure operation was an open winding in phase 1^B of channel B. Like the short-circuit condition described in Section 5.2, the defect of phase 1^B was simulated. At a certain point in time, the switch S_1 was opened (Figure 1c,d). The defect of phase 1^B in the DCBLDM occurred at $t = 0.7$ s, and in the DCSRM, at $t = 1$ s. The results of the numerical calculations are shown in Figure 13.

The failure condition caused by an open winding in one of the phases of channel B prevents further operation of machines in SCO mode. In DCO mode, after this defect occurs, a DCSR machine can continue to operate. In the case of a DCBLDC machine with delta-connected windings, this is also possible. Here, in both cases, the torque ripple increased (DCBLDCM—64% and DCSRM—165%). In the case of the DCSRM, the analyzed fault condition was identical to a defect of one or even both transistors in the power supply system. In laboratory conditions, in the case of the DCBLDCM, the detection of rotor position was based on a sensorless control algorithm. Failure to supply one of the phases of channel B resulted in errors in detection of rotor position in the defective channel (Figure 14a). Start-up of the DCBLDCM was possible with defective channel B (and with channel A not being supplied).

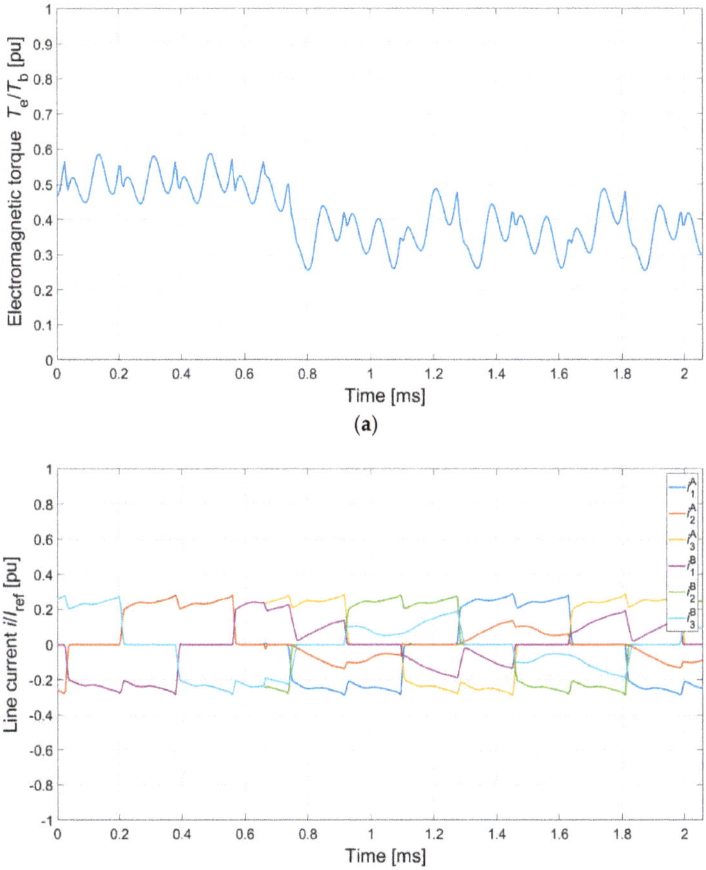

(a)

(b)

Figure 13. *Cont.*

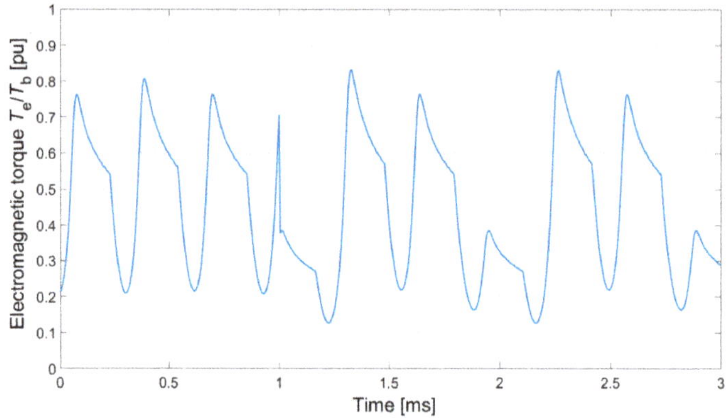

(**c**)

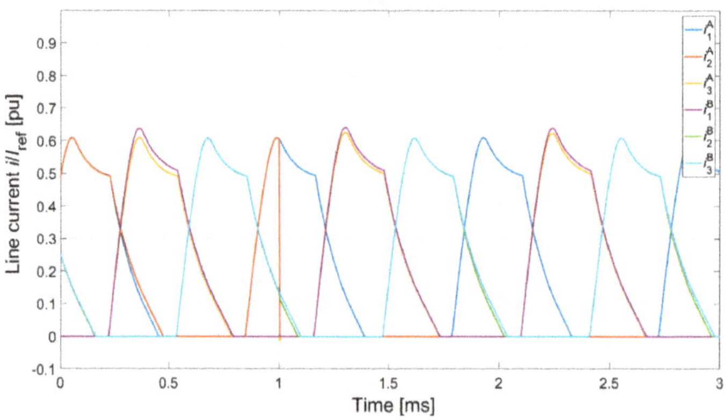

(**d**)

Figure 13. Relative electromagnetic torque and line currents in DCO mode for the DCBLDCM and the DCSRM at open phase in channel B. (**a**) Electromagnetic torque T_e for the DCBLDCM. (**b**) Line currents for the DCBLDCM. (**c**) Electromagnetic torque T_e for the DCSRM. (**d**) Line currents for the DCSRM.

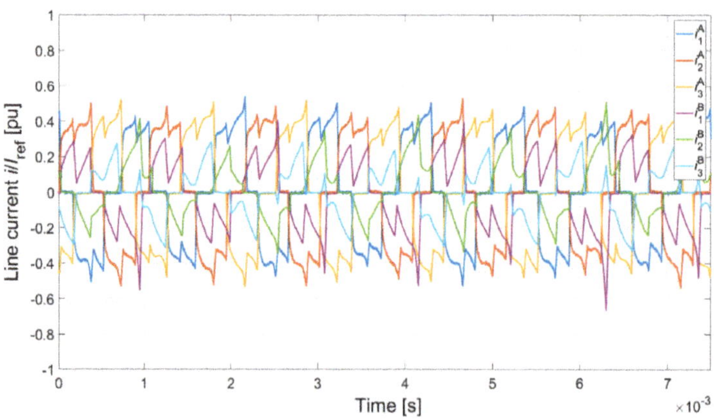

(a)

Figure 14. *Cont.*

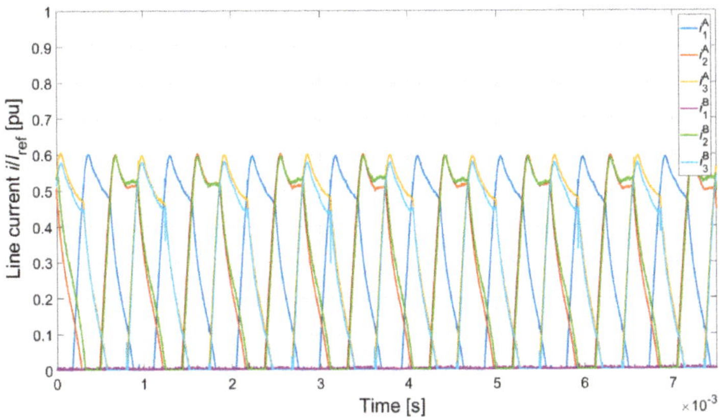

(**b**)

Figure 14. Relative line currents in DCO mode at open phase in channel B—laboratory test: (**a**) DCBLDCM; (**b**) DCSRM.

6. Conclusions

This study compared two dual-channel electronically commutated motors: a DCBLDCM and a DCSRM. However, it was not a full comparison and it was focused on selected operation modes. The analysis excluding thermal, vibroacoustic and efficiency aspects was performed. We proposed a mathematical model for a three-phase DCBLDC machine that considered the nonlinearity of the magnetization characteristics and all couplings between the channels. We presented a model of a DCSRM machine and demonstrated that the structure of its equations can be derived from the DCBLDCM model by omitting all components related to fluxes generated by permanent magnets. In the case of both designs, it is possible to omit the couplings between the channels in symmetrical control conditions. As demonstrated, nontypical operating conditions result in a much greater impact of couplings between the channels in the analyzed DCBLDCM design. The DCBLDCM solution is better in symmetrical control conditions due to higher torque-to-mass ratio (Table 1). However, the DCSRM solution is characterized by much greater tolerance and reliability. In the case of a DCSRM, a short-circuit condition in a part of the winding of one of the phases does not prevent further operation of the drive system. In the case of a DCBLDCM, on the other hand, occurrence of such a failure condition is a critical failure that can bring the drive system to a stop. Switching a DCBLDCM into SCO mode does not reduce the value of the short-circuit current. This problem is not present in the case of a DCSRM. A DCSRM switches into SCO mode without any problems and the current in the shorted winding of one channel does not influence the operation of the remaining channels of the drive system.

Author Contributions: Conceptualization, M.K., Formal analysis, J.P., Investigation, M.K., J.P., B.P. and G.P., Methodology, M.K. and J.P., Project administration, M.K. and B.P., Software, M.K. and B.P., Supervision, J.P. and P.B., Visualisation, M.K. and G.P., Writing-original draft, M.K., Writing-review & editing, J.P., P.B., B.P. and G.P.

Funding: This work is financed in part by the statutory funds of the Department of Electrodynamics and Electrical Machine Systems, Rzeszow University of Technology and in the part by Polish Ministry of Science and Higher Education under the program "Regional Initiative of Excellence" in 2019 – 2022. Project number 027/RID/2018/19, amount granted 11 999 900 PLN.

Conflicts of Interest: The authors declare no conflict of interest.

References

1. Parsa, L.; Toliyat, H.A. Five-Phase Permanent-Magnet Motor Drives. *IEEE Trans. Ind. Appl.* **2005**, *41*, 30–37. [CrossRef]

2. Levi, E.; Jones, M.; Vukosavic, S.N. A series-connected two-motor six-phase drive with induction and permanent magnet machines. *IEEE Trans. Energy Convers.* **2006**, *21*, 121–129. [CrossRef]
3. Zheng, P.; Sui, Y.; Zhao, J.; Tong, C.; Lipo, T.A.; Wang, A. Investigation of a Novel Five-Phase Modular Permanent-Magnet In-Wheel Motor. *IEEE Trans. Magn.* **2011**, *47*, 4084–4087. [CrossRef]
4. Zhou, Y.; Lin, X.; Cheng, M. A Fault-Tolerant Direct Torque Control for Six-Phase Permanent Magnet Synchronous Motor with Arbitrary Two Opened Phases Based on Modified Variables. *IEEE Trans. Energy Convers.* **2016**, *31*, 549–556. [CrossRef]
5. Radun, A.V.; Ferreira, C.A.; Richter, E. Two-channel switched reluctance starter/generator results. *IEEE Trans. Ind. Appl.* **1998**, *34*, 1026–1034. [CrossRef]
6. Chai, J.; Wang, J.; Atallah, K.; Howe, D. Performance Comparison and Winding Fault Detection of Duplex 2-Phase and 3-Phase Fault-Tolerant Permanent Magnet Brushless Machines. In Proceedings of the 2007 IEEE Industry Applications Annual Meeting, New Orleans, LA, USA, 23–27 September 2007; pp. 566–572.
7. Barcaro, M.; Nicola Bianchi, N.; Magnussen, F. Analysis and Tests of a Dual Three-Phase 12-Slot 10-Pole Permanent-Magnet Motor. *IEEE Trans. Ind. Appl.* **2010**, *46*, 2355–2362. [CrossRef]
8. Ding, W.; Liang, D.; Sui, H. Dynamic Modeling and Performance Prediction for Dual-Channel Switched Reluctance Machine Considering Mutual Coupling. *IEEE Trans. Magn.* **2010**, *46*, 3652–3663. [CrossRef]
9. Ding, W.; Liang, D. Comparison of transient and steady-state performances analysis for a dual-channel switched reluctance machine operation under different modes'. *IET Electr. Power Appl.* **2010**, *4*, 603–617. [CrossRef]
10. Barcaro, M.; Bianchi, N.; Magnussen, F. Six-Phase Supply Feasibility Using a PM Fractional-Slot Dual Winding Machine. *IEEE Trans. Ind. Appl.* **2011**, *47*, 2042–2050. [CrossRef]
11. Ding, W.; Liu, L.; Lou, J.; Liu, Y. Comparative Studies on Mutually Coupled Dual-Channel Switched Reluctance Machines with Different Winding Connections. *IEEE Trans. Magn.* **2013**, *49*, 5574–5589. [CrossRef]
12. Ding, W.; Lou, J.; Liu, L. Improved decoupled model of mutually coupled dual-channel SRM with consideration of magnetic saturation in dual-channel operation. *IET Electr. Power Appl.* **2013**, *7*, 427–440. [CrossRef]
13. Ding, W.; Liu, Y.; Hu, Y. Performance evaluation of a fault-tolerant decoupled dual-channel switched reluctance motor drive under open-circuits. *IET Electr. Power Appl.* **2014**, *8*, 117–130. [CrossRef]
14. Jiang, X.; Huang, W.; Cao, R.; Hao, Z.; Li, J.; Jiang, W. Analysis of a Dual-Winding Fault-Tolerant Permanent Magnet Machine Drive for Aerospace Applications. *IEEE Trans. Magn.* **2015**, *51*, 8114704. [CrossRef]
15. Jiang, X.; Huang, W.; Cao, R.; Hao, Z.; Jiang, W. 'Electric Drive System of Dual-Winding Fault-Tolerant Permanent-Magnet Motor for Aerospace Applications'. *IEEE Trans. Ind. Electron.* **2015**, *62*, 7322–7330. [CrossRef]
16. Mese, E.; Yasa, Y.; Akca, H.; Aydeniz, M.G.; Garip, M. Investigating Operating Modes and Converter Options of Dual Winding Permanent Magnet Synchronous Machines for Hybrid Electric Vehicles. *IEEE Trans. Energy Convers.* **2015**, *30*, 285–295. [CrossRef]
17. Bai, H.; Zhu, J.; Qin, J.; Sun, J. Fault-tolerant control for a dual-winding fault-tolerant permanent magnet motor drive based on SVPWM. *IET Power Electron.* **2017**, *10*, 509–516. [CrossRef]
18. Chen, Q.; Xu, D.; Xu, L.; Wang, J.; Lin, Z.; Zhu, X. Fault-Tolerant Operation of a Novel Dual-Channel Switched Reluctance Motor Using Two 3-Phase Standard Inverters. *IEEE Trans. App. Supercond.* **2018**, *28*, 5204205. [CrossRef]
19. Miyama, Y.; Ishizuka, M.; Kometani, H.; Akatsu, K. Vibration Reduction by Applying Carrier Phase-Shift PWM on Dual Three-Phase Winding Permanent Magnet Synchronous Motor. *IEEE Trans. Ind. Appl.* **2018**, *54*, 5998–6004. [CrossRef]
20. Zhu, J.; Bai, H.; Wang, X.; Li, X. Current Vector Control Strategy in a Dual-Winding Fault-Tolerant Permanent Magnet Motor Drive. *IEEE Trans. Energy Convers.* **2018**, *33*, 2191–2199. [CrossRef]
21. Jiang, X.; Li, Q.; Huang, W.; Cao, R. A Dual-Winding Fault-Tolerant Motor Drive System Based on the Redundancy Bridge Arm. *IEEE Trans. Ind. Electron.* **2019**, *66*, 654–662. [CrossRef]
22. Mecrow, B.C.; Jack, A.G.; Atkinson, D.J.; Green, S.R.; Atkinson, G.J.; King, A.; Green, B. Design and testing of a four-phase fault-tolerant permanent-magnet machine for an engine fuel pump. *IEEE Trans. Energy Convers.* **2004**, *19*, 671–678. [CrossRef]

23. Ursu, D.; Gradinaru, V.; Fahimi, B.; Boldea, I. Six-Phase BLDC Reluctance Machines: FEM-Based Characterization and Four-Quadrant Control. *IEEE Trans. Ind. Appl.* **2015**, *51*, 2105–2115. [CrossRef]
24. Wang, S.; Zhan, Q.; Ma, Z.; Zhou, L. Implementation of a 50-kW four-phase switched reluctance motor drive system for hybrid electric vehicle. *IEEE Trans. Magn.* **2005**, *41*, 501–504. [CrossRef]
25. Korkosz, M.; Bogusz, P.; Prokop, J. Modelling and experimental research of fault-tolerant dual-channel brushless DC motor. *IET Electr. Power Appl.* **2018**, *12*, 787–796. [CrossRef]

energies

MDPI

Article

The Fault-Tolerant Quad-Channel Brushless Direct Current Motor

Mariusz Korkosz *, Piotr Bogusz and Jan Prokop

The Faculty of Electrical and Computer Engineering, Rzeszow University of Technology, Al. Powstancow Warszawy 12, 35-959 Rzeszow, Poland; pbogu@prz.edu.pl (P.B.); jprokop@prz.edu.pl (J.P.)
* Correspondence: mkosz@prz.edu.pl; Tel.: +48-178-651-389

Received: 12 August 2019; Accepted: 20 September 2019; Published: 25 September 2019

Abstract: In this study, a permanent magnet brushless direct current machine with multi-phase windings is proposed for critical drive systems. We have named the solution, which has four-stator winding, a quad-channel permanent magnet brushless direct current (QCBLDC) motor. The stator windings are supplied by four independent power converters under quad-channel operation (QCO) mode. After a fault in either one, two, or three channels, further operation of the machine can be continued in triple-channel operation (TCO) mode, dual-channel operation (DCO) mode, or single-channel operation (SCO) mode. In this paper, a novel mathematical model is proposed for a QCBLDC machine. This model takes into account the nonlinearity of a magnetic circuit and all of the couplings between the phases within a given channel, as well as between channels. Based on numerical calculations, the static electromagnetic moment and the coupled fluxes were determined for the individual windings of the variants and work modes being analyzed. A normal work condition can be achieved in the QCO or DCO modes. For the DCO mode, an acceptable case uses a balanced magnetic pull (A and C channels supplied). The DCO A and B type work mode is comparable to the DCO A and C mode with regard to its efficiency in processing electrical energy. The vibroacoustic parameters of this mode, however, are much worse. In fault states, TCO, DCO, and SCO work modes are possible. As the number of active channels decreases, the efficiency of energy processing also decreases. In a critical situation, the motor works in overload mode (SCO mode). Laboratory tests conducted for one of the variants demonstrated that the TCO work mode is characterized by worse vibroacoustic parameters than the DCO A and C mode.

Keywords: multi-channel; quad-channel operation (QCO); triple-channel operation (TCO); dual-channel operation (DCO); single-channel operation (SCO); permanent magnet brushless direct current motor; BLDCM

---

1. Introduction

Critical drives are characterized by improved reliability and are usually intended for special applications, such as airplanes, submarines, and electric cars [1–3]. The main property of critical drives is their ability to continue operation after the occurrence of a fault state or emergency state. There are several ways to improve the reliability of these drives. The first is to use, for example, two independent drive systems [4]. In this case, full independence and separation of the drives is achieved. This is a safe solution because when one drive becomes defective, the other remains in operation. However, this solution requires more space and is heavier. Another group of solutions that has recently been developed consists of using only a one drive system using multi-channel [5–8] or multi-phase [9,10] machines. Such systems make it possible to reduce the mass of the motor and the space occupied by the drives while maintaining high reliability. Electric machines used in high-reliability critical drives include induction machines [11–13], switched reluctance machines (SRM) [8,14], permanent magnet synchronous machines (PMSM) [15–17], and brushless direct current machines with a permanent magnet (BLDCM) [7,18]. Unlike multi-phase solutions, multi-channels have clearly separated channels that are usually supplied by separate power supply sources [7,8]. In the case of multi-channel three-phase machines, each channel consists of three windings that are electrically separated from the windings of the other channels [15]. In the case of multi-phase solutions, the occurrence of an emergency state caused by a break (e.g., in one of the phases) must lead to the occurrence of an asymmetric magnetic pull. In the case of a multi-channel solution, depending on the configuration adopted, operation of the drive can continue with a balanced magnetic pull.

This article analyzes the problem of operating a quad-channel brushless direct current machine with a permanent magnet (QCBLDC) in quad-channel operation (QCO) mode. The four channels of the machine can be supplied from one, two, or four supply sources. An example of supply from two sources (DC1 and DC2) in the QCO mode is shown in Figure 1.

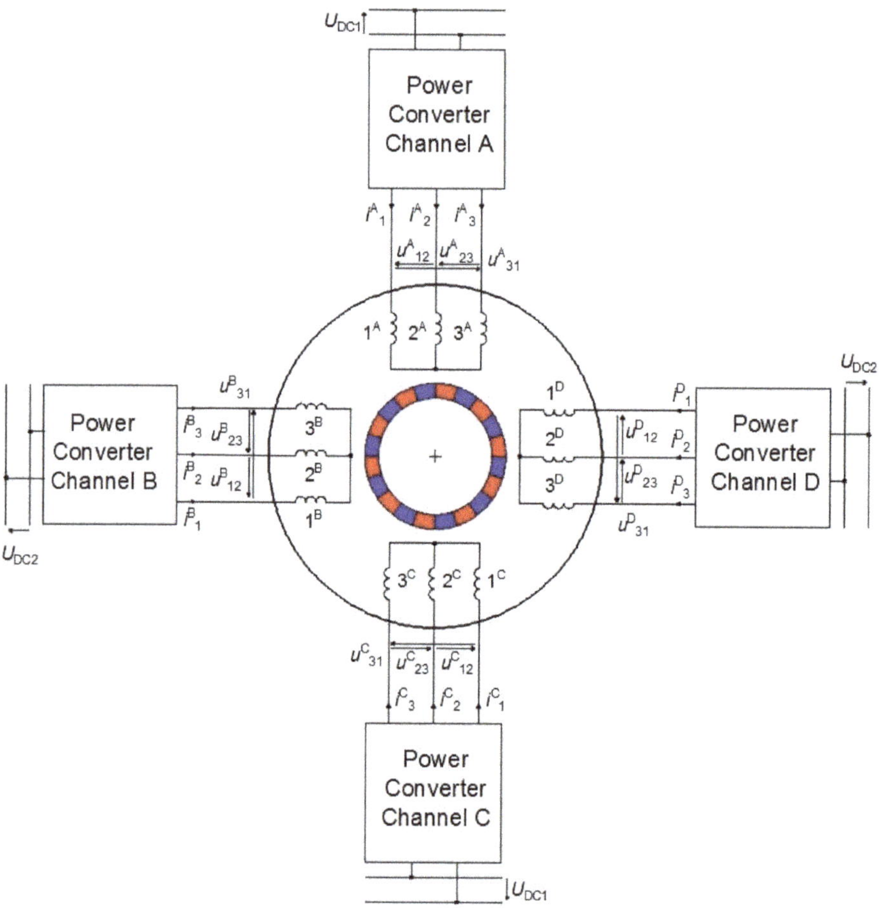

Figure 1. A scheme of a quad-channel brushless permanent magnet direct current (QCBLDC) supply system under quad-channel operation (QCO) mode with two supply sources.

In the case of an emergency state in one of the channels, the system can switch to operating in the triple-channel operation (TCO) mode, the dual-channel operation (DCO) mode, or the single-channel operation (SCO) mode. This article assumes that a QCBLDC working in the QCO mode, after the occurrence of an emergency state in one of its channels, goes into the DCO mode. The third working channel in this case works as an emergency channel for the DCO mode. In the event of operation in the TCO mode, a balanced magnetic pull can be achieved only for the selected configurations of the channel location. Consequently, this operation status will not be analyzed at this stage of the research. The research problem formulated in this article is an answer to the question: which two channels, out of the three that are not defective, work in the DCO mode? So far, the literature on this topic has not presented an analysis of the properties of a four-channel permanent magnet brushless direct current (QCBLDC) machine, and there is no solution for the aforementioned research problem.

The objective of this article is to present the results of the research conducted by the authors on the properties of a QCBLDC motor working in two modes: QCO and DCO. Knowledge of the results of such an analysis is necessary to prepare a control algorithm that, in the event of the occurrence of an emergency state in one channel, would switch the motor from the QCO mode to the DCO mode. The analysis of the properties of a QCBLDC motor is conducted based on simulations and laboratory tests. Proprietary nonlinear circuit mathematical models for the QCBLDC motor are presented for the QCO and DCO modes. As part of the simulation tests, the results of a finite element method (FEM) two-dimensional analysis of the distribution of the magnetic fluxes and the distribution of stresses on the circumference of the stator are presented. The static characteristics, the current waveforms, and the electromagnetic torque are compared for the operating modes being analyzed. The results of the simulation tests were verified by conducting laboratory tests. The conclusions drawn from the results of the tests are presented in the summary.

2. Model and Winding Configurations of a Quad-Channel BLDC Motor

Figure 2a shows a proprietary prototype of the hybrid drive from a small unmanned aerial vehicle (UAV). The electric motor of this object is a three-phase brushless motor with permanent magnets (BLDCM). The stator of the motor with its windings is shown in Figure 2b. When designing the motor, the possibility to improve the reliability of the drive's operation is taken into account. The possibility of an independent multi-channel supply for the motor is provided. The tested motor has the possibility to use a quad-channel supply or, as shown in the literature [5], a dual-channel supply. In the case of a quad-channel supply, there are two possible configurations of the stator windings. These possibilities are shown in Figure 3a (variant I) and Figure 3b (variant II). What makes these configurations different is the location of the windings of the different channels on the circumference of the stator. In variant I, the phases of each channel are staggered by 120 mechanical degrees (Figure 3a). In variant II, the windings of each channel are staggered by 30 mechanical degrees (Figure 3b). The channels in both variants are distributed as follows:

- variant I: ADCBADCBADCB,
- variant II: AAABBBCCCDDD.

Moreover, it was assumed that continuous operation was also possible using only two channels, i.e., in the DCO mode. In this case, the two remaining channels are redundant.

Regardless of the variant, there are six possible configurations of DCO: A and B, A and C, A and D, B and C, B and D, and C and D. In this paper, only two are analyzed; A and B and A and C. This analysis is shown in Table 1 and Figure 3c–f. Figure 3g,h show a phasor diagram of the induced voltages (BEMF) in the windings (1,2,3) of individual channels (A,B,C,D) for variants I and II. At the same time, under the conditions of a fully operational drive, it is possible to use all channels. This operating condition is identified as the QCO (quad-channel operation) mode. In this operating condition, each channel works with half of the required power. When analyzing the multi-channel supply, the impact of the location of the channels and the way they are supplied on the magnetic pull force was not considered. These aspects will be analyzed in subsequent stages of the research based on a coupled electromagnetic-mechanical analysis. Further, operations in odd channel number conditions (i.e., TCO and SCO) are not considered. The TCO mode may occur when one of the channels becomes defective. SCO is a condition of the critical operation of the system. This means that the remaining three channels have already become defective.

(a)

(b)

Figure 2. The prototype of a three-phase quad-channel brushless machine with a permanent magnet (QCBLDC) machine; (**a**) the hybrid drive of a small unmanned aerial vehicle; (**b**) the stator with windings.

Table 1. The selected typical operation conductions of a QCBLDC motor.

Channel	Variant I					Variant II				
	QCO	TCO	DCO		SCO	QCO	TCO	DCO		SCO
			A and B	A and C				A and B	A and C	
A	X	X	X	X	X	X	X	X	X	X
B	X	X	X	-	-	X	X	X	-	-
C	X	X	-	X	-	X	X	-	X	-
D	X	-	-	-	-	X	-	-	-	-

Table legend: QCO — quad-channel operation, TCO — triple-channel operation, DCO — dual- channel operation, SCO — single-channel operation.

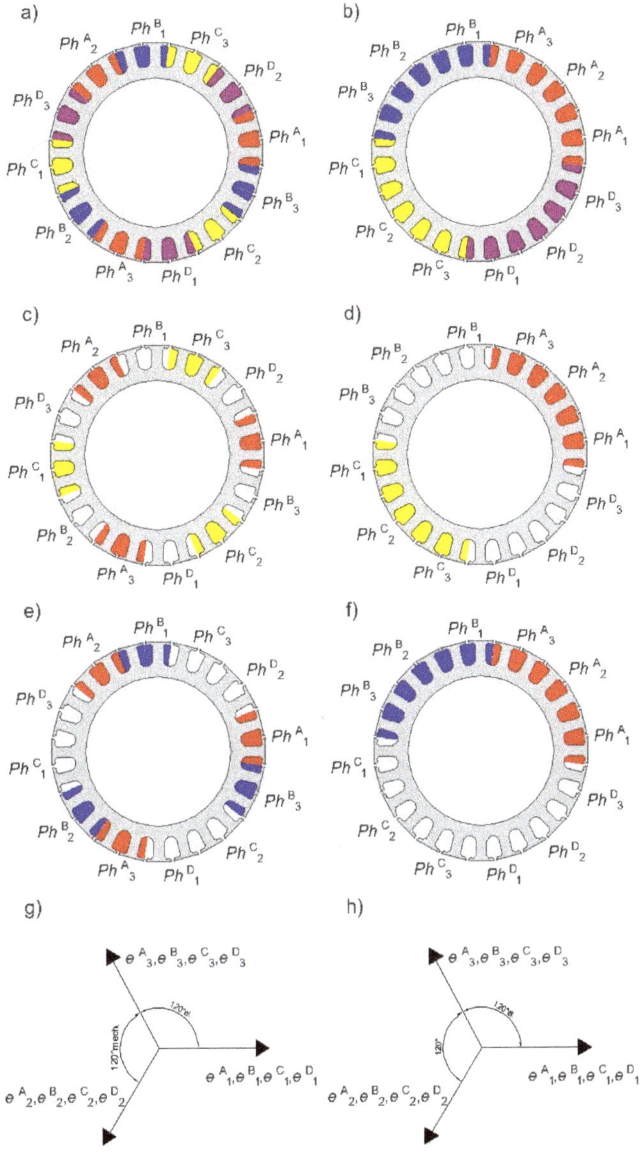

Figure 3. Scheme of the winding distribution of channels on the stator of a three-phase QCBLDC motor: variant I—quad-channel operation (QCO) (**a**), variant II—QCO (**b**), variant I—dual-channel operation (DCO) (channel A and C) (**c**), variant II—DCO (channel A and C) (**d**), variant I—DCO (channel A and B) (**e**), variant II—DCO (channel A and B) (**f**), variant I—phasor diagram (**g**), and variant II—phasor diagram (**h**).

3. Mathematical Model of the QCBLDC Motors

The subject of the mathematical modelling is the QCBLDC motor, for which the authors' circuit-based models, known as flux models, were proposed. The mathematical models are presented while taking into account the non-linearity of the magnetic circuit and the magnetic couplings between the particular phases within the given channel, as well as between the channels (A, B, C, D). Models are included for a machine that works in two modes: the quad-channel operation (QCO) mode and the dual-channel operation (DCO) mode.

The following simplifying assumptions have been adopted in the proposed mathematical model of the QCBLDC machine:

1. Symmetry of the magnetic and electric circuit structure of both the stator and rotor;
2. Decomposition of the phase fluxes into a sum of fluxes induced by phase currents (leakage and main fluxes) and fluxes from permanent magnets;
3. Simplified leakage fluxes from currents in the end-turns of windings;
4. Omitting the influence of temperature on the fluxes generated by permanent magnets and stator resistance.

3.1. Model for QCO Mode

The general structure of the mathematical model of the three-phase QCBLDC motor in QCO mode can be written in the following form:

$$\begin{bmatrix} \mathbf{u}^A \\ \mathbf{u}^B \\ \mathbf{u}^C \\ \mathbf{u}^D \end{bmatrix} = \begin{bmatrix} \mathbf{R}^A & 0 & 0 & 0 \\ 0 & \mathbf{R}^B & 0 & 0 \\ 0 & 0 & \mathbf{R}^C & 0 \\ 0 & 0 & 0 & \mathbf{R}^D \end{bmatrix} \begin{bmatrix} \mathbf{i}^A \\ \mathbf{i}^B \\ \mathbf{i}^C \\ \mathbf{i}^D \end{bmatrix} + \frac{\mathrm{d}}{\mathrm{d}t} \begin{bmatrix} \boldsymbol{\psi}^A \\ \boldsymbol{\psi}^B \\ \boldsymbol{\psi}^C \\ \boldsymbol{\psi}^D \end{bmatrix} + \begin{bmatrix} \mathbf{e}^A \\ \mathbf{e}^B \\ \mathbf{e}^C \\ \mathbf{e}^D \end{bmatrix} \tag{1}$$

$$J\frac{\mathrm{d}\omega}{\mathrm{d}t} + D\omega + T_L = T_\mathrm{e} \tag{2}$$

$$\frac{\mathrm{d}\theta}{\mathrm{d}t} = \omega \tag{3}$$

where for channels $\mathrm{k} \in (\mathrm{A,B,C,D})$, the vectors representing the phase voltages, $\mathbf{u}^\mathrm{k}$, phase currents, $\mathbf{i}^\mathrm{k}$, phase back-EMF voltages, $\mathbf{e}^\mathrm{k} = \mathbf{e}^\mathrm{k}(\theta, \mathrm{i}^\mathrm{PM})$, the flux linkages caused by the phase winding currents, $\boldsymbol{\psi}^\mathrm{k} = \boldsymbol{\psi}^\mathrm{k}(\theta, \mathbf{i}^\mathrm{A}, \mathbf{i}^\mathrm{B}, \mathbf{i}^\mathrm{C}, \mathbf{i}^\mathrm{D}, \mathrm{i}^\mathrm{PM})$, as well as the matrixes of the stator resistances, $\mathbf{R}^\mathrm{k}$, are defined as follows:

$$\mathbf{u}^\mathrm{k} = \left[\mathrm{u}_1^\mathrm{k}, \mathrm{u}_2^\mathrm{k}, \mathrm{u}_3^\mathrm{k}\right]^\mathrm{T},\ \mathbf{i}^\mathrm{k} = \left[\mathrm{i}_1^\mathrm{k}, \mathrm{i}_2^\mathrm{k}, \mathrm{i}_3^\mathrm{k}\right]^\mathrm{T},\ \mathbf{e}^\mathrm{k} = \left[\mathrm{e}_1^\mathrm{k}, \mathrm{e}_2^\mathrm{k}, \mathrm{e}_3^\mathrm{k}\right]^\mathrm{T},\ \boldsymbol{\psi}^\mathrm{k} = \left[\psi_1^\mathrm{k}, \psi_2^\mathrm{k}, \psi_3^\mathrm{k}\right]^\mathrm{T},$$
$$\mathbf{R}^\mathrm{k} = \mathrm{diag}(\mathrm{R}_1^\mathrm{k}, \mathrm{R}_2^\mathrm{k}, \mathrm{R}_3^\mathrm{k}).$$

The following symbols are used in Equations (1) to (3): θ—rotor angle position, ω—the rotor angular speed, i^PM—the permanent magnet magnetization equivalent current, J—the rotor's (and load's) moments of inertia, D—the rotor damping of the viscous friction coefficient, T_L—the load torque, T_e—the total electromagnetic torque.

The phase back-EMF vectors in Equation (1) for channels $\mathrm{k} \in (\mathrm{A,B,C,D})$ are defined as follows:

$$\mathbf{e}^\mathrm{k} = \frac{\mathrm{d}}{\mathrm{d}t} \begin{bmatrix} \psi_1^\mathrm{kPM}(\theta, \mathrm{i}^\mathrm{PM}) \\ \psi_2^\mathrm{kPM}(\theta, \mathrm{i}^\mathrm{PM}) \\ \psi_3^\mathrm{kPM}(\theta, \mathrm{i}^\mathrm{PM}) \end{bmatrix} = \omega \begin{bmatrix} \frac{\partial \psi_1^\mathrm{kPM}(\theta, \mathrm{i}^\mathrm{PM})}{\partial \theta} \\ \frac{\partial \psi_2^\mathrm{kPM}(\theta, \mathrm{i}^\mathrm{PM})}{\partial \theta} \\ \frac{\partial \psi_3^\mathrm{kPM}(\theta, \mathrm{i}^\mathrm{PM})}{\partial \theta} \end{bmatrix} \tag{4}$$

where $\psi_\mathrm{i}^\mathrm{kPM}(\theta, \mathrm{i}^\mathrm{PM})$ for $\mathrm{i} \in (1,2,3)$ are the permanent magnet fluxes linking the stator windings. The permanent magnet flux linking each stator winding of the QCBLDC motor follows the trapezoidal profile back-EMF. The real back-EMF is not a flat and ideal trapezoidal waveform. Other real back-EMF profiles can be defined in Equation (1). For example, the back-EMF waveform in the Fourier series for $\mathrm{k} \in (\mathrm{A,B,C,D})$ and $\mathrm{i} \in (1,2,3)$ is represented as

$$\mathrm{e}_\mathrm{i}^\mathrm{k} = \omega \left[a_{\mathrm{i}0}^\mathrm{k} + \sum_{\nu=1}^{\infty} \left(a_{\mathrm{i}\nu}^\mathrm{k} \sin(\nu\theta) + b_{\mathrm{i}\nu}^\mathrm{k} \cos(\nu\theta) \right) \right] \tag{5}$$

The flux linkages caused by the phase winding currents in Equation (1) for $k \in (A, B, C, D)$ can be written in the following form:

$$\psi^k = \begin{bmatrix} L_{1\sigma}^{kk} i_1^k + \sum\limits_{l=A}^{D} \left(\sum\limits_{j=1}^{3} \psi_{1j}^{kl}(\theta, i_j^l, i^{PM}) \right) \\ L_{2\sigma}^{kk} i_2^k + \sum\limits_{l=A}^{D} \left(\sum\limits_{j=1} {}^3\psi_{2j}^{kl}(\theta, i_j^l, i^{PM}) \right) \\ L_{3\sigma}^{kk} i_3^k + \sum\limits_{l=A}^{D} \left(\sum\limits_{j=1}^{3} \psi_{3j}^{kl}(\theta, i_j^l, i^{PM}) \right) \end{bmatrix} \tag{6}$$

where $k, l \in (A, B, C, D)$ is the stator channel index, $i, j \in (1, 2, 3)$ is the stator phase number, and $L_{1\sigma}^{kk}$ are the coefficients of the end-turn self-inductances. The stator flux linking (the so-called self-flux) ψ_{ii}^{kk} i-th $i \in (1, 2, 3)$ phase for the k-th channel $k \in (A, B, C, D)$ in Equation (6) is calculated based on the following dependencies:

$$\psi_{ii}^{kk}(\theta, i_i^k, i^{PM}) = \psi_i^k - L_{i\sigma}^{kk} i_i^k - \sum_{\substack{l=A \\ l \neq k}}^{D} \left(\sum_{\substack{j=1 \\ j \neq i}}^{3} \psi_{ij}^{kl}(\theta, i_j^l, i^{PM}) \right) \tag{7}$$

From non-linear dependence (Equation (7)), the phase current, i_i^k, is calculated:

$$\psi_{ii}^{kk} = \psi_{ii}^{kk}(\theta, i_i^k, i^{PM}) \Rightarrow i_i^k = i_i^k(\theta, \psi_{ii}^{kk}, i^{PM}) \tag{8}$$

The electromagnetic torque in Equation (2) can be calculated as a derivative of the total magnetic field co-energy in the air gap with respect to the rotor angle's position, θ. The expression $T_e = T_e(\theta, i^A, i^B, i^C, i^D, i^{PM})$ for electromagnetic torque for the QCO mode can be written in the following form:

$$\begin{aligned} T_e = & \sum_{k=A}^{D} \sum_{i=1}^{3} \left(\frac{\partial \psi_i^{kPM}(\theta, i^{PM})}{\partial \theta} i_i^k \right) + \sum_{k=A}^{D} \sum_{i=1}^{3} \left(\int_0^{i_i^k} \frac{\partial \psi_{ii}^{kk}(\theta, \tilde{i}_i^k, i^{PM})}{\partial \theta} d\tilde{i}_i^k \right) + \sum_{k=A}^{D} \sum_{i=2}^{3} \sum_{j=1}^{i-1} \left(\frac{\partial \psi_{ij}^{kk}(\theta, i_j^k, i^{PM})}{\partial \theta} i_i^k \right) \\ & + \sum_{k=B}^{D} \sum_{i=1}^{3} \sum_{j=1}^{3} \left(\frac{\partial \psi_{ij}^{kA}(\theta, i_j^A, i^{PM})}{\partial \theta} i_i^k \right) + \sum_{k=C}^{D} \sum_{i=1}^{3} \sum_{j=1}^{3} \left(\frac{\partial \psi_{ij}^{kB}(\theta, i_j^B, i^{PM})}{\partial \theta} i_i^k \right) + \sum_{i=1}^{3} \sum_{j=1}^{3} \left(\frac{\partial \psi_{ij}^{DC}(\theta, i_j^C, i^{PM})}{\partial \theta} i_i^D \right) \\ & + T_{cog}(\theta, i^{PM}) \end{aligned} \tag{9}$$

Electromagnetic torque (Equation (9)) is the sum of the so-called cogging torque, $T_{cog}(\theta, i^{PM})$, torques from fluxes linking permanent magnets and windings currents. The cogging torque of the permanent magnet (PM) machines, produced by magnets, can be expanded into a Fourier series:

$$T_{cog} = T_{cog}(\theta, i^{PM}) = \sum_{\nu=1}^{\infty} T_\nu(i^{PM}) \sin(\nu q \theta + \theta_0) \tag{10}$$

where $T_\nu(i^{PM})$ is the amplitude of the ν-th harmonic, q is the number of slots, and θ_0 is the initial angle.

The component's electromagnetic torque produced by the permanent magnets and currents can be acquired in the form:

$$T_e^{PM} = \sum_{k=A}^{D} \sum_{i=1}^{3} \left(\frac{\partial \psi_i^{kPM}(\theta, i^{PM})}{\partial \theta} i_i^k \right) = \frac{1}{\omega} \sum_{k=A}^{D} \sum_{i=1}^{3} \left(e_i^k i_i^k \right) \tag{11}$$

Equations (1) and (2) with Equations (4), (6), and (9) constitute the nonlinear mathematical model of the QCBLDC motors in the QCO mode.

3.2. Model for DCO Mode

The voltage Equation (1) for the DCO mode, i.e., where only channels A and B are supplied, can be written in the following form:

$$\begin{bmatrix} \mathbf{u}^A \\ \mathbf{u}^B \end{bmatrix} = \begin{bmatrix} \mathbf{R}^A & 0 \\ 0 & \mathbf{R}^B \end{bmatrix}\begin{bmatrix} \mathbf{i}^A \\ \mathbf{i}^B \end{bmatrix} + \frac{d}{dt}\begin{bmatrix} \boldsymbol{\psi}^A \\ \boldsymbol{\psi}^B \end{bmatrix} + \begin{bmatrix} \mathbf{e}^A \\ \mathbf{e}^B \end{bmatrix} \tag{12}$$

where, for channels $k \in (A, B)$, the phase back-EMFs voltages, $\mathbf{e}^k = \mathbf{e}^k(\theta, i^{PM})$, and vectors representing the flux linkages caused by phase winding currents, $\boldsymbol{\psi}^k = \boldsymbol{\psi}^k(\theta, \mathbf{i}^A, \mathbf{i}^B, \mathbf{i}^{PM})$, are defined as follows:

$$\mathbf{e}^A = \omega\begin{bmatrix} \frac{\partial\psi_1^{APM}(\theta,i^{PM})}{\partial\theta} \\ \frac{\partial\psi_2^{APM}(\theta,i^{PM})}{\partial\theta} \\ \frac{\partial\psi_3^{APM}(\theta,i^{PM})}{\partial\theta} \end{bmatrix}, \ \mathbf{e}^B = \omega\begin{bmatrix} \frac{\partial\psi_1^{BPM}(\theta,i^{PM})}{\partial\theta} \\ \frac{\partial\psi_2^{BPM}(\theta,i^{PM})}{\partial\theta} \\ \frac{\partial\psi_3^{BPM}(\theta,i^{PM})}{\partial\theta} \end{bmatrix} \tag{13}$$

$$\begin{gathered} \boldsymbol{\psi}^A = \begin{bmatrix} L_{1\sigma}^{AA} i_1^A + \sum\limits_{j=1}^{3}\left(\psi_{1j}^{AA}(\theta, i_j^A, i^{PM}) + \psi_{1j}^{AB}(\theta, i_j^B, i^{PM})\right) \\ \mathit{vspace3pt}\ L_{2\sigma}^{AA} i_2^A + \sum\limits_{j=1}^{3}\left(\psi_{2j}^{AA}(\theta, i_j^A, i^{PM}) + \psi_{2j}^{AB}(\theta, i_j^B, i^{PM})\right) \\ L_{3\sigma}^{AA} i_3^A + \sum\limits_{j=1}^{3}\left(\psi_{3j}^{AA}(\theta, i_j^A, i^{PM}) + \psi_{3j}^{AB}(\theta, i_j^B, i^{PM})\right) \end{bmatrix}, \\ \boldsymbol{\psi}^B = \begin{bmatrix} L_{1\sigma}^{BB} i_1^B + \sum\limits_{j=1}^{3}\left(\psi_{1j}^{BA}(\theta, i_j^A, i^{PM}) + \psi_{1j}^{BB}(\theta, i_j^B, i^{PM})\right) \\ L_{2\sigma}^{BB} i_2^B + \sum\limits_{j=1}^{3}\left(\psi_{2j}^{BA}(\theta, i_j^A, i^{PM}) + \psi_{2j}^{BB}(\theta, i_j^B, i^{PM})\right) \\ L_{3\sigma}^{BB} i_3^B + \sum\limits_{j=1}^{3}\left(\psi_{3j}^{BA}(\theta, i_j^A, i^{PM}) + \psi_{3j}^{BB}(\theta, i_j^B, i^{PM})\right) \end{bmatrix} \end{gathered} \tag{14}$$

The expression for the electromagnetic torque, $T_e = T_e(\theta, \mathbf{i}^A, \mathbf{i}^B, \mathbf{i}^{PM})$, with Equation (14) taken into account, can be written in the following form:

$$\begin{gathered} T_e = \sum_{i=1}^{3}\left(\frac{\partial\psi_i^{APM}(\theta,i^{PM})}{\partial\theta} i_i^A + \frac{\partial\psi_i^{BPM}(\theta,i^{PM})}{\partial\theta} i_i^B\right) + \sum_{i=1}^{3}\left(\int_0^{i_i^A} \frac{\partial\psi_{ii}^{AA}(\theta,\tilde{i}_i^A,i^{PM})}{\partial\theta} d\tilde{i}_i^A + \int_0^{i_i^B} \frac{\partial\psi_{ii}^{BB}(\theta,\tilde{i}_i^B,i^{PM})}{\partial\theta} d\tilde{i}_i^B\right) \\ + \sum_{i=2}^{3}\sum_{j=1}^{i-1}\left(\frac{\partial\psi_{ij}^{AA}(\theta,i_j^A,i^{PM})}{\partial\theta} i_i^A + \frac{\partial\psi_{ij}^{BB}(\theta,i_j^B,i^{PM})}{\partial\theta} i_i^B\right) + \sum_{i=1}^{3}\sum_{j=1}^{3}\left(\frac{\partial\psi_{ij}^{BA}(\theta,i_j^A,i^{PM})}{\partial\theta} i_i^B\right) + T_{cog}(\theta, i^{PM}) \end{gathered} \tag{15}$$

The components of the electromagnetic torque produced by permanent magnets and currents (the first component of the right side of Equation (15)) can be determined in the form:

$$T_e^{PM} = \frac{1}{\omega}\sum_{i=1}^{3}\left(e_i^A i_i^A + e_i^B i_i^B\right) \tag{16}$$

Equation (12) with Equations (13) and (14), and Equation (2) with Equation (15) constitute the mathematical model of a QCBLDC machine in the DCO mode.

4. Static Characteristics

The following assumptions were made in the FEM two-dimensional numerical calculations:
1. Symmetry of the magnetic and electric circuit structure of both the stator and the rotor;
2. Omitting the influence of temperature on fluxes generated by permanent magnets;
3. Omitting the influence of temperature on phase resistance;
4. Maintaining a constant speed in the transient analysis.

4.1. Electromagnetic Torque

The characteristics of static electromagnetic torque were determined for the QCO operation, and for both the analyzed variants (variant I and variant II) of the DCO operation, using FEM two dimensional commercial software [19].

The calculations were performed for one electrical period (36 mechanical degrees) at I = constant, supplying the phases, $Ph1$ and $Ph2$, and speed, n = 0.167 r/min. The current was changed in the range of 0 to 25 A for QCO and 0 to 50 A for DCO. The average value of the electromagnetic torque, T_{eav}, as a function of the current, I, flowing in the channel is shown in Figure 4a. Examples of the relationship between the electromagnetic torque, T_e, and the rotor position are shown in Figure 4b. The numerical calculations were verified under laboratory conditions. A laboratory stand used to determine static characteristics is shown in Figure 4c. Examples of laboratory static torque characteristics are shown in Figure 4d.

In QCO operation, the configuration type (variant I, variant II) is completely unimportant. In DCO operation, the electromagnetic torque decreases (as a result of saturation). In the operating range (to the value of the rated torque), this influence is practically negligible. In overload operation (or in emergency operation), the constant torque in the DCO mode decreases. The difference between variant I and variant II is insignificant. A slightly smaller value of torque was generated in variant II. In both variants, no difference between the A and B configuration and the A and C configuration was identified in the DCO operation. This means that the type of variant and the configuration does not influence the value of the electromagnetic torque produced. There are differences between the QCO mode and the DCO mode in the stress within the magnetic circuit of the stator. Figure 5 shows examples of the surface force density of the magnetic circuit of the stator (of magnetic origin) for the selected positions of the rotor and for variant I of the stator winding. The results obtained for variant II are similar.

In the DCO mode, there is a significant increase in the stress compared to the QCO mode. In the case of the DCO A and B supply, the distribution of the stresses is non-symmetric, which is conducive to the occurrence of vibrations in the structure. In this regard, this configuration is not recommended. However, if operation needs to continue, e.g., after the C and D channels have become defective, the motor can continue to operate with increased asymmetry of the magnetic pull.

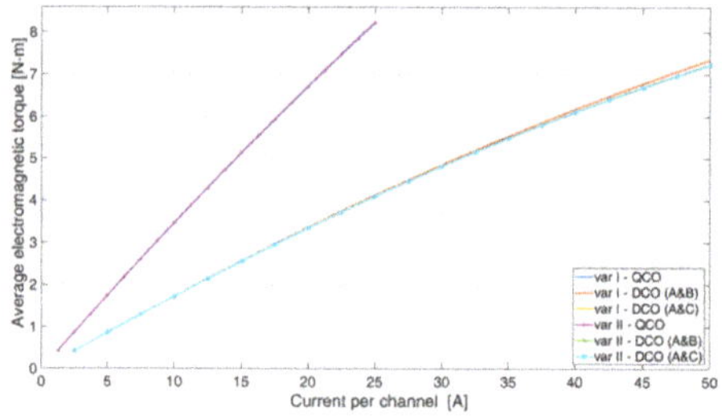

(a)

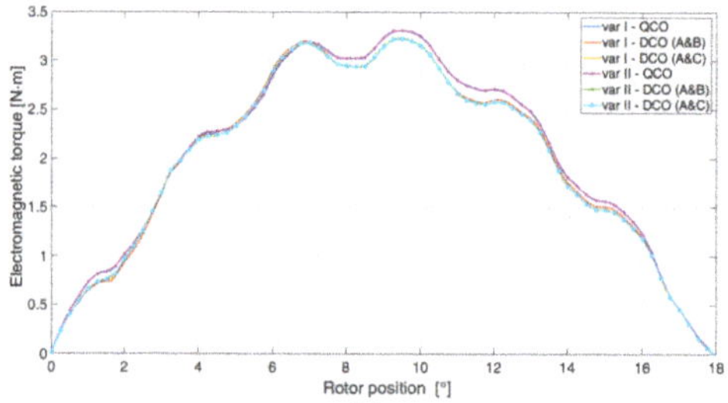

(b)

(c)

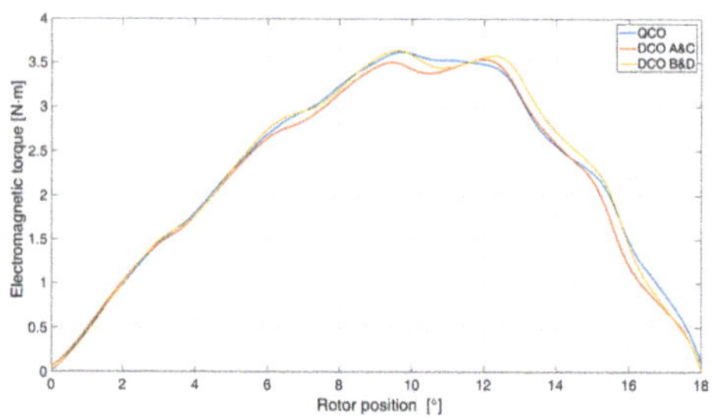

(d)

Figure 4. The static characteristics of electromagnetic torque: The average value of electromagnetic torque vs. current (**a**); electromagnetic torque vs. rotor positions—simulation tests (**b**); the stand to determine static characteristics (**c**); electromagnetic torque vs. rotor positions (laboratory test—variant II) (**d**).

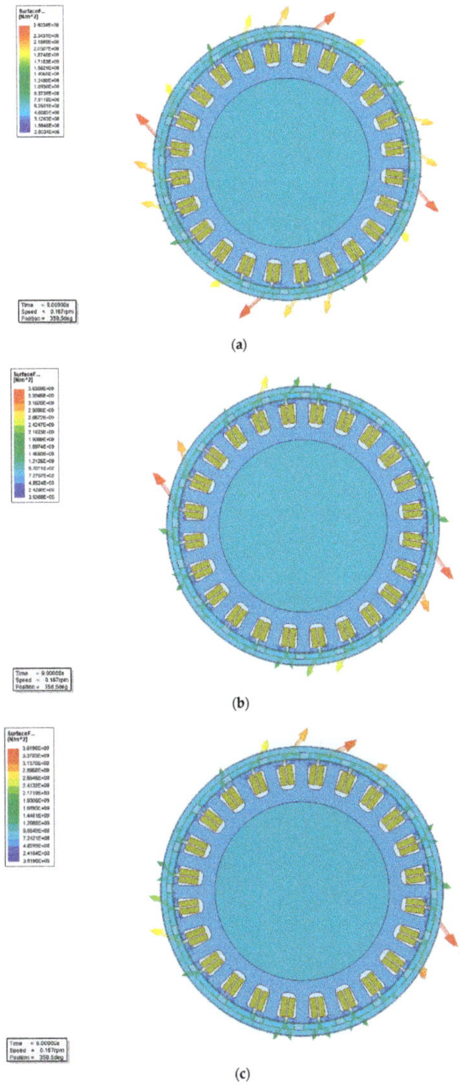

Figure 5. Surface force density: QCO (a); DCO A and C (b); DCO A and B (c)—variant I.

4.2. Flux Characteristics-Variant I

Due to the division of the windings into four channels, there are twelve flux linkages with individual phases. Figures 6–8 show the flux linkages as a function of the rotor position for variant I in all analyzed configurations.

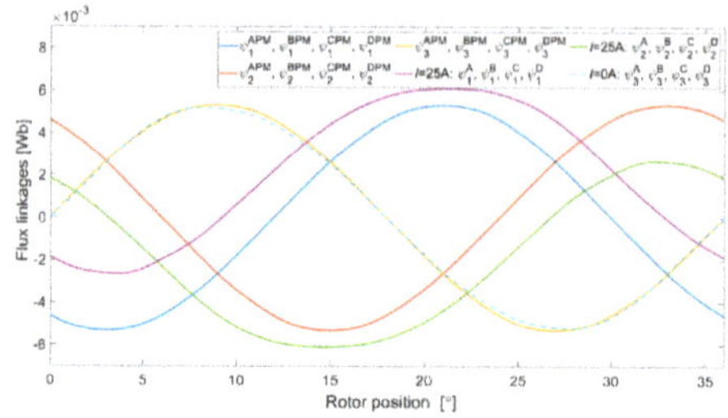

Figure 6. Flux linkages vs. rotor positions for QCO—variant I.

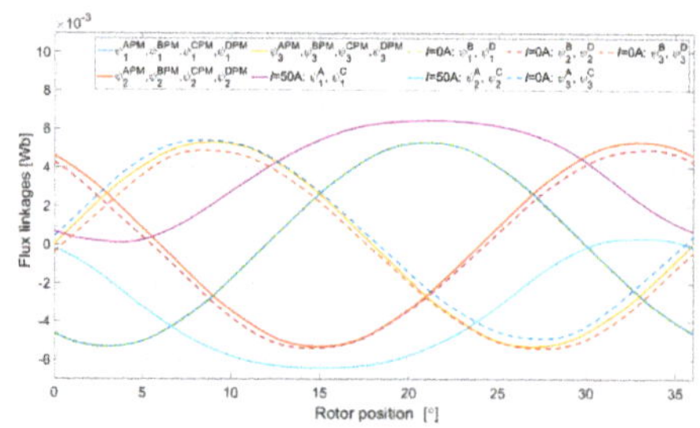

Figure 7. Flux linkages vs. rotor positions for DCO A and C—variant I.

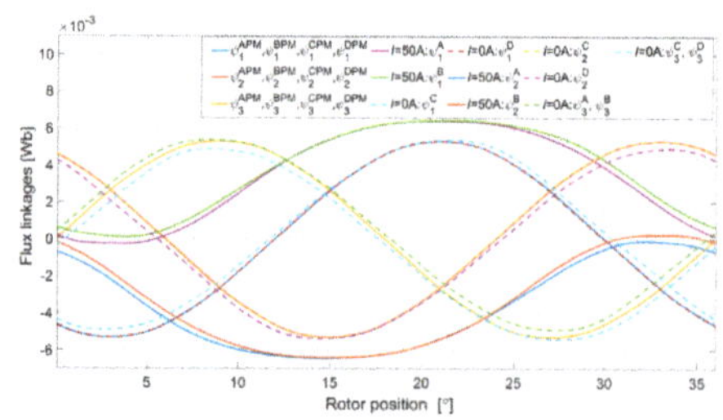

Figure 8. Flux linkages vs. rotor positions for DCO A and B—variant I.

In the case of the dual-channel supply in variant I, the A and C supply configuration (Figure 7) is more beneficial because of its minimal multi-channel magnetic coupling. The A and B configuration has a slightly greater magnetic coupling between the channels. However, the difference is not significant.

4.3. Flux Characteristics-Variant II

Figures 9 and 10 show the flux linkages for the analyzed configurations for variant II of the dual-channel supply.

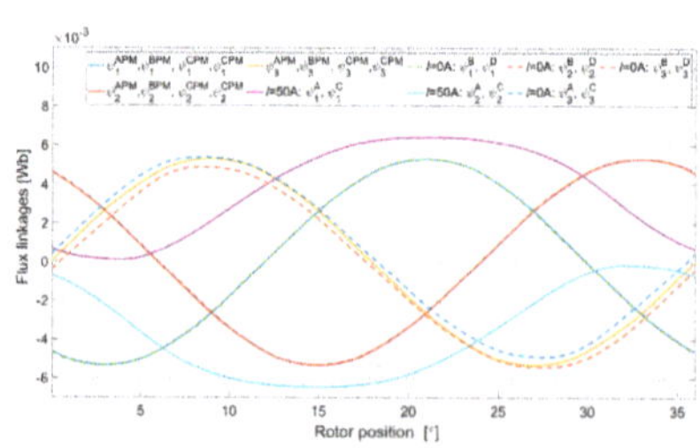

Figure 9. Flux linkages vs. rotor positions for DCO A and C—variant II.

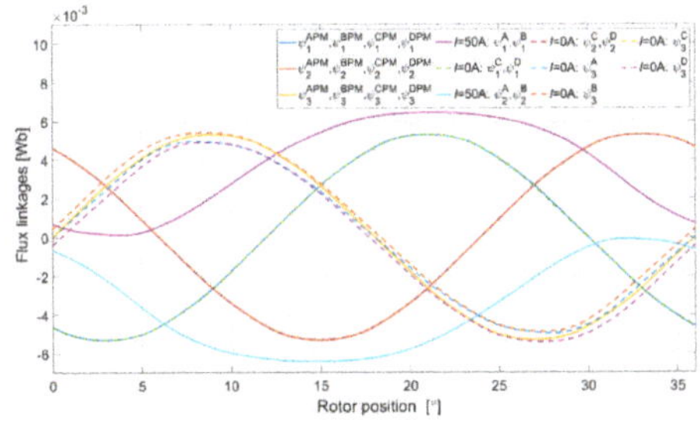

Figure 10. Flux linkages vs. rotor positions for DCO A and B—variant II.

In the case of variant II, it is not possible to indicate a more advantageous configuration due to the impact of the linkages or the impact of saturation of the magnetic circuit.

All determined characteristics were implemented in the simulation model as per the two-dimensional lookup table in the Matlab SISOTOOL system (R2019a, MathWorks, Natick, MA, USA) [20]. This has been explained in a previous paper [21].

5. Waveforms, Current, Voltage, and Electromagnetic Torque

5.1. Numerical Calculations

For the purpose of transient analysis, numerical calculations were performed for a constant rotor speed of $n = 1000$ r/min. In the calculations, it was assumed that, in the QCO mode, the value of the reference current set on the current control devices in all channels was equal to 10 A. In the dual-channel operation mode, the reference current was equal to 20 A for each of the analyzed configurations of both variants. Figures 11 and 12 show the electromagnetic torque of the motor for variant I (Figure 11) and variant II (Figure 12). Figure 13 shows the relationship between the flux linkage of the phase, *Ph*1, and the current for all cases analyzed.

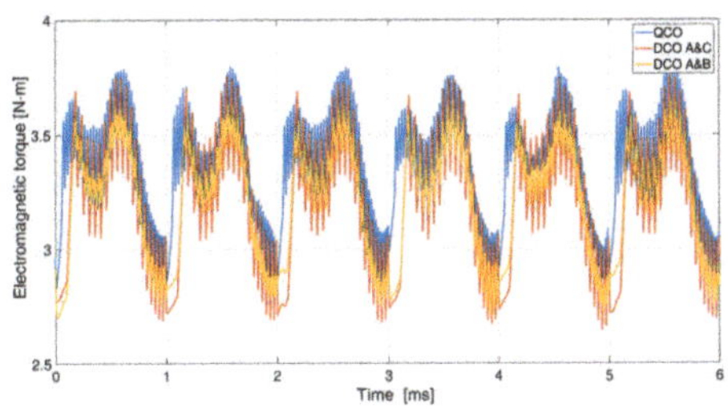

Figure 11. Waveforms of electromagnetic torque—variant I.

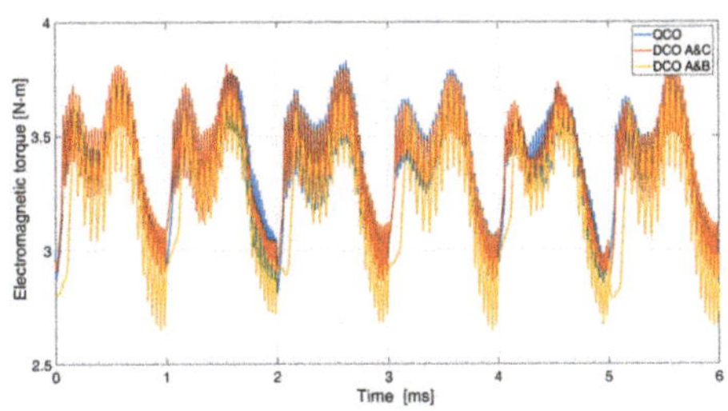

Figure 12. Waveforms of electromagnetic torque—variant II.

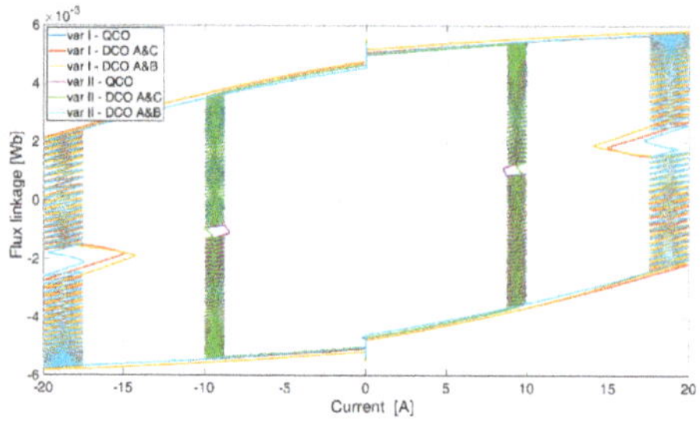

Figure 13. Flux linkages vs. rotor positions for all configurations.

Selected results of the examinations are presented in Table 2.

Table 2. The selected results of the calculations for the analyzed operating conditions.

Parameter	Variant I			Variant II		
	QCO	DCO A and C	DCO A and B	QCO	DCO A and C	DCO A and B
T_{eav} [N·m]	3.35	3.21	3.21	3.35	3.21	3.21
Torque ripple [%]	14.5	17.7	15.6	15.2	17.7	17.3
$I^A{}_{rms\_av}$ [A]	7.6	14.9	14.9	7.6	15	15
$I^B{}_{rms\_av}$ [A]	7.6	-	15.1	7.6	-	15
$I^C{}_{rms\_av}$ [A]	7.6	14.9	-	7.6	15	-
$I^D{}_{rms\_av}$ [A]	7.6	-	-	7.6	-	-
P_{Fe} [W]	10.6	7.6	7.3	10.5	7.2	7.5
P_{cu} [W]	48.5	93.2	94.7	48.5	94.5	94.5
η [%]	84.5	76.0	75.9	84.5	75.9	75.6

The variant types and configurations of the channels have little impact on the average value of the electromagnetic moment, T_{eav}. Switching to the DCO mode results in an increase in copper losses (P_{cu}), with a small reduction of iron losses (P_{Fe}). In the case of dual-channel operation, the electromagnetic torque's ripple increases slightly. The efficiency of energy processing in the DCO mode is significantly deteriorated due to increased winding losses. The highest efficiency in the DCO mode was achieved for variant I of A and C. However, in general, the differences in the energy processing efficiency for each of the two variants of the DCO mode are small.

5.2. Laboratory Test

In laboratory conditions, a quad-channel supply for a QCBLDC motor was developed. Figure 14 shows the test stand. Laboratory tests were performed only for variant II.

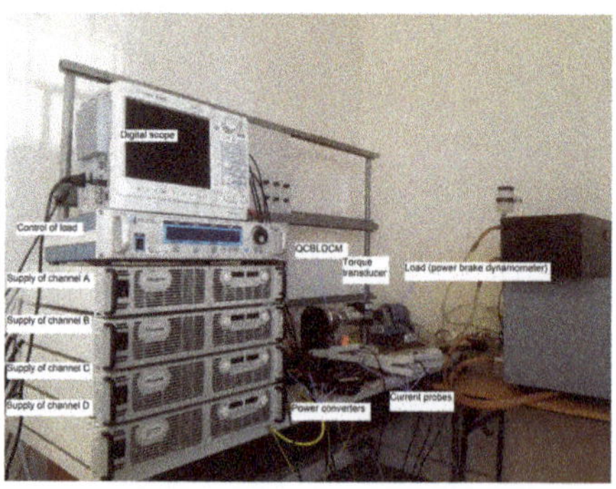

Figure 14. Stand for transient test of a QCBLDC motor.

In laboratory conditions, the current waveforms were recorded during quad-channel, triple-channel, dual-channel, and single-channel operations (Figure 16): A switch from the QCO mode was connected to the TCO mode (Figure 16a), the DCO A and B mode (Figure 16b), the DCO A and C mode (Figure 16c), and to the SCO mode (Figure 16d) of variant II (U_{dc} = 24 V, T_L = 1.2 N·m).

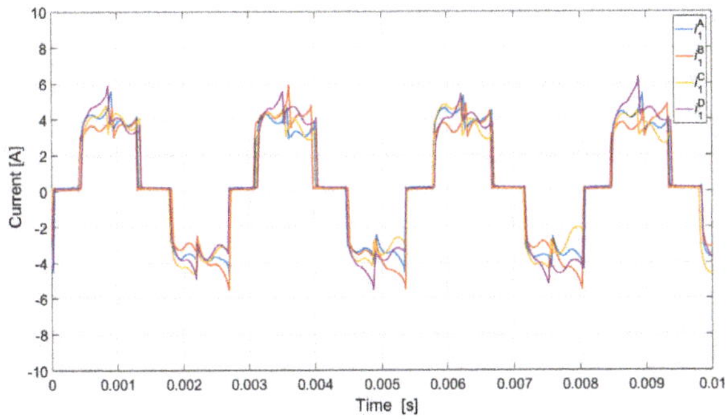

(**a**)

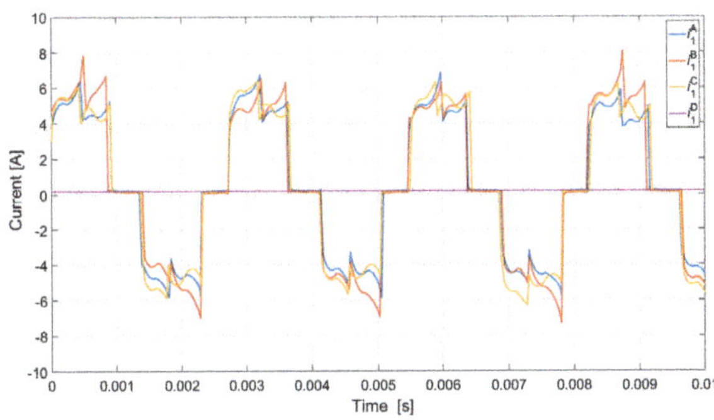

(**b**)

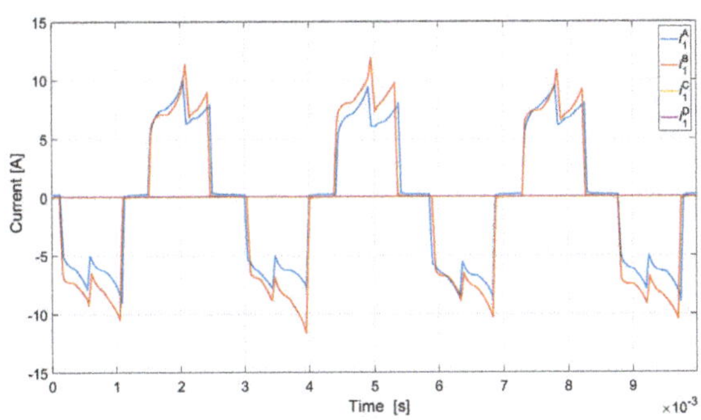

(**c**)

Figure 15. *Cont.*

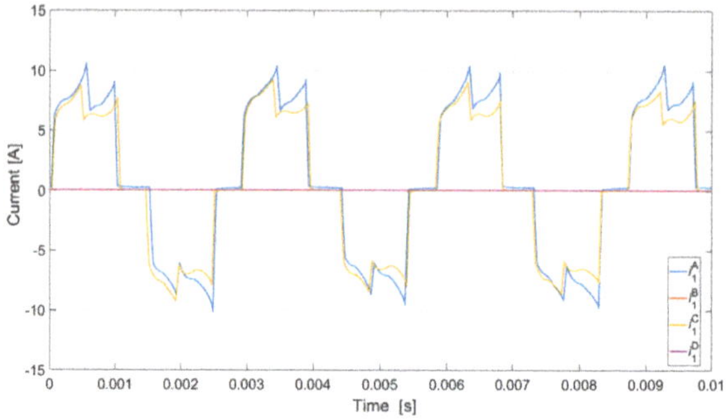

(d)

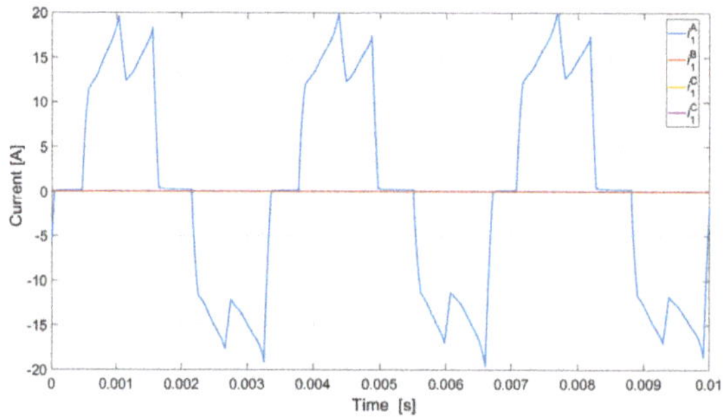

(e)

Figure 15. Waveforms of currents for QCO (**a**); TCO (**b**); DCO A and C (**c**); DCO A and B (**d**); and SCO (**e**).

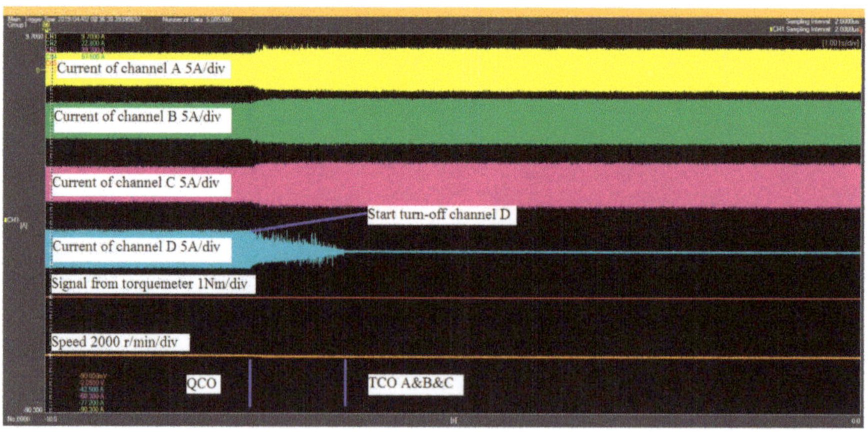

(a)

Figure 16. *Cont.*

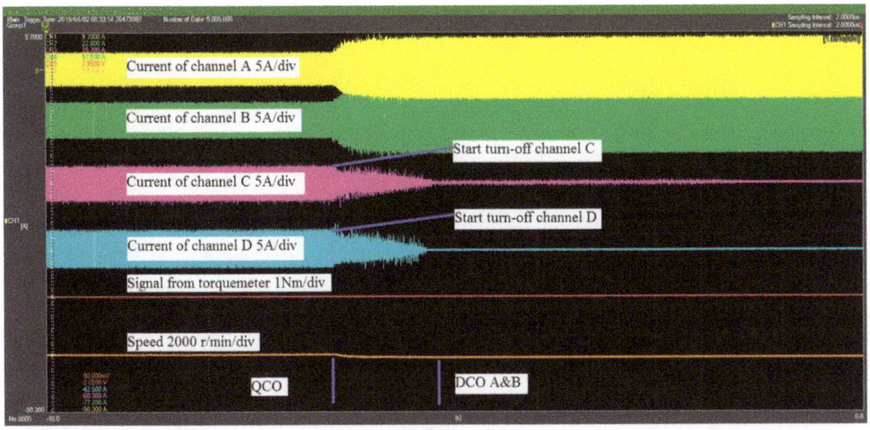

(**b**)

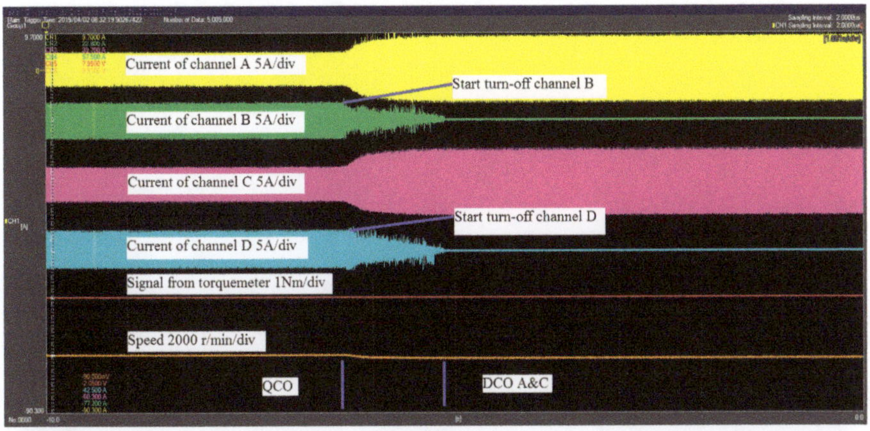

(**c**)

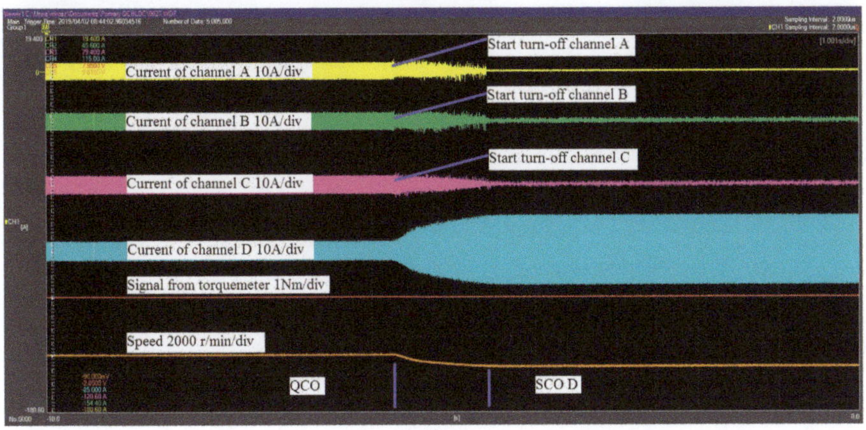

(**d**)

Figure 16. Waveforms of the currents, load torque, and speed for QCO in the TCO mode (**a**), the DCO A and B mode (**b**), the DCO A and C mode (**c**), and the SCO mode (**d**).

During the transition from the QCO mode to the TCO, DCO, and SCO modes, channel D (Figure 16a), C and D (Figure 16b), B and D (Figure 16c), and A and B and C (Figure 16d) were disconnected, and channel A and B and C (Figure 16a), A and B (Figure 16b), A and C (Figure 16c), and D (Figure 16d) started to operate with a shaft load. To achieve the same moment, the currents in the active channels must increase by 133% in the TCO mode, by 200% in the DCO mode, and by 400% in the SCO mode. This increase leads to a decrease in speed in an open-loop control system and results in an output power decrease of a few percent.

The mechanical characteristics and the general efficiency were determined for the selected work mode in a stable state (without current regulation). The load torque was changed to 4 N·m (or 2 N·m for the SCO mode). The rotational speed as a function of the load torque is shown in Figure 17a. The general efficiency as a function of the load torque is shown in Figure 17b. This was determined using the direct method ($\eta = P_{out}/P_{in}$). For a load torque of $T_L = 2$ N·m, the acceleration of vibration and the noise level were recorded. The results are given in Table 3.

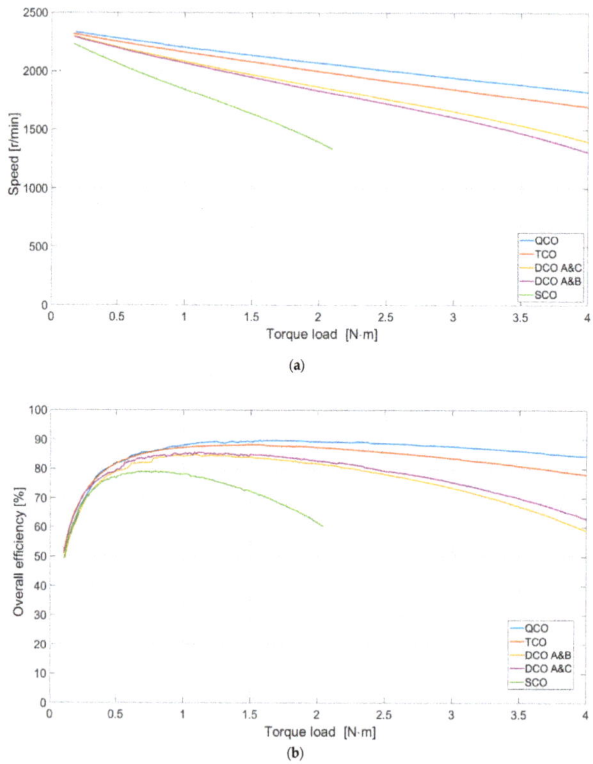

Figure 17. Speed vs. torque load (**a**), overall efficiency vs. torque load (**b**)—variant II.

Table 3. Selected results of the acceleration and the noise level.

Mode	Noise [dB]	Acceleration [m/s^2]
QCO	61.1	1.7
TCO	65.8	5.0
DCO A and B	75.1	7.5
DCO A and C	61.5	2.4
SCO	76.2	12.5

The QCO, TCO, and DCO work modes enable continuous operation of the machine. In the case of the SCO mode, the motor is usually already working in the overload range. This is a critical work condition that should enable operation of the device for a specific period of time. The QCO mode ensures the highest efficiency. In the TCO mode, the efficiency of energy processing is slightly reduced. In the DCO mode, a slightly higher efficiency is achieved in the A and C mode. This efficiency is smaller than that in the TCO mode.

The vibration and noise results presented in Table 3 indicate that the most advantageous work mode is QCO. The DCO A and C work condition was only slightly worse than QCO. The TCO work mode (regardless of the variant) was noticeably noisier than the DCO A and C mode. The DCO A and B work mode was significantly noisier than The DCO A and C mode and TCO mode. The SCO mode was comparable to the DCO A and B mode.

It is possible to maintain a constant speed after turning off one, two, or three channels under operation in a constant torque region.

In a practical layout, some differences between the channels were visible (Figure 16a–c). These differences are due to the differences between the voltages induced in the windings of the different channels. The induced voltages (BEMF) of the shut-down channels in the DCO of one phase are shown in Figure 18. These differences do not affect the reliability of the machine but, unfortunately, result in uneven loads on different channels.

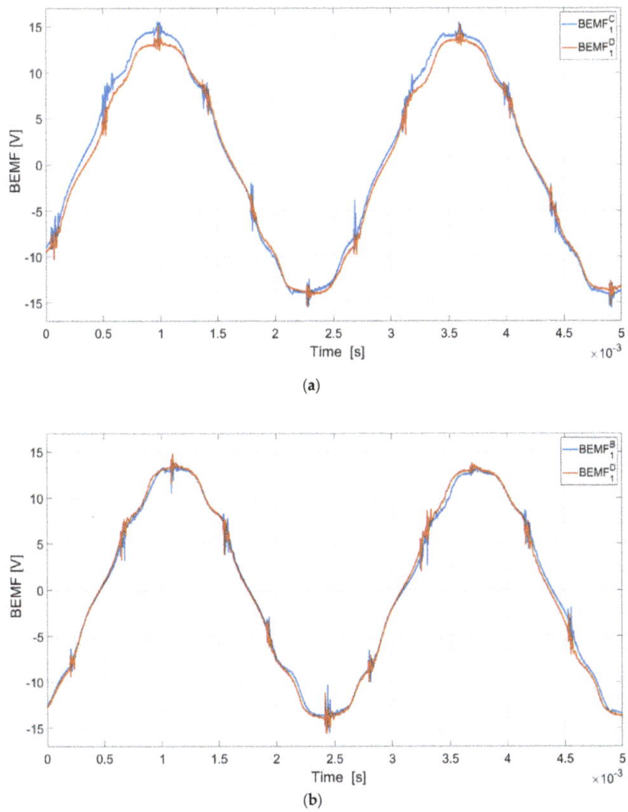

Figure 18. The induced voltages: DCO A and B mode (**a**), DCO A and C mode (**b**).

The problem of operating in TCO, DCO, and SCO modes after the occurrence of a fault state (e.g., short-circuit in the winding) was initially analyzed. Under certain conditions, further motor operation is possible. This was confirmed by preliminary laboratory tests. This will be discussed in future publications.

6. Conclusions

Reliable operation is vitally important for critical drives. This article proposes a quad-channel design for a BLDC machine (QCBLDC) that allows operation with an independent supply from four inverter systems. During fault-free operation, this machine can work in one of the following two operating modes: quad-channel operation (QCO) or dual-channel operation (DCO). In the DCO mode, there are two possible configurations, which are not significantly different from each other with regards to their electrical parameters. The two variants analyzed in this article have very similar parameters. For mechanical reasons, it is more beneficial to supply channel windings staggered by 120 mechanical degrees (variant I). For technological reasons, it is easier to manufacture variant II (shorter winding connections). Compared to the QCO mode, the DCO mode for the same duty point is characterized by a slightly lower efficiency (greater copper losses) and only a slightly higher but balanced magnetic pull (A and C mode). In the case of the DCO A and C mode, the type of variant is not important. In the case of the DCO A and B variant II, a significant increase in vibrations and noise was observed. In the case of variant II of the TCO mode, there was also an insignificant (but noticeable) increase in the vibration and noise level compared to DCO A and C. The results of both the simulation tests and the laboratory tests confirm that the suggested design for the QCBLDC motor effectively works in both the QCO and DCO modes. The problem of operating in TCO, DCO, and SCO modes after the occurrence of a fault state, and the preparation of control algorithms that will facilitate the continued operation of a QCBLDC machine in such a situation, will be discussed in future publications.

Author Contributions: Conceptualization, J.P. and M.K.; methodology, J.P. and M.K.; software, M.K.; formal analysis, J.P. and M.K.; investigation, J.P.; resources, M.K.; writing—original draft preparation, M.K.; writing—review and editing, J.P. and P.B.; visualization, M.K.; supervision, J.P.; project administration, M.K.; funding acquisition, M.K.

Funding: This work is financed in part by the statutory funds of the Department of Electrodynamics and Electrical Machine Systems, Rzeszow University of Technology and in part by the Polish Ministry of Science and Higher Education under the program "Regional Initiative of Excellence" in 2019–2022. Project number 027/RID/2018/19, amount granted 11,999,900 PLN.

Conflicts of Interest: The authors declare no conflict of interest.

References

1. Cao, W.; Mecrow, B.C.; Atkinson, G.J.; Bennett, J.W.; Atkinson, D.J. Overview of Electric Motor Technologies Used for More Electric Aircraft (MEA). *IEEE Trans. Ind. Electron.* **2012**, *59*, 3523–3531.
2. Zhang, G.; Zhang, H.; Huang, X.; Wang, J.; Yu, H.; Graaf, R. Active Fault-Tolerant Control for Electric Vehicles With Independently Driven Rear In-Wheel Motors Against Certain Actuator Faults. *IEEE Trans. Control Syst. Technol.* **2016**, *24*, 1557–1572. [CrossRef]
3. Nounou, K.; Charpentier, J.F.; Marouani, K.; Benbouzid, M.; Kheloui, A. Emulation of an Electric Naval Propulsion System Based on a Multiphase Machine under Healthy and Faulty Operating Conditions. *IEEE Trans. Veh. Technol.* **2018**, *67*, 6895–6905. [CrossRef]
4. Kwon, S.; Ha, J. Fault-tolerant operation with 1-phase open in parallel-connected motor. In Proceedings of the IEEE Applied Power Electronics Conference and Exposition (APEC), San Antonio, TX, USA, 4–8 March 2018.
5. Vaidya, J.G. Redundant Multiple Channel Electric Motors and Generators. United States Patent Number 4550267A, 29 October 1985.
6. Targo, B.A.; Lordo, R.E. Fault Tolerant Electric Machine. United States Patent Number 5929549A, 27 July 1999.
7. Korkosz, M.; Bogusz, P.; Prokop, J. Modelling and experimental research of fault-tolerant dual-channel brushless DC motor. *IET Electr. Power Appl.* **2018**, *12*, 787–796. [CrossRef]
8. Ding, W.; Liang, D. Comparison of transient and steady-state performances analysis for a dual-channel switched reluctance machine operation under different modes. *IET Electr. Power Appl.* **2010**, *4*, 603–617. [CrossRef]

9. Arafat, A.K.M.; Choi, S. Optimal Phase Advance Under Fault-Tolerant Control of a Five-Phase Permanent Magnet Assisted Synchronous Reluctance Motor. *IEEE Trans. Ind. Electron.* **2018**, *65*, 2915–2924. [CrossRef]
10. Levi, E.; Bojoi, R.; Profumo, F.; Toliyat, H.A.; Williamson, S. Multiphase induction motor drives—A technology status review. *IET Electr. Power Appl.* **2007**, *1*, 489–516. [CrossRef]
11. Bojoi, R.; Cavagnino, A.; Tenconi, A.; Vaschetto, S. Control of Shaft-Line-Embedded Multiphase Starter/Generator for Aero-Engine. *IEEE Trans. Ind. Electron.* **2016**, *63*, 641–652. [CrossRef]
12. Popescu, M.; Dorrell, D.G.; Alberti, L.; Bianchi, N.; Staton, D.A.; Hawkins, D. Thermal Analysis of Duplex Three-Phase Induction Motor Under Fault Operating Conditions. *IEEE Trans. Ind. Appl.* **2013**, *49*, 1523–1530. [CrossRef]
13. Umesh, B.S.; Sivakumar, K. Dual-Inverter-Fed Pole-Phase Modulated Nine-Phase Induction Motor Drive With Improved Performance. *IEEE Trans. Ind. Electron.* **2016**, *63*, 5376–5383. [CrossRef]
14. Ullah, S.; McDonald, S.P.; Martin, R.; Benarous, M.; Atkinso, G.J. A Permanent Magnet Assist, Segmented Rotor, Switched Reluctance Drive for Fault Tolerant Aerospace Applications. *IEEE Trans. Ind. Appl.* **2019**, *55*, 298–305. [CrossRef]
15. Wang, B.; Wang, J.; Griffo, A.; Shi, Y. Investigation Into Fault-Tolerant Capability of a Triple Redundant PMA SynRM Drive. *IEEE Trans. Power Electron.* **2019**, *34*, 1611–1621. [CrossRef]
16. Jiang, X.; Huang, W.; Cao, R.; Hao, Z.; Li, J.; Jiang, W. Electric Drive System of Dual-Winding Fault-Tolerant Permanent-Magnet Motor for Aerospace Applications. *IEEE Trans. Ind. Electron.* **2015**, *62*, 7322–7330. [CrossRef]
17. Guo, H.; Xu, J.; Chen, Y. Robust Control of Fault-Tolerant Permanent-Magnet Synchronous Motor for Aerospace Application with Guaranteed Fault Switch Process. *IEEE Trans. Ind. Electron.* **2015**, *62*, 7309–7321. [CrossRef]
18. Li, H.; Li, W.; Ren, H. Fault-Tolerant Inverter for High-Speed Low-Inductance BLDC Drives in Aerospace Applications. *IEEE Trans. Power Electron.* **2017**, *32*, 2452–2463. [CrossRef]
19. ANSYS. *Ansys Electronics Release 2019 R1*; ANSYS Inc.: Canonsburg, PA, USA, 2019.
20. MathWorks. *MATLAB R2019a*; The MathWorks Inc.: Natick, MA, USA, 2019.
21. Bogusz, P.; Korkosz, M.; Prokop, J. The analysis of flux characteristics of dual-channel BLDC machine. In Proceedings of the 2017 International Symposium on Electrical Machines (SME), Naleczow, Poland, 18–21 June 2017.

energies

MDPI

Article

Design and Implementation of a Speed-Loop-Periodic-Controller-Based Fault-Tolerant SPMSM Drive System

Tian-Hua Liu [1,*], Muhammad Syahril Mubarok [1], Muhammad Ridwan [1,2,3] and Suwarno [2]

1 Department of Electrical Engineering, National Taiwan University of Science and Technology, Taipei 106, Taiwan; syahril.elmubarok@gmail.com (M.S.M.); ridwan.bantulz@gmail.com (M.R.)
2 School of Electrical and Informatics Engineering, Institut Teknologi Bandung, Bandung, West Java 40132, Indonesia; suwarno@stei.itb.ac.id
3 PT. PLN (Persero), Jakarta 12160, Indonesia
* Correspondence: Liu@mail.ntust.edu.tw

Received: 1 June 2019; Accepted: 20 June 2019; Published: 20 September 2019

Abstract: This paper proposes a speed-loop periodic controller design for fault-tolerant surface permanent magnet synchronous motor (SPMSM) drive systems. Faulty conditions, including an open insulated-gate bipolar transistor (IGBT), a short-circuited IGBT, or a Hall-effect current sensor fault are investigated. The fault-tolerant SPMSM drive system using a speed-loop periodic controller has better performance than when using a speed-loop PI controller under normal or faulty conditions. The superiority of the proposed speed-loop-periodic-controller-based SPMSM drive system includes faster transient responses and better load disturbance responses. A detailed design of the speed-loop periodic controller and its related fault-tolerant method, including fault detection, diagnosis, isolation, and control are included. In addition, a current estimator is also proposed to estimate the stator current. When the Hall-effect current sensor is faulty, the estimated current is used to replace the current of the faulty sensor. A 32-bit digital signal processor, type TMS-320F-2808, is used to execute the fault-tolerant method and speed-loop periodic control. Measured experimental results validate the theoretical analysis. The proposed implementation of a fault-tolerant SPMSM drive system and speed-loop periodic controller design can be easily applied in industry due to its simplicity.

Keywords: periodic controller; surface permanent magnet synchronous motor; fault-tolerant system

1. Introduction

Electrical motors, including DC motors, induction motors, and permanent magnet synchronous motors, were used to for decades, enabling modern life. Electric motors and their related inverters are used to transform electric power into mechanical power. Electric motors are used in pumps, cranes, conveyors, mills, elevators, and transportation. The surface permanent magnet synchronous motor (SPMSM) became more popular due to its excellent characteristics: high-power density, high efficiency, and a simple control method [1]. The SPMSM is widely used in traction applications, including land and marine vehicles because the SPMSM does not require any brushes, and there is no slip frequency between its stator and rotor [1]. In addition, increased awareness of global warming and motivation to decrease carbon emissions further increased the attraction of electric vehicles, most of which are driven by SPMSMs, which have the benefits of high-power density, good dynamic responses, and simple control methods [2].

Failure of an SPMSM drive system can put drivers, operators, passengers, and people in the vicinity at risk of injury or even death. Failures can be divided into two main categories: motor faults and inverter faults. An inverter is far more fragile and more likely to suffer a fault than a motor due to its high PWM switching frequency, vulnerable power devices, and complicated control algorithm. Development of advanced fault-tolerant control methods is important to reduce the potential for accidents and huge financial losses incurred by them [3]. Research on advanced fault-tolerant control technology was successfully applied in motor drives, power supplies, transportation, and other industrial applications [4,5]. For example, Naidu et al. proposed fault-tolerant SPMSM drive topologies for automotive vehicles, which used X-by-wire systems to improve their safety, reliability, and performance [6]. Kontarcek et al. investigated a low-cost fault-tolerant SPMSM drive system for an open-phase fault in an SPMSM drive system, which was based on field orientation control. In addition, a prediction stator current for the next sampling interval and a new post-fault operation method of the SPMSM was investigated [7]. Jung et al. proposed a model reference adaptive technique-based diagnosis of an open-circuit fault. An observer was implemented to determine the faulty condition. Two major post-fault actions were discussed as well [8]. Cai et al. proposed a Bayesian network-based data-driven fault diagnosis methodology for three-phase inverters. Two output line-to-line voltages were measured to detect and diagnose faults, which could be used for multilevel inverter SPMSM drive systems [9]. Meinguet et al. used multiple fault indices to retrieve the most likely state of the AC drive systems. Based on the unbalance of the three-phase currents and instantaneous frequency, a fault-tolerant topology was derived [10]. Wang et al. proposed a fault-tolerant control for dual three-phase SPMSM drive systems under open-phase faults. The object of the research included two parts. The first part was to maximize the torque capability while protection was considered, and the second part was to minimize copper loss [11]. Tseng et al. proposed a fault-tolerant control for a dual-SPMSM drive system. Two simple methods, including a short-circuit fault-tolerant method and an open-circuit fault-tolerant method, were investigated. Experimental results showed that this dual-SPMSM drive system could maintain speed although one power device was open-circuited or short-circuited [12]. Wang et al. proposed a fault-tolerant control system of a parallel-voltage inverter-fed SPMSM drive system. Three fault-tolerant control strategies were proposed and compared. The proposed method not only provided smooth torque but also had less copper loss under open-circuit faults [13].

Recently, Nasiri et al. proposed a full digital current control of an SPMSM for vehicular applications. The objective of the control is to achieve a deadbeat dynamic response for the speed control of an SPMSM. The proposed method discussed a robust sensorless method; as a result, an encoder fault was allowed [14]. Bennett et al. investigated a fault-tolerant electric drive for an aircraft nose wheel steering actuator. The wheel steering actuator included two independent controllers. Each controller operated one-half of a dual three-phase SPMSM drive system. As a result, the other controller could control the aircraft nose when one controller failed [15]. Jeong proposed a fault detection and fault-tolerant control of the IPMSM drive system for electric vehicles. Once the fault was detected, the control scheme automatically reconfigured to provide post-fault operational capability [16]. Wang et al. implemented a fault-tolerant control with an active fault diagnosis for four-wheel independently driven electric ground vehicles. An adaptive control-based passive fault-tolerant controller was designed to ensure that the vehicle system was stable and tracked a desired vehicle motion when the in-wheel motor drive system failed [17]. Zhang et al. proposed an active fault-tolerant control for electric vehicles with independently driven rear in-wheel motors against actuator faults. After the fault was detected, a proper reconfigurable controller was switched on to achieve optimal post-fault performance [18]. Bolognan proposed remedial strategies against failures occurring in an inverter power device for an SPMSM drive system. Minimal redundant hardware was implemented [19]. Bai proposed a fault-tolerant control for a dual-winding SPMSM drive system based on the space vector pulse width modulation technique. The distribution of the space vector voltages was analyzed, and the vector control strategies under healthy and one-phase open-circuit faulty conditions were investigated to maintain the magnetomotive force of the SPMSM as a constant [20]. The papers mentioned above, however, only focused on the fault detection, diagnosis, and isolation [6–20]. None or only a few researchers focused on the controller design of fault-tolerant drive systems. When the SPMSM drive system is operated in normal conditions, the three-phase currents are balanced. Thus, the torque pulsations are small. However, when the SPMSM drive system is operated in faulty conditions, the three-phase currents are seriously imbalanced, causing obvious torque pulsations. As a result, the drive system in a faulty condition is very difficult to control. To solve this challenge, this paper proposes a speed-loop periodic controller to improve the dynamic responses of the drive system under an open-circuit fault or short-circuit fault. To the authors' best knowledge, the ideas of this paper are original. No previously published papers covered this issue.

This paper proposes a speed-loop periodic controller to improve the transient responses and load disturbance responses for SPMSMs under normal and faulty conditions. This paper is divided into the following sections: firstly, a fault-tolerant inverter is presented. Secondly, the fault detection, diagnosis, isolation, and control methods are discussed. The methods use a back-up leg to replace the faulty leg in the inverter. After that, a speed-loop periodic controller and a current-loop PI controller are designed to improve the dynamic responses of the SPMSM drive system, including fast transient responses and good load disturbance responses. Next, the implementation of the drive system is discussed. Finally, several experimental results and conclusions are included.

2. Fault-Tolerant SPMSM Drive System

Failure of the SPMSM drive system can be categorized into two major types: motor failures and circuit failures. Motor failures includes bearing damage, open winding, and partially short-circuited winding. Circuit failures include inverter failure, current sensor failure, and encoder failure. The inverter is the most likely location of a fault and not the motor because, compared to the SPMSM, the inverter is more fragile and more likely to be open- or short-circuited. In addition, the current sensor and its circuit malfunction easily due to the offset voltage and aging of the circuit. As a result, a fault-tolerant control method is proposed here to use the estimated current to replace the measured current. This paper only focuses on the fault-tolerant method of the inverter and sensor and not the SPMSM. In this section, fault detection, diagnosis, isolation, and control of a fault-tolerant inverter are discussed first, and then fault detection, isolation, estimation, and control of a Hall-effect sensor are investigated.

2.1. Fault Detection and Diagnosis of a Fault-Tolerant Inverter

This research covers the situation when one power switch of the inverter is open- or short-circuited. The fault-tolerant inverter drive system is shown in Figure 1, which contains six IGBTs, S_a, S'_a, S_b, S'_b, S_c, and S'_c, and two back-up IGBTs, S_t and S'_t. At the output of the inverter, six TRIACs, including T_{at}, T_{bt}, T_{ct}, T_{af}, T_{bf}, and T_{cf} are added. In addition, three high-speed fuses F_a, F_b, and F_c are inserted into the inverter and a back-up leg, including S_t and S'_t, is added as well. This paper discusses an open-circuit fault and a short-circuit fault of one leg in the inverter.

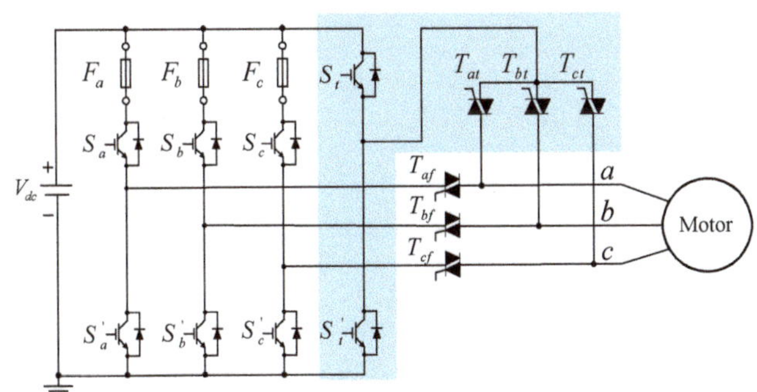

Figure 1. Fault-tolerant inverter.

A performance index is established to identify if the SPMSM drive system failed [7]. During normal operation, the square magnitude error is calculated as follows:

$$\varepsilon_n(k) \cong \left(\frac{v_d(k)}{L_s}\Delta t\right)^2 + \left(\frac{v_q(k)-e_q(k)}{L_s}\Delta t\right)^2, \tag{1}$$

where $\varepsilon_n(k)$ is the performance index under normal conditions. $v_d(k)$ is the d-axis voltage, L_s is the self-inductance, $v_q(k)$ is the q-axis voltage, $e_q(k)$ is the back-EMF, and Δt is the time of each time interval. To avoid false detection, according to the authors' experiences, 10 times the normal error vector is an adequate threshold to determine the faulty condition. The performance index in the faulty condition, $\varepsilon(k)$ can be expressed as

$$\varepsilon(k) > 10\ \varepsilon_n(k). \tag{2}$$

In Equation (2), the performance index of the SPMSM in a faulty condition, $\varepsilon(k)$, can be defined as

$$\varepsilon(k) = \Delta i_d^2(k) + \Delta i_q^2(k), \tag{3}$$

where $\varepsilon(k)$ is the performance index in a faulty condition. $\Delta i_d(k)$ and $\Delta i_q(k)$ are the current deviations in the d-axis and q-axis in a faulty condition. The DSP diagnoses the faulty condition by measuring the deviations of the d-axis and q-axis currents and then identifying whether the faulty condition occurred in either the a-phase, b-phase, or c-phase. The DSP transforms the a, b, c axis currents in the α–β axis currents, and then computes the current angle δ [21,22]. Taking the a-phase fault as an example, Figure 2a shows the b-phase and c-phase currents when the a-phase winding is open-circuited. The current can flow in either direction as shown in Figure 2b. The current may flow from the b-phase to the c-phase, which results in the current vector having a 270° angle, or the current may flow from the c-phase to the b-phase, which results in a 90° angle. The summarized results of the current angle δ when one phase is faulty are shown in Table 1. By computing the current angle δ, one can easily diagnose which phase is open. After that, an isolation and control method is executed to isolate the faulty part, and uses the back-up leg to replace the faulty leg. A fault-tolerant SPMSM drive system, thus, can be achieved.

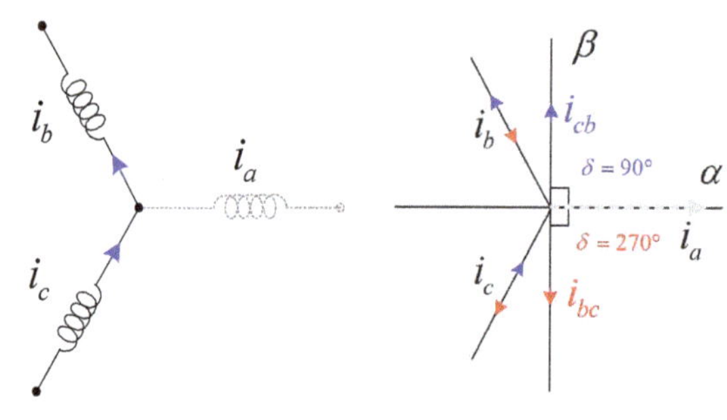

Figure 2. Output current vector of a-phase fault: (**a**) three-phase winding; (**b**) current vector.

Table 1. Current angle at different faulty phases.

Current Angle δ		**Faulty Phase**
Upper Quadrant	**Lower Quadrant**	
90°	270°	a-phase
30°	210°	b-phase
150°	330°	c-phase

2.2. Fault Detection and Control of a Current Sensor

This paper also investigates the detection and control of a fault in a one-phase current sensor. Previous research used a current estimator to evaluate the current sensor error, and an adaptive threshold was used to detect and diagnose the faulty condition [23,24]. In a faulty condition, the estimated current replaces the faulty current. The α–β axis voltages and currents were obtained using a coordinate transformation, and the estimated α–β currents in the discrete time domain can be expressed as

$$\hat{i}_\alpha(k) = \hat{i}_\alpha(k-1) + \frac{T_s}{L_s}[v_\alpha(k) - r_s i_\alpha(k) - \omega_e(k)\lambda_m \sin\theta_e(k)], \tag{4}$$

and

$$\hat{i}_\beta(k) = \hat{i}_\beta(k-1) + \frac{T_s}{L_s}\left[v_\beta(k) - r_s i_\beta(k) + \omega_e(k)\lambda_m \cos\theta_e(k)\right], \tag{5}$$

where $\hat{i}_\alpha(k)$ and $\hat{i}_\beta(k)$ are the estimated current, $v_\alpha(k)$ and $v_\beta(k)$ are the α–β axis voltages, and $i_\alpha(k)$ and $i_\beta(k)$ are α–β axis currents. $\omega_e(k)$ and $\theta_e(k)$ are the electrical speed and angle. The current waveform factor $F_x(k)$ and $\hat{F}_x(k)$ can be calculated as

$$F_x(k) = \frac{|i_x|_{\text{RMS}}(k)}{|i_x|_{\text{AVG}}(k) + e}, \tag{6a}$$

and

$$\hat{F}_x(k) = \frac{|i_x + \varepsilon_x|_{\text{RMS}}(k)}{|i_x + \varepsilon_x|_{\text{AVG}}(k) + e}, \tag{6b}$$

where e is a constant to prevent the denominator in the Equations (6a) and (6b) from reaching zero. $|i_x|_{RMS}(k)$ is the absolute value of the measured RMS current, $|i_x|_{AVG}(k)$ is the absolute value of the measured average current, and $\varepsilon_x(k)$ is the estimated error. The residual function $R_x(k)$ is obtained by computing the difference between the estimated waveform factor $\hat{F}_x(k)$ and the measured waveform factor $F_x(k)$. It can be expressed as

$$R_x(k) = \hat{F}_x(k) - F_x(k). \tag{7}$$

By substituting Equations (6a) and (6b) into Equation (7), the residual function $R_x(k)$ in the equation can be transformed into

$$R_x(k) = \frac{|i_x + \varepsilon_x|_{\text{RMS}}(k)}{|i_x + \varepsilon_x|_{\text{AVG}}(k) + e} - \frac{|i_x|_{\text{RMS}}(k)}{|i_x|_{\text{AVG}}(k) + e} = -\frac{|i_x + \varepsilon_x|_{\text{RMS}}(k)}{|i_x + \varepsilon_x|_{\text{AVG}}(k) + e} - \frac{|i_x|_{\text{RMS}}(k)}{|i_x|_{\text{AVG}}(k) + e}, \tag{8}$$

where $\varepsilon_x(k)$ is the estimated current error of the a, b, c phases. The numerator of the estimated absolute value of the RMS current $|i_x + \varepsilon_x|_{RMS}(k)$ is always lower than or equal to the total of $|i_x|_{RMS}(k)+|\varepsilon_x|_{RMS}(k)$ due to the triangular inequality rule. By using this relationship, the residual function $R_x(k)$ can be rewritten as

$$R_x(k) \leq \frac{|i_x|_{\text{RMS}}(k)}{|i_x + \varepsilon_x|_{\text{AVG}}(k) + e} + \frac{|\varepsilon_x|_{\text{RMS}}(k)}{|i_x + \varepsilon_x|_{\text{AVG}}(k) + e} - \frac{|i_x|_{\text{RMS}}(k)}{|i_x|_{\text{AVG}}(k) + e}. \tag{9}$$

The difference between the threshold value and the residual value is used to determine if a faulty condition occurred. When the system is in a steady-state condition and the current sensor is in a normal condition, the measured and estimated waveform factors are constants and the residual value is near zero. On the other hand, when the current sensor is in a faulty condition, the residual value abruptly increases due to its large error. Finally, the estimated current $\hat{i}_x(k)$ replaces the measured faulty current $i_x(k)$. However, in this paper, the estimated currents are near the measured currents only in steady-state conditions. The transient responses of the estimated currents are ignored to simplify the current estimating method.

3. Speed-Loop Periodic Controller

The speed-loop periodic controller for a fault-tolerant SPMSM in this paper is an original idea. The internal model principle states that perfect asymptotic tracking of persistent inputs can be attained by replicating the signal generator in a stable feedback loop [25]. The internal model of the inputs is a signal generator. Figure 3a shows the basic continuous s-domain structure of a periodic signal generator, which includes a delay device e^{-sT_O} and a positive feedback. According to Figure 3a and assuming $Q(s) = 1$, the transfer function of the periodic signal generator can be expressed as

$$G_{rs}(s) = \frac{u_{rc}(s)}{e(s)} = \frac{e^{-sT_O}}{1-e^{-sT_O}}, \tag{10}$$

where $G_{rs}(s)$ is the transfer function of the periodic signal generator, and e^{-sT_O} is a time-delay unit. From Equation (10), the periodic signal generator $G_{rs}(s)$ can be expanded as follows [26]:

$$G_{rs}(s) = \frac{e^{-sT_O}}{1-e^{-sT_O}} = -\frac{1}{2} + \frac{1}{sT_O} + \frac{1}{T_O}\sum_{n=1}^{\infty}\frac{2s}{s^2+(n\omega_o)^2}. \tag{11}$$

In Equation (11), the first item is a transfer function of an impulse, the second item is a transfer function of a step input, and the third item is the transfer function of the harmonics. In the real world, a low-pass filter $Q(s)$ is required to compensate for the related harmonics, and a phase-lead compensator $G_f(s)$ is used for the entire system delay compensation. To simplify the problem, assuming $Q(s)$ is 1, the classic periodic controller makes $G_{rs}(s)$ approach ∞ at poles $s = \pm jn\omega_o$. In this research, a DSP is used to execute the control algorithm; as a result, the s-domain periodic signal generator needs to convert into the z-domain periodic signal generator shown in Figure 3b. The z-domain periodic signal generator is expressed in a discrete form as follows:

$$G_{rs}(z) = k_{rc}\frac{Q(z)z^{-N}}{1-Q(z)z^{-N}}G_f(z), \tag{12}$$

where k_{rc} is a constant control gain, $Q(z)$ is a low-pass filter (LPF), $G_f(z)$ is a phase-lead compensator that compensates for the time delay, and N is the number of delay steps.

In the discrete time domain, the z^{-N} is added as shown in Figure 3b. N can be expressed as

$$N = \frac{T_0}{T_s}, \tag{13}$$

where T_0 is the fundamental period and T_s is the sampling interval of the speed-loop control system. The fundamental period T_0 determines the delay of the periodic controller in N steps. The delay steps determine the settling time of the SPMSM drive system. If the delay time is set too short, the output generates obvious overshoot but has quick responses; however, if the delay time is too long, the periodic controller has slow responses. The choice of the parameter N depends on the designer's experience. In addition, the periodic controller is added into the speed-loop PI controller in the forward loop [26], which is shown in Figure 3c. The speed-loop PI controller is used to improve the transient responses and load disturbance responses for the normal operation speed dynamics; however, the speed-loop periodical controller is used when the SPMSM drive system is faulty, which causes three-phase current imbalance. In Figure 3c, $G_p(z)$ is the transfer function of the SPMSM drive system, $G_c(z)$ is the speed-loop PI controller, and $G_{rs}(z)$ is the periodic signal generator, which is used to reduce the current harmonics. After that, the speed command ω_r^* is input into the closed-loop system. In this closed-loop system, a periodic signal output $u_{rs}(z)$ is added to the speed error $\Delta\omega_r(z)$ to generate the total input of the PI controller to control the system.

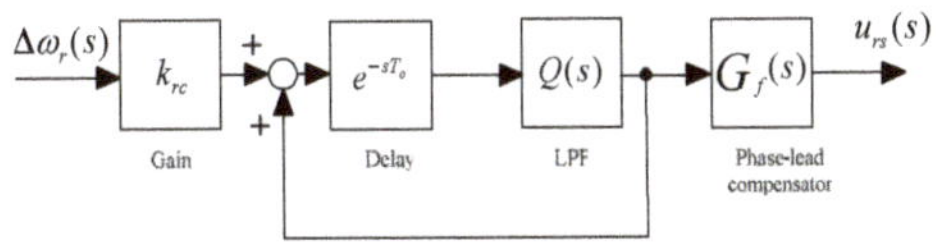

(a)

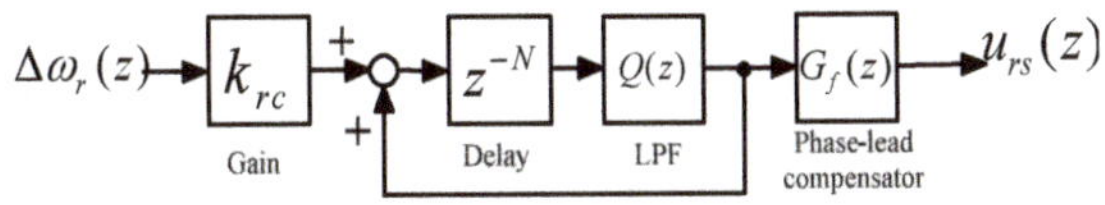

(b)

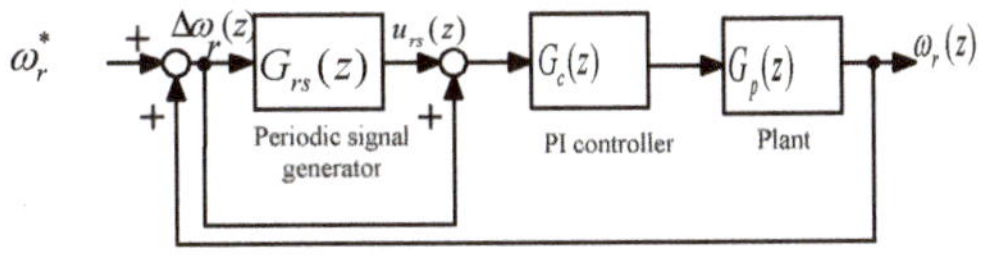

(c)

Figure 3. Periodic controller in speed-loop: (**a**) s-domain periodic controller $G_{rs}(s)$; (**b**) z-domain periodic controller $G_{rs}(z)$; (**c**) closed-loop control system.

Compared to the traditional speed-loop PI controller, the proposed method uses a periodic controller to cascade to the traditional speed-loop PI controller, which increases the gain at certain frequencies. As a result, the transient responses and load disturbance responses of the SPMSM can be effectively improved. The computation of the periodic controller is very simple, which only includes a delay operation, a low-pass filter, a positive feedback operation, and phase-lead compensation. As a result, it is easy to implement the proposed control method by using a DSP.

The parameters of the periodic controller, including a control gain k_{rc} and a phase-lead compensation $G_f(z)$, are determined by using stability analysis in the closed-loop control system. The detailed analysis and the stable condition of a closed-loop control system were previously discussed and can be expressed as follows [13]:

$$k_{rc} < \frac{2\cos(\theta_H + p\omega)}{|H(e^{j\omega})|} \text{ and } k_{rc} \geq 0, \tag{14a}$$

and

$$H(z) = \frac{G_c(z)G_p(z)}{1 + G_c(z)G_p(z)}, \tag{14b}$$

where θ_H is the phase angle of $H(z)$, ω is the frequency, p is the order of the phase-lead compensation, and $H(z)$ is the transfer function of the closed-loop control system. The control gain k_{rc} and the order p of the phase-lead compensation are determined as shown below. In the z-domain analysis, the phase-lead compensation $G_f(z)$ is commonly expressed as follows [13]:

$$G_f(z) = z^p \tag{15}$$

The characteristics of the closed-loop speed-control SPMSM drive system are discussed here. Figure 4 shows the relationship between the boundary of the phase angle and operating frequency of the closed-loop drive system. The phase lead step p includes steps 0, 1, 2, and 3, which are shown as $p = 0$, $p = 1$, $p = 2$, and $p = 3$ in Figure 4, respectively. In the physical system, the available range of the compensated phase is between −90° and 90°, which is shown as the dashed line in Figure 4. From Figure 4, to operate in the phase boundary between −90° and 90°, the maximum operating frequency is 0.4 kHz for zero-step phase-lead compensation, 5 kHz for one-step phase-lead compensation, 3.3 kHz for two-step phase-lead compensation, and 1.9 kHz for three-step phase-lead compensation. In order to obtain the widest operating frequency of the closed-loop SPMSM drive system, the one-step phase lead ($p = 1$) is selected in this research. After that, the gain k_{rc} is chosen according to the stability analysis. The stability condition is shown in Equation (14a), which shows that the gain k_{rc} needs to satisfy the inequality equation. Figure 5 shows the relationship between the maximum boundary $\frac{2\cos(\theta_H + p\omega)}{|H(e^{j\omega})|}$ and the operation frequency. In order to both satisfy Equation (14a) and obtain the widest operating frequency range, the one-step phase-lead compensation that provides a very smooth curve was chosen for this paper. By using the one-step phase-lead compensation and satisfying Equation (14a), k_{rc} was selected as 1.5 because $\frac{2\cos(\theta_H + p\omega)}{|H(e^{j\omega})|}$ was varied between 1.5 and 150 when the operating frequency varied from 0 kHz to 5 kHz.

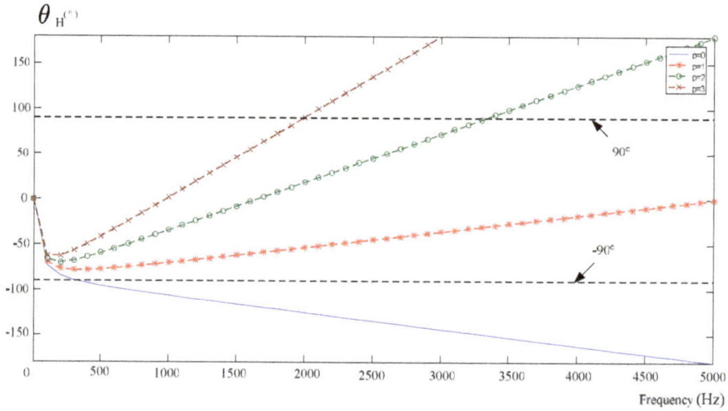

Figure 4. Compensated phase responses using a periodic controller.

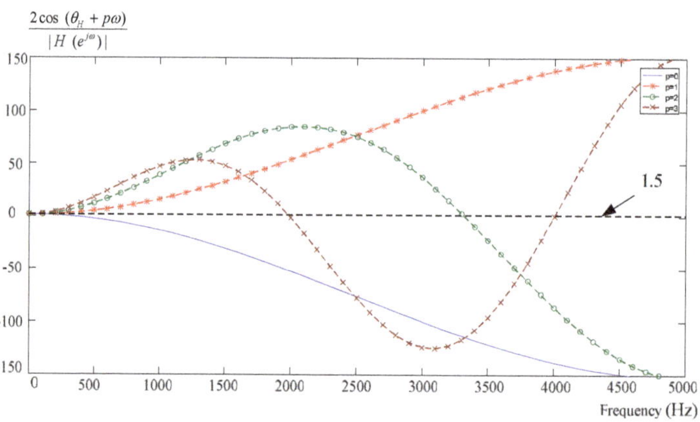

Figure 5. Boundary of control gain under different compensated phases.

The low-pass filter, LPF $Q(z)$, was designed by using finite impulse response (FIR). FIR was chosen here because it is commonly used in digital filter applications. The transfer function of an FIR LPF $Q(z)$ can be expressed as

$$Q(z) = \sum_{i=0}^{m} a_i z^{-i}. \tag{16}$$

The components of the periodic controller are shown in Figure 6. The speed error $\Delta\omega_e(z)$ is multiplied with a control gain k_{rc}, and then added to the $z^{-p}u_{rs}(z)$ to generate $e_s(z)$. A low-pass filter, $a_0 + a_1 z^{-1} + a_2 z^{-2}$, is used to reduce the high-frequency noise. After that, the output of the low-pass filter, which is $z^{-p}u_{rs}(z)$, is added to the $k_{rc}\Delta\omega_r(k)$ to obtain $e_s(z)$. Finally, $u_{rs}(z-p)$ passes through the phase-lead compensator z^p to obtain the $u_{rs}(z)$. By using k as the interval step number, the output before delay is expressed as $u_{rs}(z)z^{-p}$, and then the system error $e_s(z)$ can be transformed into

$$e_s(k) = k_{rc}\Delta\omega_r(k) + u_{rs}(k-p). \tag{17}$$

By using the LPF $Q(z)$ with a_i as the coefficient, the $u_{rc}(k)$ can be expressed as

$$u_{rs}(k) = \sum_{i=0}^{m} a_i e_s(k - N - i + p),\ i = 0, 1, 2. \tag{18}$$

The total of $u_{rs}(k)$ and $\Delta\omega_r(z)$ becomes the control input of the PI controller.

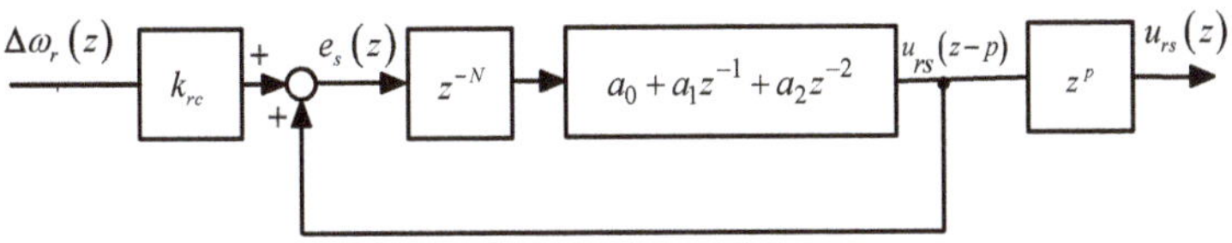

Figure 6. The proposed periodic controller.

4. Current-Loop Controller

In general, the current-loop PI controller, which is a minor-loop of the SPMSM drive system, is cascaded with the speed-loop controller. Figure 7 shows the detailed block diagram of the speed-loop controller and current-loop controller in an SPMSM drive system. First, the speed ω_r is subtracted from the speed command ω_r^* to obtain the speed error $\Delta\omega_m$. Then, the speed-loop controller is executed to generate the q-axis current command i_q^*. The d-axis current command is set at zero in this research. Next, two PI controllers are designed to compute the d-axis voltage command v_d^* from the d-axis current error, and also the q-axis voltage command v_q^* from the q-axis current error. After that, the SVPWM inverter generates a-, b-, c-axis voltages v_a, v_b, and v_c from the information of the v_d^*, v_q^*, and electrical rotor position θ_e. The a, b, c voltages are injected into the SPMSM to generate the a, b, c currents i_a, i_b, and i_c. Finally, the SPMSM rotates and reports its mechanical angle θ_m to the DSP.

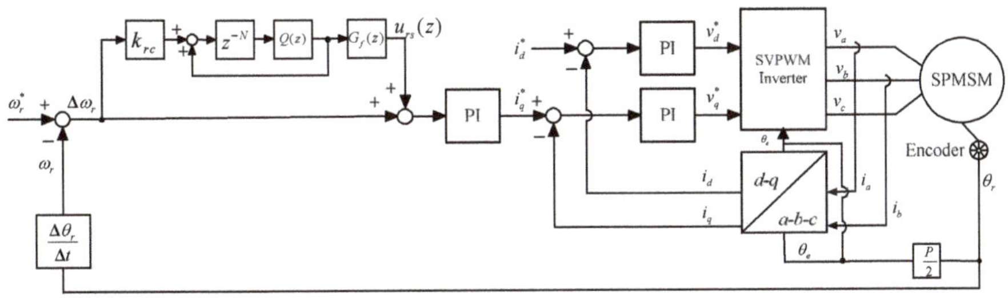

Figure 7. Detailed block diagram of speed-loop and current-loop PI controllers in an SPMSM drive system.

The SPMSM drive system returns the signals from the encoder and two Hall-effect current sensors to the DSP. The encoder detects the rotor angle θ_m, and then computes the electrical angle θ_e by multiplying pole pairs. The rotor speed ω_r is obtained by taking the difference operation from the θ_r. Two Hall-effect current sensors are used to measure the a-phase and b-phase currents i_a and i_b, and then the c-phase current i_c can be calculated because it is a three-phase balanced system. The relationship between the a, b, c currents and the d-, q-axis currents is shown below.

$$\begin{bmatrix} i_d \\ i_q \end{bmatrix} = \frac{2}{3}\begin{bmatrix} \cos(\theta_e) & \cos(\theta_e - \frac{2\pi}{3}) & \cos(\theta_e + \frac{2\pi}{3}) \\ \sin(\theta_e) & \sin(\theta_e - \frac{2\pi}{3}) & \sin(\theta_e + \frac{2\pi}{3}) \end{bmatrix}\begin{bmatrix} i_a \\ i_b \\ i_c \end{bmatrix}. \tag{19}$$

In the d-, q-axis synchronous frame, the dynamic equation of currents for an SPMSM is expressed as

$$\frac{di_d(t)}{dt} = \frac{1}{L_s}\big(v_d(t) - r_s i_d(t) + \omega_e(t) L_s i_q(t)\big), \tag{20}$$

$$\frac{di_q(t)}{dt} = \frac{1}{L_s}\big(v_q(t) - r_s i_q(t) - \omega_e(t) L_s i_d(t) - \omega_e(t)\lambda_m\big). \tag{21}$$

The dynamic equation of the speed is

$$\frac{d\omega_e(t)}{dt} = \frac{1}{J_m}(T_e - B_m\omega_e(t) - T_L), \tag{22}$$

and the electromagnetic torque is

$$T_e = \frac{3}{2}\frac{P}{2}\big(\lambda_m i_q(t)\big), \tag{23}$$

where $\frac{d}{dt}$ is the differential operator, L_s is the stator inductance, r_s is the stator resistance, λ_m is the flux linkage, J_m is the inertia, B_m is the friction coefficient, and T_L is the external load. Assuming the resistance voltage is neglected and the decoupling forward method is used, then the d-, q-axis voltage v_d^* and v_q^* can be expressed as

$$\frac{1}{L_s}v_d^* = \frac{1}{L_s}\big(v_d + \omega_e L_s i_q\big), \tag{24}$$

and

$$\frac{1}{L_s}v_q^* = \frac{1}{L_s}\big(v_q - \omega_e L_s i_d - \omega_e\lambda_m\big). \tag{25}$$

Substituting Equations (24) and (25) into Equations (20) and (21), the dynamics of the SPMSM can be rewritten as

$$\frac{di_d}{dt} = -\frac{r_s}{L_s}i_d + \frac{1}{L_s}v_d^*, \tag{26}$$

and

$$\frac{di_q}{dt} = -\frac{r_s}{L_s}i_q + \frac{1}{L_s}v_q^*. \tag{27}$$

After using the current-loop PI controllers, the d-, q-axis voltage commands, v_d^* and v_q^*, are expressed as

$$v_d^*(t) = K_P\big(i_d^*(t) - i_d(t)\big) + K_I\int_0^t \big(i_d^*(\tau) - i_d(\tau)\big)d\tau, \tag{28}$$

and

$$v_q^*(t) = K_P\big(i_q^*(t) - i_q(t)\big) + K_I\int_0^t \big(i_q^*(\tau) - i_q(\tau)\big)d\tau. \tag{29}$$

The d-axis voltage is obtained by substituting Equation (28) into Equation (24), and the q-axis voltage is obtained by substituting Equation (29) into Equation (25). Finally, the output voltages can be expressed as

$$v_d(t) = K_P\big(i_d^*(t) - i_d(t)\big) + K_I\int_0^t \big(i_d^*(\tau) - i_d(\tau)\big)d\tau - \omega_e(t)L_s i_q(t), \tag{30}$$

and

$$v_q(t) = K_P\left(i_q^*(t) - i_q(t)\right) + K_I \int_0^t \left(i_q^*(\tau) - i_q(\tau)\right)d\tau + \omega_e(t)L_s i_d + \omega_e(t)\lambda_m. \quad (31)$$

After transferring the continuous time domain into discrete time domain, one can obtain the *d*-, *q*-axis voltage commands as

$$v_d(k) = K_P\left(i_d^*(k) - i_d(k)\right) + K_I T_c \sum_{n=1}^{k}\left(i_d^*(k) - i_d(k)\right) - \omega_e(k)L_s i_q(k), \quad (32)$$

and

$$v_q(k) = K_P\left(i_q^*(k) - i_q(k)\right) + K_I T_c \sum_{n=1}^{k}\left(i_q^*(k) - i_q(k)\right) + \omega_e(k)L_s i_d(k) + \omega_e(k)\lambda_m, \quad (33)$$

where T_c is the sampling interval of the current loop. From Equations (32) and (33), a block diagram of the PI current-loop controller can be constructed as shown in Figure 8. In this research, the parameters of the PI controller were obtained by using the pole assignment technique.

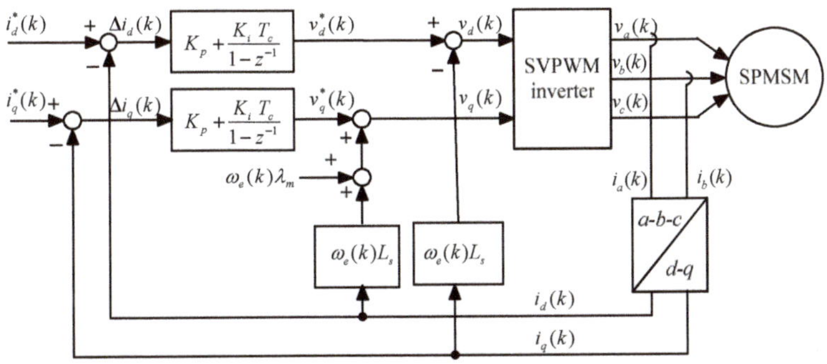

Figure 8. Current-loop PI controller.

5. Implementation

A block diagram of the implemented SPMSM drive system is shown in Figure 9a. A DSP type TMS320F2808 was used as the control center. The SPMSM drive system includes a fault-tolerant inverter, a DSP, gate drivers, current-sensing circuits, an encoder circuit, and an overcurrent protection circuit. The speed-loop PI controller includes $K_p = 0.5$ and $K_I = 0.2$, which are obtained by pole assignment with two poles, $p_1 = 0.79$ and $p_2 = 0.93$. The speed-loop periodic controller includes $k_{rc} = 1.5$, $Q(z) = 0.2 + 0.45z^{-1} + 0.2z^{-2}$, $N = 50$, and $G_f(z) = z$. The sampling interval of the speed loop is 1 ms. The current-loop PI controllers include $K_p = 12.17$ and $K_I = 0.0006$, which determine the inner-loop current dynamics. The sampling interval of the current loop is 100 μs.

The SPMSM has the following parameters: $r_s = 0.73\ \Omega$, $L_s = 1.37$ mH, $\lambda_m = 0.167$ Wb, $B_m = 0.003$ N·m·s/rad, and $K_T = 1.0$ N·m/A. Figure 9b shows a photograph of the implemented drive system, which includes an SPMSM and a dynamometer, which provides the external load for the SPMSM drive system.

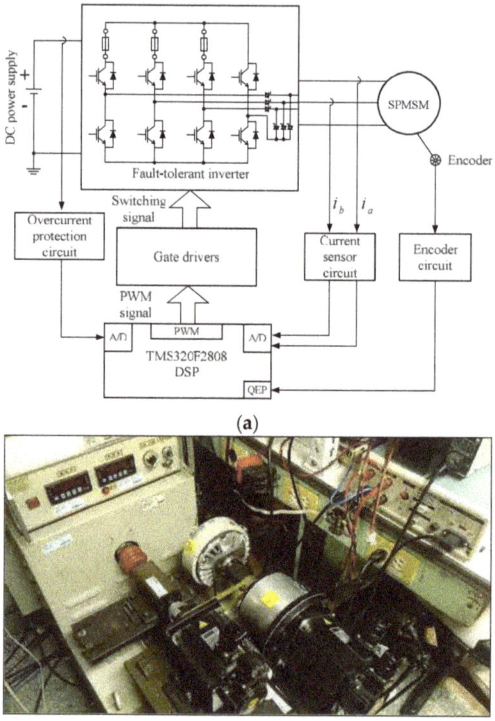

(a) (b)

Figure 9. The implemented system (**a**) block diagram, and (**b**) photograph.

6. Simulated and Experimental Results

The simulated and experimental results were measured under the following five conditions: a normal condition, an open-circuit condition, a short-circuit condition, a faulty current sensor condition, and a multiple faulty condition. The details are given below.

6.1. Normal Condition Experimental Results

Figure 10a shows the measured speed responses at 100 r/min, 300 r/min, and 500 r/min. The periodic controller has quicker transient responses than the PI controller. Figure 10b shows the measured q-axis currents. The periodic controller provides greater input power when compared to the PI controller. Figure 11a shows the speed responses at 500 r/min when an external load of 3.5 N·m was added. The periodic controller provides a lower speed drop and quicker recovery time than the PI controller. Figure 11b shows the q-axis current responses in the same case. The periodic controller shows better performance than the PI controller, including a lower overshoot and quicker recovery time when an external load is added.

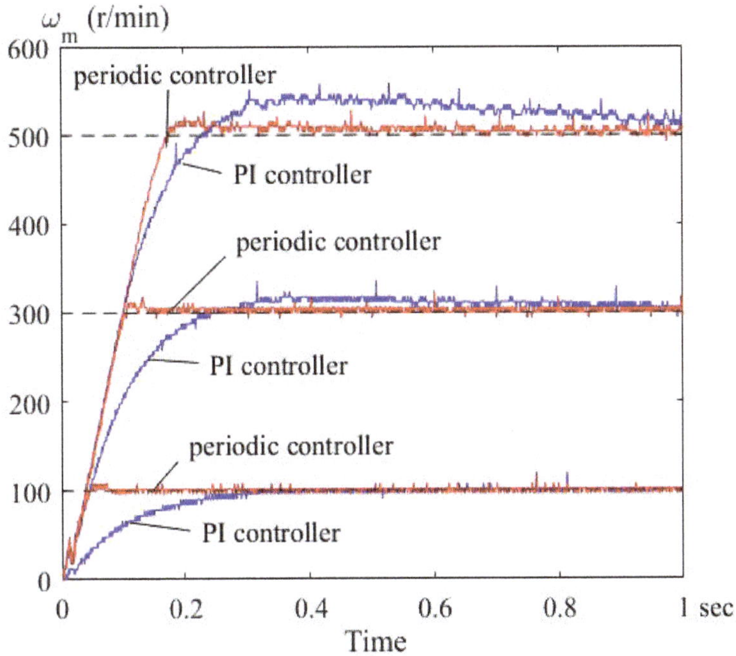

(a)

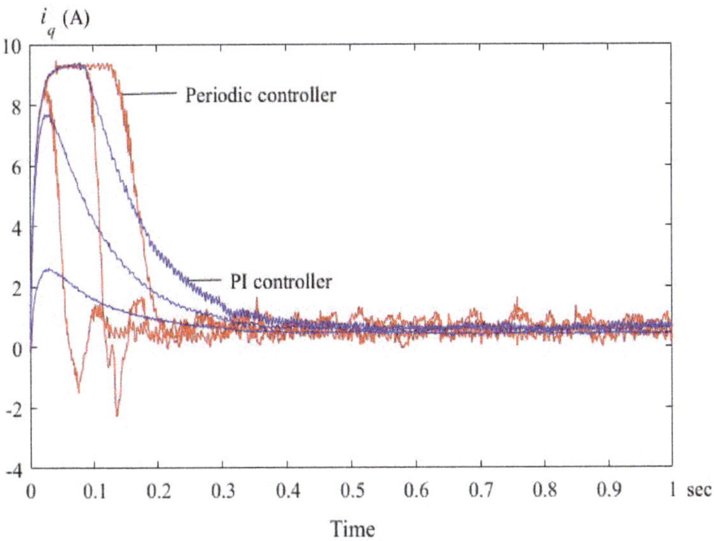

(b)

Figure 10. Measured speeds at 100 r/min, 300 r/min, and 500 r/min: (**a**) speed responses; (**b**) q-axis currents.

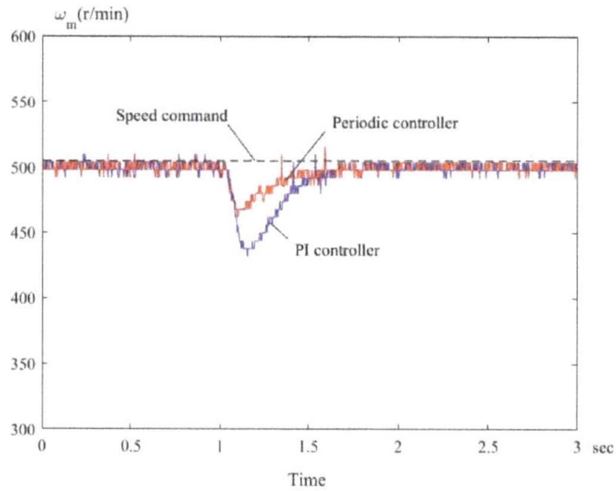

(a)

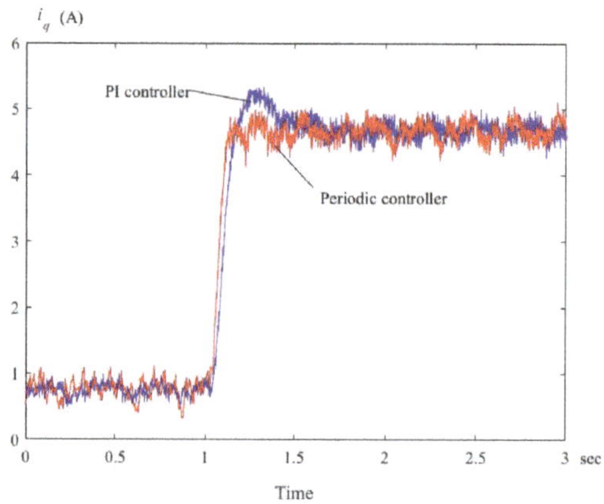

(b)

Figure 11. Measured results at 500 rpm and 3.5 N·m load: (**a**) speed responses; (**b**) *q*-axis currents.

6.2. Inverter Open-Circuit Faulty Condition Experimental Results

Figure 12a–c show the simulated results of the *a*-phase open circuit at 300 r/min without using the fault-tolerant method. The simulated results include the three-phase currents, q-axis current, and speed. Figure 13a shows the measured three-phase currents without using the proposed fault-tolerant method of the SPMSM drive system when its *a*-phase upper leg was open-circuited at 300 r/min. The faulty condition occurs at 0.15 s. A manual switch was in series with the power device. When the switch was opened, the power device was instantaneously opened. Thus, the PMSM drive system became a three-phase unbalanced drive system. Figure 13b shows the *q*-axis current response when the *a*-phase upper leg is open. The *q*-axis current oscillated due to the unbalanced three-phase currents. Figure 13c shows the measured speed response. As we can observe, in this figure, the speed varied between 485 r/min to 510 r/min when the *a*-phase upper leg was open. Figure 14a–c show the simulated results using the fault-tolerant control when the *a*-phase leg was open-circuited. The *d*-axis inductance remained the same as its nominal value, but the *q*-axis inductance was reduced to 50% of its nominal value due to the influence of saturation. The simulated results include the currents of the speed-loop PI controller, the currents of the speed-loop periodic controller, and the speed responses of speed-loop PI and periodic controllers. Figure 15a–c show the measured results of fault-tolerant control at 300 r/min when the *a*-phase was open-circuited. The periodic controller had better performance than the PI controller, including lower peak current and smaller speed variations during faulty intervals. Figure 15a shows the three-phase currents using the fault-tolerant method. Figure 15b shows the measured currents of the speed-loop periodic controller. The measured speed responses of both the PI controller and periodic controller are shown in Figure 15c. Figure 16a–c show the measured results of the fault-tolerant control when one switch of the a-phase leg was open-circuit at 1500 r/min.

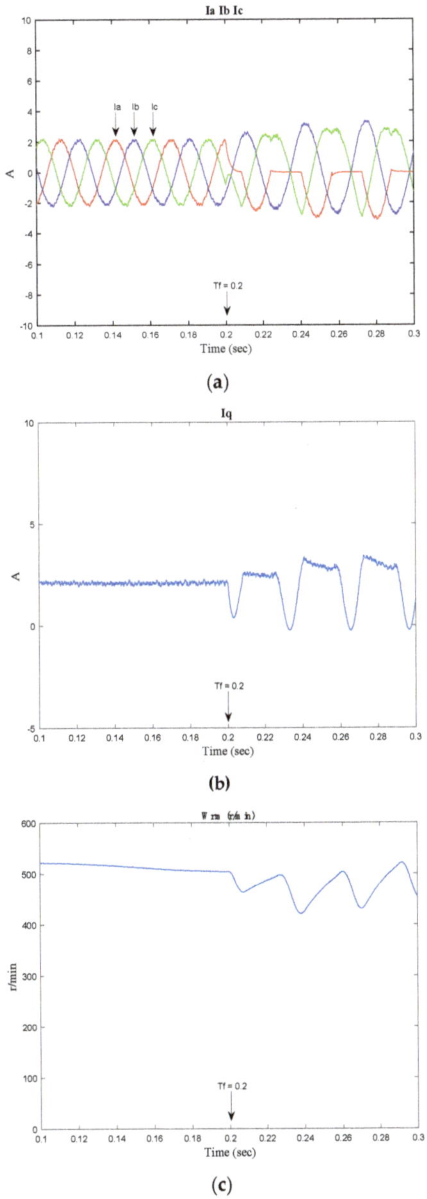

Figure 12. Simulated results of the a-phase open-circuited without using the fault-tolerant method: (**a**) three-phase currents; (**b**) q-axis current; (**c**) speed.

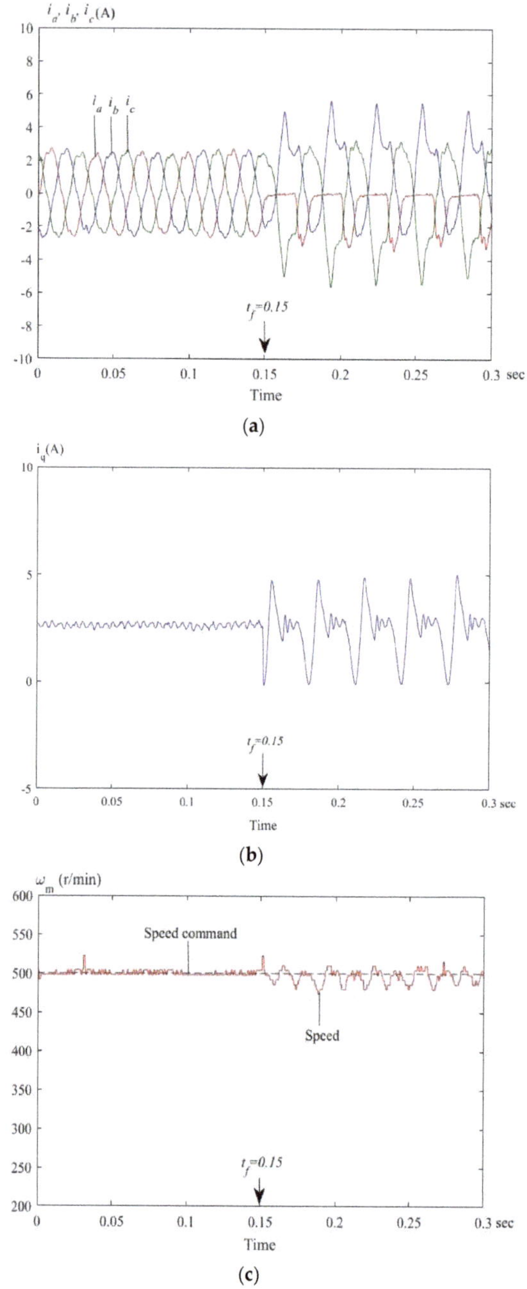

Figure 13. Measured results of the *a*-phase open-circuited without using the fault-tolerant method: (**a**) three-phase currents; (**b**) *q*-axis current; (**c**) speed.

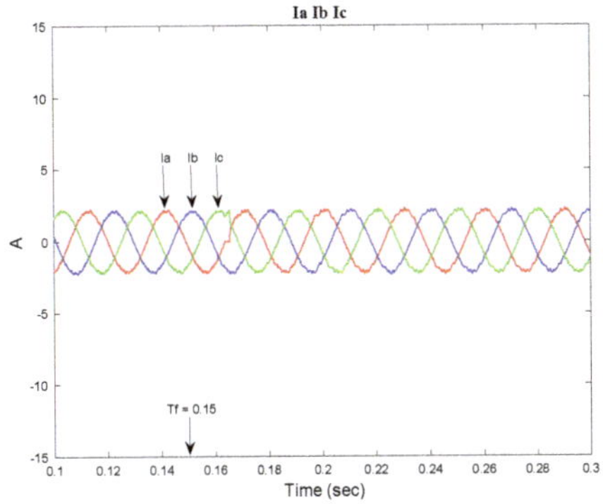

(a)

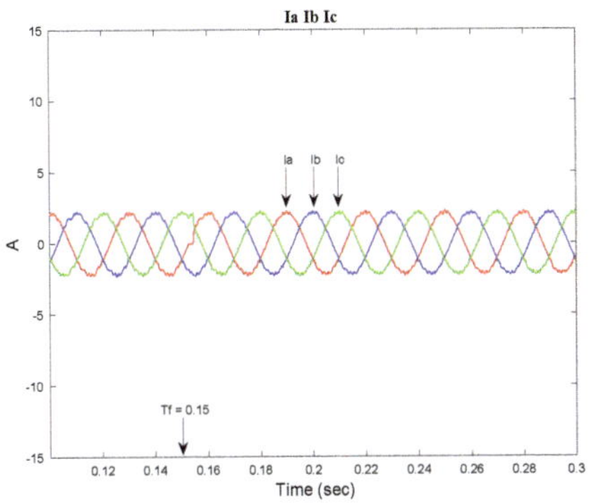

(b)

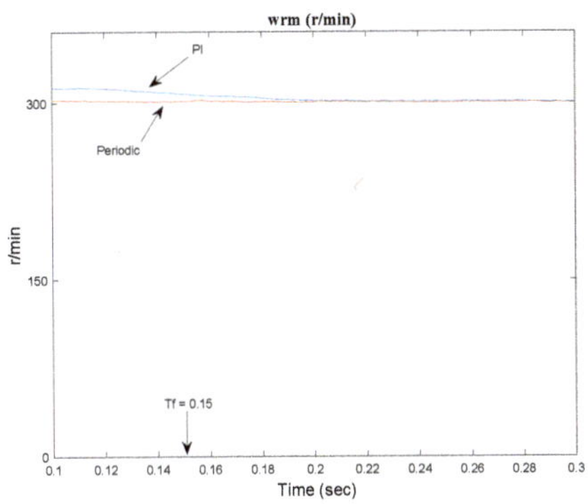

(c)

Figure 14. Simulated results of the fault-tolerant control when the *a*-phase was open-circuited in *d*–*q* inductance asymmetry conditions: (**a**) currents of the speed-loop PI controller; (**b**) currents of the speed-loop periodic controller; (**c**) speed responses.

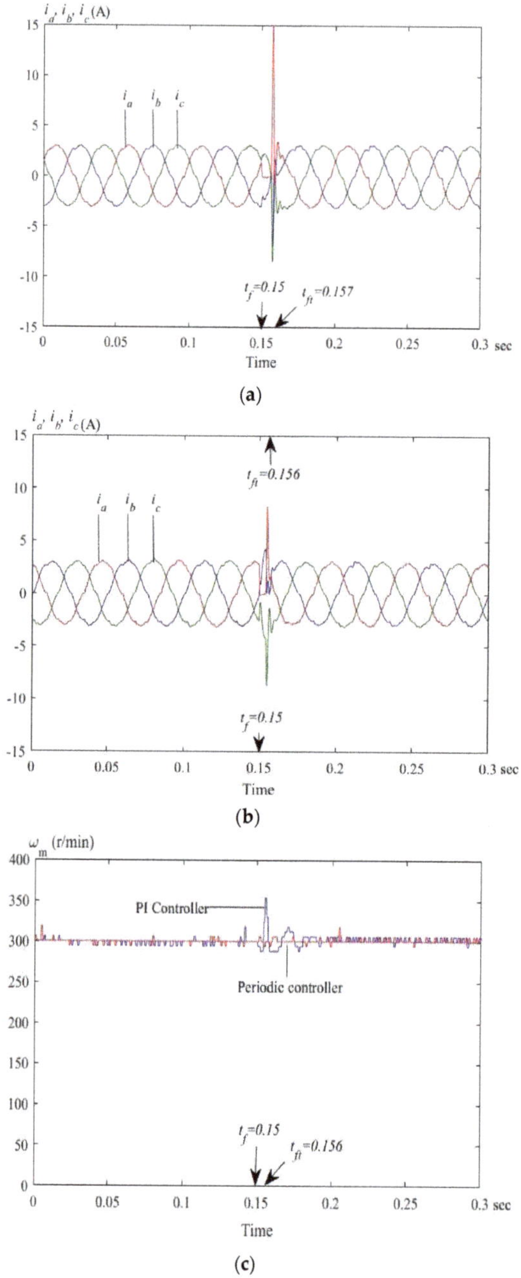

Figure 15. Measured results of the fault-tolerant control when the *a*-phase was open-circuited: (**a**) currents of the speed-loop PI controller; (**b**) currents of the speed-loop periodic controller; (**c**) speed responses.

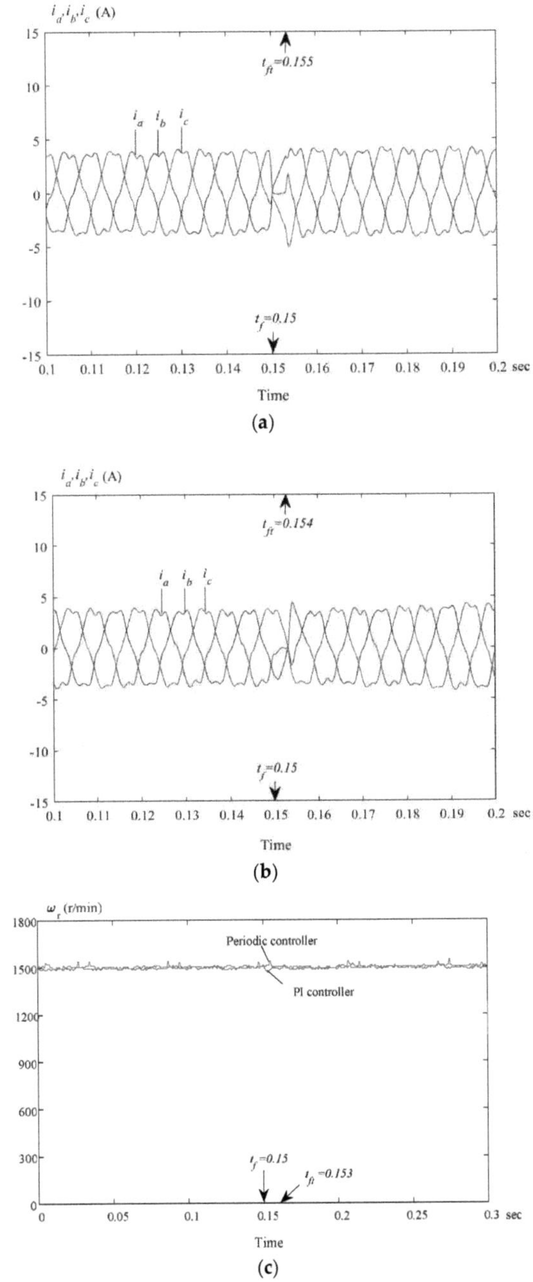

Figure 16. Measured results at 1500 r/min of the fault-tolerant control when the *a*-phase was open-circuited: (**a**) currents of the PI controller; (**b**) currents of the periodic controller; (**c**) speed-responses.

6.3. Inverter Short-Circuit Faulty Condition Experimental Results

Figure 17a–c show the simulated results of the a-phase short-circuited at 300 r/min without using the fault-tolerant method, including the three-phase currents, q-axis current, and speed response. Figure 18a–c show the measured results of the same case. Figure 18a shows the measured three-phase currents without using the fault-tolerant method. Figure 18b shows the measured q-axis current. Figure 18c shows the measured speed response that dropped quickly due to the trip of the inverter. Figure 19a–c show the simulated results of the a-phase short-circuited at 300 r/min using the fault-tolerant method. The simulated results include the current responses, speed responses, and performance index. Figure 20a–c show the measured results of the fault-tolerant control when the a-phase inverter was short-circuited at 300 r/min. The two power devices in the upper leg and lower leg were both turned on to have this leg short-circuited. Figure 20a shows the measured current responses when using a PI controller. Figure 20b shows that the measured speed variation was 80 r/min when using a speed PI controller, but it was 50 r/min when using the speed-loop periodic controller. These results show that the periodic controller has better transient response than the PI controller. Figure 20c shows the performance index before and after the fault. Yan et al. proposed a PWM voltage source inverter diagnosis method for a PMSM drive system based on a fuzzy logic approach [27]. By using the fuzzy logic diagnosis method, the DSP could identify the faulty condition in 0.09 s after the fault occurrence. Compared to Yan's method, in this paper, from Figure 20b, the short-circuit fault-tolerant control was finished in 0.01 s. As a result, this work reduced the time by approximately 89% when compared to Yan's method. The reason is that fuzzy logic is more complicated than the method proposed in this paper. Hang et al. proposed the detection and discrimination of an open-phase fault in an SPMSM drive system based on the zero-sequence voltage components [28]. For one switch open, Hang's proposed detection and discrimination method required 0.04 s. Compared to Hang's method, in this paper, from Figure 15c, the open-circuit fault-tolerant control was finished in 0.006 s. As a result, this work reduced the time by approximately 85% when compared to Hang's method. However, this paper may cause more conduction loss because six TRIACs were used to change the structure of the inverter. Figure 21a–c show the measured results of the short-circuit fault-tolerant control at one switch of the a-phase leg at 1500 r/min.

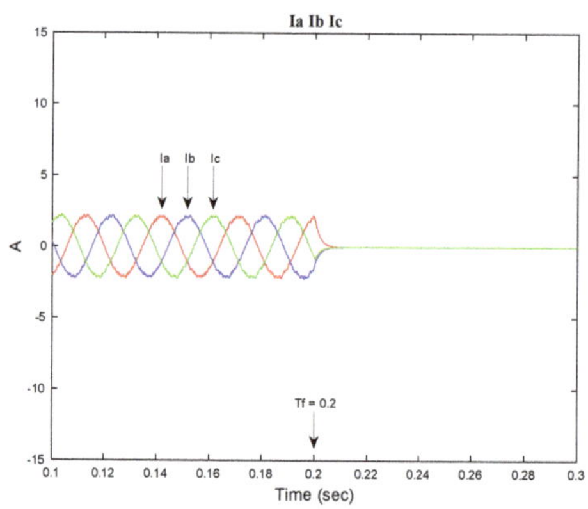

(a)

Figure 17. *Cont.*

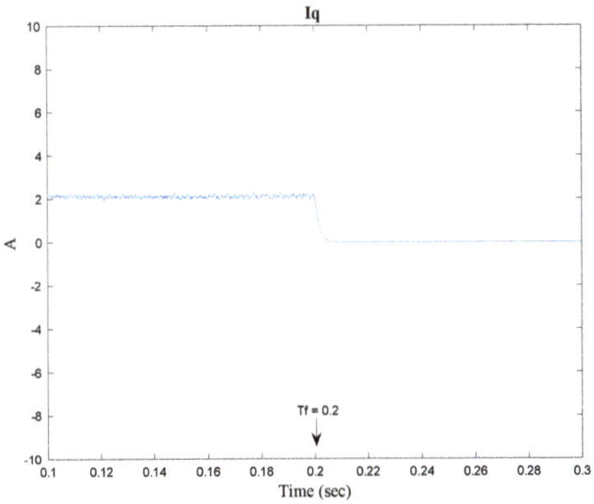

(b)

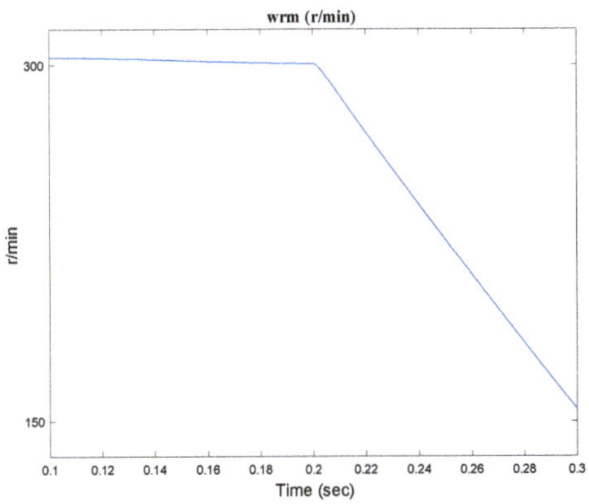

(c)

Figure 17. Simulated results when the a-phase was short-circuited without using the fault-tolerant method: (**a**) three-phase currents; (**b**) q-axis current; (**c**) speed response.

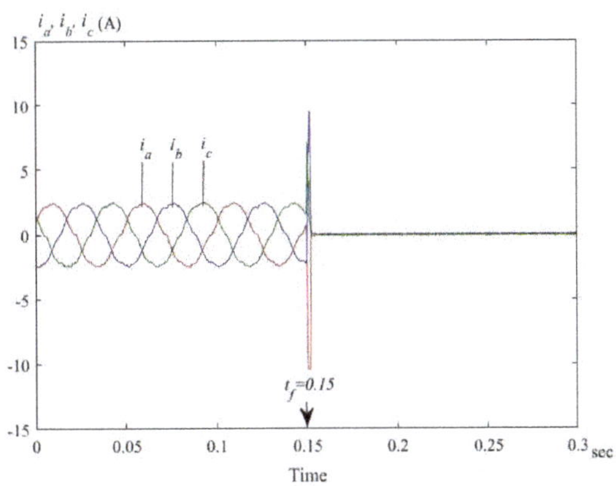

(a)

Figure 18. *Cont.*

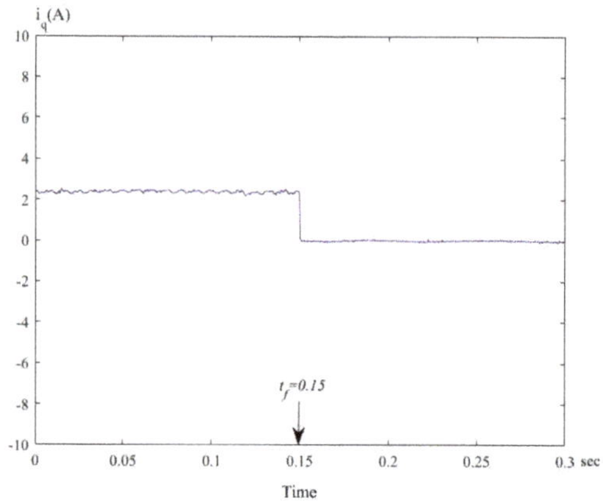

(b)

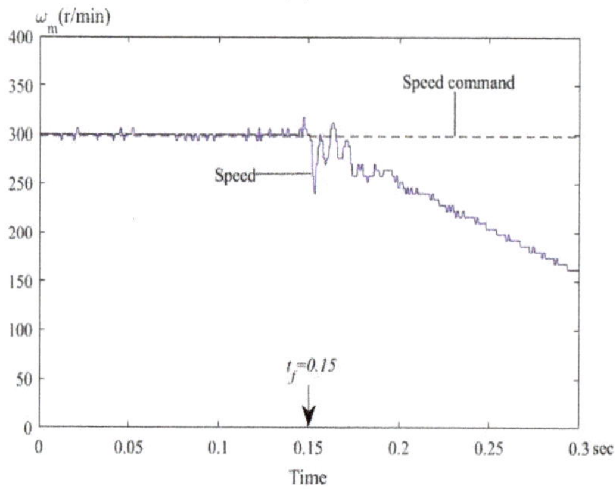

(c)

Figure 18. Measured results when the *a*-phase was short-circuited without using the fault-tolerant method: (**a**) three-phase currents; (**b**) *q*-axis current; (**c**) speed response.

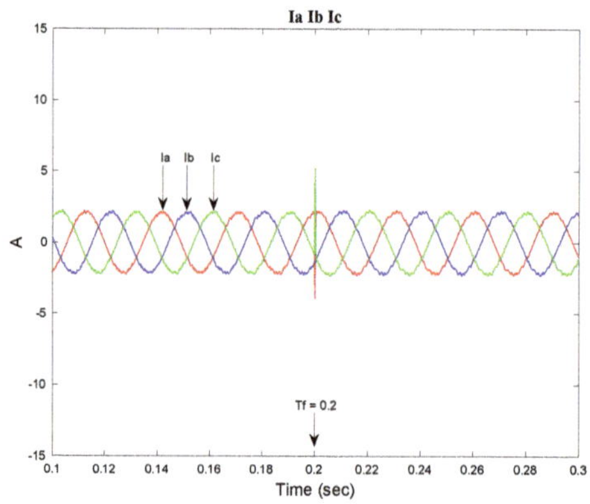

(**a**)

Figure 19. *Cont.*

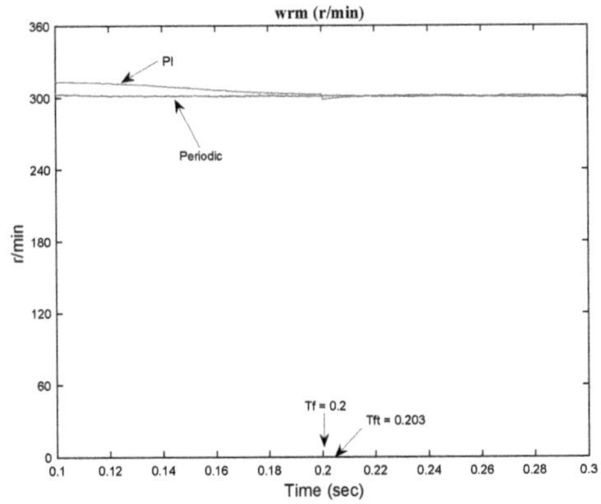

(**b**)

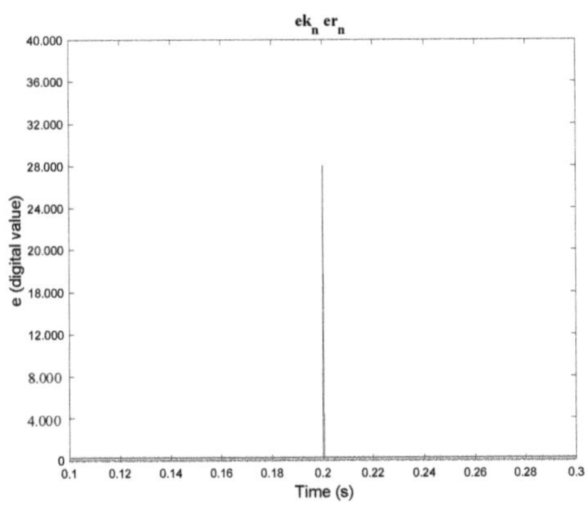

Figure 19. Simulated results of the fault-tolerant control when the *a*-phase was short-circuited: (**a**) currents of PI controller; (**b**) speed responses; (**c**) performance index.

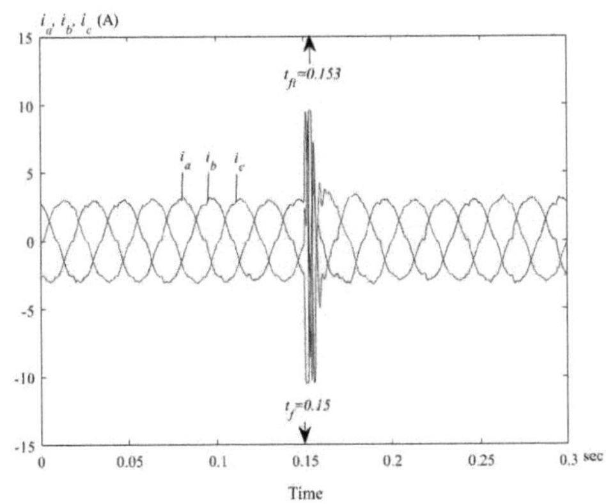

(**a**)

Figure 20. *Cont.*

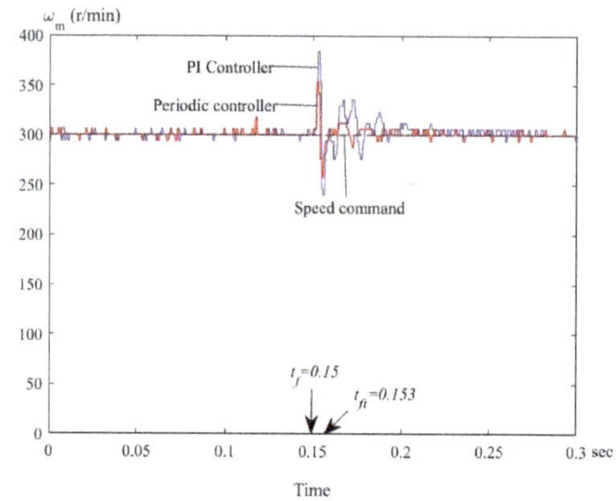

(b)

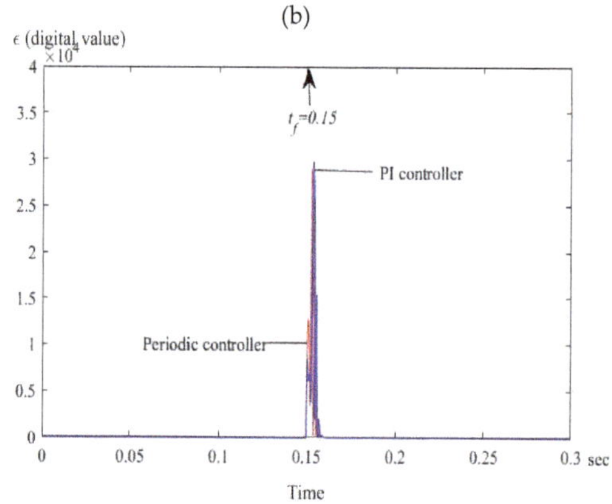

(c)

Figure 20. Measured results of the fault-tolerant control when the *a*-phase was short-circuited: (**a**) currents of the PI controller; (**b**) speed responses; (**c**) performance indexes.

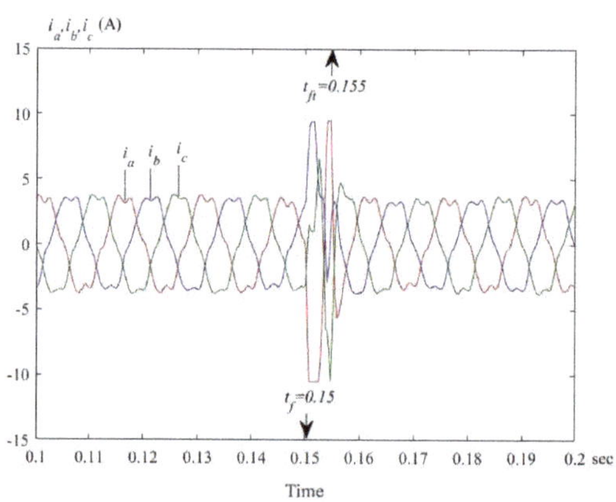

(a)

Figure 21. *Cont.*

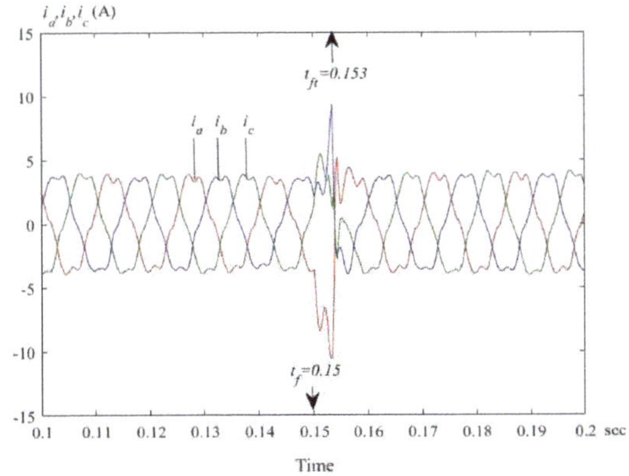

(**b**)

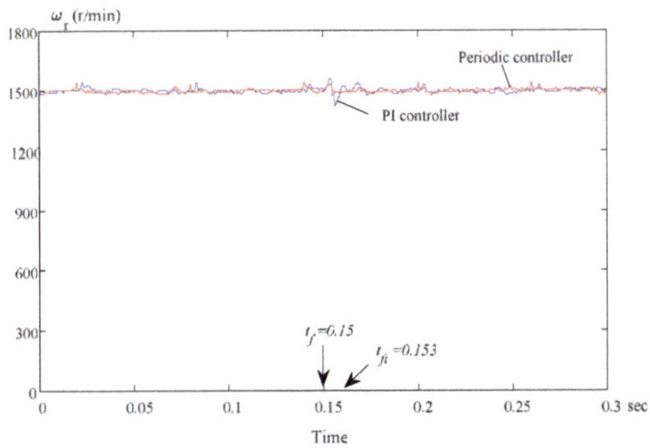

(**c**)

Figure 21. Measured results at 1500 r/min of the fault-tolerant control when the *a*-phase was short-circuited: (**a**) currents of the PI controller; (**b**) currents of the periodic controller; (**c**) speed responses.

6.4. Current Sensor Faulty Condition Experimental Results

In addition, when the *a*-phase Hall-effect current sensor is open, the *a*-phase measured current suddenly becomes zero. Then, the estimated current is used to replace the measured current. In the experiment, a manual switch was connected with the current-sensing circuit. When the switch was opened, the phase current became zero, resulting in a one-phase current fault. Figure 22a shows the measured *a*-phase current and its estimated current in normal operating conditions. As we can see in this figure, they were very close. Figure 22b shows the residual and adaptive threshold. The residual was always below its adaptive threshold because the system was operated in normal conditions. Figure 23a,b show the *b*-phase measured current and its estimated current when the *b*-phase current sensor was faulty at 0.15 s. The estimated *b*-phase current replaced the measured *b*-phase current at 0.154 s. Figure 24a–c show the measured three-phase currents when the *a*-phase current sensor was faulty. Figure 24a shows the measured three-phase currents using the PI controller. Figure 24b shows the measured three-phase current using the speed-loop periodic controller. Again, the periodic controller performed better than the PI controller. Figure 24c shows the measured speed responses using the speed-loop periodic controller and the speed-loop PI controller. The speed-loop periodic controller once again performed better than the speed-loop PI controller.

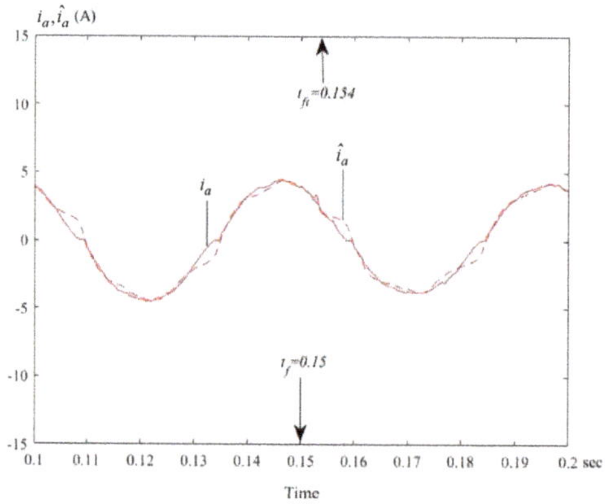

(**a**)

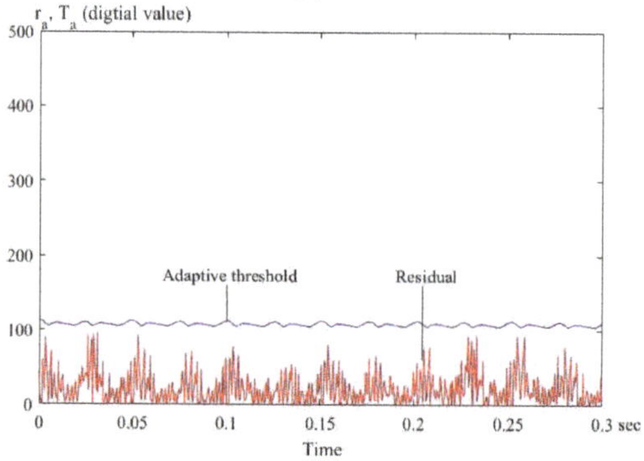

(**b**)

Figure 22. Measured results of the *a*-phase in normal operating conditions: (**a**) measured and estimated currents; (**b**) residual and adaptive threshold.

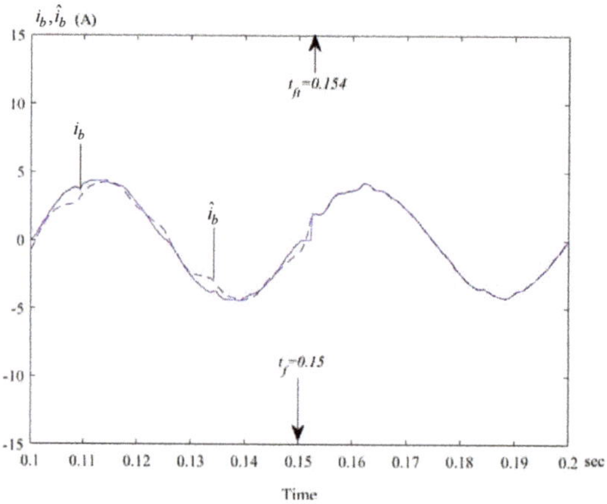

(**a**)

Figure 23. *Cont.*

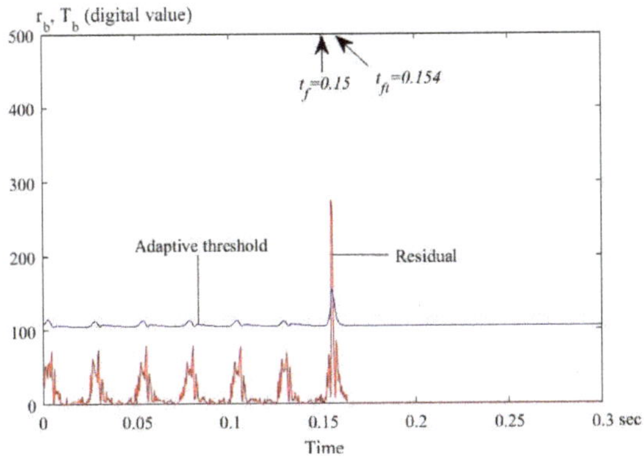

(**b**)

Figure 23. Measured results of the *b*-phase when the *b*-phase current sensor was faulty: (**a**) measured and estimated current; (**b**) residual and adaptive threshold.

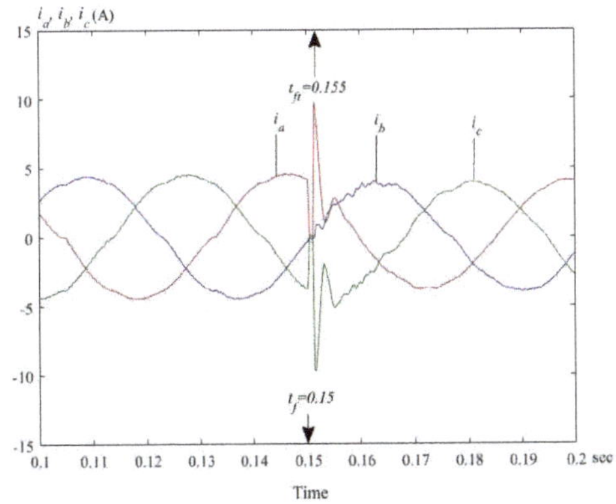

(**a**)

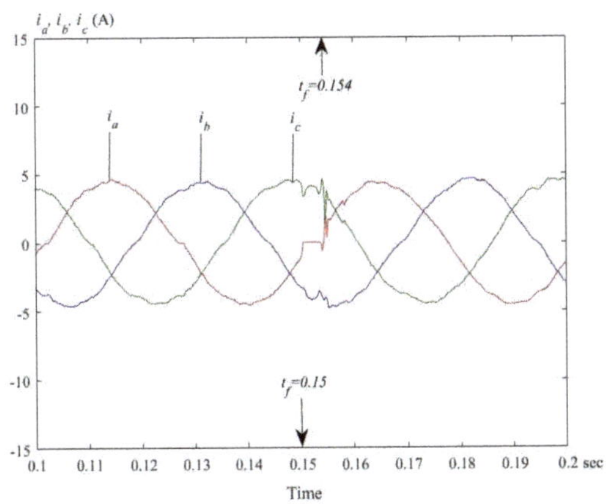

(**b**)

Figure 24. *Cont.*

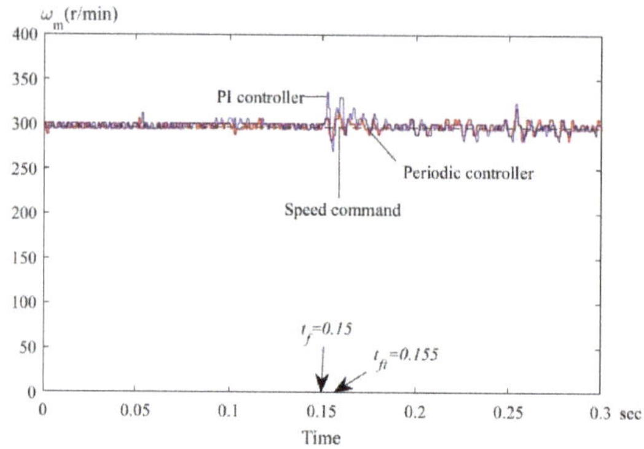

(c)

Figure 24. Measured three-phase currents of the fault-tolerant control when the *a*-phase current sensor was faulty: (**a**) current using PI controller; (**b**) current using periodic controller; (**c**) speed.

6.5. Multiple Faulty Conditions Experimental Results

Figure 25a,b show the simulated multiple faults when the *a*-phase leg was open-circuited and the *a*-phase current sensor was faulty using the periodic speed-loop controller. Figure 25a shows the simulated currents, and Figure 25b shows the simulated speed.

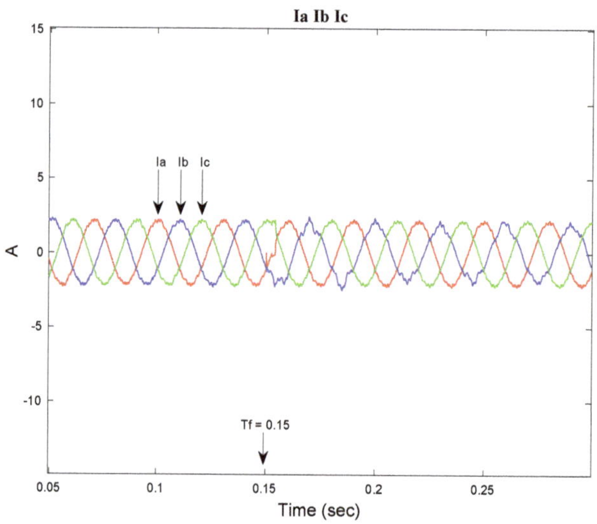

(a)

Figure 25. *Cont.*

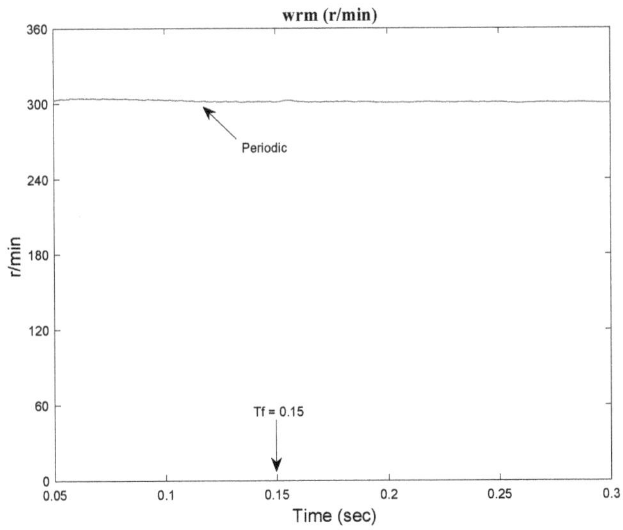

(b)

Figure 25. Simulated multiple faults when the *a*-phase leg was open-circuited and the *a*-phase current sensor was faulty using the periodic speed-loop controller: (**a**) current; (**b**) speed.

The proposed method required more computation time for a DSP. In addition, the proposed method also added two IGBTs for the back-up leg, and six TRIACs for changing the structure of the inverter. As a result, the proposed drive system required a higher cost, and generated more conduction losses. In addition, the proposed method required more CPU computation time. These were considered the overheads of the process. According to the experimental results, there were no faulty cases that the proposed design failed to detect. All faulty cases were successfully detected and controlled.

7. Conclusions

In this paper, the design of a speed-loop periodic controller for a fault-tolerant SPMSM drive system was investigated and discussed. A 32-bit DSP, TMS-320F-2808, was used to execute the speed-loop periodic controller and fault-tolerant algorithm. The detailed design procedures of the speed-loop periodic controller design were presented. The experimental results showed that the proposed periodic speed-loop controller provided better performance, including faster transient responses and better load disturbance responses, than the speed-loop PI controller under normal operating conditions and faulty conditions. The experimental results validated the theoretical analysis. The proposed method can be applied in industry due to its simplicity. This paper only focused on the faulty conditions that were clearly open- or short-circuited. Unclear faulty conditions, including resistance changing, noise interruption, overheating, and current or voltage derating of the IGBT, will be discussed in future research.

Author Contributions: Conceptualization, T.-H.L.; methodology, T.-H.L., M.S.M. and M.R.; software, M.S.M. and M.R.; hardware, M.R.; data curation, M.S.M. and M.R.; writing—review and editing, T.-H.L., M.S.M.; funding acquisition, T.-H.L.; supervision, T.-H.L. and S.

Funding: This research was funded by the Ministry of Science and Technology, Taiwan, under Grant MOST-105-2221-E-011-095-MY2.

Conflicts of Interest: The authors declare no conflict of interest.

References

1. Sul, S.K. *Control of Electric Machine Drive Systems*; John Wiley & Sons: Hoboken, NJ, USA, 2011.
2. Krause, P.; Wasynczuk, O.; Sudhoff, S.; Pekarek, S. *Analysis of Electric Machinery and Drive Systems*, 3rd ed.; John Wiley & Sons: Hoboken, NJ, USA, 2013.
3. Isermann, R. *Fault-Diagnosis Applications-Model Based Condition Monitoring: Actuators, Drives, Machinery, Plants, Sensors, and Fault-Tolerant Systems*; Springer: Berlin/Heidelberg, Germany, 2011.
4. Cardoso, A.J.M. *Diagnosis and Fault Tolerance of Electrical Machines, Power Electronics and Drives*; The Institution of Engineering and Technology: London, UK, 2018.
5. Ginart, A. *Fault Diagnosis for Robust Inverter Power Drives*; The Institution of Engineering and Technology: London, UK, 2019.
6. Naidu, M.; Gopalakrishnan, S.; Nehl, T.W. Fault-tolerant permanent motor drive topologies for automotive X-by-wire systems. *IEEE Trans. Ind. Appl.* **2010**, *46*, 1–8. [CrossRef]
7. Kontarcek, A.; Bajec, P.; Nemec, M.; Ambrozic, V.; Nedeljkovic, D. Cost-effective three-phase PMSM drive tolerant to open-phase fault. *IEEE Trans. Ind. Electron.* **2015**, *62*, 6708–6718. [CrossRef]
8. Jung, S.M.; Park, J.S.; Kim, H.W.; Cho, K.Y.; Youn, M.J. An MRAS-based diagnosis of open-circuit fault in PWM voltage-source inverters for PM synchronous motor drive systems. *IEEE Trans. Power Electron.* **2013**, *28*, 2514–2526. [CrossRef]
9. Cai, B.; Zhao, Y.; Liu, H.; Xie, M. A data-driven fault diagnosis methodology in three-phase inverters for PMSM drive systems. *IEEE Trans. Power Electron.* **2017**, *32*, 5590–5600. [CrossRef]
10. Meinquet, F.; Sandulescu, P.; Kestlyn, X.; Semail, E. A method for fault detection and isolation based on the processing of multiple diagnostic indices: Application to inverter faults in AC drives. *IEEE Trans. Veh. Technol.* **2013**, *62*, 995–1009. [CrossRef]
11. Wang, W.; Zhang, J.; Cheng, M.; Li, S. Fault-tolerant control of dual three-phase permanent-magnet synchronous machine drives under open-phase faults. *IEEE Trans. Power Electron.* **2017**, *32*, 2052–2063. [CrossRef]
12. Tseng, S.K.; Liu, T.H.; Hsu, J.W.; Ramelan, L.R.; Firmansyah, E. Fault-tolerant control for a dual-PMSM drive system. *IET J. Eng.* **2016**. [CrossRef]
13. Wang, Z.; Chen, J.; Cheng, M.; Zheng, Y. Fault-tolerant control of paralleled-voltage-source-inverter-fed PMSM drives. *IEEE Trans. Ind. Electron.* **2015**, *62*, 4749–4760. [CrossRef]
14. Nasiri, A. Full digital current control of permanent magnet synchronous motors for vehicular applications. *IEEE Trans. Veh. Technol.* **2007**, *56*, 1531–1537. [CrossRef]
15. Bennett, J.W.; Mecrow, B.C.; Atkinson, O.J.; Maxwell, C.; Benarous, M. Fault-tolerant electric drive for an aircraft nose wheel steering actuator. *IET Electr. Syst. Transp.* **2011**, *1*, 117–125. [CrossRef]
16. Jeong, Y.S.; Sul, S.K.; Schulz, S.E.; Patel, N.R. Fault detection and fault-tolerant control of interior permanent-magnet motor drive system for electric vehicle. *IEEE Trans. Ind. Appl.* **2015**, *41*, 46–51. [CrossRef]
17. Wang, R.; Wang, J. Fault-tolerant control with active fault diagnosis for four-wheel independently driven electric ground vehicles. *IEEE Trans. Veh. Technol.* **2011**, *60*, 4276–4287. [CrossRef]
18. Zhang, G.; Zhang, H.; Huang, X.; Wang, J.; Yu, H.; Graaf, R. Active fault-tolerant control for electric vehicles with independently driven rear in-wheel motors against certain actuator faults. *IEEE Trans. Control Syst. Technol.* **2016**, *34*, 1557–1571. [CrossRef]
19. Bolognani, S.; Zordan, M.; Zigliotto, M. Experimental fault-tolerant control of a PMSM drive. *IEEE Trans. Ind. Electron.* **2000**, *47*, 1134–1141. [CrossRef]
20. Bai, H.; Zhu, J.; Qin, J.; Sun, J. Fault-tolerant control for a dual-winding fault-tolerant permanent magnet motor drive based on SVPWM. *IET Power Electron.* **2016**, *10*, 509–516. [CrossRef]
21. Kontarcek, A.; Nemec, M.; Bajec, P.; Ambrozic, V. Single open-phase fault detection with fault-tolerant control of an inverter-fed permanent magnet synchronous machine. *Automatika* **2014**, *55*, 474–486. [CrossRef]
22. An, Q.T.; Sun, L.; Sun, L.Z. Current residual vector-based open-switch fault diagnosis of inverters in PMSM drive systems. *IEEE Trans. Power Electron.* **2015**, *30*, 2814–2827. [CrossRef]
23. Beng, G.F.H.; Zhang, X.; Vilathgamuwa, D.M. Sensor fault-resilient control of interior permanent-magnet synchronous motor drives. *IEEE/ASME Trans. Mechatron.* **2015**, *20*, 855–864. [CrossRef]

24. Jlassi, I.; Estima, J.O.; El Khil, S.K.; Bellaaj, N.M.; Cardoso, A.J.M. A Robust Observer-Based Method for IGBTs and Current Sensors Fault Diagnosis in Voltage-Source Inverters of PMSM Drives. *IEEE Trans. Ind. Appl.* **2017**, *53*, 2894–2905. [CrossRef]
25. Francis, B.A.; Wonham, W.M. The internal model principle of control theory. *Automatica* **1976**, *12*, 457–465. [CrossRef]
26. Wang, D. *Periodic Control of Power Electronic Converters Periodic Control of Power*; CPI Group Ltd.: Croydon, UK, 2016.
27. Yan, H.; Xu, Y.; Cai, F. PWM-VSI fault diagnosis for a PMSM drive based on the fuzzy logic approach. *IEEE Trans. Power Electron.* **2019**, *31*, 759–768. [CrossRef]
28. Hang, J.; Zhang, J.; Cheng, M.; Ding, S. Detection and discrimination of open-phase fault in permanent magnet synchronous motor drive system. *IEEE Trans. Power Electron.* **2016**, *31*, 4697–4708. [CrossRef]

energies

MDPI

Article

Performance Analysis of Synchronous Reluctance Motor with Limited Amount of Permanent Magnet

Duc-Kien Ngo [1] and Min-Fu Hsieh [2,*]

1 Department of Systems and Naval Mechatronic Engineering, National Cheng Kung University, Tainan 70101, Taiwan; eng.duckien.ngo@gmail.com
2 Department of Electrical Engineering, National Cheng Kung University, Tainan 70101, Taiwan
* Correspondence: mfhsieh@mail.ncku.edu.tw; Tel.: +866-06-275-7575 (ext. 62366)

Received: 11 August 2019; Accepted: 9 September 2019; Published: 11 September 2019

Abstract: This paper analyzes the performance of a synchronous reluctance motor (SynRM) equipped with a limited amount of a permanent magnet (PM). This is conventionally implemented by inserting PMs in rotor flux barriers, and this is often called the PM-assisted SynRM (PMa-SynRM). However, common PMa-SynRMs could be vulnerable to irreversible demagnetization. Therefore, motor performance and PM demagnetization should be simultaneously considered, and this would require the PM to be properly arranged. In this paper, various rotor configurations are carefully studied and compared in order to maximize the motor performance, avoid irreversible demagnetization and achieve higher PM utilization. Moreover, the field weakening capability is investigated and improved by regulating armature excitation. A particular rotor type with flux intensification was found to possess higher PM utilization, lower demagnetization possibility with fairly high performance. Thus, suitable rotor configurations are recommended for certain applications.

Keywords: SynRM; irreversible demagnetization; PMa-SynRM; flux intensifying

1. Introduction

The synchronous reluctance motor (SynRM), with its robustness, high overload capability and low cost, has become a popular research target for many years [1–4]. However, the relatively lower torque/power density and power factor are the inherent disadvantages of SynRMs [5–7] compared to a permanent magnet synchronous machine (PMSM). To overcome such weaknesses, a permanent magnet (PM) can be inserted into the rotor of the SynRMs with a modest volume, which leads to the birth of a type of motor called the permanent magnet assisted synchronous reluctance motor (PMa-SynRM) [8–11]. With the increasing number of related research works, the PMa-SynRM has become a popular choice in some applications and can be an alternative to a SynRM or PMSM [12–15].

Generally, the PMs inserted inside the rotor flux barriers produce a negative flux linkage along the q-axis. The q-axis inductance L_q is usually low due to the multiple flux barriers. The permanent magnet (PM) flux linkage (the flux linkage due to PM solely) promotes the rotation of the flux linkage vector, and therefore the voltage vector goes close to the current vector to increase the power factor [9]. The PM flux linkage also contributes to torque production so that the total torque increases. However, the volume/size of the added PM needs to be limited to avoid the motor becoming an interior permanent magnet synchronous motor (IPMSM) [16], which could also increase the cost. Nevertheless, the volume/size of the PM should not be too small to achieve the desired torque and power density, or to be vulnerable to irreversible demagnetization [10].

For SynRMs or PMa-SynRMs, various design possibilities can be considered, e.g., the number of flux barriers, with or without PMs, rare earth or other types of PM materials or the amount of PM employed. In an effort to standardize the design process of SynRMs and PMa-SynRMs, Bianchi et al. [11] proposed a series of steps that are synthesized from some example studies [11,17–20], where the inward PMs (near the rotor shaft) are larger than the outward ones to improve flux flows and avoid demagnetization. This PM arrangement is considered as a common trend for the PMa-SynRM rotor design.

Some motor designs presented in previous research [10,11,21–23] using either rare-earth or ferrite PMs are summarized in Table 1, including their ratio of PM-to-motor volume and torque density. It can be observed that the variety of rotor designs and PM arrangements is rich in these motors. However, the first four motors listed in Table 1 [10,11,21,22] employ a relatively large PM volume compared to the motor studied in Reference [23]. Furthermore, the PM size in the motor in Reference [23] is purposefully made identical for all the PM layers to reduce the manufacturing cost, which is different from common designs. The multiple flux barrier design allows the torque density of this motor to reach 28.1 Nm/L with PM taking only 0.95% of the motor volume by assuming sufficient cooling is applied, as shown in Table 1. However, as mentioned in Reference [16], the low PM volume and high excitation current could lead to its negligible contribution in torque production due to the low PM-torque-to-total-torque ratio and high probability of irreversible demagnetization with field-weakening applied. In addition, since the armature current I_s is far from the characteristic current I_{ch} on d-axis [24], the constant power speed range (CPSR) could become relatively low for PMa-SynRM with a little amount of PM. From the above discussions, it is necessary to propose a solution to improve the performance of this type of motor in terms of PM utilization, demagnetization resistivity and field weakening capability.

Table 1. The reference motor parameters and torque production.

Reference Motor Source	[10]	[11]	[21]	[22]	[23]
Stator diameter (mm)	150	200	125	112	160
Stack length (mm)	105	40	27	40	120
Motor volume (L)	1.856	1.257	0.331	0.394	2.413
PM volume (L)	0.066	0.077	0.013	0.009	0.023
PM-to-Motor volume ratio (%)	3.58	6.13	3.93	2.28	0.95
PM material type	Rare earth	Ferrite	Ferrite	Rare earth	Rare earth
Number of poles	4	4	4	4	4
Number of flux barriers for each pole	4	3	2	1	5
Number of PM layers	4	3	2	1	4
PM size between layers	Unequal	Unequal	Unequal	-	Equal
Torque (Nm)	17.9	12.47	1.27	4.54	67.8
Torque density (Nm/L)	9.7	9.9	3.8	11.5	28.1

Therefore, in this paper, several motor models based on a prototyped PMa-SynRM [23] with various PM arrangements using limited amount of PM are analyzed in detail. The analysis concentrates on the effect of the PM position on the magnetic distribution, inductances, torque production, torque/power-speed curves and magnetization characteristics. The armature current is also adjusted for observation on the correlation between the electrical and magnetic parameters affecting the motor performance. From the above analysis, this paper aims to achieve a high PM utilization rate to produce torque though a limited amount of PMs in a more efficient way. Demagnetization can also be avoided under high performance operations. The analysis was conducted using finite element analysis (FEA), which has been partially validated using previous experimental studies [23]. Note that differing from Reference [23], where the evaluation was only conducted for a fixed rotor structure, this paper makes a complete analysis with a sufficient number of models in order to make proper suggestions for the improvement of SynRM performance. In Reference [25], the PM volume was optimized for predetermined field-intensified PM machines. Here, in the present study, the models investigated cover not only conventional PMa-SynRMs, but also the novel flux-intensifying PMa-SynRMs [26].

In what follows, the terms flux-intensification, flux-intensifying and flied-intensified are all abbreviated as FI. In addition, by investigating over an existing prototype, the analysis can be better convincing. The comparisons can also be made to highlight the novelty of the current analysis.

This paper is organized as follows. The mathematical model and the configuration of the motor models are presented first in Section 2. Then, the investigation for the influence of PM positions on motor characteristics is carried out in Section 3, followed by the comparison of some motor models in Section 4. Section 5 presents the discussions over these investigated models. Finally, the paper is concluded by making suggestions for the design of such motors in Section 6.

2. Mathematical Model and Configuration of Investigated Motors

2.1. Mathematical Modeling of Investigated Motors

A conventional SynRM with a limited amount of PM embedded along the flux barriers, i.e., facing the physical q-axis, is called the first PM arrangement (hereafter denoted Type 1), as shown in Figure 1a. In contrast, when PM is added crossing the flux barriers, i.e., facing the d-axis, it is called the second PM arrangement (hereafter denoted Type 2), as the example illustrated in Figure 1b. For the Type 1 rotor, the flux linkage produced by the PM is arranged against the q-axis armature flux linkage, while for the Type 2 rotor, the PM flux linkage complements the d-axis armature flux linkage. The Type 2 motor can thus be called the flux-intensifying PMa-SynRM (FI-PMa-SynRM) [26].

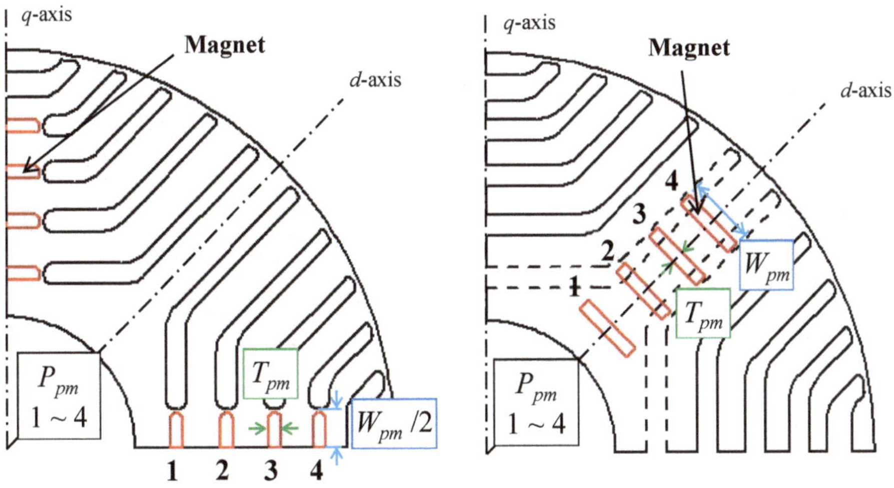

Figure 1. Rotor configurations: (**a**) The first permanent magnet (PM) arrangement (Type 1); (**b**) The second PM arrangement (Type 2).

Assuming that the iron saturation and the cross-coupling effect are neglected, the stator dynamic voltage equations for synchronous machines in the d-q frame [27,28] can be expressed as:

$$v_d = R_s i_d + \frac{d\lambda_d}{dt} - \omega\lambda_q \tag{1}$$

$$v_q = R_s i_q + \frac{d\lambda_q}{dt} + \omega\lambda_d \tag{2}$$

where the subscripts d and q represent the d- and q-axis, respectively, i_d and i_q are the currents, λ_d and λ_d are the flux linkages, R_s is the phase resistance, and ω is the electrical angular speed.

Equations (1) and (2) are general voltage equations for synchronous machines. To be applied to the two types of motors mentioned above, the flux linkages λ_d and λ_q in (1) and (2) need to be further discussed since the PMs are arranged differently in these two types of motors. For the Type 1 motors, as previously mentioned, the PMs are arranged in q-axis against the stator flux due to the q-axis current (i_d), and therefore the flux linkages in the d-q frame can be expressed as:

$$\lambda_d = L_d i_d, \lambda_q = L_q i_q - \lambda_m \tag{3}$$

where L_d and L_q are the stator inductances in the d-q frame, and λ_m is the PM flux linkage. Note that L_d and L_q do not take into account the PM flux linkage but only the flux linkage produced by i_d and i_q. For the Type 2 motors, the PMs are placed in the d-axis to complement the stator flux and thus the flux linkages are given as:

$$\lambda_d = L_d i_d + \lambda_m, \lambda_q = L_q i_q \tag{4}$$

From (3) and (4), the flux-weakening nature for the Type 1 motors and the flux-intensifying characteristics for the Type 2 can be clearly observed.

Figure 2 presents the equivalent circuits for the Type 1 [27] and Type 2 motors. As can be seen, the two types of motors have a difference in PM flux linkages. The phasor diagrams for Type 1 [9] and Type 2 are illustrated in Figure 3a,b, respectively, with the winding resistance being neglected [9,29]. Therefore, the voltage equations can be further expressed as:

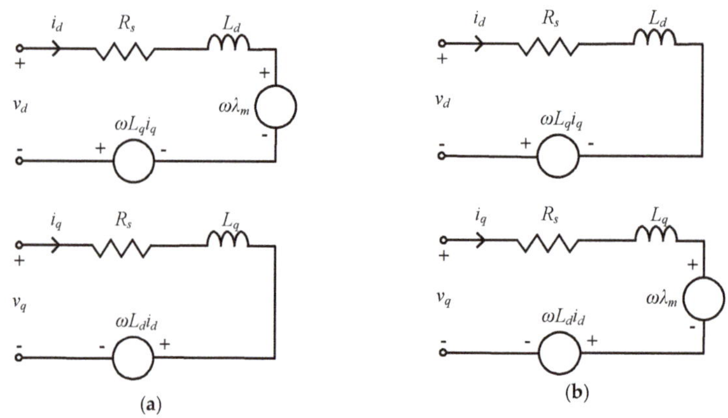

Figure 2. Equivalent circuits: (**a**) Type 1; (**b**) Type 2.

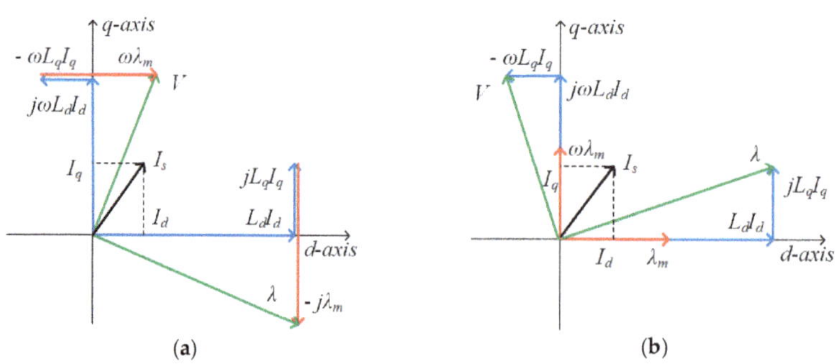

Figure 3. Phasor diagrams: (**a**) Type 1; (**b**) Type 2.

$$V = \omega\sqrt{(L_d I_d)^2 + \left(L_q I_q - \lambda_m\right)^2} \tag{5}$$

for Type 1 [30], and:

$$V = \omega\sqrt{(L_d I_d + \lambda_m)^2 + \left(L_q I_q\right)^2} \tag{6}$$

for Type 2.

Note that the "-" sign in front of λ_m in (3) and (5) indicates that the direction of this quantity is opposing $L_q I_q$, differing from the definition in [30].

The torque equations for Type 1 and Type 2 are respectively expressed as:

$$T = \frac{3N_m}{4}\left[\lambda_m I_d + (L_d - L_q) I_d I_q\right] \tag{7}$$

for Type 1 [29], and

$$T = \frac{3N_m}{4}\left[\lambda_m I_q + (L_d - L_q) I_d I_q\right] \tag{8}$$

for Type 2 [26].

Figure 4 illustrates the circle diagrams of these two types of motors. For the Type 1 motor with the first PM arrangement shown in Figures 1a, 3a and 4a indicate that the flux linkage generated by the q-axis current could cause the PM to be irreversibly demagnetized, especially for the thin PM. In contrast, for the Type 2 motor with the second PM arrangement shown in Figure 1b, the demagnetization on the PM can be avoided during maximum torque per ampere (MTPA) operation but may be possibly locally demagnetized during field weakening operation (this can be avoided by careful design), as shown in Figures 3b and 4b. This configuration may be subject to a lower power factor at low speed, but for medium and high-speed operations, the current phase advance would improve the power factor. On the other hand, inserting PM in the d-axis (flux paths) can decrease L_d and then decrease the reluctance torque so that the Type 2 motors would become closer to surface PM synchronous motors (SPMSMs). However, this requires a further investigation [31] and is not discussed here.

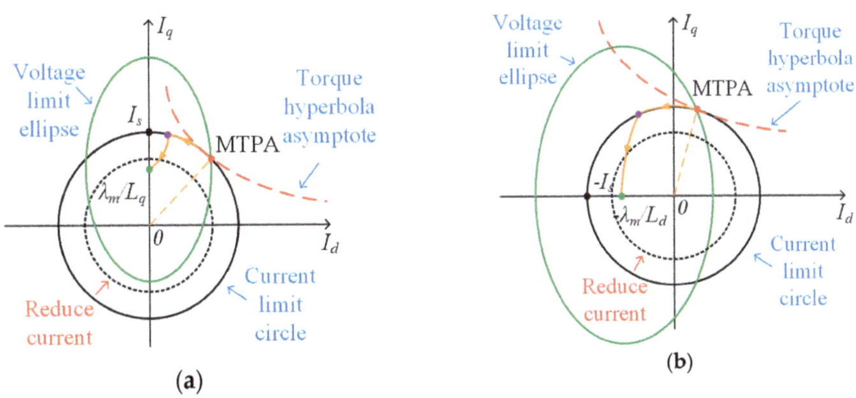

Figure 4. Circle diagrams: (**a**) Type 1; (**b**) Type 2.

The field weakening theory has been discussed in References [32,33] where ideally, the infinite speed could be achieved when the value of the armature current I_s is equal to the characteristic current I_{ch}. Practically, to increase the CPSR, I_s should be selected as close to I_{ch} as possible [32,34]. The motors in this paper have I_s greater than I_{ch} due to low the PM flux linkage by the limited PM quantity, as shown in Figure 4, where $I_{ch} = \lambda_m / L_q$ for Type 1 and $I_{ch} = \lambda_m / L_d$ for Type 2. There are two potential methods for improving the CPSR. The first method is to increase I_{ch} by using stronger or more magnet (higher λ_m) or changing the rotor configuration to reduce the q-axis or d-axis inductance for Type 1 or Type 2, respectively. However, this would lead to a redesign of the motors [35]. The second method is to reduce I_s (active reduction), as indicated in Figure 4, which however, should face a direct reduction of motor torque and power [24]. On the other hand, these motors can be operated in the maximum torque per voltage (MTPV) mode in which the current is reduced (passive reduction) when the speed increases. This would result in a partial overlap in the power-speed curves between the active and passive reductions of the excitation current in the field weakening region at high speed. This is explained in Section 4.3.

2.2. Configuration of Investigated Motors

As previously illustrated in Figure 1, for the rotor configuration of Type 1, the PMs are embedded along the flux barriers and located in the central part of each rotor pole. As indicated in Figure 1, P_{pm} is the magnet position from 1 to 4, W_{pm} is the magnet width and T_{pm} is the magnet thickness. For the Type 2 configuration, the PMs are arranged along the d-axis. Both types have the same PM positions viewed from the motor shaft (e.g., the PMs at Position 1 of Types 1 and 2 keep the same distance to the shaft). These arrangements of PM positions help to evaluate the effect of the PM directions (i.e., facing d- or q-axis). Note that, the magnetization of the PMs is all in the parallel pattern. The motor specifications and parameters are listed in Table 2, where the analysis at the peak current condition is for the purpose of exploring the capacity of the motors.

Table 2. Main specifications/parameters of models.

Parameter/ Specification	Unit	Value	Parameter/ Specification	Unit	Value
Desired peak power	kW	10	Stator diameter	mm	160
Number of phases	-	3	Rotor diameter	mm	94
Number of poles	-	4	Air-gap	mm	0.5
Number of slots	-	36	Stack length	mm	120
DC voltage	V	220	PM meterial	-	N35H
Maximum current	A	80	PM volume	mm^3	23040
Number of turns	turns	6	PM/Motor volume ratio	%	0.95

The performance of the prototype motor with Type 1 arrangement has been investigated with both the experiments and FEA simulations in Reference [23], where the results show that although the torque is high, this motor could not maintain the power and presented a low CPSR. This reduces its practical usefulness.

According to the previous discussion and the mathematical models described in the first part of this section, it is worthwhile to study the Type 2 motor as a potential alternative. Note that Section 2 mainly provides the mathematical background and briefly introduces the basic topologies of the two types of motor.

3. Comparative Analysis of Influence of PM Position

In this Section, the analyses on the effect of individual PM position on Type 1 and Type 2 motors using finite element analysis (FEA) were conducted. In these analyses, the no-load analysis aims to investigate the contribution of PM at different positions in the rotor to motor flux, and the on-load analysis is used to study the correlation between the PM position and armature excitation. No complete motor models are involved in this Section.

3.1. No-Load Operation Comparison

Base on the prototype motor in Reference [23], the PM pieces with a cross-sectional dimension of 1.5×8 mm were chosen and inserted into the PM positions from 1 to 4 (from inmost to outmost) in this study. Note that, all the PMs were fixed inside the flux barriers embracing them. In the beginning, to investigate the influence of the PM position, each time only one PM piece was placed at one of the above positions and the armature current was removed (no-load operation). The PM flux linkage and air-gap flux density B_g for each PM position of both types of motors are illustrated in Figure 5. As can be seen in Figure 5a,c, although the waveforms of the PM flux linkage for Type 1 seem to be different at different PM positions and the trend is unclear, the amplitudes are similar. On the other hand, the waveforms of B_g tend to spread out and the peak value deceases as the PM moves towards the inmost position. As shown in Figure 5b,d, all the PM flux linkage waveforms for Type 2 are basically trapezoidal and the amplitudes increase when the PM moves towards the outmost position. In addition, the waveforms of B_g are the same but the peak B_g increases when the PM moves towards the outmost position. The differences between the two types are significant. The PM position can be used on Type 1 to adjust the waveform of the PM flux linkage and both the waveform and amplitude of the air-gap flux density. For Type 2, this can adjust the amplitude of both the PM flux linkage and air-gap flux density.

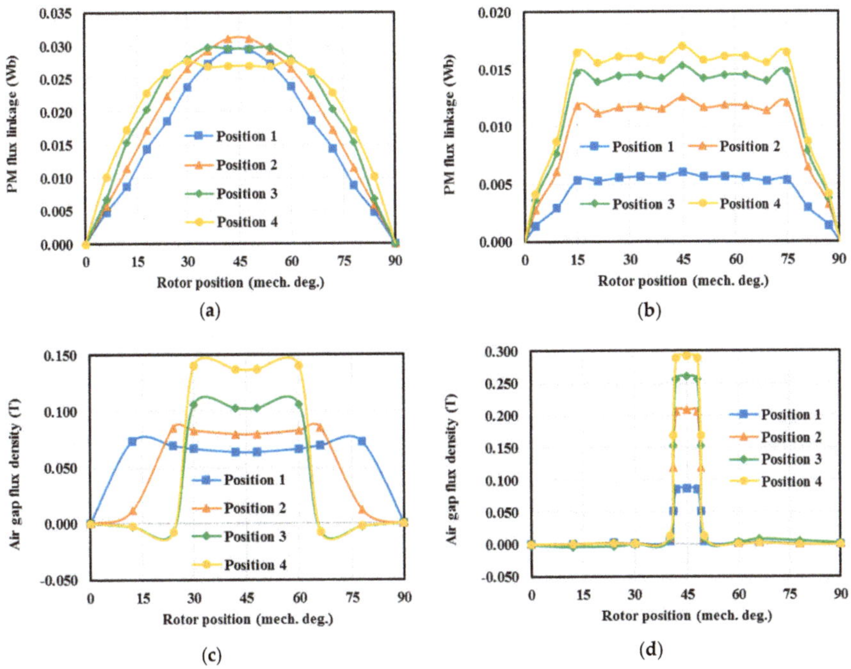

Figure 5. The PM flux linkage and air-gap flux density at no-load: (**a**) The PM flux linkage for Type 1; (**b**) the PM flux linkage for Type 2; (**c**) the air-gap flux density for Type 1; (**d**) the air-gap flux density for Type 2.

Table 3. Air-gap flux density for each PM position.

PM Position	1st	2nd	3rd	4th
Air-gap flux density for Type 1 (T)	0.056	0.060	0.067	0.075
Air-gap flux density for Type 2 (T)	0.025	0.059	0.074	0.083

The no-load peak flux density in the air gap for each PM position of both types is summarized in Table 3. For Type 1, the flux density at position 1 is the lowest (0.056 T) and that at position 4, is the highest (0.075 T). Similarly, for Type 2, the flux density at position 1 is the lowest (0.025 T) and that at position 4, is the highest (0.083 T). However, the lowest flux density for Type 2 is lower than Type 1, while the highest flux density for Type 2 is higher than Type 1. Thus, the no-load air-gap flux density for the Type 2 rotor seems to be more sensitive to positions. This is possibly because for Type 2, the PM flux is not blocked by the outer flux barriers while the blockage appears for Type 1. The revious analyses imply that for Type 1, the room for the PM flux can be more at the inward positions, i.e., larger or stronger PM [23], while for Type 2, more PM for the outward position could be used. In addition, the improvement of the PM flux linkage and air-gap flux density can be anticipated when the number of PM layers or the PM width increases, which is analyzed later.

3.2. On-Load Operation Comparison and Flux Balance Index

To fully investigate the influence of PM positions, the on-load operation is considered. Figure 6 shows the flux density distribution in the rotor with an 80 A peak current and maximum torque for each PM position (single PM each case). For Type 1, the most unbalanced flux density distribution occurs at the flux segments near the PMs (red circle) since the PM flux is obstructed by the surrounding flux barriers. The heavily unbalanced flux distribution may cause the PMs not to be utilized efficiently and lead to problems, such as local saturation, torque ripple or risk of demagnetization in motors [36]. For Type 2, the unbalance also occurs but appears to be lighter (dark blue circle) and the condition is almost the same for every PM position. This paper develops an index called the flux balance index to rate the degree of balance of flux distribution, which is given as:

$$K_u = \frac{B_u}{B_{rotor}} \cdot 100\% \tag{9}$$

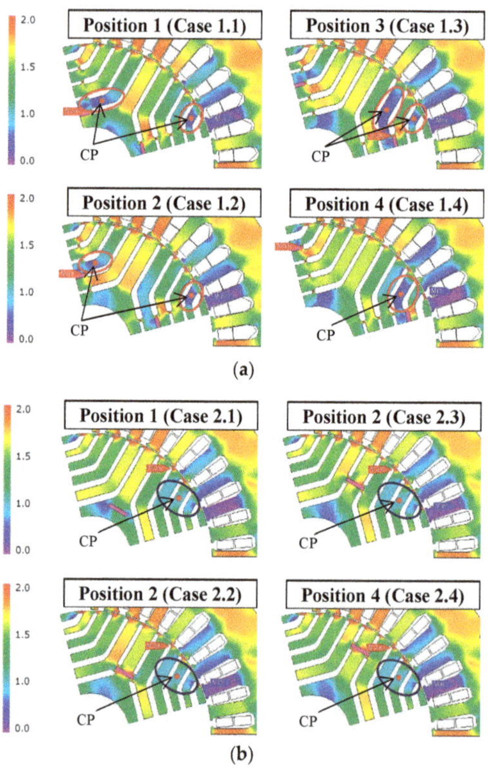

Figure 6. Comparison of the flux density distribution in rotors: (**a**) Type 1; (**b**) Type 2.

where B_u is the lowest flux density at the central point (CP) of the main unbalanced magnetic distribution zones as highlighted in Figure 6, and B_{rotor} is the average flux density in the rotor core obtained from a number of selected points (40 points in this paper) evenly spread out on the rotor core, as illustrated in Figure 7. The selection of these points only aims to represent the average flux density in the rotor without a particular criterion. Note that the determination of B_u and B_{rotor} does not take into account the singularities of the magnetic field, which do not represent the general magnetic distribution although these points are significant for saliency and the torque of motors [18,37]. The higher the flux balance index is, the better and more balanced magnetic distribution is in the rotor. The flux balance indices for various the PM positions of both types of motors are shown in Table 4. As can be seen, for Type 1, the flux balance index increases in the order of PM positions from 1 to 4 [Cases 1.1 to 1.4 in Figure 6a], indicating that the magnetic distribution would be better when the PM moves outward. In contrast, for Type 2, the unbalanced zones and the unbalanced condition do not seem to change as the PM position changes [Case 2.1 to 2.4 in Figure 6b]. This may be due to the fact that for such a configuration, the PM flux is not obstructed by the flux barriers. Furthermore, this unbalance is only caused by the armature reaction at load condition. As a result, Type 2 has a greater flux balance index than Type 1 does, meaning that the arrangement of PMs in Type 2 can be a decent choice to help improve the balance of the magnetic field.

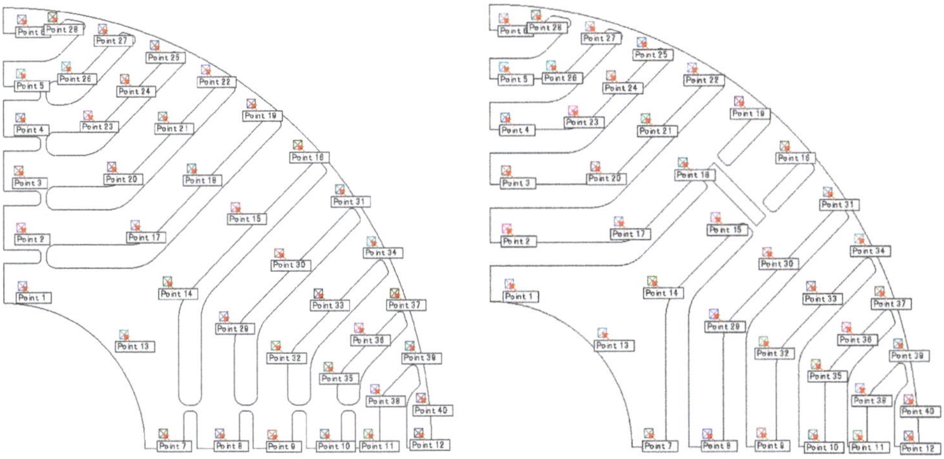

(**a**) (**b**)

Figure 7. The selected points for the rotor average flux density calculation: (**a**) Type 1; (**b**) Type 2, where the PM at position 4 is used as representative.

Table 4. The flux balance index of magnetic distribution in rotors.

PM Position	1st	2nd	3rd	4th
Lowest flux density for Type 1 (T)	0.262	0.302	0.355	0.372
Average flux density in rotor core for Type 1 (T)	1.104	1.111	1.103	1.120
Flux balance index, K_u for Type 1 (%)	23.74	27.19	32.18	33.23
Lowest flux density for Type 2 (T)	0.663	0.640	0.645	0.641
Average flux density in rotor core for Type 2 (T)	1.190	1.171	1.172	1.169
Flux balance index, K_u for Type 2 (%)	55.73	54.67	55.05	54.82

A brief comparison of the torque and torque ripple between the various PM positions of both types is shown in Figure 8. It can be seen that the torque ripple for Type 1 significantly changes with the PM positions, but this does not happen in Type 2. As previously mentioned, for Type 1, both the flux balance index and the locations of unbalance zones vary with the PM position, and the most unbalanced area occurs on the segment near the PM wherever it is placed (Figure 6a and Table 4). For Type 2, in contrast, both the flux balance index and the unbalanced zones seem to be independent of PM position (Figure 6b and Table 4). This indicates that the magnetic distribution in a Type 2 rotor is insensitive to the PM position, and this accounts for the invariant torque ripple with PM positions. In other words, for Type 1, the selection of PM position in a rotor should consider its effect on the magnetic distribution and thus torque ripple. For Type 2 motors, the torque ripple should not be a major concern for the placement of the PM. Note that, the torque output level is not affected significantly by the PM positions and motor types.

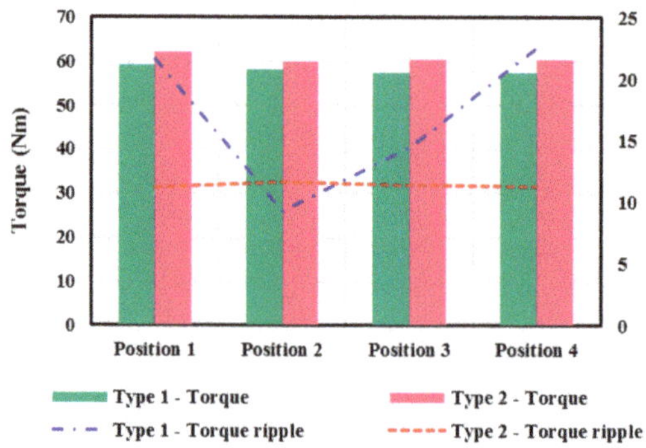

Figure 8. Torque production for each PM position.

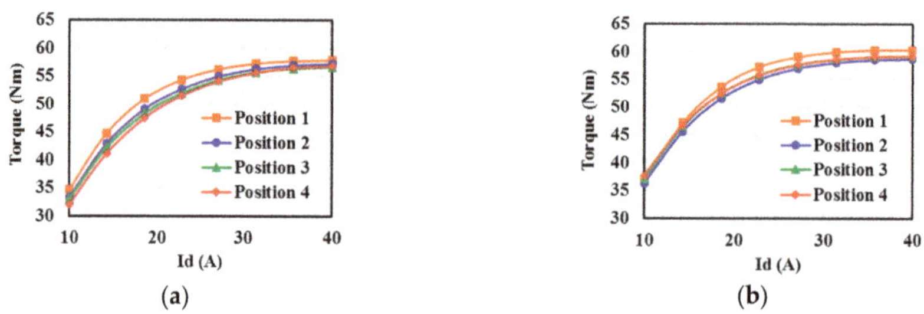

Figure 9. Torque production with Id for each PM position. (**a**) Type 1, (**b**) Type 2.

To fully compare the contribution of each PM position on torque production, the current is regulated in terms of the magnitude and phase advances with respect to the q-axis. The current I_q is kept at 70A and I_d is changed from 10 to 40 A, which considers the cross influence between the d- and q-axes and limits the armature within the peak value, i.e., 80 A. As the result shown in Figure 9, for Type 1, the torque decreases when the PM advances to position 4. In Type 2, although the highest air-gap flux density at position 4 is much higher than position 1, the torque is not much different at any positions. Besides, the highest torque can be generated by the PM at position 1 where the PM is separated from the flux barriers (no intersections), as shown in Figure 1b. However, the width of the PM is limited in this case. In Type 1, the most effective PM position in the torque production is the inward one [23] but it is limited by the possible room given in the rotor, and therefore the configuration with multiple PM layers is suitable for Type 1 to enhance the torque output. In Type 2, with the torque production at each PM position being similar and with the highest air-gap flux density occurring at position 4, this implies that the effective PM configuration is outwards, and the multiple PM layers construction may not be necessary.

In this section, small PM pieces have been used in both Type 1 and 2 motors for the comparison of the influence of PM position. Generally, the placement of these small PM pieces does not significantly change the rotor structure, especially for Type 1. However, the larger PM dimension (e.g., wider) or more pieces (e.g., multiple layers) would be more practical, and thus more variation in rotor configuration may require further assessment.

4. Comparative Analysis of Motor Characteristics

From the previous discussions, the comparison for various numbers of the PM layers as well as various the PM positions is presented. In this section, to fully investigate the performance and characteristics, six models broken down into the two types of motors discussed in Section 2 are created and shown in Figure 10, where Models 1, 2, and 3 are categorized in Type 1 and Models 4, 5, and 6 belong to Type 2. In these models, Models 1 and 4 have the PMs installed at position 1, Models 2 and 5 at position 4, and Models 3 and 6 at all the positions. As the previous analysis, the PMs are all magnetized in the parallel pattern. Note that these models can be treated as proper motors and the effect of various PM layouts (inmost, outmost and multiple layers) on motor inductances and torque production can be investigated through these rotor arrangements. All the models have the same main specifications and the PM thickness/volume. Their PM positions and dimensions are given in Table 5. Note that, the PM thickness is 1.5 mm, which is considered to be manufacturable [23,26].

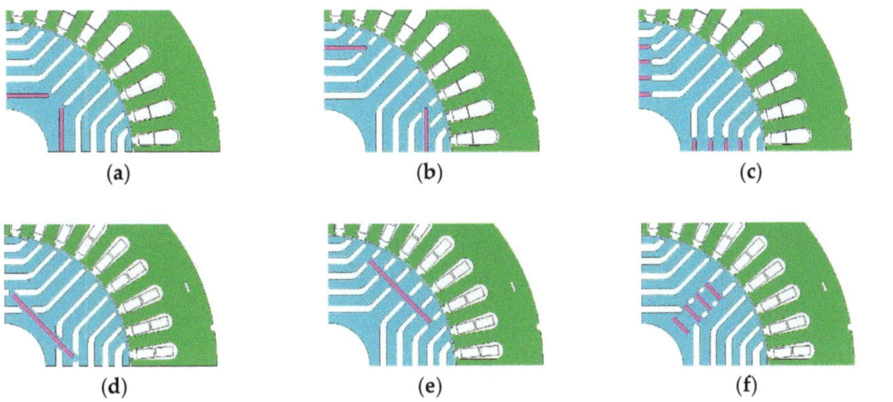

Figure 10. Configurations of motor models: (**a**) Model 1; (**b**) Model 2; (**c**) Model 3; (**d**) Model 4; (**e**) Model 5; (**f**) Model 6.

Table 5. Position and dimension of PMs.

Items	Model 1	Model 2	Model 3	Model 4	Model 5	Model 6
PM position	1	4	1, 2, 3,4	1	4	1, 2, 3, 4
PM dimenssion	1.5 × 32 mm	1.5 × 32 mm	1.5 × 8 mm	1.5 × 32 mm	1.5 × 32 mm	1.5 × 8 mm

4.1. Motor Inductances

The inductance variation with the phase angle and magnitude of the current are shown in Figure 11, where the current angle was set to be zero for Figure 11c,d. Note that, these are the motor *d*- and *q*-axis inductances (L_{dm} and L_{qm}) with the presence of PMs rather than the ones without considering the PM flux, i.e., L_d and L_q (stator inductances) in Equations (3)–(8).

First, as shown in Figure 11a,c, for the Type 1 models (Models 1, 2 and 3), the inductance L_{dm} is the highest for Model 3 but not much higher than the others. The *q*-axis inductance L_{qm} is similar for Models 1, 2 and 3, which indicates that PM position does not significantly influence the electromagnetic characteristics of the motors. Conversely, as shown in Figure 11b,d, for the Type 2 models (Models

4, 5 and 6), the PM position has a great impact on L_{dm}, with the lowest for Model 5 because of the stronger flux created by the PM that limits the armature flux linkage. The inductance L_{qm} is similar for Models 4, 5 and 6. The above analysis shows that the PM position would be the most important design key for Type 2 motors.

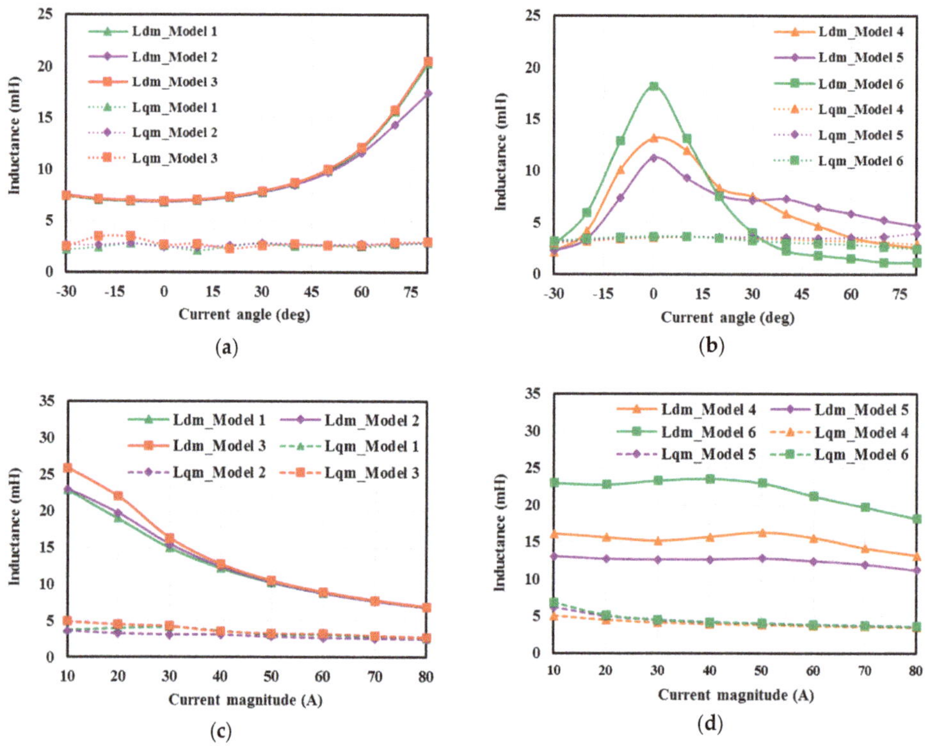

Figure 11. Inductance comparison: (**a**) Inductance versus current angle for Type 1; (**b**) Inductance versus current angle for Type 2; (**c**) Inductance versus current magnitude for Type 1; (**d**) Inductance versus current magnitude for Type 2.

Second, as the current magnitude increases, the inductance L_{dm} for the Type 1 models decreases rapidly while L_{qm} decreases gently, as shown in Figure 11c. However, as shown in Figure 11d, both L_{dm} and L_{qm} for Type 2 declines slightly with the current magnitude. This results in a small inductance difference ($L_{dm} - L_{qm}$) variation, especially for Model 5 whose L_{dm} and L_{qm} almost do not change. This may be attributed to the flux barriers that slightly brings down L_{qm}. Meanwhile, L_{dm} decreases mildly because of the alleviated *q*-axis magnetic saturation effects [38]. The insignificant inductance variation that may benefit sensorless control shows that fewer PM layers and placing the PM outwards would be more beneficial to the Type 2 motor design.

4.2. Torque Production

The torque production and torque components (i.e., PM/reluctance torques) are illustrated in Figure 12 and Table 6. To obtain the curves of torque versus the current angle using FEA, the initial rotor position can first be set up and based on that which the current is applied. Then, the current angle is regulated and swept through the prescribed range so that the output torque can be calculated for each current angle. As shown in Figure 12a, the torque production and torque components for all the models of Type 1 are almost similar. Of these cases, the multiple PM layers one, i.e., Model 3 is the best choice for the sake of its highest achieved total torque and PM torque ratio, as presented in Table 6. However, Model 1 should also be further considered since its torque is only slightly lower, taking the advantage of inmost PM arrangement, as discussed in the previous analysis. As shown in Figure 12b, the torque production and components are diverse for the models of Type 2, which agrees with the previous discussion. The case with the most outward PM and fewer PM layers, i.e. Model 5 would the best choice for its highest achieved total torque by making the most utilization of the PM compared to Models 4 and 6, as presented in Table 6. Furthermore, as presented in Table 6, the torque production of all the models of Type 1 is higher than every model of Type 2. This seems to indicate that the models of Type 1 have better torque production than Type 2. However, the high PM torque and its easy regulation by applying the various PM configurations (i.e., large PM torque difference between these investigated models) is a significant advantage of Type 2. On the other hand, for the Type 1 models, the torque is brought down to zero when the current angle is zero (i.e., I_d is zero) while for Type 2, the torque can only reduce to zero at the high current angle since the PM torque and the reluctance torque offset each other. Therefore, for the Type 2 motors, the armature excitation, i.e., the stator current I_s can be easily used to regulate the torque characteristics.

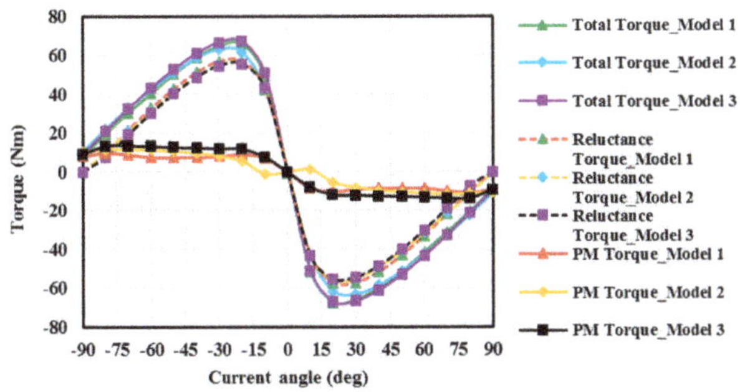

(a)

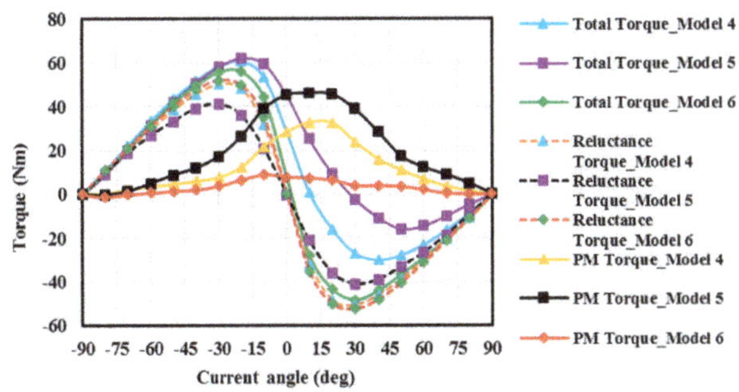

(b)

Figure 12. Torque production and components: (**a**) Type 1; (**b**) Type 2.

Table 6. Torque production and components.

Items	Model 1	Model 2	Model 3	Model 4	Model 5	Model 6
Maximum torque (Nm)	66.3	63.7	67.8	60.9	62.4	57.3
PM torque at maximum torque (Nm)	8.6	7.4	11.9	11.6	28.5	5.5
PM torque ratio (%)	13.0	11.6	17.6	19.0	45.7	9.6

4.3. Torque and Power-Speed Curves

As previously mentioned, the armature current I_s can be reduced to be close to I_{ch} to improve the speed range. The relationship between I_s and I_{ch} as I_s of Models 1, 3 and 5 varies and is presented in Table 7, where as expected, the I_s/I_{ch} ratio generally decreases as I_s decreases. Furthermore, Figures 13 and 14 show the comparison of torque-speed and power-speed curves between Models 1, 3 and 5, where a 220 V DC voltage is applied. In these curves, the rhombuses and circles denote the start- and end-points of the constant power region. The pink ones are for Model 3 and the blue ones are for Models 1 or 5. As can be seen in Figures 13 and 14, for the three models, the power-speed curves tend to converge at high speed and are partially overlapping. When I_s reduces to be close to I_{ch}, the CPSR of the motors improves but the torque and the power suffer some reduction. In Figure 13, the torque and power-speed curves of Model 1 are constantly slightly lower than Model 3 for the same I_s although the highest torque production is obtained when the PM is at the inmost position in the rotor. In contrast, although at a low speed operation, the torque production of Model 5 is lower than Model 3, but the difference decreases as I_s decreases. This results in an enhancement of the power-speed curves and CPSR for Model 5, which then surpasses Model 3, as seen in Figure 14. However, if I_s continues to reduce, e.g., I_s = 20 A, the CPSR of Model 5 cannot maintain the superiority to that of Model 3. This can be observed via the power-speed curves where there is an intersection point between these curves. The power corresponding to this point is denoted the intersection power, P_i. For such an intersection, if the required output power is greater than P_i, the CPSR of Model 5 is better than Model 3 and vice versa. Note that, as can be seen in Figure 14, this intersection point locates at the overlap of the MTPV control [33] regions of the target models so that this point no longer depends on the required current, but on the electromagnetic properties or the motor construction, i.e., PM flux linkage, inductance and I_{ch}.

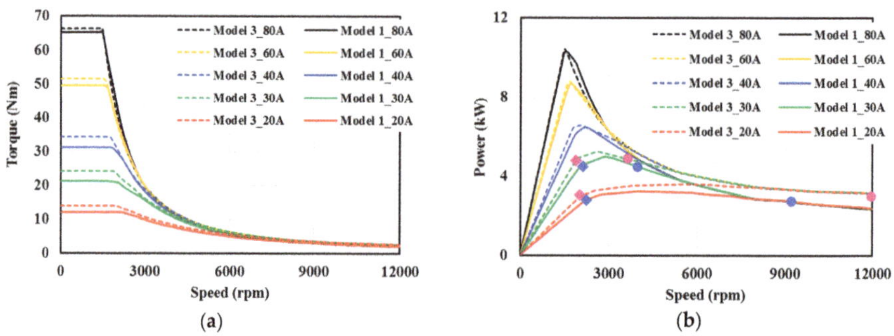

Figure 13. Field weakening comparison between Models 1 and 3: (**a**) Torque-speed curves; (**b**) Power-speed curves.

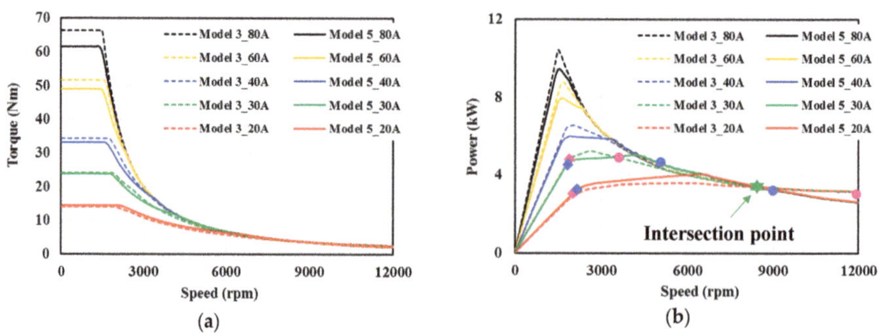

Figure 14. Field weakening comparison between Models 3 and 5: (**a**) Torque-speed curves; (**b**) Power-speed curves.

From the above discussions, with limited PM volume (less than common IPM motors), the selection of torque and power is closely related to desired motor speed range and needs to be chosen carefully by simultaneously considering the mechanical and electromagnetic characteristics of the motors and their applications. Although these motors are ideally infinite speed [16], the actual operating speed is limited. In addition, the selection of the PM positions and directions, e.g., moving PM from *q*-axis to *d*-axis in this paper, is the key to improve the performance of SynRMs with limited PM used.

Table 7. The constant power speed range (CPSR) analysis.

I_s (A)	80	60	40	30	20
I_{ch} (A) (Model 1)	18.9	17.1	13.7	10.2	9.4
I_s/I_{ch} (Model 1)	4.23	3.52	2.93	2.95	2.13
I_{ch} (A) (Model 3)	12.0	10.7	9.3	8.6	7.5
I_s/I_{ch} (Model 3)	6.67	5.59	4.28	3.50	2.68
I_{ch} (A) (Model 5)	19.7	17.2	15.1	14.2	13.7
I_s/I_{ch} (Model 5)	4.06	3.49	2.65	2.12	1.46

4.4. Demagnetization Analysis

The no-load PM flux linkage and air-gap flux density of Models 1, 3 and 5 are shown in Figure 15. It can be seen that when the PM width increases, the PM flux linkage and air-gap flux density of Type 2 (Model 5) increases significantly. This demonstrates the intensification of the magnetic field that is enabled by placing the PM along the *d*-axis instead of the *q*-axis. This can also explain the high PM torque ratio of Model 5 as presented in Table 6.

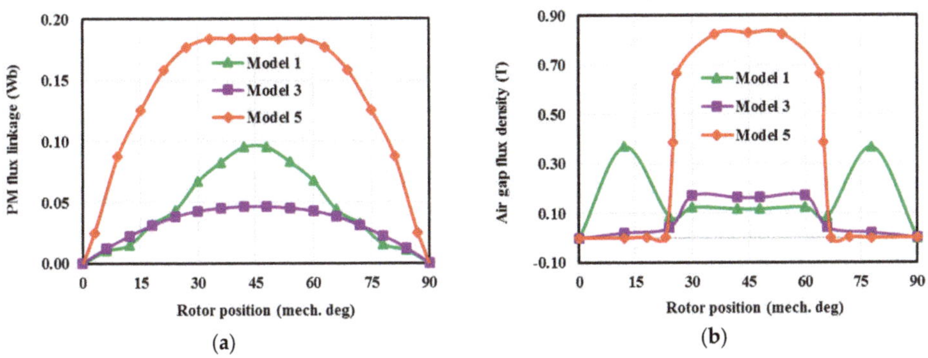

Figure 15. No-load magnetic characteristic. (**a**) PM flux linkage, (**b**) Air-gap flux density.

The distribution of flux density in the PMs of Models 1, 3 and 5 under the peak current at 120 °C are shown in Figure 16. The average flux density in the PMs at various temperatures are presented in Table 8. Note that, the demagnetization curves of N35H (PM material) are illustrated in Figure 17 [16,39]. It can be seen that the flux density in the PMs of Model 1 is very low, meaning that the PMs can be easily demagnetized. Moreover, the flux density in the PM of Model 1 is lower than Model 3 where the PMs are inserted at all positions. This implies that if only the inward PM position is used, an appropriate thickness of the PM should be carefully chosen to avoid irreversible demagnetization. The PMs in Models 3 and 5 have better demagnetization resistance compared to Model 1 since they possess better operating points, as the flux density presented in Table 8. Particularly, Model 5 possesses an advantage with an average PM flux density over 0.8 T. In contrast, the PMs in Model 3 can be locally demagnetized since their flux density are low, as shown in Figure 16. The key is the direction of the flux. In Type 1, the *q*-axis armature flux is opposing the PM flux and then reducing the flux at the *q*-axis. Instead, in Type 2, the *d*-axis armature flux and PM flux are in the same direction and then the d-axis flux is intensified. The motors that employ flux intensifying such as the Type 2 ones have been called the flux intensifying IPMSM [38,40,41]. However, in this paper, less PM is used for the rotor with a dominant reluctance torque, and therefore they should still be considered as a kind of SynRMs. They are named the FI-PMa-SynRM in this paper [26].

Overall, as investigated in this section, the layout with the PMs crossing the flux barriers have a decreased possibility to be demagnetized.

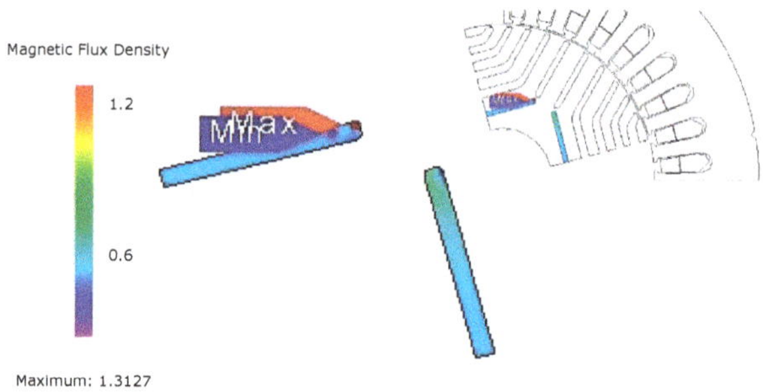

(a)

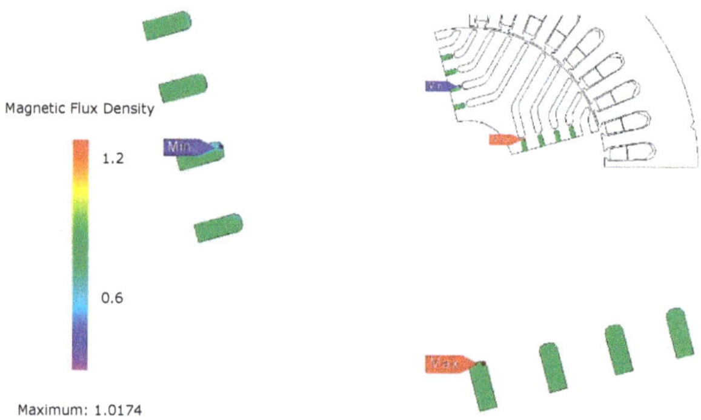

(b)

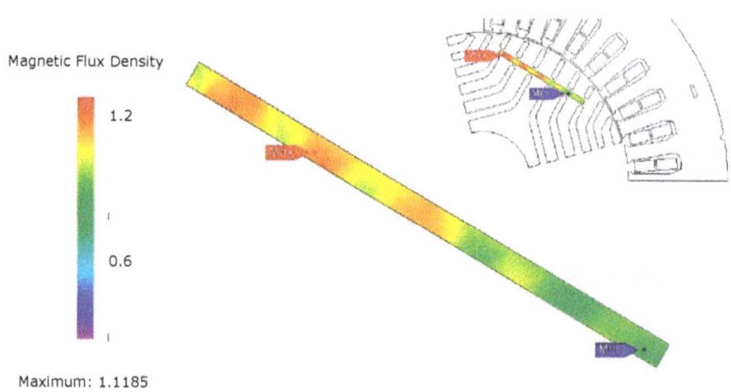

(c)

Figure 16. Flux density in PM at the 120 °C of temperature: (**a**) Model 1; (**b**) Model 3; (**c**) Model 5.

Table 8. Average flux density in PM.

Temperature	20 °C	90 °C	105 °C	120 °C	130 °C	155 °C
Model 1	0.403 T	0.354 T	0.342 T	0.327 T	0.273 T	0.133 T
Model 3	0.711 T	0.614 T	0.595 T	0.577 T	0.563 T	0.508 T
Model 5	0.912 T	0.869 T	0.860 T	0.852 T	0.845 T	0.828 T

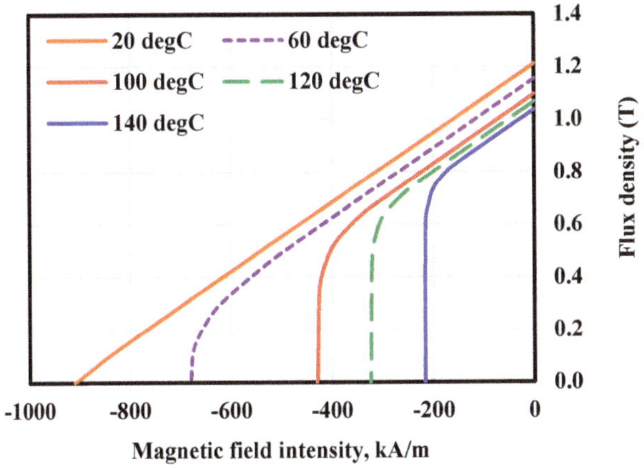

Figure 17. Demagnetization curves of N35H.

5. Discussion

Based on the previous analyses, some brief summaries are listed as follows.

- For the effect of the PM position, the outward PMs generally produce greater air-gap flux density but torque production does not exactly have the same trend.
- The PM position and its arrangement in the *d*- or *q*-axis have a great impact on flux distribution in the rotor where the *d*-axis PM arrangement possesses a higher flux balance index.
- The PM position has a greater impact on motor inductance of the Type 2 motor than the Type 1 one and the effect is approximately linear.
- Model 3 (conventional multiple-layer PMa-SynRM) has the highest torque production while Model 5 (FI-PMa-SynRM) has the most utilization of PM.
- Model 5 (FI-PMa-SynRM) is the best choice for demagnetization resistance while Model 1 is the worst one.
- For SynRM with a limited PM amount, the reduction of armature current I_s leads to an increase of CPSR but a trade-off with torque reduction should be considered.

The experimental studies were reported in [23], where the simulations were conducted using the same software package (JMAG). Since the focus of this paper is on the analysis and comparison of several types of SynRM rotors, therefore the experiments are not provided here.

6. Conclusions

In this paper, the analyses for six models of SynRMs with two different categories of PM layouts have been conducted, and their performance and electromagnetic characteristics have been comprehensively compared. From these analyses, it can be observed that the layout with PMs being arranged along the *q*-axis or embedded into the flux barriers has better torque production capability. For the other layout with the PM facing the *d*-axis or across the flux barriers, the advantages of SynRMs using a limited PM amount can be maintained and the inherent drawbacks, such as irreversible demagnetization, can be overcome. These indicate that, for SynRMs with a limited amount of PM added and placed along the *d*-axis would better make use of the PM. In addition, the PM arrangement at the inward position is a decent choice for SynRM with the *q*-axis PM, but the PM dimension should be calculated carefully to avoid irreversible demagnetization.

Author Contributions: D.-K.N. conceived and conducted the research and wrote the paper. M.-F.H. suggested the research topic, guided D.-K.N. to complete the research and helped edit and finalize the paper. M.-F.H. also provided the laboratory space and facilities for this study.

Funding: This work is supported by the Ministry of Science and Technology, Taiwan under project contracts MOST 108-2622-8-006-014 and 107-2622-E-006-005-CC2.

Acknowledgments: The authors would like to thank Thanh Anh Huynh for his help in this work. JSOL is also acknowledged for providing JMAG software.

Conflicts of Interest: The authors declare no conflicts of interest.

References

1. Lipo, T. Synchronous reluctance machines—A viable alternative for AC drives. *Electr. Mach. Power Syst.* **1991**, *19*, 659–671. [CrossRef]
2. Ozcelik, N.G.; Dogru, U.E.; Imeryuz, M.; Ergene, L.T. Synchronous reluctance motor vs. induction motor at low-power industrial applications: Design and comparison. *Energies* **2019**, *12*, 2190. [CrossRef]
3. Ibrahim, M.N.; Sergeant, P.; Rashad, E.M. Synchronous reluctance motor performance based on different electrical steel grades. *IEEE Trans. Magn.* **2015**, *51*, 1–4. [CrossRef]
4. Reddy, P.B.; El-Refaie, A.M.; Galioto, S.; Alexander, J.P. Design of synchronous reluctance motor utilizing dual-phase material for traction applications. *IEEE Trans. Ind. Appl.* **2017**, *53*, 1948–1957. [CrossRef]
5. Cai, S.; Jin, M.-J.; Hao, H.; Shen, J.-X. Comparative study on synchronous reluctance and PM machines. *COMPEL Int. J. Comput. Math. Electr. Electron. Eng.* **2016**, *35*, 607–623. [CrossRef]
6. Guan, Y.; Zhu, Z.Q.; Afinowi, I.A.A.; Mipo, J.C.; Farah, P. Design of synchronous reluctance and permanent magnet synchronous reluctance machines for electric vehicle application. *COMPEL Int. J. Comput. Math. Electr. Electron. Eng.* **2016**, *35*, 586–606. [CrossRef]
7. Carraro, E.; Degano, M.; Morandi, M.; Bianchi, N. PM synchronous machine comparison for light electric vehicles. In Proceedings of the IEEE International Electric Vehicle Conference (IEVC), Florence, Italy, 17–19 December 2014; pp. 1–8.
8. Ibrahim, M.N.F.; Rashad, E.; Sergeant, P. Performance comparison of conventional synchronous reluctance machines and PM-assisted types with combined star-delta winding. *Energies* **2017**, *10*, 1500. [CrossRef]
9. Bianchi, N.; Fornasiero, E.; Soong, W. Selection of PM flux linkage for maximum low-speed torque rating in a PM-assisted synchronous reluctance machine. *IEEE Trans. Ind. Appl.* **2015**, *51*, 3600–3608. [CrossRef]
10. Guglielmi, P.; Boazzo, B.; Armando, E.; Pellegrino, G.; Vagati, A. Permanent magnet minimization in PM-assisted synchronous reluctance motors for wide speed range. *IEEE Trans. Ind. Appl.* **2013**, *49*, 31–41. [CrossRef]
11. Bianchi, N.; Mahmoud, H.; Bolognani, S. Fast synthesis of permanent magnet assisted synchronous reluctance motors. *IET Electr. Power App.* **2016**, *10*, 312–318. [CrossRef]
12. Niazi, P.; Toliyat, H.A.; Cheong, D.-H.; Kim, J.-C. A low-cost and efficient permanent-magnet-assisted synchronous reluctance motor drive. *IEEE Trans. Ind. Appl.* **2007**, *43*, 542–550. [CrossRef]
13. Lee, J.H.; Jang, Y.J.; Hong, J.P. Characteristic analysis of permanent magnet-assisted synchronous reluctance motor for high power application. *J. Appl. Phys.* **2005**, *97*, 10Q503. [CrossRef]
14. Huynh, T.A.; Hsieh, M.F. Performance analysis of permanent magnet motors for electric vehicles (EV) traction considering driving cycles. *Energies* **2018**, *11*, 1385. [CrossRef]
15. Cai, H.; Guan, B.; Xu, L. Low-Cost ferrite PM-Assisted synchronous reluctance machine for electric vehicles. *IEEE Trans. Ind. Electron.* **2014**, *61*, 5741–5748. [CrossRef]
16. Huynh, T.A.; Hsieh, M.-F. Comparative study of PM-assisted SynRM and IPMSM on constant power speed range for EV applications. *IEEE Trans. Magn.* **2017**, *53*, 1–6. [CrossRef]
17. Vagati, A. Synchronous Reluctance Electrical Motor Having a Low Torque-Ripple Design. U.S. Patent 5,818,140, 6 October 1998.

18. Ferrari, M.; Bianchi, N.; Fornasiero, E. Rotor saturation impact in synchronous reluctance and PM assisted reluctance motors. In Proceedings of the IEEE Energy Conversion Congress and Exposition (ECCE), Denver, CO, USA, 15–19 September 2014; pp. 1235–1242.
19. Binns, K.J.; Lawrenson, P.J.; Trowbridge, C.W. *The Analytical and Numerical Solution of Electric and Magnetic Fields*; Wiley: Chichester, UK, 1992.
20. Moghaddam, R.R. Synchronous Reluctance Machine (SYNRM) in Variable Speed Drives (VSD) Applications. Ph.D. Thesis, KTH Royal Institute of Technology, Stockholm, Sweden, 2011.
21. Morimoto, S.; Sanada, M.; Takeda, Y. Performance of PM-Assisted synchronous reluctance motor for high-efficiency and wide constant-power operation. *IEEE Trans. Ind. Appl.* **2001**, *37*, 1234–1240. [CrossRef]
22. Imamura, K.; Sanada, M.; Morimoto, S.; Inoue, Y. Improvement of characteristics by flux barrier shape and magnet thickness of IPMSM with Dy-free rare-earth magnet. In Proceedings of the 15th European Conference on Power Electronics and Applications (EPE), Lille, France, 2–6 September 2013; pp. 1–10.
23. Huynh, T.A.; Hsieh, M.-F.; Shih, K.-J.; Kuo, H.-F. An investigation into the effect of PM arrangements on PMa-SynRM performance. *IEEE Trans. Ind. Appl.* **2018**, *54*, 5856–5868. [CrossRef]
24. Pellegrino, G.; Vagati, A.; Guglielmi, P. Design tradeoffs between constant power speed range, uncontrolled generator operation, and rated current of IPM motor drives. *IEEE Trans. Ind. Appl.* **2011**, *47*, 1995–2003. [CrossRef]
25. Prins, M.H.A.; Kamper, M.J. Design optimisation of field-intensified permanent magnet machine. In Proceedings of the IEEE International Conference on Electrical Machines (ICEM), Berlin, Germany, 2–5 September 2014; pp. 117–123.
26. Ngo, D.-K.; Hsieh, M.-F.; Huynh, T.A. Torque enhancement for a novel flux intensifying PMa-SynRM using surface-inset permanent magnet. *IEEE Trans. Magn.* **2019**, *55*, 8106108. [CrossRef]
27. Pillay, P.; Krishnan, R. Modeling of permanent magnet motor drives. *IEEE Trans. Ind. Electron.* **1988**, *35*, 537–541. [CrossRef]
28. Morimoto, S.; Takeda, Y.; Hatanaka, K.; Tong, Y.; Hirasa, T. Designand control system of inverter-driven permanent magnet synchronous motors for high torque operation. *IEEE Trans. Indus. Appl.* **1993**, *29*, 1150–1155. [CrossRef]
29. Baek, J.; Bonthu, S.S.R.; Choi, S. Design of five-phase permanent magnet assisted synchronous reluctance motor for low output torque ripple applications. *IET Electr. Power App.* **2016**, *10*, 339–346. [CrossRef]
30. Lee, J.H.; Kim, J.C.; Hyun, D.S. Effect analysis of magnet on Ld and Lq inductance of permanent magnet assisted synchronous reluctance motor using finite element method. *IEEE Trans. Magn.* **1999**, *35*, 1199–1202. [CrossRef]
31. Ngo, D.-K.; Hsieh, M.-F. Analysis of flux intensifying effect on synchronous motors applied to electric scooter. In Proceedings of the IEEE Vehicle Power and Propulsion Conference (IEEE-VPPC), Hanoi, Vietnam, 14–17 October 2019.
32. Jahns, T.M. Flux-weakening regime operation of an interior permanent-magnet synchronous motor drive. *IEEE Trans. Ind. Appl.* **1987**, *23*, 681–689. [CrossRef]
33. Soong, W.L.; Miller, T.H.E. Field-weakening performance of brushless synchronous AC motor drives. *IEE Proc. Electr. Power Appl.* **1994**, *141*, 331–340. [CrossRef]
34. Jahns, T.M.; Kliman, G.B.; Neumann, T.W. Interior permanent-magnet synchronous motors for adjustable-speed drives. *IEEE Trans. Ind. Appl.* **1986**, *22*, 738–747. [CrossRef]
35. Barcaro, M.; Bianchi, N.; Magnussen, F. Permanent-magnet optimization in permanent-magnet-assisted synchronous reluctance motor for a wide constant-power speed range. *IEEE Trans. Ind. Electron.* **2012**, *59*, 2495–2502. [CrossRef]
36. Kim, W.-H.; Kim, K.-S.; Kim, S.-J.; Kang, D.-W.; Go, S.-C.; Chun, Y.-D.; Lee, J. Optimal PM design of PMA-SynRM for wide constant-power operation and torque ripple reduction. *IEEE Trans. Magn.* **2009**, *45*, 4660–4663. [CrossRef]
37. Ferrari, M.; Bianchi, N.; Doria, A.; Fornasiero, E. Design of synchronous reluctance motor for hybrid electric vehicles. *IEEE Trans. Ind. Appl.* **2015**, *51*, 3030–3040. [CrossRef]
38. Zhu, X.; Yang, S.; Du, Y.; Xiang, Z.; Xu, L. Electromagnetic performance analysis and verification of a new flux-intensifying permanent magnet brushless motor with two-layer segmented permanent magnets. *IEEE Trans. Magn.* **2016**, *52*, 1–4. [CrossRef]

39. N35H—Arnold Magnetic Technologies. Available online: https://www.arnoldmagnetics.com/wp-content/uploads/2017/11/N35H-151021.pdf (accessed on 23 July 2019).
40. Zhu, X.; Wu, W.; Yang, S.; Xiang, Z.; Quan, L. Comparative design and analysis of new type of flux-intensifying interior permanent magnet motors with different q-axis rotor flux barriers. *IEEE Trans. Energy Convers.* **2018**, *33*, 2260–2269. [CrossRef]
41. Limsuwan, N.; Shibukawa, Y.; Reigosa, D.D.; Lorenz, R.D. Novel design of flux-intensifying interior permanent magnet synchronous machine suitable for self-sensing control at very low speed and power conversion. *IEEE Trans. Ind. Appl.* **2011**, *47*, 2004–2012. [CrossRef]

energies

MDPI

Article

A Novel Rotor Eddy Current Loss Estimation Method for Permanent Magnet Synchronous Machines with Small Inductance and a Conductive Rotor Sleeve

Le Pei, Liyi Li *, Qingbo Guo, Rui Yang and Pengcheng Du

Department of Electrical Engineering, Harbin Institute of Technology, Harbin 150080, China; 15b906021@hit.edu.cn (L.P.); aza1aza2@163.com (Q.G.); hit_yangrui@163.com (R.Y.); doooo_2010@163.com (P.D.)
* Correspondence: liliyi1024@163.com

Received: 18 July 2019; Accepted: 28 September 2019; Published: 30 September 2019

Abstract: Typically, permanent magnet synchronous machines (PMSMs) with small inductance can achieve a higher power density and higher power factor. Thus, in many industrial applications, more and more PMSMs are being designed with small inductance. Compared with traditional PMSMs, current harmonics in small inductance PMSMs are much more abundant, and the amplitudes are usually high. These current harmonics will cause large eddy current losses (ECLs) on the rotor, making the estimation of ECLs necessary in the design stage. Currently, ECL estimation methods are usually based on frequency order information, which cannot tell the travelling direction of the harmonic magneto-wave, resulting in the inaccuracy of the estimation. This article proposes a novel estimation method based on the mechanism of the formation of space-vector pulse width modulation (PWM), which considers both the frequency order and travelling direction of the harmonic wave, resulting in the improvement of the accuracy. Besides this, by using double Fourier analysis (DFA) instead of traditional fast Fourier analysis (FFA), the predicted frequencies of the current harmonics are more accurate and free of the troubles caused by traditional FFA-based methods. Simulation study and experiments are conducted to show the effectiveness of the proposed method.

Keywords: double Fourier analysis; current spectrum decomposition; eddy current loss; permanent magnet machine design

---

1. Introduction

Permanent magnet synchronous machines (PMSMs) with small inductance usually have good performance in terms of their power density, dynamic performance and overload capability; thus, they have been a trend in academia and industry. However, small inductance also introduces many disadvantages, such as large current harmonics, high rotor temperature rises, or noise and vibration problems. One key problem of the rotor eddy current loss attenuation, which relates to the rotor temperature rise, is unique to these types of machine because of the poor heat dissipation environment of the rotor and large current harmonics under voltage-source inverter (VSI) supply. Researchers found that a conductive sleeve over the rotor can reduce the eddy current losses (ECLs) significantly, and a great deal of research has been conducted into the performance of different configurations and structure parameters [1–4]. Meanwhile, continuous efforts have been devoted to the harmonic suppression algorithm of the currents produced by VSI [5–9]. Moreover, the rotor ECL estimation under different control strategies has become an important step for the lifetime guarantee and safety of the PMSMs when designing small inductance PMSMs.

Regarding the estimation method of rotor ECLs, there are mainly two categories: the finite-element method (FEM) and analytical model method. The first one is highly developed and accurate but is time-consuming and needs extensive computation resources. The latter one is fast and provides much information on the improvement of the inverter-motor system; however, it is less accurate than FEM. Thus, analytical methods are usually used in the design process, and FEMs are used to verify the results.

A large number of researchers have devoted themselves to the improvement of analytical methods, mainly concentrating on analytical motor models. Analytical motor models can be roughly divided into two categories: methods considering the reaction field, and methods without considering the reaction field. The first ones have the ability to take the saturation effect, slot opening effect, etc. into consideration, under the assumption that the eddy current field is of the resistance type. Two-dimensional models were developed with polar coordinates [10–13] or Cartesian coordinates [14], and the permeability of the stator and rotor cores is supposed to be infinite [15,16]. In [17], the time harmonics are discussed, and an extended model was developed. The slotting effect on the eddy current can be modeled using conformal mapping [18] or subdomain models [19,20], and the saturation effect is usually handled using modulation functions [21–23] or magnet equivalent circuits (MEC) [24]. Finite magnet dimensions were solved by adding a coefficient to infinite magnet dimensions in [25,26]. In [27], the authors used Carter's theory and surface impedance to calculate the ECLs, which is simple and has a low computational burden. When the eddy current field cannot be considered to be of the resistance type, which is the case when the thickness of conductor is thicker than the skin depth, the reaction field has to be considered. In [28], the eddy current reaction field was considered in a slotless PMSM. In [29], a 2D model based on Cartesian coordinate was proposed to calculate the losses in the magnet, while in [30], a 2D model based on polar coordinates was proposed. Generally, methods without consideration of the reaction field are more capable of handling complex structures with low-frequency harmonics, while methods which consider the reaction field are more suitable for simple structures with high-frequency harmonics.

The accuracy achieved by the analytical method depends on two aspects: the degree of conformity of the predicted stator current with the actual stator current, and the degree of conformity of the simplified model with the actual motor structure. This article aims to improve the method by reducing the disconformity induced by mismatched currents. For this aspect, traditional methods use the fast Fourier transform (FFT) to decompose the voltage pulses generated by the inverter and then calculate current harmonics based on the frequency spectrum of voltage pulses. These methods are fast, but under non-integer carrier wave ratios, the spectrum leakage problem deteriorates the accuracy of predicted stator current. With the development of simulation technology, many researchers use simulation software such as Simulink from Mathworks to obtain simulated current waveforms, then decompose current waveforms to obtain the frequency spectrum of the stator current. The usage of simulation software greatly improves the accuracy of the prediction of stator current; however, the process is complex, time-consuming, and reveals fewer insights into the difference between various working conditions.

This article proposes a novel ECL estimation method with improved accuracy for PWM inverters. Based on the mechanism of the formation of PWM waveforms, a new decomposition method of PWM voltage pulses is proposed to avoid the problem of spectrum leakage. With the new decomposition method, the phase sequence of different harmonics can be retrieved, which cannot be done with simple FFT or simulation. Regarding the conversion of voltage spectrum into current spectrum, the effect of different voltage components on the motor is discussed. Finally, based on the information provided by the new decomposition method, the direction of the travelling wave caused by subharmonics is carefully studied, and a new algorithm for the calculation of ECLs with different travelling wave direction judgement methods is given.

As the most widely used PWM strategy, space-vector PWM (SVPWM) is adopted as the study object. An analytical orthogonal model of the air-gap is used to ensure accuracy while keeping the model simple, and phase inductance is assumed to be constant.

2. Problem Statement

A typical structure of a small inductance permanent magnet synchronous machine is shown in Figure 1a. The stator is composed of a stator iron core and stator windings, and the rotor is composed of a retaining sleeve, permanent magnet and rotor yoke.

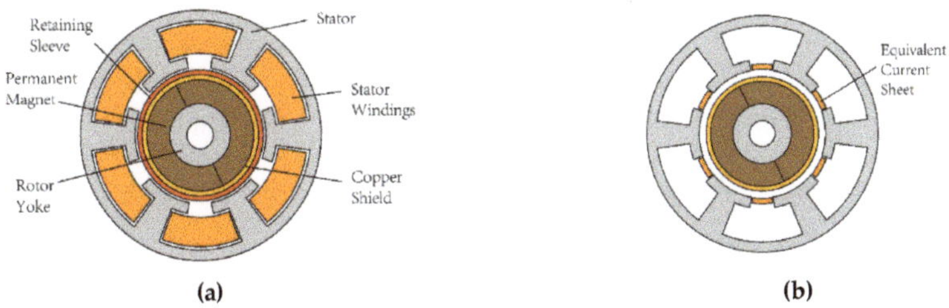

Figure 1. Structure of small inductance permanent magnet synchronous machine (PMSM) (**a**) and simplified model for the calculation of eddy current losses (**b**).

Due to the relatively high permeability in the stator iron core, the windings can be expressed as current sheets at the slot openings. The retaining sleeve is used to enhance the rotor mechanical strength and made of glass fiber. When considering eddy current losses, the retaining sleeve is often considered as part of the air gap because of its low conductivity. The copper shield, which has a high conductivity, is used to reduce the eddy current losses in permanent magnets. Additionally, the simplified model used for eddy current losses estimation is shown in Figure 1b.

Further, the model can be expanded into Cartesian coordinates for convenience in calculation. Typically, a small inductance PMSM Cartesian coordinate model for the calculation of ECLs consists of six layers: the outermost layer is the stator, whose magnet permeability is considered as infinity; next is the current sheet layer, representing the current of the stator, whose thickness is zero; the third layer is the air-gap, whose thickness is d_0; then, the copper sleeve used to resist eddy currents, whose thickness is d_r; inside the copper sleeve is the permanent magnet layer, whose thickness is d_p; and the innermost layer is the rotor yoke, whose thickness is d_z. These layers are depicted in Figure 2.

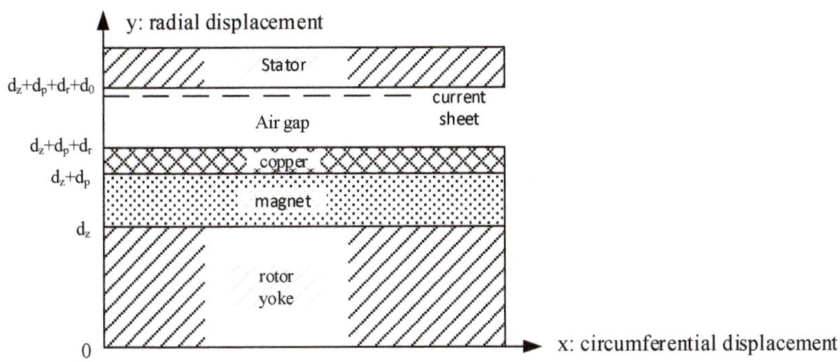

Figure 2. Model structure of a high-speed PMSM for eddy current loss (ECL) calculation.

To calculate the rotor ECLs, the spectrum of phase voltage of the small inductance PMSM is first calculated using time-FFT, with its base frequency being the fundamental current frequency, and the current harmonics are derived using the amplitudes of voltage harmonics divided by the reactance of each phase; then, the space harmonics caused by the winding structure in the motor are calculated using space-FFT, with its base period being a pole pitch of the motor. This process can be expressed as in Equation (1):

$$\begin{gathered} U_n = \frac{1}{T}\int_0^T u(t) \cdot e^{jn\omega_b t} dt \\ I_n = \frac{U_n}{jn\omega_b L_s} \end{gathered} \tag{1}$$

where $u(t)$ is the phase voltage, ω_b is the base frequency, n is the order of the current harmonics, L_s is the phase inductance, and I_n is the amplitude of current harmonics.

Once current harmonics and space harmonics are obtained, their impacts on the motor can be represented by a sum of travelling current sheets. If the order of the current harmonics is n, and the order of space harmonics is v, there would be a travelling current sheet $cs(i,x)$ close to the inner surface of the stator, and that can be expressed as in Equation (2) [19,20]:

$$\begin{gathered} \text{If the current sheet travels forwards :} \\ cs(i,x) = \sum_v K_{sov} K_{dpv} \frac{2}{\tau} \sin(-\frac{v\pi}{2}) N \cdot \sum_n I_n \sin(n\omega_b t - v\frac{\pi}{\tau}x) \\ \text{If the current sheet travels back :} \\ cs(i,x) = \sum_v K_{sov} K_{dpv} \frac{2}{\tau} \sin(-\frac{v\pi}{2}) N \cdot \sum_n I_n \sin(n\omega_b t + v\frac{\pi}{\tau}x) \end{gathered} \tag{2}$$

where K_{sov} and K_{dpv} are coefficients accounting for the slot opening effect and winding distribution effect, respectively. The direction of the travelling current sheet depends on the space harmonic order as well as the current harmonic order. Then, the ECL caused by each current sheet is calculated individually, then summed together to form the overall eddy current loss J_{Loss}. To do so, the vector potential of the motor A_z should be calculated. The governing law in the object small inductance PMSM is listed in Equation (3):

$$\frac{\partial^2 A_z}{\partial x^2} + \frac{\partial^2 A_z}{\partial y^2} = jn\omega_b \mu_i \sigma_i A_z \tag{3}$$

where i denotes different layers; e.g., i = 1 denotes the airgap, i = 2 denotes the copper shield, etc. In different layers, σ_i and μ_i have different values, depending on the material. For example, in the airgap, σ_1 is 0, and μ_1 is the permeability of the vacuum. Boundary conditions are given in Equations (4)~(6):

$$\frac{1}{\mu_1}\frac{\partial}{\partial y} A_z = cs(i,x)\Big|_{y=dz+dp+dr+d0} \tag{4}$$

$$\left.\begin{aligned} \mu_i \frac{\partial A_z}{\partial y} &= \mu_{i+1}\frac{\partial A_z}{\partial y} \\ A_{zi} &= A_{z(i+1)} \end{aligned}\right|_{1<i<6} , \ y \text{ on the border line of layer } i \text{ and } i+1 \tag{5}$$

$$A_z = 0|_{y=0} \tag{6}$$

Finally, once the vector potential A_z is obtained, Poynting's Theorem is used to calculate the loss:

$$\begin{aligned} E &= -\frac{\partial A_z}{\partial t} = jn\omega_b \cdot [C_0 \sinh(d_r + d_p + d_z) + D_0 \cosh(d_r + d_p + d_z)] \\ H &= \frac{1}{\mu_1} \cdot \frac{\partial A_z}{\partial y} = \frac{v\pi}{\mu_1 \tau}[C_0 \sinh(d_r + d_p + d_z) + D_0 \cosh(d_r + d_p + d_z)] \\ J_{Loss} &= 2\tau L_{ef} \cdot \frac{1}{2} E \cdot H^* \end{aligned} \tag{7}$$

where C_0 and D_0 are related to the specific vector potential of the motor. Detailed calculations are shown in Appendix A.

The problem relates to the way PWM voltage waves are formed: according the results in [30], the PWM modulation process can be seen as a two-variable controlled process. Taking space-vector

PWM as an example, the modulation process can be viewed as a line crossing the area formed by a modulating wave, as shown in Figure 3.

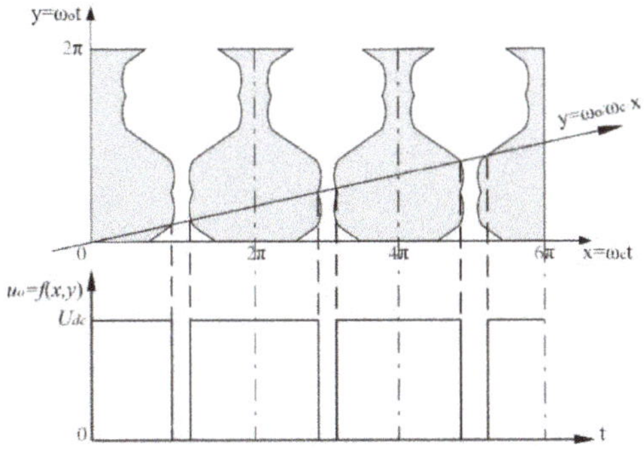

Figure 3. Graphical representation of space-vector PWM modulation process.

In Figure 3, when the line is in the shaded area, the SVPWM modulator outputs a positive voltage; otherwise, it outputs a negative voltage. The borderline of the shaded area and unshaded area is the shape of the modulation waveform. According to [31], the main frequency components of this process result in multiples of carrier wave frequency and sums of the multiples-of-carrier-wave-frequency and multiples-of-modulation-wave-frequency. The decomposition can be expressed as in Equation (8):

$$\begin{aligned} f(t) = \frac{A_{00}}{2} + \sum_{n=1}^{\infty}\left[A_{0n}\cos(\omega_0 t + \theta_0) + B_{0n}\sin(\omega_0 t + \theta_0)\right] + \\ \sum_{m=1}^{\infty}\left[A_{m0}\cos(\omega_c t + \theta_c) + B_{m0}\sin(\omega_c t + \theta_c)\right] + \\ \sum_{m=1}^{\infty}\sum_{\substack{n=-\infty \\ (n\neq 0)}}^{\infty}\left\{\begin{array}{l} A_{mn}\cos[m(\omega_c t + \theta_c) + n(\omega_o t + \theta_o)] \\ +B_{mn}\sin[m(\omega_c t + \theta_c) + n(\omega_o t + \theta_o)] \end{array}\right\} \end{aligned} \tag{8}$$

where ω_0 is the modulation wave frequency, and ω_c is the carrier wave frequency. However, ω_c is chosen so that the switching losses of power devices are within a reasonable range and need not be exact multiples of ω_0. This made it difficult for all harmonic components to find a common base. If the fundamental current frequency continues to act as the base for time-FFT analysis, spectrum leakage is unavoidable, making the eddy current loss calculation inaccurate. Furthermore, while traditional methods only consider the frequencies of harmonic currents, it will be seen in Section 5 that only one part of the harmonic order pair determines the travelling direction of magneto waves. This error is especially serious in high-speed applications because the modulation wave frequency is relatively high compared to conventional machines. To solve these problems, improved methods for the decomposition of PWM voltage and a novel estimation algorithm to improve the ECL estimation accuracy with different travelling wave direction judgement methods are proposed in this article.

3. SVPWM Working Principle and Expression Deduction of Modulation Wave Per Phase

SVPWM works based on the volt–second balance principle. As depicted in Figure 4, there are six effective vectors and two zero vectors in the stationary plane, and we divide the plane into six sectors. Under SVPWM, the target space vector is synthesized using two neighboring effective vectors and zero vectors.

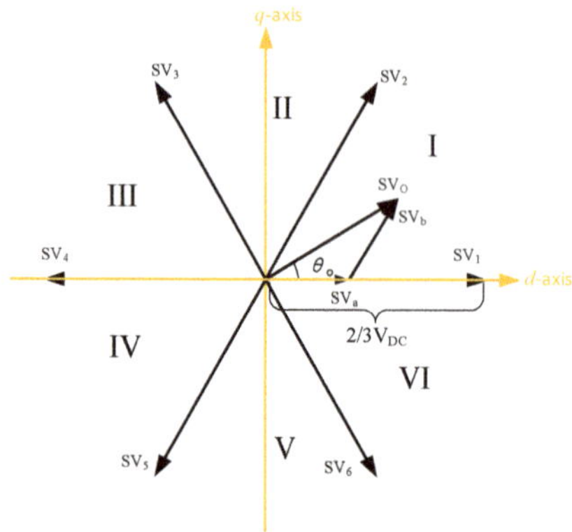

Figure 4. Synthesis of target space vector under SVPWM modulation.

Suppose the target space vector lies in sector I in Figure 4; the synthesis principle can then be expressed as in Equation (9).

$$\overline{SV_O} = \frac{T_{SV1}}{\Delta T/2}\overline{SV_1} + \frac{T_{SV2}}{\Delta T/2}\overline{SV_2} \tag{9}$$

If target space vector is expressed in complex form, as in Equation (10),

$$\overline{SV_O^*} = U_o e^{j\theta_o}, \tag{10}$$

then the length of the two effective space vectors of which the target space vector is composed can be derived as in Equation (11):

$$\begin{array}{l} \left|\overline{SV_a}\right| = \frac{T_{SV1}}{\Delta T/2}\left|\overline{SV_1}\right| = \frac{U_O}{\sqrt{3}/2}\sin(60^\circ - \theta_o) \\ \left|\overline{SV_b}\right| = \frac{T_{SV2}}{\Delta T/2}\left|\overline{SV_2}\right| = \frac{U_O}{\sqrt{3}/2}\sin\theta_o \end{array} \tag{11}$$

The action time of two effective vectors can be derived by solving Equation (11), and the result is Equation (12):

$$\begin{array}{ll} T_{SV1} & = \frac{\sqrt{3}}{2}\frac{U_o \cdot \Delta T}{V_{DC}}\sin(60^\circ - \theta_o) \\ & = \frac{\sqrt{3}}{2}\frac{U_o \cdot \Delta T}{V_{DC}}\cos(\theta_o + \frac{\pi}{6}) \\ T_{SV2} & = \frac{\sqrt{3}}{2}\frac{U_o \cdot \Delta T}{V_{DC}}\sin\theta_o \\ & = \frac{\sqrt{3}}{2}\frac{U_o \cdot \Delta T}{V_{DC}}\cos(\theta_o - \frac{\pi}{2}) \end{array} \tag{12}$$

Finally, the action times of the effective space vectors in other sectors are derived in a similar way, as listed in Table 1.

Table 1. Action time of effective space vectors under SVPWM.

$\theta_o = \omega_o t$	Active Space Vectors	Effect Time Span
$0 < \theta_o \leq \pi/3$	$\overline{SV_1} = A\overline{BC}$ $\overline{SV_2} = AB\overline{C}$	$\frac{\sqrt{3}}{2}\frac{U_o \cdot \Delta T}{V_{DC}}\cos(\theta_o + \frac{\pi}{6})$ $\frac{\sqrt{3}}{2}\frac{U_o \cdot \Delta T}{V_{DC}}\cos(\theta_o - \frac{\pi}{2})$
$\pi/3 < \theta_o \leq 2\pi/3$	$\overline{SV_2} = AB\overline{C}$ $\overline{SV_3} = \overline{A}B\overline{C}$	$\frac{\sqrt{3}}{2}\frac{U_o \cdot \Delta T}{V_{DC}}\cos(\theta_o - \frac{\pi}{6})$ $\frac{\sqrt{3}}{2}\frac{U_o \cdot \Delta T}{V_{DC}}\cos(\theta_o - \frac{5\pi}{6})$
$2\pi/3 < \theta_o \leq \pi$	$\overline{SV_3} = \overline{A}B\overline{C}$ $\overline{SV_4} = \overline{A}BC$	$\frac{\sqrt{3}}{2}\frac{U_o \cdot \Delta T}{V_{DC}}\cos(\theta_o - \frac{\pi}{2})$ $\frac{\sqrt{3}}{2}\frac{U_o \cdot \Delta T}{V_{DC}}\cos(\theta_o - \frac{7\pi}{6})$

Table 1. Cont.

$\theta_o = \omega_o t$	Active Space Vectors	Effect Time Span
$\pi < \theta_o \le 4\pi/3$	$\overline{SV_4} = \overline{A}BC$ $\overline{SV_5} = \overline{AB}C$	$\frac{\sqrt{3}}{2}\frac{U_o \cdot \Delta T}{V_{DC}}\cos(\theta_o - \frac{5\pi}{6})$ $\frac{\sqrt{3}}{2}\frac{U_o \cdot \Delta T}{V_{DC}}\cos(\theta_o - \frac{3\pi}{2})$
$4\pi/3 < \theta_o \le 5\pi/3$	$\overline{SV_5} = \overline{AB}C$ $\overline{SV_6} = A\overline{B}C$	$\frac{\sqrt{3}}{2}\frac{U_o \cdot \Delta T}{V_{DC}}\cos(\theta_o - \frac{7\pi}{6})$ $\frac{\sqrt{3}}{2}\frac{U_o \cdot \Delta T}{V_{DC}}\cos(\theta_o - \frac{11\pi}{6})$
$5\pi/3 < \theta_o \le 2\pi$	$\overline{SV_6} = A\overline{B}C$ $\overline{SV_1} = A\overline{BC}$	$\frac{\sqrt{3}}{2}\frac{U_o \cdot \Delta T}{V_{DC}}\cos(\theta_o - \frac{3\pi}{2})$ $\frac{\sqrt{3}}{2}\frac{U_o \cdot \Delta T}{V_{DC}}\cos(\theta_o - \frac{\pi}{6})$

The expression of the modulation wave can then be obtained. Taking phase A as example, the expression is listed in Table 2.

Table 2. Phase A modulation wave expression under SVPWM.

$\theta_o = \omega_o t$	$f_A(\theta_o)$
$0 < \theta_o \le \pi/3$	$\frac{\sqrt{3}}{2}\frac{U_o \cdot \Delta T}{V_{DC}}\cos(\theta_o - \frac{\pi}{6})$
$\pi/3 < \theta_o \le 2\pi/3$	$\frac{3}{2}\frac{U_o \cdot \Delta T}{V_{DC}}\cos(\theta_o)$
$2\pi/3 < \theta_o \le \pi$	$\frac{\sqrt{3}}{2}\frac{U_o \cdot \Delta T}{V_{DC}}\cos(\theta_o + \frac{\pi}{6})$
$\pi < \theta_o \le 4\pi/3$	$\frac{\sqrt{3}}{2}\frac{U_o \cdot \Delta T}{V_{DC}}\cos(\theta_o - \frac{\pi}{6})$
$4\pi/3 < \theta_o \le 5\pi/3$	$\frac{3}{2}\frac{U_o \cdot \Delta T}{V_{DC}}\cos(\theta_o)$
$5\pi/3 < \theta_o \le 2\pi$	$\frac{\sqrt{3}}{2}\frac{U_o \cdot \Delta T}{V_{DC}}\cos(\theta_o + \frac{\pi}{6})$

4. Refined SVPWM Frequency Spectrum Structure

Once the expression of the modulation wave is derived, the frequency spectrum structure of the voltage pulses produced by SVPWM can also be derived using double Fourier analysis (DFA). Generally, the pulses produced by SVPWM can be decomposed into a series of sine wave bands, namely the base band, carrier wave bands and sideband harmonics. This can be expressed as in Equation (13):

$$\begin{aligned} f(t) = \frac{A_{00}}{2} + &\sum_{n=1}^{\infty} [A_{0n}\cos(\omega_0 t + \theta_0) + B_{0n}\sin(\omega_0 t + \theta_0)] + \\ &\sum_{m=1}^{\infty} [A_{m0}\cos(\omega_c t + \theta_c) + B_{m0}\sin(\omega_c t + \theta_c)] + \\ \sum_{m=1}^{\infty} &\sum_{\substack{n=-\infty \\ (n \neq 0)}}^{\infty} \left\{ \begin{array}{l} A_{mn}\cos[m(\omega_c t + \theta_c) + n(\omega_o t + \theta_o)] \\ +B_{mn}\sin[m(\omega_c t + \theta_c) + n(\omega_o t + \theta_o)] \end{array} \right\} \end{aligned} \tag{13}$$

The coefficients can be obtained using Table 2, shown as Equation (14):

$$A_{mn} + jB_{mn} = \frac{1}{2\pi^2}\sum_{i=1}^{6}\int_{y_s(i)}^{y_e(i)}\int_{x_r(i)}^{x_f(i)} 2V_{dc}e^{j(mx+ny)}dxdy \tag{14}$$

Suppose the modulation ratio $M = U_o/V_{DC}$; then, the integral limits of the coefficients are listed as shown in Table 3.

Table 3. Integral limits for modulation wave per phase under SVPWM.

i	$y_s(i)$	$y_e(i)$	$x_r(i)$	$x_f(i)$
1	0	$\pi/3$	$-\frac{\pi}{2}\left[1+\frac{\sqrt{3}}{2}M\cos(\theta_o-\frac{\pi}{6})\right]$	$\frac{\pi}{2}\left[1+\frac{\sqrt{3}}{2}M\cos(\theta_o-\frac{\pi}{6})\right]$
2	$\pi/3$	$2\pi/3$	$-\frac{\pi}{2}\left[1+\frac{3}{2}M\cos(\theta_o)\right]$	$\frac{\pi}{2}\left[1+\frac{3}{2}M\cos(\theta_o)\right]$
3	$2\pi/3$	π	$-\frac{\pi}{2}\left[1+\frac{\sqrt{3}}{2}M\cos(\theta_o+\frac{\pi}{6})\right]$	$\frac{\pi}{2}\left[1+\frac{\sqrt{3}}{2}M\cos(\theta_o+\frac{\pi}{6})\right]$
4	π	$4\pi/3$	$-\frac{\pi}{2}\left[1+\frac{\sqrt{3}}{2}M\cos(\theta_o-\frac{\pi}{6})\right]$	$\frac{\pi}{2}\left[1+\frac{\sqrt{3}}{2}M\cos(\theta_o-\frac{\pi}{6})\right]$
5	$4\pi/3$	$5\pi/3$	$-\frac{\pi}{2}\left[1+\frac{3}{2}M\cos(\theta_o)\right]$	$\frac{\pi}{2}\left[1+\frac{3}{2}M\cos(\theta_o)\right]$
6	$5\pi/3$	2π	$-\frac{\pi}{2}\left[1+\frac{\sqrt{3}}{2}M\cos(\theta_o+\frac{\pi}{6})\right]$	$\frac{\pi}{2}\left[1+\frac{\sqrt{3}}{2}M\cos(\theta_o+\frac{\pi}{6})\right]$

Through cumbersome simplification, the analytical solution of the voltage harmonics is shown in Equation (15):

$$\begin{aligned} f_A(t) = \frac{A_{00}}{2} + \sum_{n=1}^{\infty}\left[A_{0n}\cos(\omega_0 t+\theta_0) + B_{0n}\sin(\omega_0 t+\theta_0)\right] + \\ \sum_{m=1}^{\infty}\left[A_{m0}\cos(\omega_c t+\theta_c) + B_{m0}\sin(\omega_c t+\theta_c)\right] + \\ \sum_{m=1}^{\infty}\sum_{\substack{n=-\infty \\ (n\neq 0)}}^{\infty}\left\{\begin{array}{l} A_{mn}\cos[m(\omega_c t+\theta_c)+n(\omega_o t+\theta_o)] \\ +B_{mn}\sin[m(\omega_c t+\theta_c)+n(\omega_o t+\theta_o)] \end{array}\right\} \end{aligned} \tag{15}$$

where

$$A_{mn}+jB_{mn} = \frac{8V_{dc}}{m\pi^2}\left[\begin{array}{l} \frac{\pi}{6}\sin\left([m+n]\frac{\pi}{2}\right)\left\{J_n\left(m\frac{3\pi}{4}M\right)+2\cos n\frac{\pi}{6}\cdot J_n\left(m\frac{\sqrt{3}\pi}{4}M\right)\right\} \\ +\frac{1}{n}\sin m\frac{\pi}{2}\cos n\frac{\pi}{2}\sin n\frac{\pi}{6}\left\{J_0\left(m\frac{3\pi}{4}M\right)-J_0\left(m\frac{\sqrt{3}\pi}{4}M\right)\right\}\Big|_{n\neq 0} \\ +\sum_{\substack{k=1 \\ k\neq -n}}^{\infty}\left[\begin{array}{l} \frac{1}{n+k}\sin\left([m+k]\frac{\pi}{2}\right)\cos\left([n+k]\frac{\pi}{2}\right)\sin\left([n+k]\frac{\pi}{6}\right) \\ \cdot\left\{J_k\left(m\frac{3\pi}{4}M\right)+2\cos\left([2n+3k]\frac{\pi}{6}\right)J_k\left(m\frac{\sqrt{3}\pi}{4}M\right)\right\} \end{array}\right] \\ +\sum_{\substack{k=1 \\ k\neq n}}^{\infty}\left[\begin{array}{l} \frac{1}{n-k}\sin\left([m+k]\frac{\pi}{2}\right)\cos\left([n-k]\frac{\pi}{2}\right)\sin\left([n-k]\frac{\pi}{6}\right) \\ \cdot\left\{J_k\left(m\frac{3\pi}{4}M\right)+2\cos\left([2n-3k]\frac{\pi}{6}\right)J_k\left(m\frac{\sqrt{3}\pi}{4}M\right)\right\} \end{array}\right] \end{array}\right] \tag{16}$$

in which J_n denotes the n-th order Bessel function.

It can be noticed that, in Equation (16), $B_{mn} = 0$. This implies that the ideal modulation process does not change the phase of the modulation wave. Furthermore, when $m \geq 1$, $n \neq 3k$, A_{mn} does not have to be zero; however, when $m = 0$, $n \neq 3k$, A_{mn} does have to be zero. This implies that an ideal modulation process does introduce new frequency components into the modulated voltage waveform.

For phase B and C, the expressions are shown as in Equations (17) and (18),

$$\begin{aligned} f_B(t) = \frac{A_{00}}{2} + \sum_{n=1}^{\infty}\left[A_{0n}\cos(\omega_0 t+\theta_0-\tfrac{2\pi}{3}) + B_{0n}\sin(\omega_0 t+\theta_0-\tfrac{2\pi}{3})\right] + \\ \sum_{m=1}^{\infty}\left[A_{m0}\cos(\omega_c t+\theta_c) + B_{m0}\sin(\omega_c t+\theta_c)\right] + \\ \sum_{m=1}^{\infty}\sum_{\substack{n=-\infty \\ (n\neq 0)}}^{\infty}\left\{\begin{array}{l} A_{mn}\cos[m(\omega_c t+\theta_c)+n(\omega_o t+\theta_o-\frac{2\pi}{3})] \\ +B_{mn}\sin[m(\omega_c t+\theta_c)+n(\omega_o t+\theta_o-\frac{2\pi}{3})] \end{array}\right\} \end{aligned} \tag{17}$$

$$
\begin{aligned}
f_C(t) = \tfrac{A_{00}}{2} + \sum_{n=1}^{\infty}\Big[A_{0n}\cos(\omega_0 t + \theta_0 + \tfrac{2\pi}{3}) + B_{0n}\sin(\omega_0 t + \theta_0 + \tfrac{2\pi}{3})\Big] + \\
\sum_{m=1}^{\infty}[A_{m0}\cos(\omega_c t + \theta_c) + B_{m0}\sin(\omega_c t + \theta_c)] + \\
\sum_{m=1}^{\infty}\sum_{\substack{n=-\infty \\ (n\neq 0)}}^{\infty}\left\{\begin{array}{c} A_{mn}\cos[m(\omega_c t + \theta_c) + n(\omega_o t + \theta_o + \tfrac{2\pi}{3})] \\ +B_{mn}\sin[m(\omega_c t + \theta_c) + n(\omega_o t + \theta_o + \tfrac{2\pi}{3})] \end{array}\right\}
\end{aligned} \tag{18}
$$

where the coefficient $A_{mn}+jB_{mn}$ is the same as phase A.

When calculating harmonic currents, the amplitude of current can be calculated using Equation (19):

$$
i_x(t) = \sum_m \sum_n \left. \frac{\left\{\begin{array}{c} A_{mn}\cos[m(\omega_c t + \theta_c) + n(\omega_o t + \theta_o)] \\ +B_{mn}\sin[m(\omega_c t + \theta_c) + n(\omega_o t + \theta_o)] \end{array}\right\}}{(m\omega_c + n\omega_o)L_s} \right|_{x=A,B,C} ; \tag{19}
$$

however, the respective phase of voltage harmonics must be considered. From Equations (17) and (18), the following can be seen:

1. When $n = 3k$, the coefficients in $f_a(t)$, $f_b(t)$ and $f_c(t)$ are the same. Thus, the harmonic voltage does not produce a respective resulting harmonic current.
2. Switching harmonics does not exist in phase currents; i.e., the voltage harmonics bearing the form in Equation (20)

$$
\sum_{m=1}^{\infty}[A_{m0}\cos(\omega_c t + \theta_c) + B_{m0}\sin(\omega_c t + \theta_c)] \tag{20}
$$

do not produce current harmonics. This is because the coefficients of such voltage harmonics are the same in all three phases.

3. Voltage harmonics with $m \pm n$ being even do not produce current harmonics either. The reasons for this are as follows. Components bearing the form in Equation (21)

$$
\frac{\pi}{6}\sin\!\left([m+n]\frac{\pi}{2}\right)\left\{J_n\!\left(m\frac{3\pi}{4}M\right) + 2\cos n\frac{\pi}{6}\cdot J_n\!\left(m\frac{\sqrt{3}\pi}{4}M\right)\right\} \tag{21}
$$

will not produce current harmonics because $\sin\!\left([m+n]\frac{\pi}{2}\right)$=0. Components bearing the form in Equation (22)

$$
\left.\frac{1}{n}\sin m\frac{\pi}{2}\cos n\frac{\pi}{2}\sin n\frac{\pi}{6}\left\{J_0\!\left(m\frac{3\pi}{4}M\right) - J_0\!\left(m\frac{\sqrt{3}\pi}{4}M\right)\right\}\right|_{n\neq 0} \tag{22}
$$

will not produce current harmonics because $\sin m\frac{\pi}{2}\cos n\frac{\pi}{2} = 0$ when m and n are both even or odd, which is a synonym of $m \pm n$ being even. Components bearing the form in Equation (23)

$$
\sum_{\substack{k=1 \\ k\neq -n}}^{\infty}\left[\begin{array}{c} \frac{1}{n+k}\sin\!\left([m+k]\frac{\pi}{2}\right)\cos\!\left([n+k]\frac{\pi}{2}\right)\sin\!\left([n+k]\frac{\pi}{6}\right) \\ \cdot\left\{J_k\!\left(m\frac{3\pi}{4}M\right) + 2\cos\!\left([2n+3k]\frac{\pi}{6}\right)J_k\!\left(m\frac{\sqrt{3}\pi}{4}M\right)\right\} \end{array}\right] \tag{23}
$$

will not produce current harmonics because $\sin\left([m+k]\frac{\pi}{2}\right)\cos\left([n+k]\frac{\pi}{2}\right)=0$ when m and n are both even or odd, which is another synonym of $m \pm n$ being even. Components bearing the form in Equation (24)

$$\sum_{\substack{k=1 \\ k \neq n}}^{\infty} \left[\begin{array}{c} \frac{1}{n-k}\sin\left([m+k]\frac{\pi}{2}\right)\cos\left([n-k]\frac{\pi}{2}\right)\sin\left([n-k]\frac{\pi}{6}\right) \\ \cdot\left\{J_k\left(m\frac{3\pi}{4}M\right)+2\cos\left([2n-3k]\frac{\pi}{6}\right)J_k\left(m\frac{\sqrt{3}\pi}{4}M\right)\right\} \end{array} \right] \tag{24}$$

will not produce current harmonics; the reason for this is similar.

5. ECL Estimation Algorithm with Different Travelling Current Sheet Direction Judgement Methods

The travelling direction of current sheets partly depends on the space harmonics and partly on current harmonics. For simplicity, we suppose that space harmonics are only caused by a non-sinusoidal property of stator winding. Other sources of space harmonics, including non-constant permeability, are neglected.

We take a two-pole, 12-slot permanent magnet synchronous machine as the study object. The cross-section of this machine is shown in Figure 5.

Figure 5. Structure of target permanent magnet synchronous machine.

The winding structure of phase A can be expressed as in Figure 6.

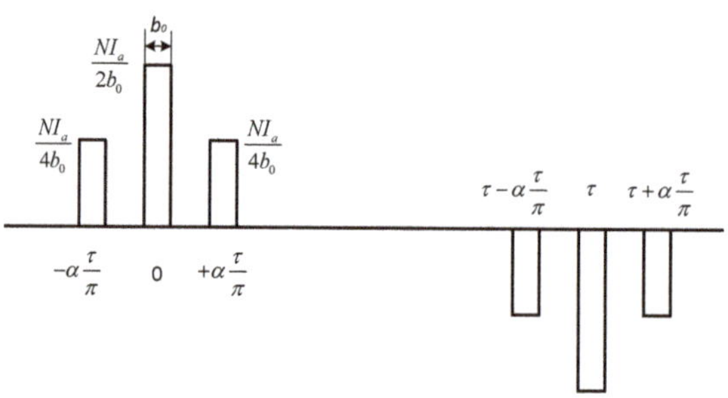

Figure 6. Winding structure of phase A.

In Figure 6,
α denotes the electrical angle between adjacent slot opening;
N denotes the winding turns of a single phase;
b_0 denotes the width of one slot opening;
τ denotes the pole pitch; and

I_a denotes the current in a phase winding.

Thus, the current sheet of phase A can be expressed as in Equation (25):

$$cs_a(I_a, x) = \sum_v \frac{\sin\left(v\frac{\pi b_0}{2\tau}\right)}{v\frac{\pi b_0}{2\tau}} \cdot \frac{1}{2}[1+\cos(v\cdot\alpha)]\cdot\sin\left(-\frac{v\pi}{2}\right)\cdot\frac{2}{\tau}N\cdot I_a\cdot\sin\left(v\frac{\pi}{\tau}x\right) \tag{25}$$

Similarly, the current sheet expressions of phase B and phase C are listed in Equations (26) and (27):

$$cs_b(I_b, x) = \sum_v \frac{\sin\left(v\frac{\pi b_0}{2\tau}\right)}{v\frac{\pi b_0}{2\tau}} \cdot \frac{1}{2}[1+\cos(v\cdot\alpha)]\cdot\sin\left(-\frac{v\pi}{2}\right)\cdot\frac{2}{\tau}N\cdot I_b\cdot\sin\left(v\frac{\pi}{\tau}x-\frac{2}{3}v\pi\right) \tag{26}$$

$$cs_c(I_c, x) = \sum_v \frac{\sin\left(v\frac{\pi b_0}{2\tau}\right)}{v\frac{\pi b_0}{2\tau}} \cdot \frac{1}{2}[1+\cos(v\cdot\alpha)]\cdot\sin\left(-\frac{v\pi}{2}\right)\cdot\frac{2}{\tau}N\cdot I_c\cdot\sin\left(v\frac{\pi}{\tau}x+\frac{2}{3}v\pi\right) \tag{27}$$

As mentioned in Section 4, the current waveforms can be expressed as in Equation (19). Then, the overall travelling current sheet can be derived by adding three current sheets:

$$\begin{aligned} cs_{tot}(i,x) &= cs_a(i,x)+cs_b(i,x)+cs_c(i,x) \\ &= \sum_v K_{sov}K_{dpv}\tfrac{2}{\tau}\sin\left(-\tfrac{v\pi}{2}\right)N\cdot \\ &\sum_m\sum_n \frac{I_{mn}}{2}\cdot\left\{\begin{array}{c} \sin(m\omega_c t+n\omega_o t+v\frac{\pi}{\tau}x+\varphi)+ \\ \sin(m\omega_c t+n\omega_o t-v\frac{\pi}{\tau}x+\varphi)+ \\ \sin\left(m\omega_c t+n\omega_o t+v\frac{\pi}{\tau}x-\frac{2}{3}(v+n)\pi+\varphi\right)+ \\ \sin\left(m\omega_c t+n\omega_o t-v\frac{\pi}{\tau}x+\frac{2}{3}(v-n)\pi+\varphi\right)+ \\ \sin\left(m\omega_c t+n\omega_o t+v\frac{\pi}{\tau}x+\frac{2}{3}(v+n)\pi+\varphi\right)+ \\ \sin\left(m\omega_c t+n\omega_o t-v\frac{\pi}{\tau}x-\frac{2}{3}(v-n)\pi+\varphi\right) \end{array}\right\} \end{aligned} \tag{28}$$

where

$$\begin{aligned} K_{sov} &= \frac{\sin\left(v\frac{\pi b_0}{2\tau}\right)}{v\frac{\pi b_0}{2\tau}} \\ K_{dpv} &= \tfrac{1}{2}[1+\cos(v\alpha)] \\ \varphi &= m\cdot\theta_c+n\cdot\theta_o+\arctan\tfrac{A_{mn}}{B_{mn}} \end{aligned} \tag{29}$$

It can be seen that only the current harmonic order n has an influence on the direction of the travelling wave, while current harmonic order m has nothing to do with the travelling wave direction. Further, this can be divided into two situations:

1. (v+n) can be divided by three, while (v-n) cannot:

$$cs_{tot}(i,x) = \sum_v K_{sov}K_{dpv}\frac{2}{\tau}\sin\left(-\frac{v\pi}{2}\right)N\cdot\sum_m\sum_n\frac{3}{2}I_{mn}\cdot\sin\left(m\omega_c t+n\omega_o t+v\frac{\pi}{\tau}x+\varphi\right) \tag{30}$$

In this situation, the travelling wave travels backwards with regard to the rotor. Suppose the velocity of the outer rotor of the permanent magnet synchronous machine is v_r; then, the travel velocity v_{tr} of the travelling current sheet is expressed as in Equation (31):

$$v_{tr} = \frac{m\cdot\omega_c+n\cdot\omega_o}{v\cdot\pi}\tau+v_r \tag{31}$$

2. (v-n) can be divided by three, while (v+n) cannot:

$$cs_{tot}(i,x) = \sum_v K_{sov}K_{dpv}\frac{2}{\tau}\sin\left(-\frac{v\pi}{2}\right)N\cdot\sum_m\sum_n\frac{3}{2}I_{mn}\cdot\sin\left(m\omega_c t+n\omega_o t-v\frac{\pi}{\tau}x+\varphi\right) \tag{32}$$

In this situation, the travelling wave travels forwards with regard to the rotor. Suppose the velocity of the outer rotor of the synchronous machine is v_r; then, the travel velocity v_{tr} of the travelling current sheet is expressed as in Equation (33):

$$v_{tr} = \frac{m \cdot \omega_c + n \cdot \omega_o}{v \cdot \pi} \tau - v_r \tag{33}$$

Finally, a novel ECL estimation algorithm can be developed. The key steps are shown in Figure 7.

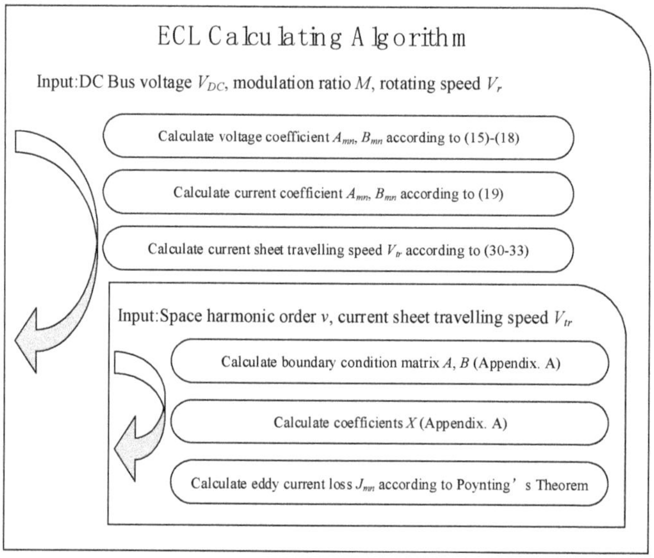

Figure 7. Novel ECL calculation algorithm.

6. Simulation Study and Experiments

A 2 kW surface-mounted permanent magnet synchronous machine is studied. Its main characteristics are listed in Table 4:

Table 4. Main parameters for the object PMSM.

Parameter	Value
Poles/slots	2/12
Stator inner diameter	36 mm
Rotor outer diameter	30.2 mm
Air gap flux density	0.4 T
Effective length	50 mm
Magnet material	N40EH
Phase inductance	48 μH

The winding structure is shown in Figure 8, and relevant parameters are listed in Table 5.

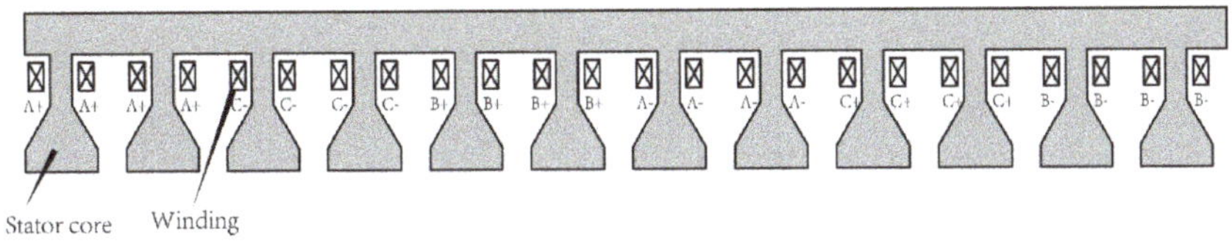

Figure 8. Winding structure of the object small inductance PMSM.

Table 5. Structure parameters for the object PMSM.

Parameter	Value
Width of slot opening	0.8 mm
Winding turns of a single phase	12
Electrical angle between adjacent slot opening	30°
Pole pitch	56.6 mm

To highlight the performance of the proposed algorithm, the switching frequency of the inverter is chosen at 10 kHz, which is relatively low compared with the phase inductance and harmonic currents, which would be large as a consequence. Firstly, the voltage and current spectrum prediction methods are verified using a current probe and an oscilloscope. Secondly, the conventional method for calculating the ECLs is conducted against the newly proposed method. Because experimental verification of ECLs is subject to heat conduction, we are unable to separate the mechanic friction loss and other interferences, making it very hard to be accurate; thus, both methods are verified by comparison with FEM results. The FEM software chosen is Flux 2D from Altair, which is widely used in the motor design industry.

The experimental platform is shown in Figure 9.

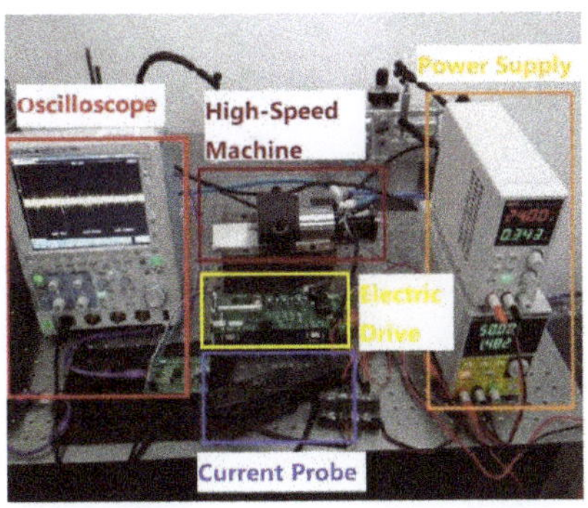

Figure 9. Experimental platform for the verification of current spectrum prediction.

The electric drive is set to the sensorless vector control work mode. To verify the performance of the proposed method under integer carrier wave ratios and non-integer carrier wave ratios, the base current frequency is chosen at 200 Hz and 240 Hz, with a carrier wave ratio of 50 and 46.667, respectively. Under the SVPWM method, the phase voltage spectrum is shown in Figure 10, and the predicted current distribution is shown in Figure 11.

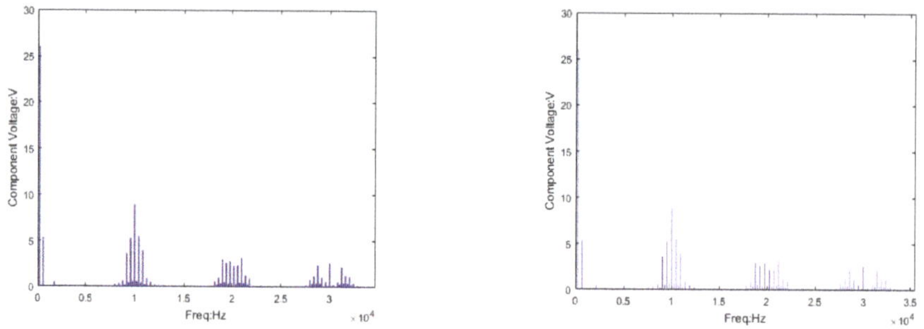

Figure 10. Phase voltage spectrum at a base frequency of 200 Hz (left) and 240 Hz (right) under SVPWM modulation. The DC bus voltage is 30 V.

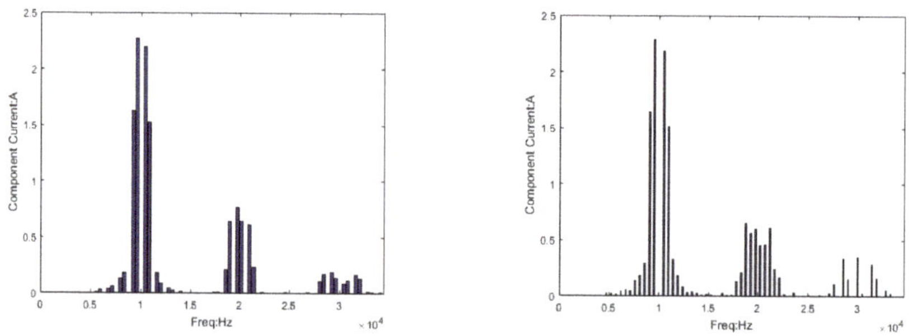

Figure 11. Phase harmonic current prediction of 200 Hz (left) and 240 Hz (right) under SVPWM modulation. The DC bus voltage is 30 V and the phase inductance is 48 μH.

It can be seen in Figure 11 that, when the base current frequency changed from 200 Hz to 240 Hz, the diversity of the current frequency components and their amplitudes increased greatly. The experimental results are shown in Figure 12.

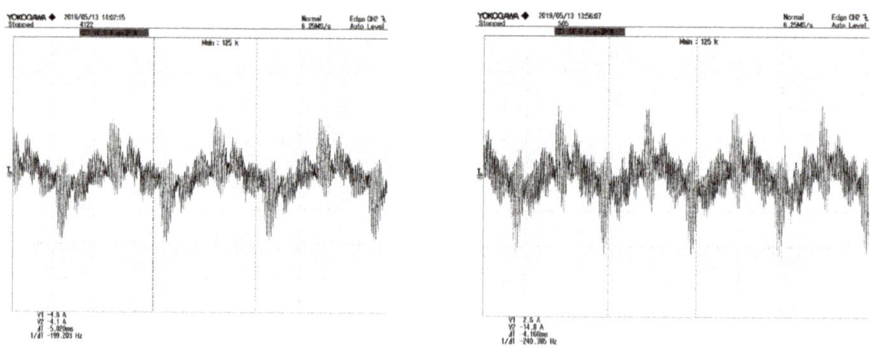

(a) (b)

Figure 12. Experimental current waveform under sensorless vector control mode. The base current frequency is 200 Hz in (**a**) and 240 Hz in (**b**).

To make a comparison of the predictive performance of the newly proposed method (DFA) and traditional analysis method (FFT), analyses by both methods are given in Figure 13. Nevertheless, a spectrum analysis of the experimental current waveform is also performed and used as the standard of performance.

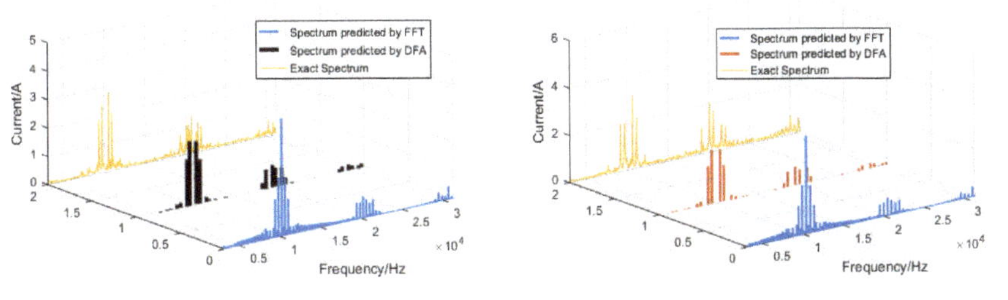

(a) **(b)**

Figure 13. Performance comparison of current spectrum prediction by newly proposed method (double Fourier analysis (DFA)) and traditional fast Fourier transform (FFT). The base current frequency is 200 Hz in (**a**) and 24 0Hz in (**b**). The experimental current spectrum served as the standard.

From Figure 13, it can be seen that traditional methods using FFT decomposition results cannot tell the phase differences among different harmonic components. Thus, traditional methods mispredict some frequency components that do not exist under real conditions. The method proposed in this article, however, fits the experimental results well, proving the effectiveness of the novel method.

The impact of different travelling directions of travelling current sheets on the prediction accuracy of ECLs are verified using FEM software. Firstly, the traditional method based on FFT and the method based on simulation which consider the travelling direction as relative to harmonic frequency are conducted. Then, the method proposed in this article, which uses different travelling direction judgement methods, is conducted, and finally all methods are compared to the results of the FEM software with experimental current waveforms. The results are shown in Figures 14 and 15.

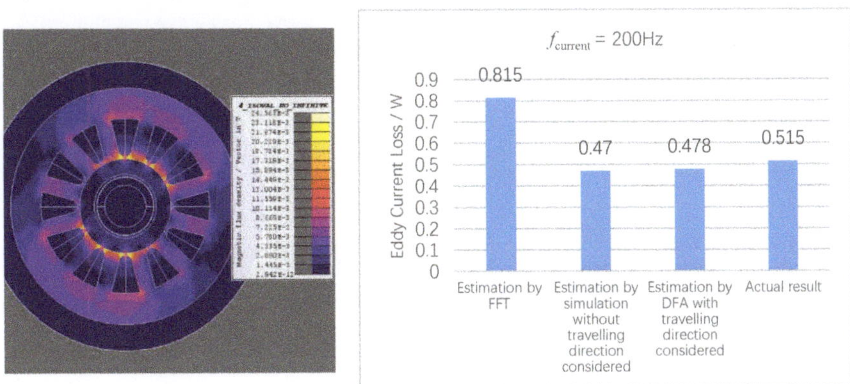

(a) **(b)**

Figure 14. Eddy current losses estimated with different methods. The base current frequency is 200 Hz. The FEM software calculation is shown in (**a**), and data are listed in (**b**).

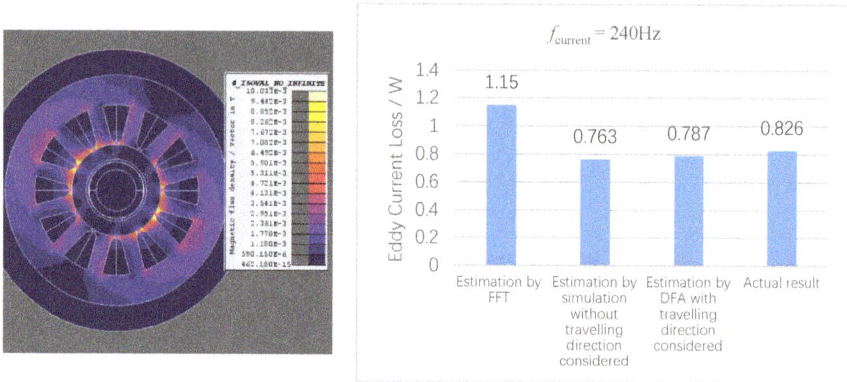

(a) (b)

Figure 15. Eddy current losses estimated with different methods. The base current frequency is 240 Hz. The FEM software calculation is shown in (**a**), and data are listed in (**b**).

From Figures 14 and 15, it can be seen that calculation methods based on FFT usually get larger results than reality, while methods based on simulation and DFA obtain results close to the actual result. However, with different travelling direction judgement methods, the ECL estimation scheme proposed in this article showed better accuracy, with improvements up to 1.9%, which proved its effectiveness.

7. Conclusions

In this article, a novel rotor ECL estimation method with improved accuracy for small inductance permanent magnet synchronous machines under SVPWM supply is proposed. A double Fourier analysis-based SVPWM harmonics decomposition algorithm is proposed to replace traditional FFT-based or simulation-based methods, which enables the phase sequence detection of current harmonics. Based on this extra information, a new method of magneto-wave travelling direction judgement is proposed. Regarding the transformation of voltage harmonics to current harmonics, an elimination method by the analysis of the coefficients of different harmonic current components make the misprediction phenomenon disappear compared to FFT-based methods. Finally, these amendments are applied to the 2D analytical motor model, and simulations and experiments showed that the proposed algorithm achieved better accuracy compared with traditional estimation methods after these modifications.

Author Contributions: Conceptualization, L.P. and Q.G.; methodology, L.P.; software, L.P.; validation, R.Y., P.D. and Q.G.; formal analysis, L.P.; investigation, L.L.; resources, L.L.; writing—original draft preparation, L.P.; writing—review and editing, L.L.

Funding: This research was funded by the National Natural Science Foundation of China, grant No. 51577038.

Conflicts of Interest: The authors declare no conflict of interest.

Appendix A Details of the Analytical Model for the Eddy Current Loss Estimation

Due to the small slot opening, the magnet field strength can be viewed as parallel to the stator inner surface. Thus, the windings can be modeled as current sheets at slot openings. The current sheet of phase A can be expressed as in Equation (A1):

$$cs_a(i,x) = \sum_{v} \frac{\sin\left(v\frac{\pi b_0}{2\tau}\right)}{v\frac{\pi b_0}{2\tau}} \cdot \frac{1}{2}[1+\cos(v\cdot\alpha)]\cdot\sin(-\frac{v\pi}{2})\cdot\frac{2}{\tau}N\cdot i_a(t)\cdot\sin(v\frac{\pi}{\tau}x) \quad \text{(A1)}$$

Similarly, the current sheet of phase B and phase C can be expressed as in Equations (A2) and (A3):

$$cs_b(i,x) = \sum_v \frac{\sin\left(v\frac{\pi b_0}{2\tau}\right)}{v\frac{\pi b_0}{2\tau}} \cdot \frac{1}{2}[1+\cos(v\cdot\alpha)]\cdot\sin(-\frac{v\pi}{2})\cdot\frac{2}{\tau}N\cdot i_b(t)\cdot\sin(v\frac{\pi}{\tau}x-\frac{2}{3}v\pi) \tag{A2}$$

$$cs_c(i,x) = \sum_v \frac{\sin\left(v\frac{\pi b_0}{2\tau}\right)}{v\frac{\pi b_0}{2\tau}} \cdot \frac{1}{2}[1+\cos(v\cdot\alpha)]\cdot\sin(-\frac{v\pi}{2})\cdot\frac{2}{\tau}N\cdot i_c(t)\cdot\sin(v\frac{\pi}{\tau}x+\frac{2}{3}v\pi) \tag{A3}$$

where b_0 is the slot opening width, τ is the pole pitch, and α is the winding distribution angle.

Time harmonics in phase currents can be decomposed as in Equation (A4):

$$\begin{aligned} i_a(t) &= \sum_n I_n\cos(n\omega_b t) \\ i_b(t) &= \sum_n I_n\cos(n\omega_b t - n\cdot\tfrac{2}{3}\pi) \\ i_c(t) &= \sum_n I_n\cos(n\omega_b t + n\cdot\tfrac{2}{3}\pi) \end{aligned} \tag{A4}$$

We define $K_{sov} = \frac{\sin\left(v\frac{\pi b_0}{2\tau}\right)}{v\frac{\pi b_0}{2\tau}}$, $K_{dpv} = \frac{1}{2}[1+\cos(v\alpha)]$, and the total travelling wave current sheet can be expressed as in Equation (A5):

$$\begin{aligned} cs_{tot}(i,x) &= cs_a(i,x)+cs_b(i,x)+cs_c(i,x) \\ &= \sum_v K_{sov}K_{dpv}\tfrac{2}{\tau}\sin(-\tfrac{v\pi}{2})N\cdot \\ &\quad \sum_n \tfrac{I_n}{2}\cdot\left\{\begin{array}{l} \sin(n\omega_b t+v\frac{\pi}{\tau}x)-\sin(n\omega_b t-v\frac{\pi}{\tau}x)+ \\ \sin[n\omega_b t+v\frac{\pi}{\tau}x-(v+n)\frac{2}{3}\pi]-\sin[n\omega_b t-v\frac{\pi}{\tau}x-(n-v)\frac{2}{3}\pi]+ \\ \sin[n\omega_b t+v\frac{\pi}{\tau}x+(v+n)\frac{2}{3}\pi]-\sin[n\omega_b t-v\frac{\pi}{\tau}x+(n-v)\frac{2}{3}\pi] \end{array}\right\} \end{aligned} \tag{A5}$$

The travelling direction can be determined using the law depicted in Equation (A6): (v+n) can be divided by three, while (v-n) cannot: current sheet travels forwards:

$$cs_{tot}(i,x) = \sum_v K_{sov}K_{dpv}\frac{2}{\tau}\sin(-\frac{v\pi}{2})N\cdot\sum_n\frac{3}{2}I_n\sin(n\omega_b t - v\frac{\pi}{\tau}x) \tag{A6}$$

(v-n) can be divided by three, while (v+n) cannot: current sheet travels backwards:

$$cs_{tot}(i,x) = \sum_v K_{sov}K_{dpv}\frac{2}{\tau}\sin(-\frac{v\pi}{2})N\cdot\sum_n\frac{3}{2}I_n\sin(n\omega_b t + v\frac{\pi}{\tau}x) \tag{A7}$$

Now, consider the domaining physical rules in the motor. Since the rotor yoke is laminated, the conductivity is relatively small. For convenience, the conductivity of rotor yoke is reduced to zero. By assuming the vector potential $A_z = 0$ at the rotor shaft, the following can be obtained:

$$\begin{gathered} \textit{In the air gap}: \nabla^2 A_z = 0, \\ \textit{in the copper sleeve}: \nabla^2 A_z = j\omega\sigma_2\mu_2, \\ \textit{in the permanent magnets}: \nabla^2 A_z = j\omega\sigma_3\mu_3, \\ \textit{in the rotor yoke}: \nabla^2 A_z = 0. \end{gathered} \tag{A8}$$

The boundary conditions can be expressed as in Equation (A9):

$$\begin{aligned} H_x &= cs_{tot}(i,x)|_{y=d_0+d_r+d_p+d_z} \\ H_{x\_airgap} &= H_{x\_copper}|_{y=d_r+d_p+d_z} \\ A_{z\_airgrap} &= A_{z\_copper}|_{y=d_r+d_p+d_z} \\ H_{x\_copper} &= H_{x\_magnet}|_{y=d_p+d_z} \\ A_{z\_copper} &= A_{z\_magnet}|_{y=d_p+d_z} \\ H_{x\_magnet} &= H_{x\_iron}|_{y=d_z} \\ A_{z\_magnet} &= A_{z\_iron}|_{y=d_z} \\ A_z &= 0|_{y=0} \end{aligned} \tag{A9}$$

A solution can be found by assuming $A_z(x,y,t) = A_z(y)\cos(v\frac{\pi}{\tau}x - \omega t)$, which carries the form in Equation (A10):

$$A_{zi}(x,y,t) = [C_i \sinh(\beta_i y) + D_i \cosh(\beta_i y)]\cos(v\frac{\pi}{\tau}x - \omega t) \tag{A10}$$

where $\beta_i = \sqrt{(v\frac{\pi}{\tau})^2 + j\mu_i\sigma_i\omega}$. The subscript i denotes different parts of the motor; e.g., $i = 1$ when in an air-gap, $i = 2$ when in the copper sleeve, etc. The whole boundary condition group can be expressed as a matrix multiplication $Ax = B$, where $A =$

$$\begin{bmatrix} \frac{v\pi}{\mu_0\tau}\cosh[\frac{v\pi}{\tau}(d_0+d_r+d_p+d_z)] & \frac{v\pi}{\mu_0\tau}\sinh[\frac{v\pi}{\tau}(d_0+d_r+d_p+d_z)] & 0 & 0 & 0 & 0 & 0 \\ \sinh[\frac{v\pi}{\tau}(d_r+d_p+d_z)] & \cosh[\frac{v\pi}{\tau}(d_r+d_p+d_z)] & -\sinh[\beta_2(d_r+d_p+d_z)] & -\cosh[\beta_2(d_r+d_p+d_z)] & 0 & 0 & 0 \\ \frac{v\pi}{\mu_0\tau}\cosh[\frac{v\pi}{\tau}(d_r+d_p+d_z)] & \frac{v\pi}{\mu_0\tau}\sinh[\frac{v\pi}{\tau}(d_r+d_p+d_z)] & -\frac{\beta_2}{\mu_2}\cosh[\beta_2(d_r+d_p+d_z)] & -\frac{\beta_2}{\mu_2}\sinh[\beta_2(d_r+d_p+d_z)] & 0 & 0 & 0 \\ 0 & 0 & \sinh[\beta_2(d_p+d_z)] & \cosh[\beta_2(d_p+d_z)] & -\sinh[\beta_3(d_p+d_z)] & -\cosh[\beta_3(d_p+d_z)] & 0 \\ 0 & 0 & \frac{\beta_2}{\mu_2}\cosh[\beta_2(d_p+d_z)] & \frac{\beta_2}{\mu_2}\sinh[\beta_2(d_p+d_z)] & -\frac{\beta_3}{\mu_3}\cosh[\beta_3(d_p+d_z)] & -\frac{\beta_3}{\mu_3}sinh[\beta_3(d_p+d_z)] & 0 \\ 0 & 0 & 0 & 0 & sinh(\beta_3\cdot d_z) & \cosh(\beta_3\cdot d_z) & -sinh(\beta_4\cdot d_z) \\ 0 & 0 & 0 & 0 & \frac{\beta_3}{\mu_3}\cosh(\beta_3\cdot d_z) & \frac{\beta_3}{\mu_3}sinh(\beta_3\cdot d_z) & -\frac{\beta_4}{\mu_4}\cosh(\beta_4\cdot d_z) \end{bmatrix},$$

$$x = \begin{bmatrix} C_0 \\ D_0 \\ C_2 \\ D_2 \\ C_3 \\ D_3 \\ C_4 \end{bmatrix}, \text{ and } B = \begin{bmatrix} cs_{tot} \\ 0 \\ 0 \\ 0 \\ 0 \\ 0 \\ 0 \end{bmatrix}.$$

After the solution of A_z is found, the eddy current loss of the rotor J_{Loss} is expressed as Equation (A11):

$$\begin{aligned} E &= j\omega \cdot [C_0 \sinh(d_r + d_p + d_z) + D_0 \cosh(d_r + d_p + d_z)] \\ H &= \frac{v\pi}{\mu_0 \tau}[C_0 \sinh(d_r + d_p + d_z) + D_0 \cosh(d_r + d_p + d_z)] \\ J_{Loss} &= 2\tau L_{ef} \cdot \frac{1}{2} E \cdot H^* \end{aligned} \quad (A11)$$

References

1. Hannon, B.; Sergeant, P.; Dupré, L. Evaluation of the Rotor Eddy-Current Losses in High-Speed PMSMs With a Shielding Cylinder for Different Stator Sources. *IEEE Trans. Magn.* **2019**, *55*, 1–10. [CrossRef]
2. Riemer, B.; Lebmann, M.; Hameyer, K. Rotor design of a high-speed permanent magnet synchronous machine rating 100,000 rpm at 10 kW. In Proceedings of the 2010 IEEE Energy Conversion Congress and Exposition, Atlanta, GA, USA, 12–16 September 2010; pp. 3978–3985.
3. Weili, L.; Hongbo, Q.; Xiaochen, Z.; Ran, Y. Influence of copper plating on electromagnetic and temperature fields in a high-speed permanent-magnet generator. *IEEE Trans. Magn.* **2012**, *48*, 2247–2253. [CrossRef]
4. Li, W.; Qiu, H.; Zhang, X.; Cao, J.; Yi, R. Analyses on electromagnetic and temperature fields of superhigh-speed permanent-magnet generator with different sleeve materials. *IEEE Trans. Ind. Electron.* **2014**, *61*, 3056–3063. [CrossRef]
5. Li, L.; Czarkowski, D.; Liu, Y.; Pillay, P. Multilevel space vector PWM technique based on phase-shift harmonic suppression. In Proceedings of the Fifteenth Annual IEEE Applied Power Electronics Conference and Exposition (Cat. No. 00CH37058), New Orleans, LA, USA, 6–10 February 2000; Volume 1, pp. 535–541.
6. Yu, F.; Zhang, X.; Wang, S. Five-phase permanent magnet synchronous motor vector control based on harmonic eliminating space vector modulation. In Proceedings of the 2005 International Conference on Electrical Machines and Systems, Nanjing, China, 27–29 September 2005; Volume 1, pp. 392–396.
7. Zhou, C.; Yang, G.; Su, J. PWM strategy with minimum harmonic distortion for dual three-phase permanent-magnet synchronous motor drives operating in the overmodulation region. *IEEE Trans. Power Electron.* **2015**, *31*, 1367–1380. [CrossRef]
8. Li, X.; Chen, G. An Approach to Harmonic Suppression Based on Triple Harmonics Injection with Passive Circuit. *Autom. Electr. Power Syst.* **2007**, *14*, 61–65.
9. Zhang, Y.; Li, Y.W. Investigation and suppression of harmonics interaction in high-power PWM current-source motor drives. *IEEE Trans. Power Electron.* **2014**, *30*, 668–679. [CrossRef]
10. Zhu, Z.Q.; Ng, K.; Schofield, N.; Howe, D. Analytical prediction of rotor eddy current loss in brushless machines equipped with surface-mounted permanent magnets. I. Magnetostatic field model. In Proceedings of the Fifth International Conference on Electrical Machines and Systems, Shenyang, China, 18–20 August 2001; Volume 2, pp. 806–809.
11. Zhu, Z.Q.; Ng, K.; Schofield, N.; Howe, D. Analytical prediction of rotor eddy current loss in brushless machines equipped with surface-mounted permanent magnets. II. Accounting for eddy current reaction field. In Proceedings of the Fifth International Conference on Electrical Machines and Systems, Shenyang, China, 18–20 August 2001; Volume 2, pp. 810–813.
12. Fang, D.; Nehl, T.W. Analytical modeling of eddy-current losses caused by pulse-width-modulation switching in permanent-magnet brushless direct-current motors. *IEEE Trans. Magn.* **1998**, *34*, 3728–3736. [CrossRef]
13. Bellara, A.; Bali, H.; Belfkira, R.; Amara, Y.; Barakat, G. Analytical Prediction of Open-Circuit Eddy-Current Loss in Series Double Excitation Synchronous Machines. *IEEE Trans. Magn.* **2011**, *47*, 2261–2268. [CrossRef]
14. Paradkar, M.; Bocker, J. 2D analytical model for estimation of eddy-current loss in the magnets of IPM machines considering the reaction field of the induced eddy currents. In Proceedings of the 2015 IEEE International Electric Machines & Drives Conference (IEMDC), Coeur d'Alene, ID, USA, 10–13 May 2015; pp. 1096–1102.
15. Atallah, K.; Howe, D.; Mellor, P.; Stone, D. Rotor loss in permanent-magnet brushless AC machines. *IEEE Trans. Ind. Appl.* **2000**, *36*, 1612–1618.
16. Wang, J.; Atallah, K.; Chin, R.; Arshad, W.M.; Lendenmann, H. Rotor Eddy-Current Loss in Permanent-Magnet Brushless AC Machines. *IEEE Trans. Magn.* **2010**, *46*, 2701–2707. [CrossRef]
17. Ishak, D.; Zhu, Z.; Howe, D. Eddy-current loss in the rotor magnets of permanent-magnet brushless machines having a fractional number of slots per pole. *IEEE Trans. Magn.* **2005**, *41*, 2462–2469. [CrossRef]

18. Markovic, M.; Perriard, Y. A simplified determination of the permanent magnet (PM) eddy-current losses due to slotting in a PM rotating motor. In Proceedings of the International Conference on Electrical Machines and Systems (ICEMS), Wuhan, China, 17–20 October 2008; pp. 309–313.
19. Wu, L.J.; Zhu, Z.Q.; Staton, D.; Popescu, M.; Hawkins, D. Analytical model for predicting magnet loss of surface-mounted permanent magnet machines accounting for slotting effect and load. *IEEE Trans. Magn.* **2012**, *48*, 107–117. [CrossRef]
20. Wu, L.J.; Zhu, Z.Q.; Staton, D.; Popescu, M.; Hawkins, D. Analytical modeling of eddy current loss in retaining sleeve of surface-mounted PM machines accounting for influence of slot opening. In Proceedings of the 2012 IEEE International Symposium on Industrial Electronics, Hangzhou, China, 28–31 May 2012; pp. 611–616.
21. Rahideh, A.; Korakianitis, T. Analytical magnetic field distribution of slotless brushless permanent magnet motors—Part I. Armature reaction field, inductance and rotor eddy-current loss calculations. *IET Electr. Power Appl.* **2012**, *6*, 628. [CrossRef]
22. Yamazaki, K.; Kanou, Y. Rotor Loss Analysis of Interior Permanent Magnet Motors Using Combination of 2D and 3D Finite Element Method. *IEEE Trans. Magn.* **2009**, *45*, 1772–1775. [CrossRef]
23. Fang, Z.X.; Zhu, Z.Q.; Wu, L.J.; Xia, Z.P. Simple and accurate analytical estimation of slotting effect on magnet loss in fractional-slot surface-mounted PM machines. In Proceedings of the 2012 XXth International Conference on Electrical Machines, Marseille, France, 2–5 September 2012; Volume 37, pp. 464–470.
24. Jaffar, M.Z.M.; Husain, I. Path permeance based analytical inductance model for IPMSM considering saturation and slot leakage. In Proceedings of the 2017 IEEE International Electric Machines and Drives Conference (IEMDC), Miami, FL, USA, 21–24 May 2017; pp. 1–7.
25. Wang, J.; Papini, F.; Chin, R.; Arshad, W.M.; Lendenmann, H. Computationally efficient approaches for evaluation of rotor eddy current loss in permanent magnet brushless machines. In Proceedings of the 2009 International Conference on Electrical Machines and Systems, Tokyo, Japan, 15–18 November 2009.
26. Bode, C.; Canders, W.R. Advanced calculation of eddy-current losses in PMSM with tooth windings. In Proceedings of the XIX International Conference on Electrical Machines, Rome, Italy, 6–8 September 2010; Volume 1, pp. 1–6.
27. Pyrhonen, J.; Jussila, H.; Alexandrova, Y.; Rafajdus, P.; Nerg, J. Harmonic Loss Calculation in Rotor Surface Permanent Magnets—New Analytic Approach. *IEEE Trans. Magn.* **2012**, *48*, 2358–2366. [CrossRef]
28. Dubas, F.; Rahideh, A. Two-Dimensional Analytical Permanent-Magnet Eddy-Current Loss Calculations in Slotless PMSM Equipped With Surface-Inset Magnets. *IEEE Trans. Magn.* **2014**, *50*, 54–73. [CrossRef]
29. Martin, F.; Zaim, M.E.H.; Tounzi, A.; Bernard, N. Improved Analytical Determination of Eddy Current Losses in Surface Mounted Permanent Magnets of Synchronous Machine. *IEEE Trans. Magn.* **2014**, *50*, 1–9.
30. Arumugam, P.; Hamiti, T.; Gerada, C. Estimation of Eddy Current Loss in Semi-Closed Slot Vertical Conductor Permanent Magnet Synchronous Machines Considering Eddy Current Reaction Effect. *IEEE Trans. Magn.* **2013**, *49*, 5326–5335. [CrossRef]
31. Holmes, D.G.; Lipo, T.A. *Pulse Width Modulation for Power Converters: Principles and Practice*; John Wiley & Sons: Hoboken, NJ, USA, 2003; pp. 215–257.

energies

MDPI

Article

Magnetic Field Characteristics and Stator Core Losses of High-Speed Permanent Magnet Synchronous Motors

Dajun Tao [1,2,*], Kai Liang Zhou [1], Fei Lv [1], Qingpeng Dou [1], Jianxiao Wu [1,2], Yutian Sun [1] and Jibin Zou [3]

1 Harbin Electric Machinery Company Limited, Harbin 150040, Heilongjiang Province, China; zhoukailiang929@163.com (K.L.Z.); lvf3ycj5658@163.com (F.L.); dqpcynic@163.com (Q.D.); zcb8386@126.com (J.W.); sunyutian@hec-china.com (Y.S.)

2 National Engineering Research Center of Large Electric Machines and Heat Transfer Technology, Harbin University of Science and Technology, Harbin 150080, Heilongjiang Province, China

3 Harbin Institute of Technology, Harbin 150001, Heilongjiang Province, China; zoujibin@hit.edu.cn

* Correspondence: tao.dj@163.com; Tel.: +86-136-1364-9258

Received: 4 January 2020; Accepted: 20 January 2020; Published: 22 January 2020

Abstract: This study focuses on the core losses in the stator region of high-speed permanent magnet synchronous motors, magnetic field characteristics in the load region, and variations in iron losses caused by changes in these areas. A two-pole 120 kW high-speed permanent magnet synchronous motor is used as the object of study, and a two-dimensional transient electromagnetic field-variable load circuit combined calculation model is established. Based on electromagnetic field theory, the electromagnetic field of the high-speed permanent magnet synchronous motor under multi-load conditions is calculated using the time-stepping finite element method. The magnetic field distribution of the high-speed permanent magnet synchronous motor under a multi-load condition is obtained, and the variations in iron core losses in different parts of the motor under multi-load conditions are further analyzed. The calculation results show that most of the stator iron core losses are dissipated in the stator yoke. The stator yoke iron loss under the no-load condition exceeds 70% of the total stator iron core loss. The stator yoke iron loss under rated operation conditions exceeds 50% of the total stator iron core loss. The stator loss under rated load operation conditions is higher than that under no-load operation. These observations are sufficient to demonstrate that the running status of high-speed motors is closely related to the stator iron losses, which have significance in determining the reasonable yoke structure of high-speed and high-power motors and the cooling methods of motor stators.

Keywords: high-speed permanent synchronous motor; magnetic field characteristic; iron loss; stator structure

1. Introduction

With the increasing market demand for high-efficiency motors for driving high-speed loads, high-speed permanent magnet synchronous motors (HPMSMs) have attracted considerable attention owing to their high power density and high efficiency [1–3]. In recent years, with the development of power electronics technology, inverters have been widely used in motors [4], allowing the motors to operate over a wide range of frequencies. Hu and Gu proposed an adaptive robust three-step control method to eliminate the influence of cogging torque and model uncertainty on the tracking control of a dc motor when its speed varies nonperiodically. This method provides new ideas for various types of motors [5]. However, when the inverter is in operation, many harmonic currents are introduced, and these harmonic currents will affect the internal magnetic field of the motor [6].

The additional losses associated with these harmonics in HPMSMs are greatly increased. This inverter characteristic will reduce the efficiency of a motor, and this reduction is especially pronounced in HPMSMs. Another problem inherent in the addition of harmonic currents is that the eddy current loss generated in the permanent magnets will cause the local temperature of the permanent magnets to rise along with the required increase in power density for high-speed operation that increases the heat load per unit volume. These two sources of additional heat generation present greater difficulties for the heat dissipation of the motor and will even cause local irreversible demagnetization to occur, which will affect the service life and reliability of the motor [7,8].

The losses caused by harmonics when the inverter is operated are called additional losses, and they contain four parts: ① additional losses in the windings. When the higher order harmonic current passes through the winding, the current density on its cross-section is distributed to the outer surface of the conductor, thus, the equivalent resistance increases and a skin effect is generated. The resulting loss increase is called additional winding loss; ② additional core iron losses. The additional iron loss is the increase in iron loss caused by the magnetic field generated by the harmonic currents alternating in the stator and rotor cores; ③ eddy current losses of the permanent magnet. The magnetic fields generated by the harmonic currents are not synchronized with the rotor rotation speed, so that the harmonic magnetic field alternates in the permanent magnet, causing eddy current loss to occur; ④ the eddy current losses of surrounding structures. The magnetic field generated by the harmonic currents will alternate in the surrounding metal structure, causing eddy current loss. High-speed motors are usually driven by high-frequency inverters, and the voltage output by the inverters is rich in higher-order harmonics [9]. The iron loss caused by high-frequency power supplies accounts for a large proportion of the high-speed motor losses. Accurate analysis of the magnetic field characteristics of high-speed motors operating under multiple operating conditions of high-frequency inverter power supplies can provide important support for this study and the analysis of motor iron losses, as well as providing the basis for the accurate calculation of motor core iron losses.

Many scholars, both foreign and domestic, have performed detailed research on motor iron loss [10–20]. Much of this research has been concerned with analytical methods and modulation ratios. Sun Ming's research [10] established an analytical model of the no-load air-gap magnetic field of the axial flux permanent magnet motor, carried out the analytical calculations, and compared the results with finite element calculation results. Some researchers [11,12] analyzed the root cause of rotor eddy current loss, and calculated this loss using the finite element method. The authors in [13] studied the influence of the modulation ratio and carrier ratio on stator losses of permanent magnet synchronous generators by using the two-dimensional field-circuit coupling time-step finite element method. A study [14] deduced the analytical algorithm of the motor core loss under the power supply of a Pulse Width Modulation (PWM) inverter and determined the relationship between the modulation ratio and the iron loss from the perspective of the analytical formula. The results showed that the larger the modulation ratio, the smaller the iron loss. There are also many studies detailing the relationship between fundamental waves and harmonics. The authors in [15] studied the distribution characteristics of the fundamental wave and harmonic iron loss in the stator and rotor core of asynchronous motors, analyzed the waveforms of the magnetic flux density at different positions of the iron core over time during no-load operation, and obtained the different areas of the iron core loss distribution of iron consumption. The authors in [16] performed a harmonic analysis of the magnetomotive force of the motor and used the two-dimensional finite element method to calculate the eddy current losses of the stator and the rotor core and the eddy current losses in the permanent magnet. Yamazaki built a time-step finite element model of a high-speed asynchronous motor and analyzed the iron core loss caused by higher harmonic magnetic fields [17]. Stators and rotors of different materials have specified reference values for the study of iron loss. Many scholars are also studying these various materials and their properties. Denis [18] carried out experiments on a permanent magnet synchronous motor using a nanocrystalline magnetic material as a stator. The results show that compared with an equivalent motor using a conventional non-oriented silicon steel stator core, the nanocrystalline stator reduced total iron loss to 64% from 75%. Okamoto [19] used a stator core of amorphous magnetic material instead of a stator core of non-oriented steel. The core loss of the motor in this study was reduced by about 50%. The numerical calculation and experimental test data were compared, and the accuracy and reliability of the results was verified. Guo introduced the core loss calculation of the magnetic flux change in a permanent magnet transverse flux machine with a soft magnetic composite stator core and a low carbon steel rotor yoke, based on a modified core loss model and finite element magnetic field analysis. The calculation of the motor core loss is consistent with the measured value on the prototype [20]. After analyzing these literature sources, it can be found that many scholars have performed in-depth research on the calculation models and methods of high-speed permanent magnet motors and high-speed induction motors. However, few studies have detailed calculations of the losses in various areas of the stator. There is still much work to be done to study the magnetic field characteristics of the HPMSMs under heavily loaded conditions and the resulting changes in iron loss.

In this study, a 120 kW high-speed permanent magnet synchronous motor is used as the research object. By establishing a two-dimensional transient electromagnetic field-variable load circuit joint calculation model and using the time-step finite element method, the magnetic field distribution characteristics of the motor under rated operating conditions, no-load conditions, and different transition states from no-load operation to normal conditions can be determined. Additionally, regular research can be performed and iron loss distribution in each structural area of the stator can be quantitatively analyzed. Then, the effects of different load conditions on the losses in each structural area of the stator core can be compared and analyzed, providing a reference for more efficient operation and structural design of HPMSMs in various fields.

2. Establishment of High-Speed Permanent Magnet Synchronous Motor Model

2.1. Motor Parameters and Physical Model

The prototype parameters of the 120 kW high-speed permanent magnet synchronous motor studied in this paper are shown in Table 1. This type of motor is mainly used to drive high-speed fans using a direct drive mechanism to achieve efficient operation of the fan system. The physical model of the motor is shown in Figure 1.

Table 1. Parameters of prototype motor.

Parameter	Value	Parameter	Value
Rated power/kW	120	Air gap length/mm	4
Rated voltage/V	380	Iron core length/mm	112
Rated frequency/hz	400	Number of poles	2
Stator outer diameter/mm	200	Number of stator slots	24
Rotor outer diameter/mm	94	Connection method	Y connection

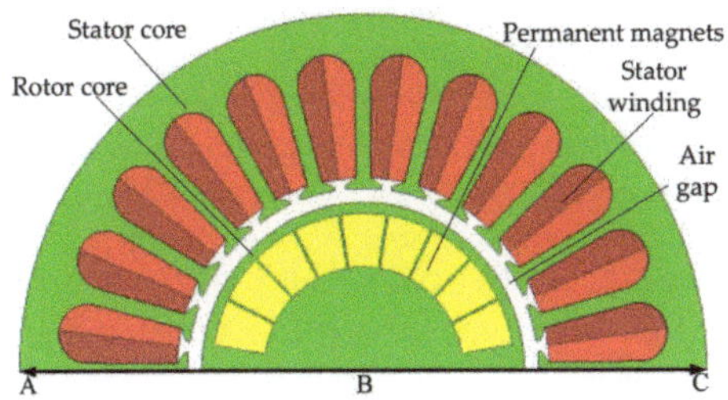

Figure 1. Solution domain physical model.

2.2. Solving Equations and Boundary Conditions

The boundary conditions are also marked in Figure 1. The vector magnetic potential A is used to analyze the magnetic field of the motor. A has only an A_Z component. A has no x- or y-axis components and also satisfies the nonlinear Poisson equation. The boundary value problem is as follows:

$$\begin{cases} \frac{\partial}{\partial x}\left(\frac{1}{\mu}\frac{\partial A_Z}{\partial x}\right) + \frac{\partial}{\partial y}\left(\frac{1}{\mu}\frac{\partial A_Z}{\partial y}\right) = -J_Z + \sigma\frac{dA_Z}{dt} \\ A_Z\big|_{\overline{AC}} = 0 \\ A_Z\big|_{\overline{AB}} = A_Z\big|_{\overline{CB}} \end{cases} \tag{1}$$

where A_Z is the z-axis component of the vector magnetic potential, $\overline{AC}$, $\overline{AB}$, and $\overline{CB}$ are the outer boundary of the stator, J_Z is the source current density, μ is the permeability of the material, and σ is the conductivity of the material.

2.3. Determination of Stator Core Loss Area

Figure 2 is a cloud diagram of the magnetic induction intensity distribution when the motor is running at no-load. The magnetic density distribution in each area of the motor structure directly affects the core loss distribution [21]. To thoroughly investigate the core loss distribution in different areas of the motor, the stator core is divided into a stator yoke, a stator tooth root, a stator tooth body, and a stator tooth top according to the small diagram marked by the arrow in Figure 2. Then, the iron consumption in each area can be calculated separately.

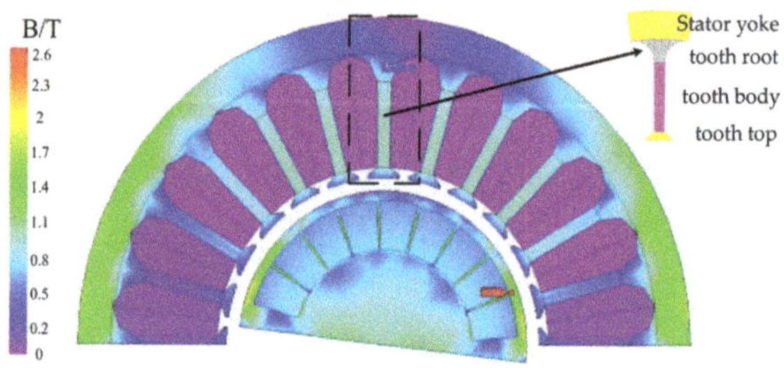

Figure 2. Magnetic flux density distribution of motor during no-load working condition.

2.4. Determination of Study Location

In order to systematically study the comprehensive magnetic flux properties of the motor, the following components mush be considered: the magnetic flux density of the 1/3 position of the stator teeth of the motor, the magnetic flux density at the center of the air gap, the flux density in the middle region between the permanent magnet and the rotor edge, the magnetic flux density in the region of the rotor magnetic bridge, the magnetic flux density of the stator yoke, the magnetic flux density of the stator tooth position, and the magnetic flux density of various areas of the stator in multi-load states. Figure 3 shows the seven arcs that correspond to the above areas in order to extract the magnetic density. The stable moment value is extracted in the vector direction of the magnetic density. The positions A, B, C, D, E, and F are centered around the rotor axis in the following arrangement: the length from the position of 1/3 tooth of the stator to the center of the circle, the length from the center of the air gap to the center of the circle, the length from the center of the area between the permanent magnet and the rotor edge to the center of the circle, the length from the center of the rotor magnetic isolation bridge to the center of the circle, and the length from the center of the stator yoke to the center of the circle, respectively. The length from the center of the stator tooth to the center of the circle is the radius where an arc can be drawn with an angle of 180° in a turn. Position G is a straight line drawn from the inner center of the stator to the outer center of the stator. The magnetic density of each of the above locations is derived.

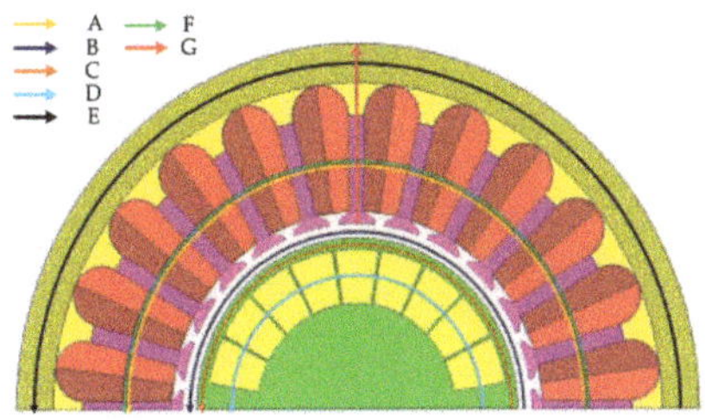

Figure 3. Study position of the model.

2.5. Numerical Calculation of Core Loss

Commonly used iron core loss calculation models include the Steinmetz model and the iron core loss separation model. This study uses the iron loss separation model to analyze the iron loss of high-speed motors.

The iron core loss separation model is also called the constant coefficient trinomial model [15], in which the iron loss is composed of three parts: hysteresis loss, classical eddy current loss, and abnormal eddy current loss. The model is expressed as follows:

$$\begin{aligned} p_{fe} &= p_h + p_c + p_e \\ &= k_h f {B_m}^2 + k_c \tfrac{1}{T}\int \left(\tfrac{dB}{dt}\right)^2 dt + k_e \tfrac{1}{T}\int \left(\tfrac{dB}{dt}\right)^{1.5} dt (W/kg) \end{aligned} \tag{2}$$

where p_{fe} is core loss, p_h is hysteresis loss, p_c is classical eddy current loss, and p_e is abnormal eddy current loss. kh is the hysteresis coefficient, f is the alternating frequency of the magnetic flux density, Bm is the magnetic density value, kc is the classical eddy current coefficient, ke is the abnormal eddy current coefficient, and B is the magnetic flux density.

In steady state operation, the hysteresis loss mainly depends on two factors, the area surrounded by the hysteresis ring and the alternating frequency of the magnetic flux density. The classical eddy current loss and the abnormal eddy current loss depend on the rate of change of the magnetic flux density. The material properties given by the material manufacturer are obtained using a numerical fitting method. The core loss per unit mass obtained using Formula (2) is multiplied by the core mass to obtain the overall core loss.

3. Magnetic Field Characteristics and Loss Analysis under Multiple Operating Conditions

3.1. Motor Electromagnetic Field Characteristics at No-Load

In this paper, a 120 kW HPMSM is used as a test example and the time-step finite element method is used to analyze the magnetic density distribution of the no-load state operation. The magnetic flux density and magnetic field line distribution of the motor at no-load are shown in Figure 4.

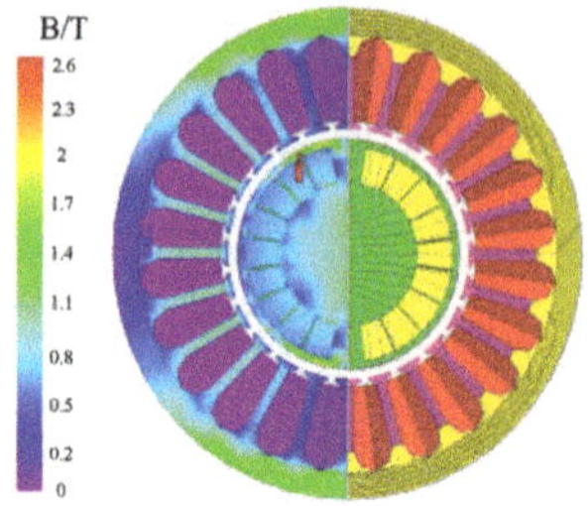

Figure 4. Motor magnetic field distribution during no-load operation.

The magnetic density waveforms extracted according to the positions A, B, C, and D are shown in Figure 5.

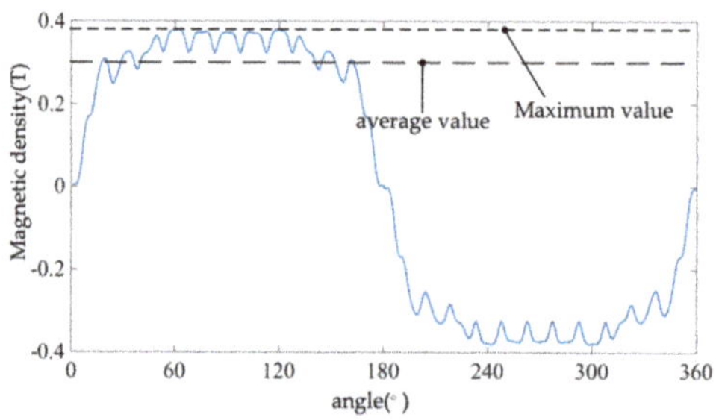

(**a**) air gap center

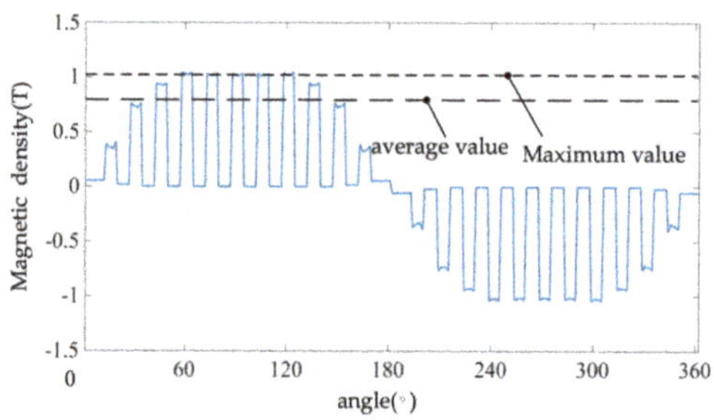

(**b**) 1/3 tooth of stator

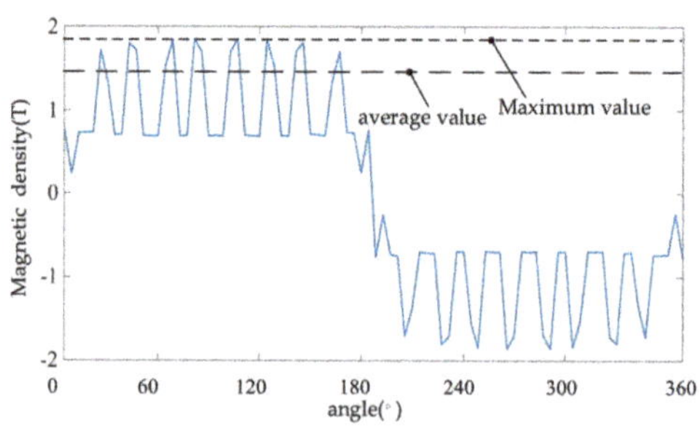

(**c**) part at 1/2 of the magnetic isolation bridge of the rotor

Figure 5. *Cont.*

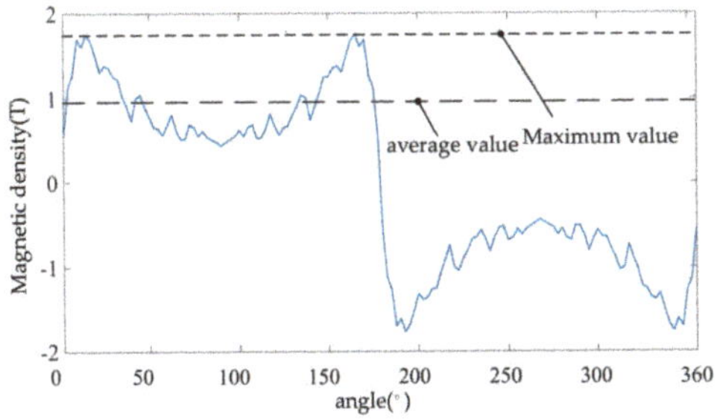

(**d**) central portion of the area between the permanent magnet and the rotor edge

Figure 5. Magnetic flux density in each region of the motor during no-load operation.

Analysis can be obtained for the following:

(1) The part of the stator teeth 1/3 teeth is just in the tooth body area. Due to the cogging effect, under no-load operating conditions, the magnetic field waveform distortion of 1/3 of the stator teeth and 1/2 of the magnetic isolation bridge is more serious than in the air gap position, and it is also more serious than in the area between the permanent magnet and the rotor edge. In comparison, the magnetic density waveform of the air gap position is closer to a sine wave.

(2) Under no-load operating conditions, the average magnetic flux density of the air gap is about 0.3 T, and the maximum value is close to 0.4 T. The average magnetic density value of 1/3 of the stator teeth is about 0.8 T, and the maximum value is just over 1 T. The average value of the magnetic flux density when the maximum magnetic flux density of the rotor magnetic isolation bridge is approximately 1.5 T and has not yet exceeded 2 T. The average magnetic density of the central part of the area between the permanent magnet and the rotor edge is approximately 1 T, and the maximum value is close to 1.8 T.

Fourier decomposition is performed on the separated air gap radial magnetic density, and the distribution of the harmonic amplitudes with their orders is shown in Figure 6.

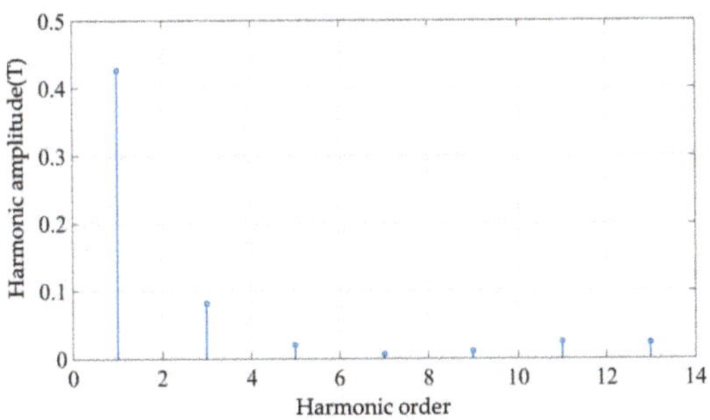

Figure 6. Harmonic distribution of air gap tangential flux density.

It can be seen from the analysis of Figure 6 that the harmonic amplitudes of the even harmonics are very small and may be neglected. The odd harmonics from 1 to 13 are separated from the total characteristic waveform. The distribution of the fundamental wave, each odd harmonic wave, and the air gap radial magnetic density in a period is shown in Figure 7.

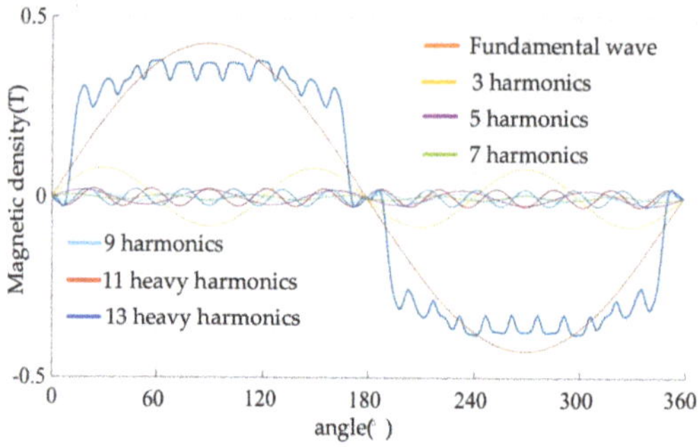

Figure 7. Distribution of each odd harmonic in a period.

The harmonic amplitudes of the 1st to 13th odd waves are extracted and the magnitude of the amplitude of each odd wave is compared to the amplitude of the fundamental wave, as shown in Table 2.

Table 2. Magnitude of each harmonic and its proportion in the fundamental.

Harmonic Order	Harmonic Amplitude/T	Proportion/%
One	0.4265	100
Three	0.0813	19.06
Fives	0.0204	4.78
Seven	0.0068	1.59
Nine	0.0109	2.56
Eleven	0.0246	5.76
Thirteen	0.0233	5.46

According to the analysis in Table 2, under the no-load operating conditions, as the harmonic order increases, the proportion of the amplitude of the odd wave to the amplitude of the fundamental wave is generally reduced. Specifically, the amplitude of the third harmonic is the highest compared to the amplitude of the fundamental wave. The remaining odd harmonic amplitudes account for a small proportion of the fundamental amplitude.

3.2. Motor Iron Loss Distribution at No-Load

Table 3 shows the iron core loss of each area of the stator and rotor under no-load operating conditions.

Table 3. Stator core loss distribution of motor during no-load operation.

Divide Area	Iron Loss/W	Proportion/%
Stator tooth top	8.02	2.42
Stator tooth body	48.21	14.56
Stator tooth root	31.71	9.58
Stator yoke	243.08	73.43
Rotor core	1	0.3

According to the analysis in Table 3, under no-load operating conditions, the iron core loss of the stator yoke accounts for the largest proportion of the total iron core loss of the motor, 73.43%, and the iron core loss in the stator tooth body and stator tooth root area account for 14.56% and 9.58%, respectively. The stator core top core loss accounts for 2.42% of the total motor core loss and the rotor iron core loss is very small accounting for the smallest proportion of the total motor iron core loss, only 0.03%. Thus, the stator iron core loss accounts for the main part of the total core iron loss of the motor, with a proportion is as high as 99.7%. The distribution of the motor iron core loss under no-load operating conditions is shown in Figure 8.

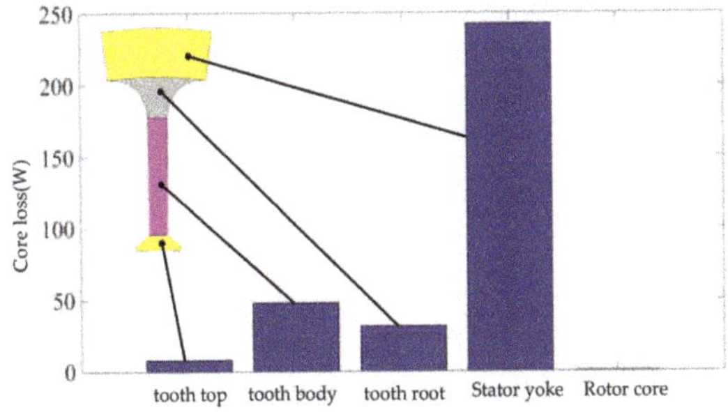

Figure 8. Stator core loss distribution in each region during rated operation.

3.3. *Analysis of the Internal Magnetic Field Distribution of a Motor under Multiple Load Conditions*

A 120 kW HPMSM is again used as a test example and the time-step finite element method is used to analyze the magnetic density distribution of the multi-load state operation. Figure 9 shows the different magnetic flux densities in different areas of the motor under multiple load conditions.

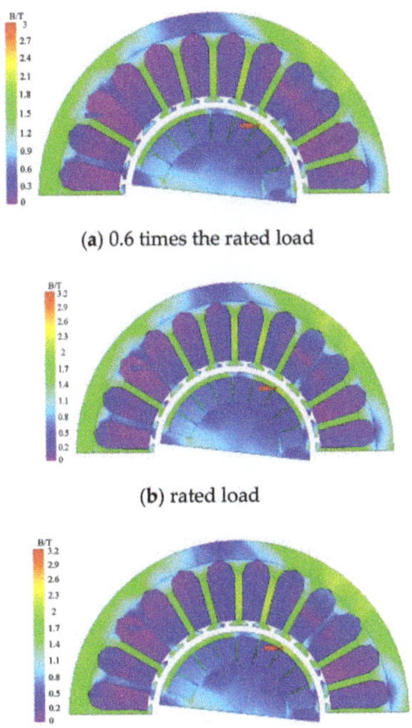

(**a**) 0.6 times the rated load

(**b**) rated load

(**c**) 1.1 times the rated load

Figure 9. Magnetic induction intensity distribution of motor in multi-load condition.

Using the analysis position determined in Figure 3, the magnetic density waveforms of position E and position F can be obtained as shown in Figure 10, and a comparative analysis can be found in the following:

(1) Due to the influence of the cogging effect, the magnetic field waveform distortion of the stator tooth body position is more serious than the stator yoke position under any operating conditions.
(2) Because the power supply contains a large number of higher harmonic components, the internal magnetic flux density waveform of the motor is severely distorted, and due to the skin effect, the high-frequency harmonics have the greatest influence on the magnetic density of the stator tooth body position.
(3) When the load changes, the magnetic density of the tooth body area changes compared with the yoke area in a shorter and more dramatic period. Because the harmonic component of the tooth body area is larger under no-load operating conditions, the average magnetic densities of the stator tooth body and stator yoke areas are the largest. During this time, the maximum values of the magnetic density are approximately 1 T and 1.5 T, and the average values are approximately 0.8 T and 0.9 T, respectively. At 1.1 times rated load, the average magnetic densities of the stator tooth body and stator yoke areas are the smallest. During this time, the maximum magnetic density is approximately 0.7 T and 0.6 T, and the average values are approximately 0.35 T and 0.4 T, respectively.
(4) At 1.1 times the rated load, the magnetic densities of the stator tooth body and stator yoke are approximately equal to the densities at the rated load. Thus, it is clear that the magnetic density has reached saturation when running at rated load.

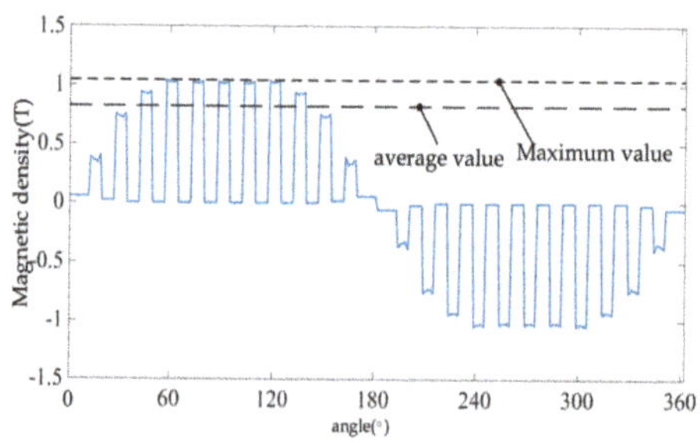

(**a**) stator tooth body under no-load operation condition

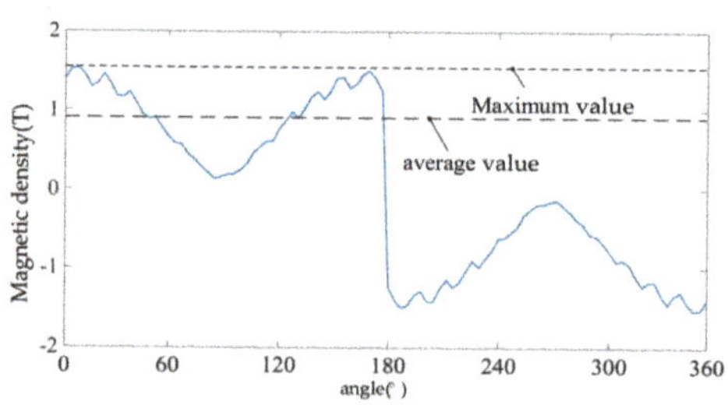

(**b**) stator yoke under no-load operation condition

Figure 10. *Cont.*

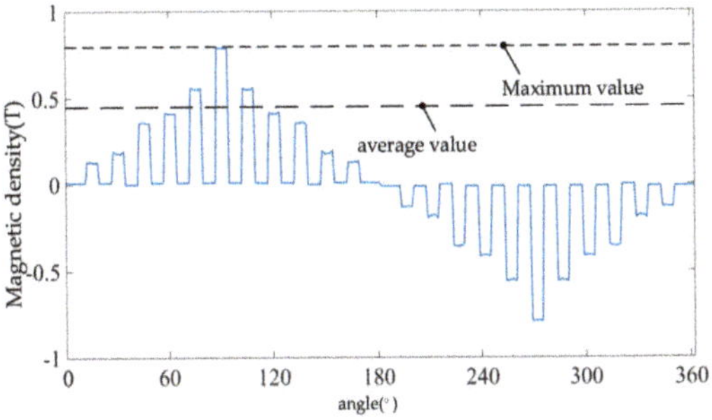

(**c**) stator tooth body under the operating condition of 0.6 times the rated load

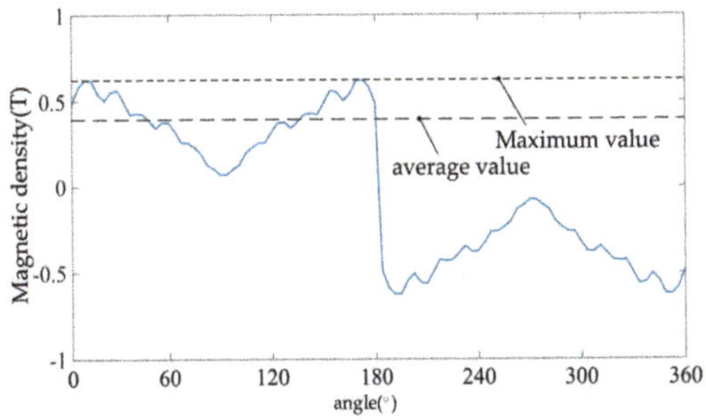

(**d**) stator yoke under the operating condition of 0.6 times the rated load

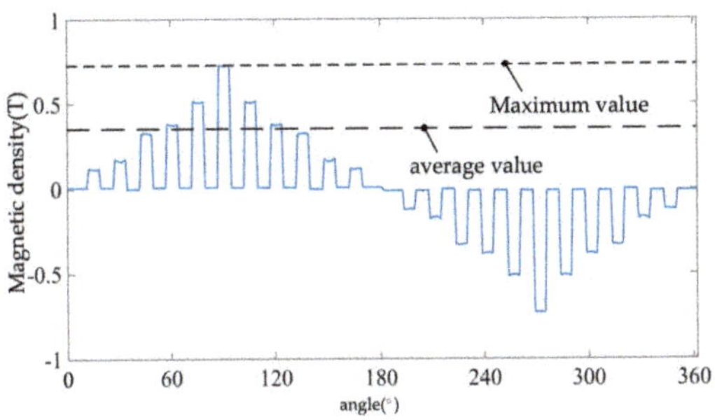

(**e**) stator tooth body under rated load operation condition

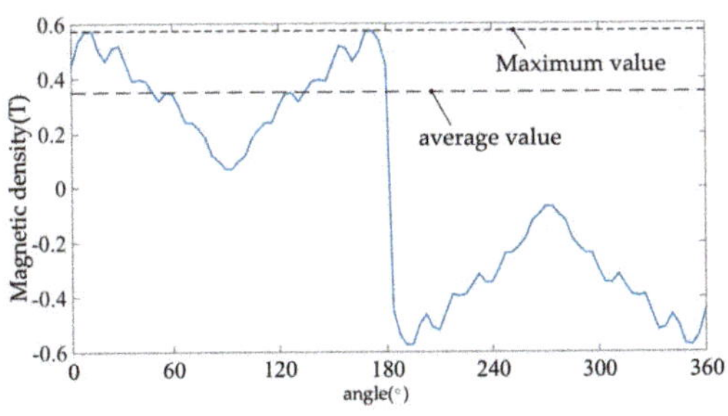

(**f**) stator yoke under rated load operating conditions

Figure 10. *Cont.*

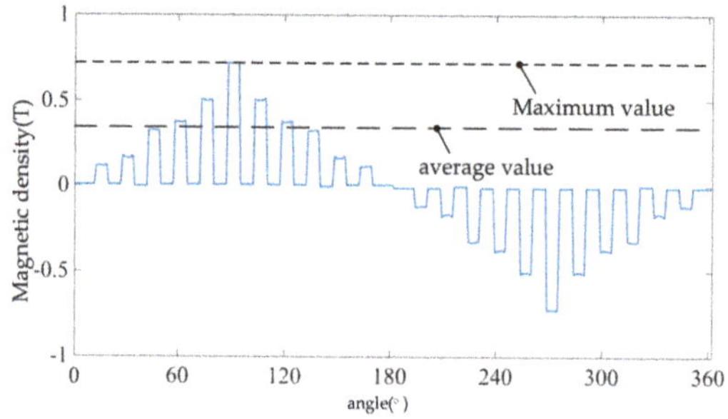

(**g**) stator tooth body under the operating condition of 1.1 times the rated load

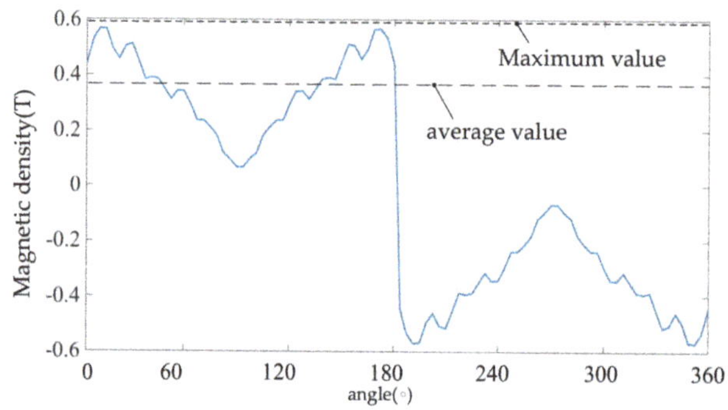

(**h**) stator yoke under the operating condition of 1.1 times the rated load

Figure 10. Magnetic flux waveforms of stator teeth and yoke under four operating conditions.

In order to explain the change of magnetic density more accurately, according to position G, the average value of the magnetic density of the center position of the four areas of the stator core under the following four operating conditions were recorded: no-load, 0.6 times rated load, rated load, and 1.1 times rated load. These results are shown in Table 4.

Table 4. Average magnetic density of each area on the stator side under four operating conditions.

Operating Conditions	Tooth Top/T	Tooth Body/T	Tooth Root/T	Yoke/T
no-load	0.7618	0.965	0.5376	0.1808
0.6 times load	0.5446	0.7385	0.4277	0.1747
rated load	0.5023	0.6811	0.3944	0.1611
1.1 times the rated load	0.4974	0.6745	0.3906	0.1596

According to the analysis in Table 4, the average value of the magnetic flux density in each area of the stator is the largest during no-load operation and the average value of the magnetic flux density in each area of the stator is the smallest when running at 1.1 times the rated load. Under all four operating conditions, the average value of the magnetic density of the tooth body is the largest among the four areas of the stator. When the load is increased, the average value of the magnetic density of each area decreases and the magnetic density of the tooth body area has the largest change compared to the other three positions. Comparing the no-load operating condition with 1.1 times the rated load operating condition, the average value of the stator tooth position magnetic density under no-load operating conditions increases by approximately 0.3 T, and the average magnetic density of the yoke is the smallest. The magnetic density of the yoke area has the smallest change compared to the other three positions, and the average magnetic flux density of the yoke is relatively unchanged under the four operating conditions.

3.4. Research on the Relationship between Load Condition and Stator Core Loss

If the waveform when the motor is stable and the iron core loss waveform when stable is extracted, then the average value of the extracted portion can be used as the core loss calculation value. The rotor iron core loss is very small when the motor is running, so we will not study it below. Table 5 shows the iron core loss distribution data for different positions of the stator.

Table 5. Stator iron core loss distribution in each region under four kinds of operating conditions.

Operating Conditions	Tooth Top/W	Tooth Body/W	Tooth Root/W	Yoke/W
no-load	8.02	48.21	31.71	243.08
0.6 times load	38.79	235.56	108.32	443.14
rated load	48.98	278.838	133.49	529.77
1.1 times the rated load	51.27	286.62	138.34	547.83

The relationship between the iron consumption of each area of the motor stator and the load is shown in Figure 11.

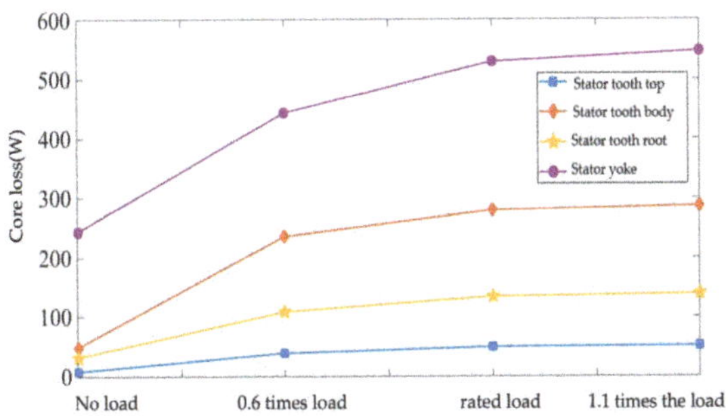

Figure 11. Stator iron core loss changes in each region under four types of operating conditions.

It can be seen from the analysis shown in Figure 11 that with the increase of the load, the core loss of the stator tooth top, tooth body, tooth root, and yoke parts also increase. Specifically, when the no-load operating condition is increased to 1.1 times the rated load operating condition, the increase of the core losses of the stator tooth root and stator yoke are 336.27% and 125.37%, respectively, which are smaller than those of the stator tooth top and the stator tooth body, which are 539.28% and 494.52%, respectively. From the no-load operating condition to 0.6 times the rated load operating condition, the total increase of the stator core loss is the largest. The increase of the core loss in the four areas of the stator tooth top, tooth body, tooth root, and yoke are 383.67%, 388.61%, 241.6%, and 82.3%, respectively. When the rated load operating condition is changed to 1.1 times the rated load operating condition, the total increase of the stator core loss is the smallest, and the increase of the core loss in the four areas of the stator is the smallest. In particular, the stator tooth top and the tooth root both increase less than 5 W, which is almost unchanged.

Through the research in this paper, the loss distribution ratio of each area of the stator is determined. The main part of the stator iron loss of the motor is the stator yoke iron loss. With the calculation of the motor temperature field, the distribution of the stator heat source can be more accurately understood and will make the temperature calculation of HPMSMs more accurate. Figure 12 shows a flowchart of the motor iron loss-temperature field transition.

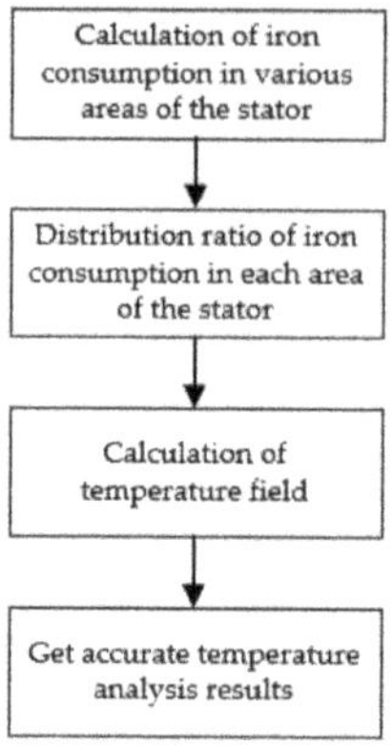

Figure 12. Flow diagram of motor iron loss-temperature field transition.

4. Conclusions

In this study, a 120 kW high-speed permanent magnet synchronous motor was used as a test example. A two-dimensional transient electromagnetic field-variable load circuit joint calculation model was established, and the time-step finite element method was used to analyze the magnetic field distribution characteristics and laws of the motor under rated operating conditions, no-load conditions, and transitory operating conditions. The four operating conditions studied were no-load, 0.6 times rated load, rated load, and 1.1 times rated load. Each state was quantitatively analyzed to determine the iron loss distribution in each structural area of the stator, and the impact of different load conditions on the losses in each structural area of the stator core of the motor was then compared and analyzed, leading to the following conclusions:

(1) The distribution principle of the stator core loss is that the stator yoke accounts for the largest proportion of the stator core loss, followed by the stator tooth body and stator tooth root, and the stator tooth top accounts for the smallest proportion.
(2) The stator core loss under the no-load operating condition is the least, and the stator core loss under the 1.1 times rated load operating condition is the most, which is particularly evident in the stator yoke. The core loss of the stator tooth has the smallest change; however, as the load increases, the growth rate of the core loss of the stator tooth top is the largest, and the growth rate of the core loss of the stator yoke is the smallest. Under the operating conditions of rated load and 1.1 times rated load, the difference between their stator core losses is not large, particularly the difference at the top of the stator teeth, which is the smallest.
(3) Comparing the average values of the magnetic flux density in the four areas on the stator, the magnetic flux density of the tooth body is larger under the no-load operating condition, and the maximum value is close to 1 T. Comparing the rated load operating conditions with 1.1 times the rated load operating conditions, the average value of the magnetic flux density at the center position of each area on the stator side under the rated load operating conditions is relatively close and the saturation of magnetic flux density is greater.

Author Contributions: Conceptualization, K.L.Z. and Y.S.; Methodology, F.L.; Software, Q.D.; Validation, D.T., K.L.Z. and F.L.; Formal Analysis, J.W.; Investigation, J.Z.; Resources, D.T.; Data Curation, K.L.Z.; Writing—Original Draft Preparation, D.T.; Writing—Review & Editing, D.T. and K.L.Z.; Visualization, Y.S.; Supervision, F.L.; Project Administration, D.T. All authors have read and agreed to the published version of the manuscript.

Funding: This research was funded by the National Natural Science Foundation of China (grant nos. 51777048 and 51407050).

Acknowledgments: This work was supported in part by the National Natural Science Foundation of China (grant nos. 51777048 and 51407050). The authors would like to thank the anonymous reviewers for their valuable comments and suggestions that strengthened this paper.

Conflicts of Interest: All of our authors declare together that there is no conflict of interest.

References

1. Zhang, Z.; Geng, W.; Lu, J. Research status and development of stator coreless permanent magnet motor technology. *Proc. CSEE* **2018**, *38*, 582–600.
2. Zhang, F.; Du, G.; Wang, T.; Liu, G. Review of Development and Design of High -Speed Motors. *Trans. China Electrotech. Soc.* **2016**, *31*, 1–18.
3. Dong, J.; Huang, Y.; Jin, L.; Lin, H.; Yang, H. Review of design and analysis techniques for high-speed permanent magnet motors. *Proc. CSEE* **2014**, *34*, 4640–4653.
4. Zhu, B.; Bai, B.; He, H. Effect of Inverter Parameters on the Eddy Current Losses in iNduction Motor Fed by PWM Inverter. In Proceedings of the International Conference on Electrical Machines and Systems, Wuhan, China, 17–20 October 2008; pp. 4240–4243.
5. Hu, Y.; Gu, W. Adaptive robust triple-step control for compensating cogging torque and model uncertainty in a dc motor. *IEEE Trans. Syst. Man Cybern. Syst.* **2019**, *49*, 2396–2405. [CrossRef]
6. Yamazaki, K.; Abe, A. Loss investigation of interior permanent-magnet motors considering carrier harmonics and magnet eddy currents. *IEEE Trans. Magn.* **2009**, *41*, 659–665. [CrossRef]
7. Fouladgar, J.; Chauveau, E. The influence of the harmonics on the temperature of electrical machines. *IEEE Trans. Magn.* **2005**, *41*, 1644–1647. [CrossRef]
8. Han, J. *Study on the Influence of PWM Inverter Power Supply on Core Loss of Asynchronous Motor*; Beijing Jiaotong University: Beijing, China, 2011.
9. Yu, J.; Li, L.; Du, P.; Zhang, J. Ripple-current suppression method for high-speed permanent magnet synchronous motor based on harmonic injection PWM. *J. Electr. Eng.* **2017**, *12*, 1–10.
10. Sun, M.; Tang, R.; Han, X.; Tong, W. Analysis and Modeling for Open Circuit Air Gap Magnetic Field Prediction in Axial Flux Permanent Magnet Machine. *Proc. CSEE* **2018**, *38*, 1525–1533.
11. Kong, X.; Wang, F.; Xing, J. Losses calculation and temperature field analysis of high speed permanent magnet machines. *Trans. China Electrotech. Soc.* **2012**, *27*, 166–173.
12. Zou, J.; Li, J.; Xu, Y.; Wei, Y. Influences of drive strategies on the loss of permanent magnet brushless direct current motor. *Trans. China Electrotech. Soc.* **2011**, *26*, 43–47.
13. Han, L.; Xie, L.; Zhang, G. Influence of converter parameters on stator losses of permanent magnet synchronous generator. *Electr. Mach. Control* **2010**, *14*, 75–81.
14. Huang, P.; Hu, Q.; Cui, Y.; Huang, Y. Analytical calculation of the iron losses of electric machine fed by PWM inverter. *Proc. CSEE* **2007**, *27*, 19–23.
15. Zhao, H.; Luo, Y.; Liu, X.; Wang, R.H.; Chen, W. Analysis on no-load iron losses distribution of asynchronous motors with time-stepping finite element method. *Proc. CSEE* **2010**, *30*, 99–106.
16. Han, S.; Cui, S.; Wang, T.; Chan, C.; Zhang, X. Application of fractional-slot concentrated winding permanent magnet machines in modular cascade machine System. *Trans. China Electrotech. Soc.* **2013**, *28*, 9–16.
17. Yamazaki, K.; Suzuki, A.; Ohto, M.; Takakura, T. Harmonic Loss and Torque Analysis of High-Speed Induction Motors. *IEEE Trans. Ind. Appl* **2012**, *48*, 933–941. [CrossRef]
18. Denis, N.; Inoue, M.; Fujisaki, K.; Itabashi, H.; Yano, T. Iron Loss Reduction in Permanent Magnet Synchronous Motor by Using Stator Core Made of Nanocrystalline Magnetic Material. *IEEE Trans. Magn.* **2017**, *53*, 1–6. [CrossRef]
19. Okamoto, S.; Denis, N.; Kato, Y.; Ieki, M.; Fujisaki, K. Core Loss Reduction of an Interior Permanent-Magnet Synchronous Motor Using Amorphous Stator Core. *IEEE Trans. Ind. Appl.* **2016**, *52*, 2261–2268. [CrossRef]
20. Guo, Y.; Zhu, J.; Lu, H.; Li, Y.; Jin, J. Core Loss Computation in a Permanent Magnet Transverse Flux Motor with Rotating Fluxes. *IEEE Trans. Magn.* **2014**, *50*, 1–4. [CrossRef]
21. Ding, S.; Li, G.; Feng, H.; Li, Y.; Deng, Y. Numerical calculation of loss under load variation for driving asynchronous motor. *Electr. Mach. Control* **2013**, *17*, 36–41.

energies

MDPI

Review

Review of Time and Space Harmonics in Multi-Phase Induction Machine

Vladimir Kindl [1,*], Radek Cermak [1], Zelmira Ferkova [2] and Bohumil Skala [1]

1 Faculty of Electrical Engineering, University of West Bohemia, 306 14 Pilsen, Czech Republic; radek@rice.zcu.cz (R.C.); skalab@kev.zcu.cz (B.S.)

2 Faculty of Electrical Engineering and Informatics, Technical University of Kosice, 040 01 Košice, Slovakia; zelmira.ferkova@tuke.sk

* Correspondence: vkindl@kev.zcu.cz

Received: 13 December 2019; Accepted: 15 January 2020; Published: 19 January 2020

Abstract: Modern multiphase electric machines take advantage of additional degrees of freedom for various purposes, including harmonic current injection to increase torque per ampere. This new approach introduces a non-sinusoidal air gap flux density distribution causing additional technical problems and so the conventional assumptions need to be revised. The paper presents a methodology for synthesis of air gap magnetic field generated by a symmetrically distributed multiphase windings including the rotor field reaction due to the machine's load. The proposed method is suitable either for single-layer or double layer windings and can be adopted either for full-pitched or chorded winding including slots effects. The article analyses the air gap flux density harmonic content and formulates conclusions important to multiphase induction motors. It also discusses effects of time harmonic currents and illustrates the principle of changing number of pole-pairs typical for harmonic currents being injected to increase torque.

Keywords: multiphase; induction; motor; space harmonics; time harmonics; injection

1. Introduction

Over the past few years the modern industry has recorded a huge and intensive development in power electronics and drives and has brought many technical upgrades not only in public transportation. This rapid technological progress results in less industrial energy consumption and improves environmental issues. One of the most frequently discussed topics relating to the traffic environment [1–3] is replacing conventional combustion engine vehicles with fully electric (battery) vehicles (EVs). The initial concept of EV [4,5] comes from the previous experience with hybrid electric vehicles (HEVs), which later evolved into the popular plug-in hybrids (PHEVs). No matter what type of EV is considered, they are mostly powered by standard ac three-phase electric motors [6].

More than 80% of electric vehicles currently use rare earth permanent magnet synchronous motors (PMSM) allowing vehicle manufacturers to increase the efficiency [7,8] compared to traditional induction motors, especially at lower motor speeds. The improved efficiency brings either an increase in transport range or a reduction in battery size (and also the cost) for a given motor weight and vehicle specification. Compared to typical city-driving vehicles, powerful highway vehicles are designed to operate at higher speeds, which somewhat reduces the relative performance increase gained by rare earth magnets and reduces their advantage over induction machines [9]. Moreover, the rising cost and complicated political situation on the market with rare-earth magnets is another very important reason [10] why manufactures start choosing induction motors [11] as main drive units for their electrical vehicles. This tendency can be observed especially when talking about higher performance and luxury vehicles, such as e-buses, locomotives, or vehicles from Audi or Tesla.

The modern e-mobility trends increasingly focus on multiphase variable-speed motor drives [12] since they could provide the transportation with numerous traction and economic advantages [13]. Probably the first mention of a multiphase electric drive dates back to 1969, where the author of [14] proposed the concept of a five-phase inverter-fed induction motor lately extended to a six-phase double-star induction motor, referred in [15,16].

This early interest in multiphase machines was mainly initiated by the possibility of torque ripple reduction, higher reliability and higher fault tolerance. Another strong argument is that for a given motor power, the input power per phase is reduced, resulting in lower demands on the inverter power electronics components. Moreover, the winding I^2R losses are inversely proportional to the square of winding distribution factor, hence the higher number of phases may significantly increase the overall motor efficiency.

Maximum theoretical value of winding loss reduction is determined in [17] as 8.8%. Moreover, as shown in [18–22], the multiphase motors may be powered with additional time harmonics injected into the winding to decrease the input current (rms value) while keeping the same torque, which is particularly relevant to the traction battery applications.

The main disadvantage lies in limited slots number available for given stator diameter, the greater number of phases, the lower the number of slots per pole per phase and consequently higher magnitudes of space harmonics. This may be even more problematic in case of outer-rotor machines with narrow-shaped slots.

No matter what number of phases is considered, electric motors are always accompanied by non-linearities and parasitic effects, usually connected with harmonics [23–26], that cause ripple of the input current and the torque, produce noise and increase losses.

These harmonics are often produced by dead times in pulse width modulation (PWM), supply voltage unbalance, magnetic circuit saturation, non-sinusoidal winding distribution, lamination slotting and some other non-linearities and asymmetries.

According to the nature of their origin, we can classify these harmonics as time harmonics and space harmonics [27], while their mutual interaction cannot be neglected [28]. This study describes/recapitulates occurrence and behavior of time/space harmonics in multiphase induction machines.

2. Space Harmonics in the Air Gap Magnetic Field

Any symmetric m-phase ($m \in \mathbb{Z}$) induction machine has a space displacement between any two successive stator phases equal to $2\pi/m$. The stator winding is designed as sinusoidally distributed as possible and is fed with balanced m-phase sinusoidal currents. The combined effect is equivalent to having the same winding excited with a constant current and rotating at the stator frequency (rotating field established). Ideally, when the number of stator slots Q_1 approaches infinity $(Q_1 \to \infty)$, and no iron core saturation will appear, the winding forms a sinusoidal magnetic field (or magneto-motive force, mmf) in the air gap δ. However, the practical windings are placed into the finite number of slots and the machines' core always experiences saturation; therefore, the resulting magneto-motive force has rather stepped than sinusoidal curve. This complex curve can be described with Fourier series of mmf waves called space harmonics. The orders of these harmonics are usually marked with symbol ν.

Synthesis of Air Gap Magnetic Field Formed by Symmetrically Distributed Windings

As shown in [29], a hypothetical single-coil winding, fed by a time-varying sinusoidal current, produces mmf having rectangular waveform according to (1). The equation considers constant air gap permeance independent of angular position (even air gap) and the iron core made of steel having infinite relative permeability.

$$H_x(\alpha) = \frac{2i}{\pi\delta} \sum_{\nu=1}^{\infty} \frac{1}{\nu} sin\left(\nu\frac{\alpha_y}{2}\right) cos(\nu[\alpha - (\xi + (x-1))\alpha_1]),\ \alpha_y = \beta\pi \quad (1)$$

In (1), $\beta \in <0;1>$ represents the shortening of the winding coil pitch (chorded winding), i is the current content in the slot, α is mechanical angle measured in the air gap periphery, ξ is position of the coil group origin given in slots number and $\alpha_1 = 2\pi/Q_1$.

For symmetric m-phase winding distributed in Q_1 stator slots, creating $2p$ magnetic poles, we can introduce $q = Q_1/2pm$ as the number of slots per phase per pole. Hence, the field produced by respective coils in group of q coils can be described using (2). For further analysis, it is reasonable to consider only the basic two-pole winding, therefore we always assume that $2p = 2$.

$$\begin{aligned}
H_1(\alpha) &= \tfrac{2i}{\pi\delta}\textstyle\sum_{v=1}^{\infty}\tfrac{1}{v}sin\left(v\tfrac{\alpha_y}{2}\right)cos(v\alpha-\xi\alpha_1)\\
H_2(\alpha) &= \tfrac{2i}{\pi\delta}\textstyle\sum_{v=1}^{\infty}\tfrac{1}{v}sin\left(v\tfrac{\alpha_y}{2}\right)cos(v[\alpha-(\xi+1)\alpha_1])\\
H_3(\alpha) &= \tfrac{2i}{\pi\delta}\textstyle\sum_{v=1}^{\infty}\tfrac{1}{v}sin\left(v\tfrac{\alpha_y}{2}\right)cos(v[\alpha-(\xi+2)\alpha_1])\\
&\vdots\\
H_q(\alpha) &= \tfrac{2i}{\pi\delta}\textstyle\sum_{v=1}^{\infty}\tfrac{1}{v}sin\left(v\tfrac{\alpha_y}{2}\right)cos(v[\alpha-(\xi+(q-1))\alpha_1])
\end{aligned} \tag{2}$$

The summation of particular fields (2) gives resulting mmf (3) generated by a group of q coils corresponding to the coil group of one stator phase.

$$H_{group}(\alpha) = \frac{2i}{\pi\delta}\sum_{v=1}^{\infty}\left[\frac{1}{v}sin\left(v\frac{\alpha_y}{2}\right)\sum_{k=1}^{q}[cos(v[\alpha-(\xi+(k-1))\alpha_1])]\right] \tag{3}$$

Simplifying (3) we obtain more useful Equation (4).

$$H_{group}(\alpha) = \frac{2qi}{\pi\delta}\sum_{v=1}^{\infty}\left[\frac{1}{v}sin\left(v\frac{\alpha_y}{2}\right)\frac{sin\left(q\frac{\alpha_1}{2}v\right)}{q\,sin\left(\frac{\alpha_1}{2}v\right)}cos\left(\alpha v - v(q-1+2\xi)\frac{\alpha_1}{2}\right)\right] \tag{4}$$

From (4), it is easy to find the resulting mmf waveform generated by any symmetric m-phase distributed winding designed with integer q. We consider twice the number of mathematical phases $m' = 2m$, therefore $\xi = (k-1)2q$ is substituted for the "plus" phases (A, B, C, D, E, ...), and $\xi = mq + (k-1)2q$ is substituted for the "minus" phases (A', B', C', D', E', ...). The input current with angular frequency ω_1 flowing through the k-th stator phase is defined as (5):

$$i_k(t) = I_{km}sin\left(\omega_1 t - \frac{k-1}{m}2\pi\right) \tag{5}$$

Hence by combination (4) with (5) we obtain (6):

$$H_{m-phase}(\alpha) = \frac{4q}{\pi\delta}I_{km}\sum_{v=1}^{\infty}\left[\begin{array}{c}\frac{1}{v}sin\left(v\frac{\alpha_y}{2}\right)sin\left(v\frac{mq\alpha_1}{2}\right)\frac{sin\left(q\frac{\alpha_1}{2}v\right)}{q\,sin\left(\frac{\alpha_1}{2}v\right)}\\ \sum_{k=1}^{m}\left[sin\left(v\frac{2\alpha+(1-k\;4q)\alpha_1}{2}\right)sin\left(\omega_1 t-\frac{k-1}{m}2\pi\right)\right]\end{array}\right] \tag{6}$$

Equation (6) can be further modified into (7) to calculate with given coil turns number per slot N carrying the input current $I\sqrt{2}$. For single-layer winding, $N/2$ must be used instead of N.

$$H_{m-phase}(\alpha) = 4\sqrt{2}\frac{NIq}{\pi\delta}\sum_{v=1}^{\infty}\left[\begin{array}{c}\frac{1}{v}sin\left(v\frac{\alpha_y}{2}\right)sin\left(v\frac{mq\alpha_1}{2}\right)\frac{sin\left(q\frac{\alpha_1}{2}v\right)}{q\,sin\left(\frac{\alpha_1}{2}v\right)}\\ \sum_{k=1}^{m}\left[sin\left(v\frac{2\alpha+(1-k\;4q)\alpha_1}{2}\right)sin\left(\omega_1 t-\frac{k-1}{m}2\pi\right)\right]\end{array}\right] \tag{7}$$

As an example, for the three-phase winding, (7) results in (8),

$$H_{3-phase}(\alpha) = 4\sqrt{2}\frac{NIq}{\pi\delta}\sum_{v=1}^{\infty}\left[\frac{1}{v}sin\left(v\frac{\alpha_y}{2}\right)sin\left(v\frac{mq\alpha_1}{2}\right)\frac{sin\left(q\frac{\alpha_1}{2}v\right)}{q\,sin\left(\frac{\alpha_1}{2}v\right)}\left[sin\left(v\frac{2\alpha+(1-4q)\alpha_1}{2}\right)sin(\omega_1 t) + \right.\right.$$
$$\left.\left. sin\left(v\frac{2\alpha+(1-8q)\alpha_1}{2}\right)sin\left(\omega_1 t-\frac{2}{3}\pi\right) + sin\left(v\frac{2\alpha+(1-12q)\alpha_1}{2}\right)sin\left(\omega_1 t-\frac{4}{3}\pi\right)\right]\right] \quad (8)$$

For the five-phase winding we have (9),

$$H_{5-phase}(\alpha) = 4\sqrt{2}\frac{NIq}{\pi\delta}\sum_{v=1}^{\infty}\left[\frac{1}{v}sin\left(v\frac{\alpha_y}{2}\right)sin\left(v\frac{mq\alpha_1}{2}\right)\frac{sin\left(q\frac{\alpha_1}{2}v\right)}{q\,sin\left(\frac{\alpha_1}{2}v\right)}\left[sin\left(v\frac{2\alpha+(1-4q)\alpha_1}{2}\right)sin(\omega_1 t) + \right.\right.$$
$$sin\left(v\frac{2\alpha+(1-8q)\alpha_1}{2}\right)sin\left(\omega_1 t-\frac{2}{5}\pi\right) + sin\left(v\frac{2\alpha+(1-12q)\alpha_1}{2}\right)sin\left(\omega_1 t-\frac{4}{5}\pi\right) + \quad (9)$$
$$\left.\left. sin\left(v\frac{2\alpha+(1-16q)\alpha_1}{2}\right)sin\left(\omega_1 t-\frac{6}{5}\pi\right) + sin\left(v\frac{2\alpha+(1-20q)\alpha_1}{2}\right)sin\left(\omega_1 t-\frac{8}{5}\pi\right)\right]\right]$$

And the seven-phase winding will generate field according to (10):

$$H_{7-phase}(\alpha) = 4\sqrt{2}\frac{NIq}{\pi\delta}\sum_{v=1}^{\infty}\left[\frac{1}{v}sin\left(v\frac{\alpha_y}{2}\right)sin\left(v\frac{mq\alpha_1}{2}\right)\frac{sin\left(q\frac{\alpha_1}{2}v\right)}{q\,sin\left(\frac{\alpha_1}{2}v\right)}\left[sin\left(v\frac{2\alpha+(1-4q)\alpha_1}{2}\right)sin(\omega_1 t) + \right.\right.$$
$$sin\left(v\frac{2\alpha+(1-8q)\alpha_1}{2}\right)sin\left(\omega_1 t-\frac{2}{7}\pi\right) + sin\left(v\frac{2\alpha+(1-12q)\alpha_1}{2}\right)sin\left(\omega_1 t-\frac{4}{7}\pi\right) +$$
$$sin\left(v\frac{2\alpha+(1-16q)\alpha_1}{2}\right)sin\left(\omega_1 t-\frac{6}{7}\pi\right) + sin\left(v\frac{2\alpha+(1-20q)\alpha_1}{2}\right)sin\left(\omega_1 t-\frac{8}{7}\pi\right) + \quad (10)$$
$$\left.\left. sin\left(v\frac{2\alpha+(1-24q)\alpha_1}{2}\right)sin\left(\omega_1 t-\frac{10}{7}\pi\right) + sin\left(v\frac{2\alpha+(1-28q)\alpha_1}{2}\right)sin\left(\omega_1 t-\frac{12}{7}\pi\right)\right]\right]$$

The air gap flux density distribution is then obtained by applying (11).

$$B_{m-phase}(\alpha) = \mu_0 H_{m-phase}(\alpha) \quad (11)$$

Previous approach assumes a uniform air gap δ, which makes the analyzed air gap field corresponding to the mmf waveform generated by the winding. The stator and the rotor surfaces are slotted in a practical machine, and therefore, the air gap permeance varies along with the machine periphery and generates additional flux waves. Hence, the air gap appears to be slightly wider than its real mechanical size. The widening of the air gap is traditionally considered via Carter's factor k_c [30]. As proposed in [29], this can be considered by introducing a fictive air gap (12),

$$\delta(\alpha) = \frac{1}{f_1(\alpha)} + \frac{1}{f_2(\alpha)} - \delta_0 \quad (12)$$

where δ_0 represents the initially assumed (even) air gap, and the functions $f_1(\alpha)$ and $f_2(\alpha)$ introduce the stator and the rotor slotting, respectively (13).

$$f_1(\alpha) = a_0 - \sum_{v=1}^{\infty} a_v \cos(vQ_1\alpha),\ a_0 = \frac{1}{k_{c1}\delta_0}$$
$$f_2(\alpha) = b_0 - \sum_{v=1}^{\infty} b_v \cos(vQ_2\alpha),\ b_0 = \frac{1}{k_{c2}\delta_0} \quad (13)$$

The Carter factors k_{c1} and k_{c2}, important to (13), are determined from given stator/rotor slot pitches t_{d1} and t_{d2} using (14),

$$k_{c1} = \frac{t_{d1}}{t_{d1}-\gamma_1\delta_0},\ k_{c2} = \frac{t_{d2}}{t_{d2}-\gamma_2\delta_0} \quad (14)$$

where $\gamma_{1,2}$ comes from (15).

$$\gamma_{1,2} = \frac{4}{\pi}\left[\frac{b_{01,02}}{2\delta_0}atan\left(\frac{b_{01,02}}{2\delta_0}\right) - ln\sqrt{1+\left(\frac{b_{01,02}}{2\delta_0}\right)^2}\right] \tag{15}$$

Parameters b_{01} and b_{02} represent the slots opening of the stator and the rotor, respectively. To complete substitution into (13), we only need to calculate the values of a_v and b_v according to (16),

$$\begin{aligned} a_v &= \frac{4\beta_1}{\pi\delta_0 v}\left[\frac{1}{2} + \frac{\left(\frac{b_{01}}{t_{d1}}v\right)^2}{0.78-2\left(\frac{b_{01}}{t_{d1}}v\right)^2}\right]sin\left(1.6\pi\frac{b_{01}}{t_{d1}}v\right) \\ b_v &= \frac{4\beta_2}{\pi\delta_0 v}\left[\frac{1}{2} + \frac{\left(\frac{b_{02}}{t_{d2}}v\right)^2}{0.78-2\left(\frac{b_{02}}{t_{d2}}v\right)^2}\right]sin\left(1.6\pi\frac{b_{02}}{t_{d2}}v\right) \end{aligned} \tag{16}$$

where β_1 and β_2 come from (17).

$$\beta_1 = \frac{1}{2} - \frac{1}{2b_{01}}\left[b_{01}\left(\frac{1}{k_{c1}} - 1\right)\right],\ \beta_2 = \frac{1}{2} - \frac{1}{2b_{02}}\left[b_{02}\left(\frac{1}{k_{c2}} - 1\right)\right] \tag{17}$$

However, by substitution (12) into (7), we find that for some chosen q the "tooth-to-tooth" synchronization of both waves is not fully observed, and therefore, it is necessary to introduce a correction factor (angular displacement) for one of the coordinate systems. Hence, Equation (7) should be rewritten into (18).

$$H_{m-phase}(\alpha) = 4\sqrt{2}\frac{NIq}{\pi\delta(\alpha)}\sum_{v=1}^{\infty}\left[\begin{array}{c}\frac{1}{v}sin\left(v\frac{\alpha_y}{2}\right)sin\left(v\frac{mq\alpha_1}{2}\right)\frac{sin\left(q\frac{\alpha_1}{2}v\right)}{q\ sin\left(\frac{\alpha_1}{2}v\right)} \\ \sum_{k=1}^{m}\left[sin\left(v\frac{2\alpha+(1-k\ 4q)\alpha_1-\alpha_{shift}}{2}\right)sin\left(\omega_1 t - \frac{k-1}{m}2\pi\right)\right]\end{array}\right] \tag{18}$$

For any full-pitch winding with odd q, we select $\alpha_{shift} = 2\pi/Q_1$ in (17), and analogously, for winding with even q, we apply $\alpha_{shift} = 0$. In case of chorded winding, we use either $\alpha_{shift} = 2\pi/Q_1$ for $q = 1, 2, 5, 6, 9, 10, \ldots$ or $\alpha_{shift} = 0$ for $q = 3, 4, 7, 8, 11, 12, \ldots$. To obtain the flux density distribution, formula $B_{m-phase}(\alpha) = \mu_0 H_{m-phase}(\alpha)$. must be applied to all equations relating to the air gap magnetic field strength distribution.

For the better understanding, we will analyze the air gap magnetic field for three various "fictive" single-layer (or double-layer) windings, considering slotless (7) and slotted (18) motor geometry. This situation is close to no-load motor operation (zero torque and zero rotor field). First, the flux density generated by a three-phase, full-pitched, winding having $Q_1 = 30$, $Q_2 = 22$, $2p = 2$ and $q = 5$ is shown in Figure 1. The left-side figure presents the flux density distribution as it depends on the air gap angular position, and the right-side figure shows the resulting frequency spectrum. While the red curve depicts the field considering smooth air gap with no slots present on the stator or the rotor, the blue curve shows the situation for slotted lamination.

Second, the flux density formed by a five-phase, full-pitched, winding having $Q_1 = 30$, $Q_2 = 22$, $2p = 2$ and $q = 3$ is shown in Figure 2.

Finally, the flux density formed by a seven-phase, full-pitched, winding having $Q_1 = 28$, $Q_2 = 22$, $2p = 2$ and $q = 2$ is shown in Figure 3.

The comparison between discussed windings shows that the motor having more phases can form a smoother magnetic field (lower content of harmonics) than the less phases motor even if it has similar slots number. Hence, this motor can generate less torque ripple, distinguishes lower THDi and can reduce noise. For air gap flux density spectrum, we can formulate several conclusions generally valid for any distributed multiphase winding having integer q.

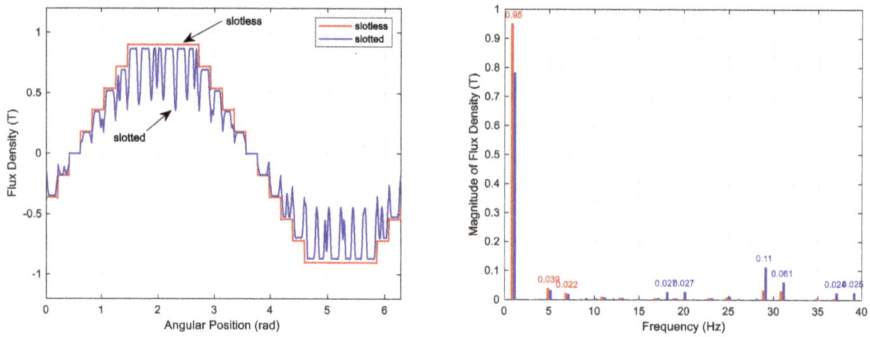

(a) (b)

Figure 1. (**a**) Air gap field generated by 3-phase winding; field distribution (**left**), (**b**) frequency spectrum (**right**).

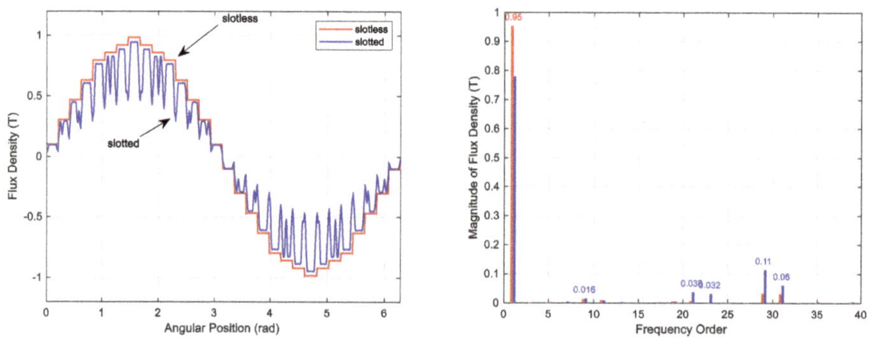

(a) (b)

Figure 2. (**a**) Air gap field generated by 5-phase winding; field distribution (**left**), (**b**) frequency spectrum (**right**).

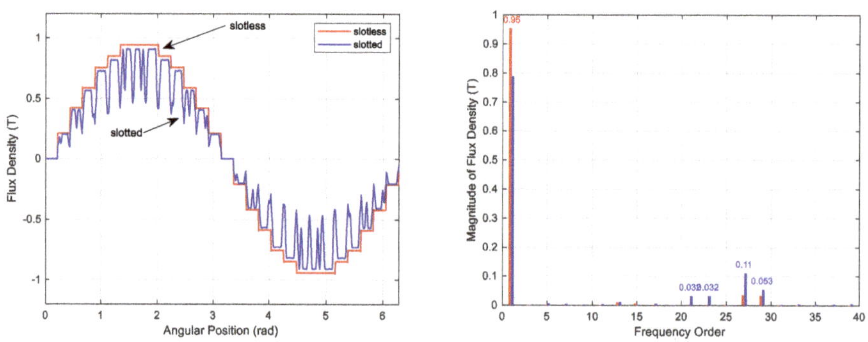

(a) (b)

Figure 3. (**a**) Air gap field generated by 7-phase winding; field distribution (**left**), (**b**) frequency spectrum (**right**).

First, the spectrum includes all harmonics orders calculated from (19). When the operator "+" is used, the relevant harmonic generates mmf travelling the air gap together with the fundamental harmonic but ν-times slower, and to the contrary, the operator "−" gives harmonics producing waves traveling the air gap in opposite direction (also ν-times slower).

$$\nu = 2mc \pm 1,\ c \in \mathbb{Z} \tag{19}$$

Second, the spectrum also includes frequency orders called "step" harmonics obtained from (20). They are present in the spectrum mainly due to the fact that the winding is placed in a finite number of stator slots. The stator "slot" harmonics have therefore the same orders as have the "step" harmonics, and hence, they both overlap in the frequency spectrum. The "step" harmonics have the winding factor of the same size as the fundamental harmonic.

$$\nu_{1step} = \nu_{1slot} = c\frac{Q_1}{p} \pm 1,\ c \in \mathbb{Z}. \tag{20}$$

Besides the "step" harmonic orders, the air gap flux density includes also the rotor "slot" harmonics (21).

$$\nu_{2slot} = c\frac{Q_2}{p} \pm 1,\ c \in \mathbb{Z} \tag{21}$$

Third, as shown in (18) the winding factor considering straight rotor bars is still given by (22),

$$k_{w\nu} = sin\left(\nu\pi\frac{\beta}{2}\right)\frac{sin\left(\nu\frac{\pi}{m'}\right)}{q\ sin\left(\nu\frac{\pi}{m'q}\right)} \tag{22}$$

Hence the magnitude of ν-th harmonics can be calculated from fundamental harmonic using (23).

$$B_\nu = \frac{B_1}{\nu}k_{w\nu} \tag{23}$$

According to previous results, the harmonic orders of the "step" and the stator "slot" harmonics interact and modify the original order (20) magnitudes. Considering only the interaction between the very first "step" and "slot" harmonic orders calculated from (20) when $c = 1$, then the resulting magnitude of (24),

$$\nu = \frac{Q_1}{p} - 1 \tag{24}$$

Becomes, according to (25),

$$B_\nu = B_{1step}\left[\frac{a_1}{2a_0}\frac{Q_1 - p}{p} + 1\right] \tag{25}$$

Moreover, analogously, for (26),

$$\nu = \frac{Q_1}{p} + 1 \tag{26}$$

We obtain (27):

$$B_\nu = B_{1step}\left[\frac{a_1}{2a_0}\frac{Q_1 + p}{p} - 1\right] \tag{27}$$

Into (25) and (27), we substitute from (13) for a_0 and from (16) for a_1 with applying $\nu = 1$. This gives us a rough estimate of analyzed harmonic amplitudes.

To demonstrate validity of the method, we can analyze (using FEA) the air gap magnetic field of real 3 kW five-phase induction motor corresponding to the case study shown in Figure 2. The motor geometry with magnetic distribution is shown on the left side of Figure 4. The right side of the same figure shows the resulting air gap flux density calculated using FEA and compared to given analytical approach. In order to compare comparable, the FEA considers no-load state (zero torque) and magnetic core composed of linear steel having very high relative permeability ($\mu_r = 10^4$).

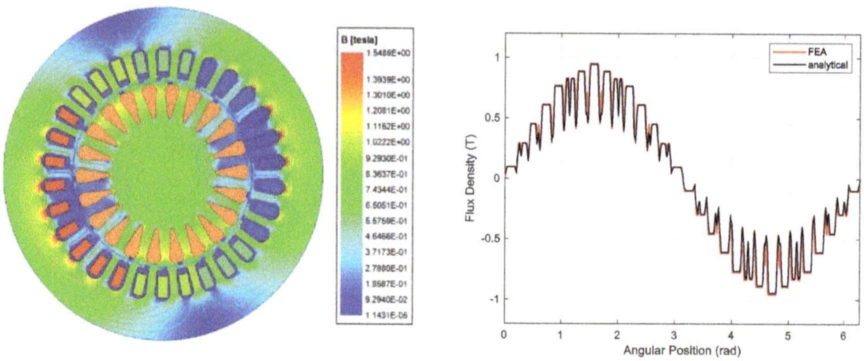

(a) (b)

Figure 4. (**a**) No-load flux density distribution inside the machine, (**b**) air gap flux density.

For the no-load operation, the proposed analytical method of air gap space harmonics prediction works well, but it may fail when the machine operates under load condition. In this case, the induced voltage generates a current flowing through the rotor, which generates its own magnetic field. This rotor field interacts with the original stator field, which in turn produces torque. As the air gap flux density combines both the stator and the rotor magnetic fields, the resulting curve is deformed as compared to the no-load air gap magnetic field. An example of the situation is shown in Figure 5. The motor (from Figure 4) operates under load (3 kW), and its magnetic field is calculated using nonlinear (steel with $\mu_r = f(I)$) transient analysis to reach as high accuracy as possible. Figure 5 on the left side represents the instantaneous distribution of the magnetic field in the motor cross-section, and the right frame shows the corresponding air gap flux density curve.

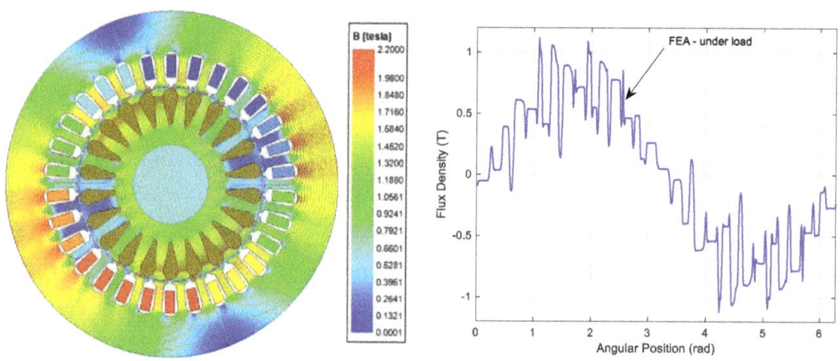

(a) (b)

Figure 5. (**a**) Flux density distribution inside the machine under load, (**b**) air gap flux density.

The no-load and the loaded operation states are compared graphically in Figure 6 by composing the two curves (taken from Figures 4 and 5) in one graph. The red line (taken from Figure 4) shows the flux density curve corresponding to the no-load operation, and the blue curve (taken from Figure 5) shows the magnetic field corresponding to the operation under load.

As obvious from the right side of Figure 6, the frequency spectrum of the loaded flux density includes harmonics that have not been predicted yet, and which are mainly caused by the rotor field presence and by saturation of the motor magnetic core.

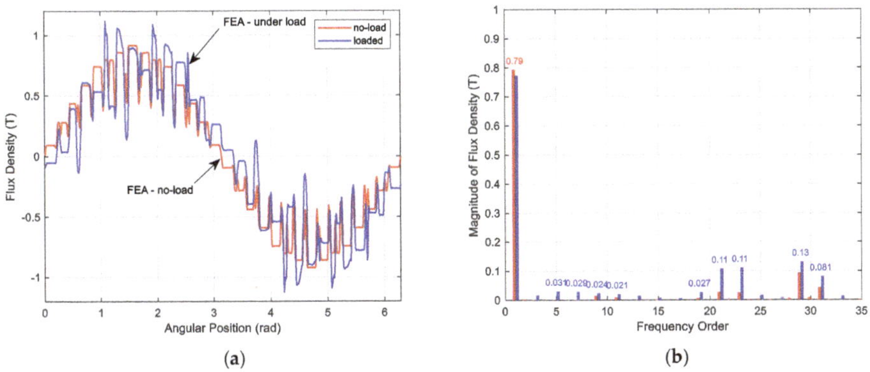

Figure 6. (**a**) Comparison between no-load and loaded operational state; field distribution (**b**) from FEA, frequency spectrum (right).

The saturated motor flats the air gap magnetic field according to the BH curve of steel used for the motor construction. As the field is a periodic odd function, the spectrum will contain group of "saturation" harmonics (28), even if the stator and rotor are considered to be slot-less.

$$v_{saturation} = 2(c-1)+1,\ c \in \mathbb{Z} \tag{28}$$

Although we classify these harmonics as space harmonics, their speed and direction do not follow the previous rules. Analyzing the air gap flux density, we find that the resulting (flattened) mmf travels the air gap with a constant shape (slots are not considered), which means that the harmonics created by the saturation must travel at the same speed and in the same direction as the fundamental wave. As a result, these harmonics contribute to the machine's useful torque.

Since these harmonics travel through the air gap synchronously with the fundamental harmonic, their slip must be the same as slip of the fundamental harmonic. Currents induced into the rotor bars will therefore include harmonics corresponding to (28).

Rotor time harmonics can be calculated either by analytical approach based on methodology described in [31] or by using FEA. The authors of [31] represented a squirrel-cage rotor by a star-connected winding with transformation of the end rings into stars. From pre-calculated saturation harmonics [30,31], they derived the induced voltage per leg of the proposed equivalent diagram and calculated the harmonic currents using transformed bar resistance and reactance.

An example relating to motor in Figure 5 is seen in Figure 7, where the left side represents the time dependency of the current (calculated from FEA) flowing through the bar, and the right side shows its frequency spectrum.

Analogously to (1), we can find the magnetic field (29) generated by the *k*th single rotor bar carrying the current i_k.

$$B_{bar-k}(\alpha) = i_k \frac{\mu_0}{\pi\delta(\alpha)} \sum_{v=1}^{\infty} \frac{1}{v} sin(v[\alpha-(k-1)\alpha_1]) \tag{29}$$

A symmetrically manufactured rotor, having bars evenly distributed along the air gap, generates magnetic field of opposite direction to the field formed by the stator. Considering sinusoidal rotor current distribution, Equation (30) describes the resulting rotor field. Here, I_{2m} represents the bar current magnitude.

$$B_{cage}(\alpha) = \frac{\mu_0}{\pi\delta(\alpha)} I_{2m} \sum_{k=1}^{Q_2} \left[sin\left(\omega t - p\frac{k-1}{Q_2}2\pi\right) \sum_{v=1}^{\infty} \frac{1}{v} sin(v[\alpha-(k-1)\alpha_1]) \right] \tag{30}$$

As shown in Figure 7, the bar current includes, besides the fundamental harmonic component, also an amount of additional time harmonics deforming the previously assumed sinusoidal current wave. Based on this, (30) should be rewritten in (31) to consider n rotor time harmonics.

$$B_{cage}(\alpha) = \frac{\mu_0}{\pi\delta(\alpha)} \sum_{k=1}^{Q_2} \left[\sum_{\mu=1}^{n} \left[I_\mu sin\left(\mu \left[\omega t - p\frac{k-1}{Q_2} 2\pi - \varphi_{i\mu} \right] - \varphi_{shift} \right) \right] \sum_{\nu=1}^{\infty} \frac{1}{\nu} sin(\nu[\alpha - (k-1)\alpha_1]) \right] \quad (31)$$

New parameter $\varphi_{i\mu}$ represents the phase shift of μth time harmonic component, and φ_{shift} is the angle measured between the stator and the rotor mmfs obtained analyzing the machines equivalent circuit [32,33]. Final magnetic field curve (see Figure 8) is then given as the difference between the stator and the rotor fields, i.e., $B_{m-phase}(\alpha) - B_{cage}(\alpha)$.

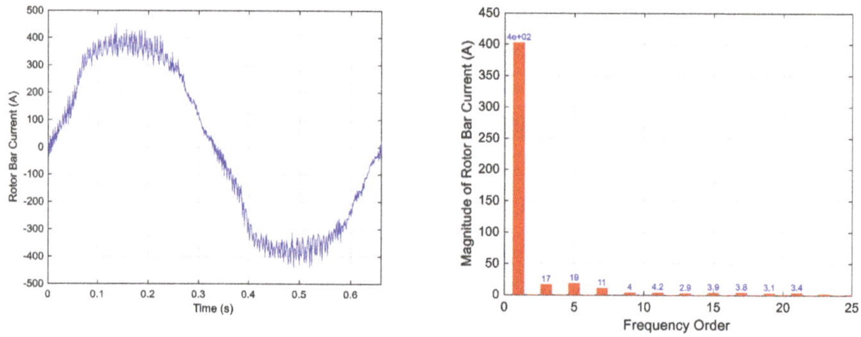

(a) (b)

Figure 7. (**a**) Time dependency of bar current (**left**) and (**b**) its harmonic spectrum (**right**).

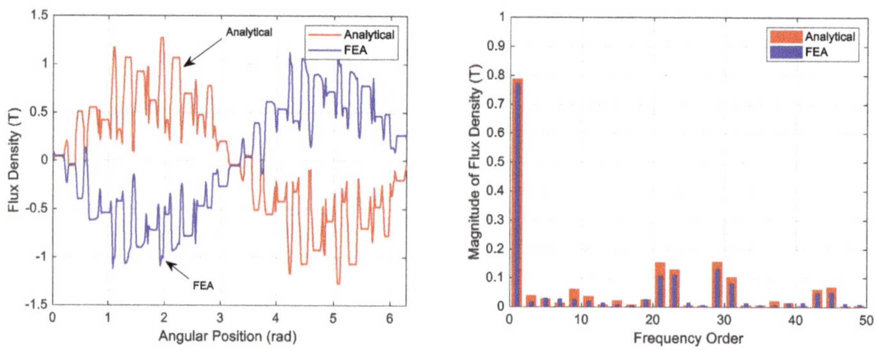

(a) (b)

Figure 8. (**a**) Final air gap flux density for motor operating under load condition; flux density distribution (**left**)—blue line inverted, (**b**) frequency spectrum (**right**).

The left side of Figure 8 compares the flux density waveform obtained using proposed analytical approach (red line) to the one derived from FEA (blue line). To ensure better clarity, the blue line is inverted. The right side of Figure 8 shows comparison between harmonic spectrum corresponding to both waveforms.

The results show good agreement between the proposed method and FEA and is therefore an effective way to air gap flux density prediction in any multiphase induction motor having distributed winding with integer slots number per pole per phase.

3. Time Harmonics in Multiphase Winding

For common three-phase industrial line-started induction machines, the significant source of non-harmonic voltage is mainly the power supply imbalance. Multiphase motors are usually powered from frequency converters used for easy and energy efficient speed control. The inverters (especially the voltage-types) generate voltage having either rectangular or pulsed shape. Thus, its spectrum contains the amount of harmonics μ dependent on the load and the pulse width modulation settings. Non-harmonic power supply can cause parasitic torque ripples, vibrations, increased noise, increased voltage stress of the insulation system and also can generate higher I^2R winding losses due to harmonic currents.

Sometimes, particularly when talking about multiphase motors, the time harmonics are injected into the power supply in order to increase torque per ampere. The extra torque is obtained due to the fact that the flux distribution in the air gap is flattened so that the saturation can be avoided for a wider operational range.

3.1. Harmonics Creating 2p Number of Pole Pairs

Operational properties of significant time harmonics and their influence on the motor can be analyzed composing a time-varying phasor diagram showing the magnetic field (32).

$$F_{m-phase\ \mu} = F_{m-\mu} \sum_{k=1}^{m} \left[cos\left(\alpha - \frac{k-1}{m} 2\pi\right) cos\left(\mu\omega_1 t - \mu \frac{k-1}{m} 2\pi\right)\right] \quad (32)$$

Sine wave current, represented by the second multiplier in the summation, flowing in each of m stationary coils, represented by the first multiplier, produces m sine varying magnetic fields perpendicular to the rotation axis. The m magnetic fields add as vectors to produce a single rotating magnetic field $F_{m-phase\ \mu}$.

For example, the fundamental harmonic component working in three-phase motor generates positive sequence field with amplitude (33).

$$\begin{gathered} F_{3-phase\ 1} = F_{m1}\tfrac{1}{2}[cos(\alpha-\omega_1 t) + cos(\alpha+\omega_1 t)] + F_{m1}\tfrac{1}{2}[cos(\alpha-\omega_1 t) + \\ cos\left(\alpha+\omega_1 t - \tfrac{4}{3}\pi\right)\Big] + F_{m1}\tfrac{1}{2}\Big[cos(\alpha-\omega_1 t) + cos\left(\alpha+\omega_1 t - \tfrac{8}{3}\pi\right)\Big] = \cdots = \\ F_{m1}\tfrac{3}{2}cos(\alpha-\omega_1 t) \end{gathered} \quad (33)$$

Similarly, we can derive (34) for the 5th and 7th harmonic components obtaining the negative and the positive sequence waves, respectively.

$$\begin{gathered} F_{3-phase\ 5} = F_{a5} + F_{b5} + F_{c5} = \cdots = F_{m5}\tfrac{3}{2}cos(\alpha + 5\omega_1 t) \\ F_{3-phase\ 7} = F_{a7} + F_{b7} + F_{c7} = \cdots = F_{m7}\tfrac{3}{2}cos(\alpha - 7\omega_1 t). \end{gathered} \quad (34)$$

Performing analyses for randomly chosen motors and harmonics, e.g., the five-phase motor fed by 5th harmonics and seven-phase motor fed by 7th harmonics, we get (35).

$$\begin{gathered} F_{5-phase\ 5} = F_{m5}\ \tfrac{5}{2}cos(\alpha + 5\omega_1 t) \\ F_{7-phase\ 7} = F_{m5}\ \tfrac{7}{2}cos(\alpha - 7\omega_1 t) \end{gathered} \quad (35)$$

Normally, the literature recognizes the harmonic content in the voltage curve of the inverter following rule (36),

$$\mu = 2mc \pm 1,\ c \in \mathbb{Z} \quad (36)$$

However, in case of harmonic injection drives, the converter can generate any harmonic we need; therefore, (36) could be rewritten in more general (37).

$$\mu = mc \pm 1,\ c \in \mathbb{Z} \quad (37)$$

Based on the previous analyses, we may conclude that by applying the operator "+" on (37), the resulting harmonics create waves traveling the air gap together with the fundamental one, but their speed is μ-times higher. Moreover, to the contrary, the operator "−" gives harmonics producing waves passing through the air gap again μ-times faster but now in opposite direction. Under perfectly balanced conditions, all these harmonics produce rotating magnetic fields having constant amplitudes (38) and angular speeds in time, so they form circular shaped fields.

$$F_{m-phase\ \mu} = F_{m\mu} \frac{m}{2} \tag{38}$$

3.2. Harmonics Creating Higher Number of Pole Pairs than 2p

For any harmonic order originating from (39), the application of (32) gives always zero value. In addition, all phase currents (related to the analyzed harmonic) are in phase with each other and therefore generate zero sequence (non-rotating) field.

$$\mu = mc,\ c \in \mathbb{Z} \tag{39}$$

When injecting subharmonic ($\mu \notin \mathbb{Z}$), the resulting magnetic field changes the amplitude and angular speed in time, which deforms originally circular field into elliptical. In extreme cases (40), the field takes the pulsating form having zero translation speed against the air gap and generates losses, produces noise and causes homopolar magnetic saturation.

$$\mu = \frac{m}{2} + m(c-1),\ c \in \mathbb{Z}. \tag{40}$$

The situation becomes more complicated when studying integer harmonic orders that are not predicted, neither by (37) nor (39). Substituting into (32) we get zero values even though they are harmonics generating fields with different phase shift relative to each other. Therefore, they produce magnetic field passing through the air gap with nonzero speed. For five-phase motor, we find those harmonics from (41),

$$\mu = m(c-1) \pm 2,\ c \in \mathbb{Z} \tag{41}$$

However, the seven-phase motor includes, besides (41), also (42).

$$\mu = m(c-1) \pm 3,\ c \in \mathbb{Z} \tag{42}$$

Hence, in case of nine-phase motor we must apply (41) to (43).

$$\mu = m(c-1) \pm 4,\ c \in \mathbb{Z} \tag{43}$$

All these harmonics produce a magnetic field with higher number of pole pairs than generated by the fundamental harmonic. This is due to the specific "time" re-assembling of the stator phases (μ-multiplication of individual phase shifts) while keeping the same winding mechanical distribution. This in turn forms a new winding with completely new Görges diagram [30] producing higher number of pole-pairs. Figure 9 shows the example of the 3rd harmonic (magnitude equals to the fundamental) injected into the five-phase motor discussed previously in Figure 5 ($Q_1 = 30$, $Q_2 = 22$, $2p = 2$, $q = 3$).

As it forms $3 \times 2p$ number of poles, the speed must be the same as that developed by the fundamental harmonic. Similar behavior can be observed also for the 3rd and 5th current harmonic in seven-phase motor and for the 3rd, 5th and 7th harmonic in nine-phase motor (see left side of Figure 10).

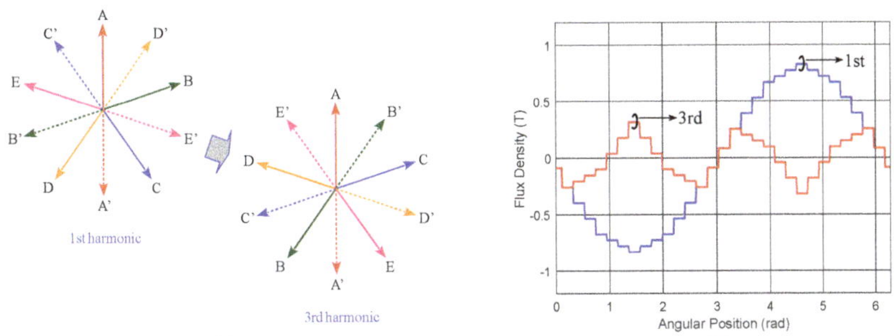

(a) (b)

Figure 9. (**a**) Stator phases re-assembling, (**b**) mmf for the fundamental and the 3rd harmonics.

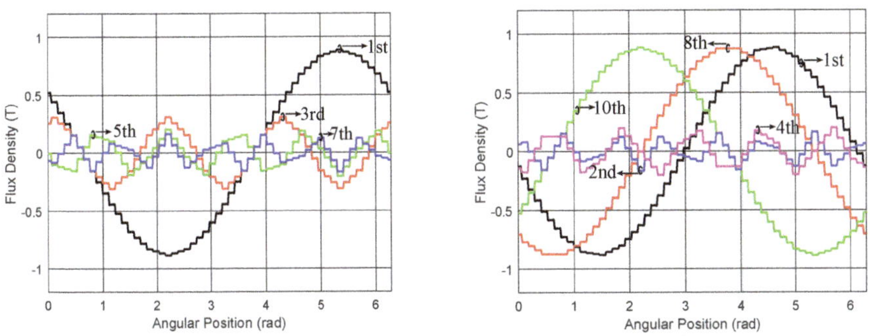

(a) (b)

Figure 10. Time harmonics acting in multiphase winding: (**a**) harmonics suitable for injection, (**b**) harmonics inappropriate for injecting.

Amplitudes of individual harmonics could be derived from new Görges diagram. As these amplitudes are defined by half ampere-turns corresponding to one magnetic pole, we may calculate them from the ampere-turns that belong to one pole pitch. These specific "pole-creating" harmonics generate in the air gap space harmonics having the same order ($\mu = \nu$), so we can consider the pole pitch to be μ-times shorter as compared to the fundamental. Hence, only (44) coil turns contribute to the final value of individual harmonic amplitude.

$$q\frac{m}{2\mu}N \tag{44}$$

With multiplication (44) by the mean value of the stator current I_μ we obtain the resulting mmf from (45).

$$H_{\mu-mag} = \frac{q}{\delta_0}\frac{m}{\mu}N\frac{I_\mu}{\pi}\sqrt{2},\ \mu = \nu \tag{45}$$

Considering the winding factor of μ-th harmonic component (still we assume $\mu = \nu$), the magnitude of the air gap flux density is then (46).

$$B_{\mu-mag} = \mu_0\frac{q}{\delta_0}\frac{m}{\pi}N\frac{k_{w\nu}}{\mu}I_\mu\sqrt{2},\ \mu = \nu \tag{46}$$

As demonstrated in the right side of the figure, other harmonics (excluding (37) and (39)) may create either circular or the elliptical fields traveling in different directions and speeds.

Table 1 shows the same overview but extended by nine-phase machines. Column "poles" gives number of magnetic poles created by the particular harmonic component, column "sequence" shows the traveling direction (or field character) and column "speed" specifies the velocity relative to the fundamental wave. The field character shows whether the mmf travels against the air gap ("+" or "−") or pulsates ("puls.") with zero speed.

Table 1. Time harmonics effect overview.

	Five-Phase Winding			Seven-Phase Winding			Nine-Phase Winding		
μ	Poles	Sequence	Speed	Poles	Sequence	Speed	Poles	Sequence	Speed
1	$2p$	+	1	$2p$	+	1	$2p$	+	1
2	$3 \times 2p$	−	2/3	$5 \times 2p$	−	2/5	$7 \times 2p$	−	2/7
2.5	$3 \times 2p$	puls.	0						
3	$3 \times 2p$	+	1	$3 \times 2p$	+	1	$3 \times 2p$	+	1
3.5				$3 \times 2p$	puls.	0			
4	$2p$	−	4	$3 \times 2p$	−	4/3	$5 \times 2p$	−	4/5
4.5							$5 \times 2p$	puls.	0
5	$5 \times 2p$	zero	0	$5 \times 2p$	+	1	$5 \times 2p$	+	1
6	$2p$	+	6	$2p$	−	6	$3 \times 2p$	−	6/3
7	$3 \times 2p$	−	7/3	$7 \times 2p$	zero	0	$7 \times 2p$	+	1
7.5	$3 \times 2p$	puls.	0						
8	$3 \times 2p$	+	8/3	$2p$	+	8	$2p$	−	8
9	$2p$	−	9	$5 \times 2p$	−	9/5	$9 \times 2p$	zero	0
10	$5 \times 2p$	zero	0	$3 \times 2p$	+	10/3	$2p$	+	10

Following this, we can construct Figures 11–13 to give graphical demonstration of the harmonics' operational influence.

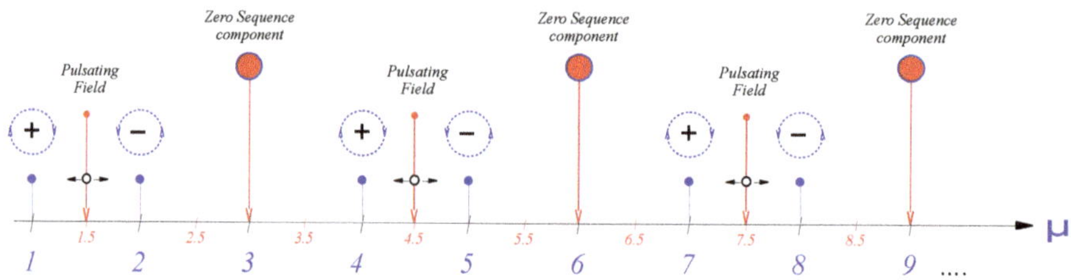

Figure 11. Time harmonics effect demonstration in three-phase machines.

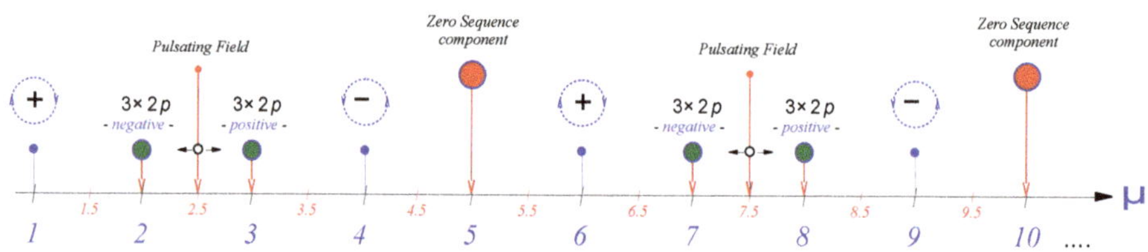

Figure 12. Time harmonics effect demonstration in five-phase machines.

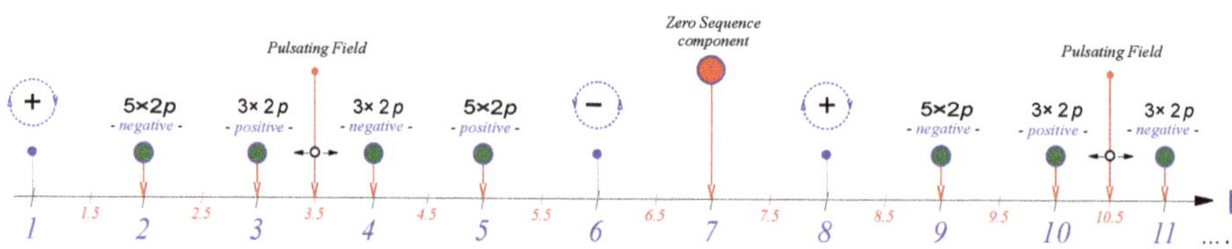

Figure 13. Time harmonics effect demonstration in seven-phase machines.

As obvious, there is an amount of harmonics that can create magnetic field traveling the air gap with the same speed and direction that travels the fundamental wave (color marked). These harmonics are usually being injected in order to increase the torque.

4. Conclusions

From the research results, it is evident that a motor having a higher number of phases can produce a more sinusoidal magnetic field than the motor having a smaller number of phases, even if it has a similar number of slots. Hence, by increasing the number of phases, the torque harmonics decrease. For distributed multiphase winding with integer q, we can conclude following rules.

First, the spectrum of the air gap flux density includes harmonic orders obtained from (19). These harmonics generate fields traveling the air gap in a different direction (according to operator used) and speed (ν-times lower than fundamental).

Second, as the windings are physically placed in slots, the spectrum includes significant orders called "step" harmonics (20) having the same winding factor as calculated for fundamental harmonic. The real "slot" harmonics are obtained also from (20), and therefore, their orders overlap (interact) with "step" harmonics. As a consequence, this interaction modifies original magnitudes of (20), and resulting air gap flux density is influenced.

Third, as shown in (18), the winding factor considering straight rotor bars is still given by (22), hence the magnitude of ν-th harmonics can be calculated from fundamental harmonic using (23).

Fourth, the injection of time harmonics can produce magnetic field having even more than $2p$ magnetic pole-pairs. For any harmonic order originating from (39), the application of (32) produces zero value, therefore zero sequence (non-rotating) field.

Fifth, considering subharmonics ($\mu \notin \mathbb{Z}$), the resulting magnetic field deforms its originally circular shape into an elliptical one. In extreme case (40), the field pulsates in only one axis forming a kind of homopolar field. When studying integer harmonic orders that are not predicted, neither by (37) nor (39), the substitution into (32) gives zero values even though they are harmonics establishing field waves passing through the air gap win non-zero speed. All these harmonics produce magnetic field having more magnetic pole-pairs than $2p$. Some of them (e.g., 3rd, 5th and 7th harmonic in nine-phase motor) can travel the air gap with the same speed as developed by the fundamental harmonic and may be therefore injected in order to increase the torque.

Author Contributions: Conceptualization, V.K. and R.C.; methodology, V.K.; software, V.K.; finite element analyses, R.C. and Z.F.; formal analysis, V.K.; writing—original draft preparation, V.K.; writing—review and editing, V.K., R.C., Z.F. and B.S.; visualization, V.K.; supervision, Z.F. and B.S. All authors have read and agreed to the published version of the manuscript.

Funding: This research was funded by the Ministry of Education, Youth and Sports of the Czech Republic under the project OP VVV Electrical Engineering Technologies with High-Level of Embedded Intelligence CZ.02.1.01/0.0/0.0/18_069/0009855 and by funding program of the University of West Bohemia number SGS-2018-009. This research was also funded by Slovak Research and Development Agency under the contract No. APVV-18-0436.

Conflicts of Interest: The authors declare no conflicts of interest.

References

1. Tie, S.F.; Tan, C.W. A review of energy sources and energy management system in electric vehicles. *Renew. Sustain. Energy Rev.* **2013**, *20*, 82–102. [CrossRef]
2. Chau, K.T.; Wong, Y.S.; Chan, C.C. An overview of energy sources for electric vehicles. *Energy Convers. Manag.* **1999**, *40*, 1021–1039. [CrossRef]
3. Su, W.; Eichi, H.; Zeng, W.; Chow, M.Y. A survey on the electrification of transportation in a smart grid environment. *IEEE Trans. Ind. Inform.* **2012**, *8*, 1–10. [CrossRef]
4. Emadi, A.; Lee, Y.J.; Rajashekara, K. Power electronics and motor drives in electric, hybrid electric, and plug-in hybrid electric vehicles. *IEEE Trans. Ind. Electron.* **2008**, *55*, 2237–2245. [CrossRef]
5. Rajashekara, K. Present status and future trends in electric vehicle propulsion technologies. *IEEE J. Emerg. Sel. Top. Power Electron.* **2013**, *1*, 3–10. [CrossRef]

6. Chan, C.C. The state of the art of electric, hybrid, and fuel cell vehicles. *Proc. IEEE* **2007**, *95*, 704–718. [CrossRef]
7. Zhu, Z.Q.; Howe, D. Electrical machines and drives for electric, hybrid, and fuel cell vehicles. *Proc. IEEE* **2007**, *95*, 746–765. [CrossRef]
8. Pellegrino, G.; Vagati, A.; Boazzo, B.; Guglielmi, P. Comparison of induction and PM synchronous motor drives for EV application including design examples. *IEEE Trans. Ind. Appl.* **2012**, *48*, 2322–2332. [CrossRef]
9. Popescu, M.; Goss, J.; Staton, D.A.; Hawkins, D.; Chong, Y.C.; Boglietti, A. Electrical vehicles—Practical solutions for power traction motor systems. *IEEE Trans. Ind. Appl.* **2018**, *54*, 2751–2762. [CrossRef]
10. Yang, Z.; Shang, F.; Brown, I.P.; Krishnamurthy, M. Comparative study of interior permanent magnet, induction, and switched reluctance motor drives for EV and HEV applications. *IEEE Trans. Transp. Electrif.* **2015**, *1*, 245–254. [CrossRef]
11. Boldea, I.; Tutelea, L.N.; Parsa, L.; Dorrell, D. Automotive electric propulsion systems with reduced or no permanent magnets: An overview. *IEEE Trans. Ind. Electron.* **2014**, *61*, 5696–5711. [CrossRef]
12. Kowal, A.; Arahal, M.R.; Martin, C.; Barrero, F. Constraint satisfaction in current control of a five-phase drive with locally tuned predictive controllers. *Energies* **2019**, *12*, 2715. [CrossRef]
13. Resa, J.; Cortes, D.; Marquez-Rubio, J.F.; Navarro, D. Reduction of induction motor energy consumption via variable velocity and flux references. *Electronics* **2019**, *8*, 740. [CrossRef]
14. Zamani, J.S. Computer simulation of converter fed synchronous machine drives containing space harmonics. In Proceedings of the International Conference on Power Electronics, Drives and Energy Systems for Industrial Growth, New Delhi, India, 8–11 January 1996; pp. 944–950.
15. Fu, J.-R.; Lipo, T.A. Disturbance-free operation of a multiphase current-regulated motor drive with an opened phase. *IEEE Trans. Ind. Appl.* **1994**, *30*, 1267–1274. [CrossRef]
16. Zhao, Y.; Lipo, T.A. Modeling and control of a multi-phase induction machine with structural unbalance. *IEEE Trans. Energy Convers.* **1996**, *11*, 578–584. [CrossRef]
17. Levi, E.; Bojoi, R.; Profumo, F.; Toliyat, H.A.; Williamson, S. Multiphase induction motor drives—A technology status review. *IET Electr. Power Appl.* **2007**, *1*, 489–516. [CrossRef]
18. Toliyat, H.A.; Lipo, T.A. Analysis of concentrated winding induction machines for adjustable speed drive applications-experimental results. *IEEE Trans. Energy Convers.* **1994**, *9*, 695–700. [CrossRef]
19. Abdel-Khalik, A.S.; Masoud, M.I.; Williams, B.W. Improved flux pattern with third harmonic injection for multiphase induction machine. *IEEE Trans. Power Electron.* **2012**, *27*, 1563–1578. [CrossRef]
20. Duran, M.J.; Salas, F.; Arahal, M.R. Bifurcation analysis of five-phase induction motor drives with third harmonic injection. *IEEE Trans. Ind. Electron.* **2008**, *55*, 2006–2014. [CrossRef]
21. Lyra, R.O.C.; Lipo, T.A. Torque density improvement in a six-phase induction motor with third harmonic current injection. *IEEE Trans. Ind. Appl.* **2002**, *38*, 351–1360. [CrossRef]
22. Abdel-Khalik, A.S.; Masoud, M.I.; Ahmed, S.; Massoud, A.M. Effect of current harmonic injection on constant rotor multiphase induction machine stators: A comparative study. *IEEE Trans. Ind. Appl.* **2012**, *48*, 2002–2013. [CrossRef]
23. Ferkova, Z.; Kindl, V. Influence of skewed squirrel cage rotor with intermediate ring on magnetic field of air gap in induction machine. *Electr. Eng.* **2017**, *23*, 26–30. [CrossRef]
24. Orság, O.; Rusnok, S.; Sobota, P.; Kačor, P. Influence of rotor slot shape on the parameters of induction motor. In Proceedings of the 2017 IEEE International Conference on Environment and Electrical Engineering and the 2017 IEEE Industrial and Commercial Power Systems Europe (EEEIC/I&CPS Europe), Milan, Italy, 6–9 June 2017; pp. 1–6.
25. Bernat, P.; Kacor, P. Operational non-contact diagnostics of induction machine based on stray electromagnetic field. *Commun. Sci. Lett. Univ. Zilina* **2015**, *17*, 89–94.
26. Alger, P.L. *The Nature of Polyphaser Induction Machines*; Wiley: Hoboken, NJ, USA, 1951.
27. Neto, L.M.; Camacho, J.R.; Salerno, C.H.; Alvarenga, B.P. Analysis of a three-phase induction machine including time and space harmonic effects: The a, b, c reference frame. *IEEE Trans. Energy Convers.* **1999**, *14*, 80–85. [CrossRef]
28. Stincescu, R.B.; Viarouge, P.; Cros, J.; Kamwa, I. A general approach of space and time harmonics interactions in induction motors. In Proceedings of the IEEE International Electric Machines and Drives Conference, Seattle, WA, USA, 3–12 May 1999; pp. 366–368. [CrossRef]

29. Heller, B.; Hamata, V. *Harmonic Field Effects in Induction Machines*; Elsevier Science Ltd.: Amsterdam, The Netherlands, 1977; ISBN 9780444998569.
30. Liao, Y.; Lipo, T.A. Effect of saturation third harmonic on the performance of squirrel-cage induction machines. *Electr. Mach. Power Syst.* **1994**, *22*, 155–171. [CrossRef]
31. Lee, C.H. Saturation 597–harmonics of polyphase induction machines. *Trans. Am. Inst. Electr. Eng. Part III Power Appar. Syst.* **1961**, *80*, 603. [CrossRef]
32. Pyrhönen, J.; Jokinen, T.; Hrabovcova, V. *Design of Rotating Electrical Machines*, 2nd ed.; John Wiley & Sons Ltd.: Chichester, UK, 2014; ISBN 978-1-118-58157-5.
33. Laksar, J.; Sobra, J.; Veg, L. Numerical calculation of the effect of the induction machine load on the air gap magnetic flux density distribution. In Proceedings of the 2017 18th International Scientific Conference on Electric Power Engineering (EPE), Kouty nad Desnou, Czech Republic, 17–19 May 2017; pp. 1–6. [CrossRef]

energies

MDPI

Article

Acoustic Noise Computation of Electrical Motors Using the Boundary Element Method

Sabin Sathyan [1,*], Ugur Aydin [1] and Anouar Belahcen [1,2]

1 Department of Electrical Engineering and Automation, Aalto University, PO Box, 15500 Helsinki, Finland; ugur.aydin@aalto.fi (U.A.); anouar.belahcen@aalto.fi (A.B.)

2 Department of Electrical Tallinn University of Technology Power Engineering and Mechatronics, 19086 Tallinn, Estonia

* Correspondence: sabin.sathyan@aalto.fi; Tel.: +358-40662-8668

Received: 11 October 2019; Accepted: 31 December 2019; Published: 3 January 2020

Abstract: This paper presents a numerical method and computational results for acoustic noise of electromagnetic origin generated by an induction motor. The computation of noise incorporates three levels of numerical calculation steps, combining both the finite element method and boundary element method. The role of magnetic forces in the production of acoustic noise is established in the paper by showing the magneto-mechanical and vibro-acoustic pathway of energy. The conversion of electrical energy into acoustic energy in an electrical motor through electromagnetic, mechanical, or acoustic platforms is illustrated through numerical computations of magnetic forces, mechanical deformation, and acoustic noise. The magnetic forces were computed through 2D electromagnetic finite element simulation, and the deformation of the stator due to these forces was calculated using 3D structural finite element simulation. Finally, boundary element-based computation was employed to calculate the sound pressure and sound power level in decibels. The use of the boundary element method instead of the finite element method in acoustic computation reduces the computational cost because, unlike finite element analysis, the boundary element approach does not require heavy meshing to model the air surrounding the motor.

Keywords: acoustics; boundary element method; electric machines; finite element method; induction motors; magneto-mechanics; modeling; noise; vibro-acoustics

1. Introduction

The acoustic noise in electric motors is a phenomenon of a complex nature and origin. The first kind, electromagnetic vibration and noise, is produced by magnetic forces, magnetostrictive expansion of the core laminations, eccentricity, phase unbalance, slot openings, and magnetic saturation. The second cause of noise is mechanical and is associated with mechanical assembly, in particular the bearings. The third major group is aerodynamic noise, which is due to the flow of ventilating air through or over the motor. These three sources are illustrated in Figure 1. A detailed review on the different forms of vibration and noise in electrical motors can be found in a paper by Vijayraghavan et al. [1]. One form of energy conversion happening in an electrical motor is from electrical energy to acoustic energy. The supply current interacts with the material to produce a magnetic field, which in turn produces magnetic forces. These forces excite the stator core and frame in the corresponding frequency range and produce mechanical vibrations. As a consequence of vibrations, the surface of the stator yoke and frame deforms with frequencies corresponding to the frequencies of forces. These stator and frame vibrations cause the surrounding medium of air to excite and vibrate and finally generate acoustic pressure variations (and thereby noise).

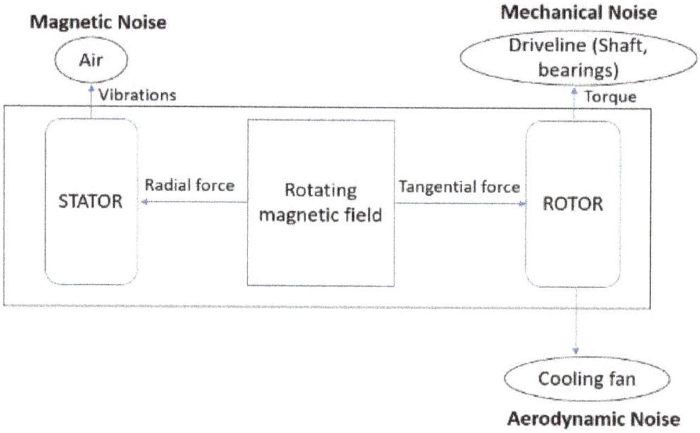

Figure 1. Generation of noise of different origins in a rotating electrical machine.

There have been various studies in the literature pertaining to the analysis of the vibrations and noise of electrical motors. Early stages of noise studies used analytical models and combined numerical and analytical methods. Belmans et al. carried out studies on the analytical formulation of acoustic noise in induction motors and authenticated the findings with experimental results [2]. They employed the rotating field theory with the Maxwell tensor method for calculating the frequency components produced by a motor that connect it to the airgap flux density time harmonics produced by the supply. The inference from their study was that high noise levels may be anticipated when one of the frequencies of the electromagnetically excited forces becomes the same as a natural frequency of the stator. They also developed a computerized model using finite element calculations and modal analysis that predicts the frequency components expected in the audible noise of a three-phase induction motor [3]. Cameron et al. have done measurement-based studies on vibrations and noise on reluctance motors and established that the stator deformation due to radial magnetic forces is the leading electromagnetic cause of noise [4]. Besnerais et al. demonstrated the impact of a Pulse Width Modulation (PWM) supply and switching frequencies on the magnetic noise of induction machines using analytical models [5]. Their approach was based on mechanical and acoustic 2D-ring stator models to compute the effect of winding space harmonics and PWM time harmonics in noise production [6]. Later on, Besnerais developed a multiphysical simulation tool for fast calculation of acoustic noise based on analytical and semi-analytical methods [7]. Their platform incorporated different models such as a permeance/magneto motive force (mmf) model, a subdomain model, and a finite element model. The semi-analytical models proved to be faster than the fully finite element models and had the same accuracy level, according to their evaluations. Devillers et al. studied the effect of tangential magnetic forces on vibrations and acoustic noise using a fast subdomain method to calculate Maxwell stress distribution and an electromagnetic vibration synthesis technique [8]. The same team later developed an experimental benchmark set-up for magnetic noise and a vibration analysis of electrical machines [9]. Fakam et al. coupled finite element structural analysis with an analytical tool to compute and compare the electromagnetic noise between surface permanent magnet and interior permanent magnet rotor topologies of a synchronous machine [10]. The same approach of a combined structural finite element method (FEM) and analytical methods was formulated by Islam et al. for computing sound power levels in synchronous motors [11].

With the improvement in numerical computational tools, the use of the boundary element method (BEM) or a combined FEM–BEM in noise computations became widespread. Moreover, these methods can give more accurate results in acoustic calculations. Juhl et al. developed a numerical method based on the BEM to calculate acoustic noise using the Helmholtz integral equation [12]. They created BEM-based numerical software for calculating sound fields on the exterior of bodies of three-dimensional shape or axisymmetric geometries [13]. Wang et al. used the BEM for computing sound power radiating from induction motors and a coupled structural FEM and acoustic BEM in their work [14]. Herrin et al. formulated a high-frequency BEM and compared it to the Rayleigh approximations method. They concluded that the high-frequency BEM is the more robust method [15]. Roivanen has employed different methods such as the BEM, high-frequency BEM, and plate approximation method combined with broad analytical, numerical, and experimental studies for calculating and comparing the sound power levels of electrical motors [16]. Neves et al. [17] presented a study on the coupling between magnetic forces, vibrations, and noise in a switched reluctance motor using FEM and BEM and gathered relatively good results. However, there were some mismatches between measured and simulated results that have been attributed to noise of aerodynamic origin. Furlan et al. [18] followed the same approach of a joint FEM–BEM-based analysis of a permanent magnet Direct Current (DC) electric motor. They simulated all three models, including an electromagnetic analysis in 3D, which could increase the memory requirements and computation time. In this paper, an electromagnetic simulation was done in 2D, and magnetic forces were taken from a 2D model and put into a 3D structural model. Schlensok et al. [19] presented an acoustic simulation of an induction machine with a squirrel-cage rotor, where 3D models were employed for electromagnetic, structural, and acoustic simulations. The electromagnetically excited structure- and air-borne noise of the motor was described in detail in their study. Deng et al. [20] investigated noise in an axial flux permanent magnet motor using electromagnetic and structural FEM and acoustic BEM; however, their approach also included a time-consuming complete 3D scheme. Järvenpää et al. [21] proposed a fast boundary element simulation of noise, where they imposed the surface velocity of a structural FEM model as a source of a fast BEM. This method facilitates the efficient modeling of large acoustic problems, although the computational cost is relatively high in this approach. Besides the computation of sound pressure in pascals and sound levels in dB surrounding the motor (shown in some previous studies), this paper also presents a far-field sound level calculation, which portrays the directivity of sound around the motor.

Inspired by the studies and findings in the literature, this paper investigated the aspect of vibro-acoustics in an electrical motor and formulated a practical and effective approach for acoustic noise calculation. The computational methodology and results successfully present a numerical technique for computing acoustic noise generated by an induction motor. The FEM- and BEM-based model describes how various quantities in different domains in an electrical motor can be calculated and used for acoustic analysis. An extensive numerical analysis of the magneto-mechanical and vibro-acoustic characteristics of a high-speed induction motor used for industrial applications is an original element of this study. Moreover, this paper paves a vivid path for researchers by detailing and clarifying the intricacies related to the preparation of a vibro-acoustic model of an electric motor by explaining both the theoretical and numerical implementation facets of the model. Although many studies have been carried out in this field of research, this paper brings an original contribution through a fully numerical analysis, the implementation of which is explained in detail.

2. Computational Methodology

This section explains our computational methods, including the main equations and numerical simulation stages employed for calculating the magnetic, mechanical, and acoustic quantities of the motor. Numerical modeling of noise generation in an electrical motor involves three models: first, modeling of the electromagnetic forces; then, modeling of the structural deformation and vibration behavior; and finally, modeling of the consequent acoustic response of the motor, as depicted in Figure 2.

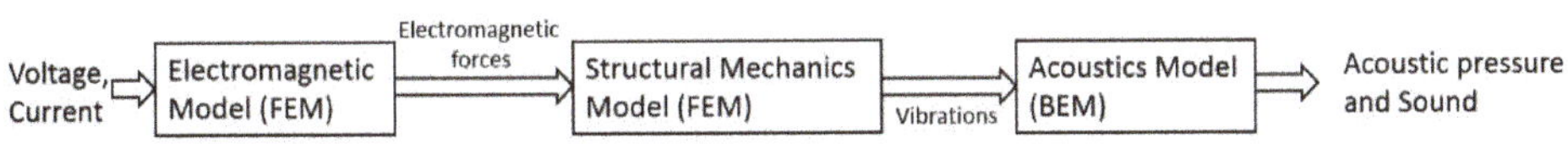

Figure 2. Numerical simulation stages.

2.1. Electromagnetics

The Maxwell equations of the magnetic field problem are solved numerically using 2D FEM. The Maxwell stress tensor method gives the magnetic torque exerted on a ferromagnetic region by integrating the magnetic stress over a surface around it [22]. Equation (1) is

$$T_{em} = \oint_S \boldsymbol{r} \times \boldsymbol{\tau} \cdot dS = \oint_S \boldsymbol{r} \times \left\{ \frac{1}{\mu_0} (\boldsymbol{B} \cdot \boldsymbol{n}) \boldsymbol{B} - \frac{1}{2\mu_0} \boldsymbol{B}^2 \boldsymbol{n} \right\} dS, \tag{1}$$

where τ is the Maxwell stress tensor, $\boldsymbol{r}$ is a vector representing the distance from the integration point to the torque axis, $\boldsymbol{B}$ is the magnetic flux density, μ_0 is the permeability of the vacuum, and $\boldsymbol{n}$ is the normal unit vector of the integration surface dS. In an electrical machine, the integration surface is chosen as the outer surface of the rotor or any cylindrical surface in the airgap.

The magnetic force can be calculated from the Maxwell stress tensor by the volume integral $\int_V \nabla . \tau dV$. This volume integral can be reduced to the closed surface integral over a surface S, and the force formula becomes [23]

$$F = \oint_S \left\{ \frac{1}{\mu_0} (\boldsymbol{B} \cdot \boldsymbol{n}) \boldsymbol{B} - \frac{1}{2\mu_0} \boldsymbol{B}^2 \boldsymbol{n} \right\} dS = \oint_S \left(\frac{1}{2\mu_0} ({B_n}^2 - {B_t}^2) \boldsymbol{n} + \frac{1}{\mu_0} B_n B_t \boldsymbol{t} \right) dS, \tag{2}$$

where $\boldsymbol{t}$ represents the outward unit vector tangential to the differential surface dS. The quantity inside the integral is usually interpreted as surface force density or traction. In this study, we follow the same interpretation and compute this force density on a surface located in the inner radius of the stator.

2.2. Structural Mechanics

Magnetic forces are the excitation parameters for the structural simulation of the stator core of the motor. The forces are fed to the stator elastic model as a body load in the simulation. The elastic model is represented by

$$\rho \frac{\partial^2 \boldsymbol{d}}{\partial t^2} - \nabla \cdot \tau = f, \tag{3}$$

where ρ is the mass density, $\boldsymbol{d}$ is the vector of displacements, and f is the given volume force. After computing the displacements, a discrete Fourier transformation is performed using fast-Fourier transform (FFT), where the time-dependent solution is transformed from times to frequencies in the frequency domain.

2.3. Acoustics

The BEM used in this study is based on the direct collocation method [13], which deals directly with acoustic variables (sound pressure and particle velocity) and boundary conditions. The multiphysics coupling that combines FEM-and BEM-based physics is employed for coupling the results of solid mechanics finite element physics to acoustics boundary element physics. In the case of an electric motor, the vibrating stator boundary can be used as the acoustics FEM–BEM boundary to couple the acceleration from FEM computation to the BEM interface. This approach allows for modeling in an FEM–BEM framework using the strength of each formulation effectively. Acoustics physics solves the Helmholtz equation for constant-valued material properties and uses the pressure as the dependent variable. The wave equation can be solved in the frequency domain for one frequency at a time. The governing Helmholtz equation for a boundary element interface is given by

$$-\frac{1}{\rho_c} \nabla^2 p_t - \frac{{k_{cq}}^2}{\rho_c} p_t = 0, \tag{4}$$

$$k_{cq}{}^2 = \left(\frac{\omega}{c_c}\right)^2 \text{and } p_t = p + p_b \tag{5}$$

where p_t is the total acoustic pressure, K_{cq} is the wave number, ρ_c is the density, ω is the angular frequency, and c_c is the speed of sound in air.

The governing equations and boundary conditions are formulated using the total pressure p_t with a scattered-field formulation. In the presence of a background pressure field defining a background pressure wave p_b, the total acoustic pressure p_t is the sum of the pressure solved for p, which is then equal to the scattered pressure p_s and the background pressure wave. The equations then contain information about both the scattered field and the background pressure field.

The benefit of the boundary element method is that only boundaries need to be meshed, and the degrees of freedom (DOFs) solved for are restricted to the boundaries. This is beneficial for handling complex geometries, as it introduces easiness in numerical computations. However, the BEM procedure results in fully populated or dense matrices, compared to the sparse system matrices in the FEM. Hence, the BEM is more expensive in terms of memory requirements per DOF than the FEM is, but it has fewer DOFs. Assembling and solving these can be very demanding. In this context, when solving acoustic models of small and medium size, the FEM will often be faster than solving the same problem with the BEM. This could be interpreted as one limitation of the BEM in small-sized computational models. However, in the case of the FEM, when the geometries are complex or two structures are far apart, large air domains need to be meshed. This is costly in terms of the numerical computational facet, as the frequency is increased.

The acoustic pressure computation problem involves solving for small acoustic pressure variations p in the surrounding medium of a sound source on top of the stationary background pressure p_0. In mathematical terms, this can be interpreted as a linearization of the dependent variables around the stationary quiescent values. The fluid flow problems in a compressible lossless fluid can be analyzed using the three governing equations, viz., the mass conservation equation or the continuity equation, the momentum conservation equation or Euler's equation, and the energy equation or the entropy equation. They are given by

$$\frac{\partial \rho}{\partial t} + \nabla \cdot (\rho \boldsymbol{u}) = \boldsymbol{M}, \tag{6}$$

$$\frac{\partial \boldsymbol{u}}{\partial t} + (\boldsymbol{u} \cdot \nabla)\boldsymbol{u} = -\frac{1}{\rho}\nabla p + \boldsymbol{F}, \tag{7}$$

$$\frac{\partial s}{\partial t} + \nabla \cdot (s\boldsymbol{u}) = 0, \tag{8}$$

where ρ is the total mass density, p is the total pressure, $\boldsymbol{u}$ is the velocity field, s is the entropy, and $\boldsymbol{M}$ and $\boldsymbol{F}$ are the possible source terms representing the body forces, if any. In conventional pressure acoustics scenarios, all thermodynamic processes are assumed to be isentropic in nature, which means the processes are both reversible and adiabatic. The small-parameter expansion is executed on a stationary fluid ($\boldsymbol{u}_0$ = 0) of density ρ_0 (kg/m^3) and at pressure p_0 (Pa) such that $p = p_0 + p_1$ with $p_1 \ll p_0$, $\rho = \rho_0 + \rho_1$ with $\rho_1 \ll \rho_0$, $\boldsymbol{u} = 0 + \boldsymbol{u}_1$ with $\boldsymbol{u}_1 \ll c$, and $s = s_0 + s_1$.The small acoustic variations are represented by the variables with subscript 1. Inserting these values in the governing equations gives

$$\frac{\partial \rho_1}{\partial t} + \nabla \cdot (\rho_0 \boldsymbol{u}_1) = \boldsymbol{M}, \tag{9}$$

$$\frac{\partial \boldsymbol{u}_1}{\partial t} = -\frac{1}{\rho_0}\nabla p_1 + \boldsymbol{F}, \tag{10}$$

$$\frac{\partial p_1}{\partial t} = c_s{}^2\left(\frac{\partial \rho_1}{\partial t} + \boldsymbol{u}_1 \cdot \nabla \rho_0\right), \tag{11}$$

where c_s is the isentropic speed of sound. The pressure time differential in the last equation is derived from the entropy equation. If the material parameters are constant, the last equation reduces to

$$p_1 = {c_s}^2 \rho_1. \tag{12}$$

This expression of acoustic pressure gives a condition that needs to be fulfilled for the linear acoustic equations to hold:

$$|p_1| \ll {c_s}^2 \rho_0. \tag{13}$$

Finally, the transient wave equations for pressure waves in a lossless medium can be obtained by rearranging Equations (9)–(11) and dropping the subscripts:

$$\frac{1}{\rho c^2}\frac{\partial^2 p}{\partial t^2} + \nabla \cdot [-\frac{1}{\rho}(\nabla p - q_d)] = Q_m, \tag{14}$$

where the source term Q_m is a monopole domain source corresponding to a mass source and q_d is a dipole domain source representing a domain force source. The speed of sound (c) and the density (ρ) may in general be space-dependent. The combination term ρc^2 is called the adiabatic bulk modulus (K_s) with unit Pa, which is related to the adiabatic compressibility coefficient $\beta_s = 1/K_s$.

In the frequency domain, the Helmholtz equation can be written as

$$\nabla \cdot [-\frac{1}{\rho_c}(\nabla p_t - q_d)] - \frac{{k_{cq}}^2 p_t}{\rho_c} = Q_m. \tag{15}$$

Acoustics problems mostly encompass simple harmonic waves such as sinusoidal waves. In numerical computations, to model acoustic–structure interactions, a structural analysis can be coupled to acoustics by imposing the acceleration as a source in the boundaries of the structure in the form of normal acceleration, specified as

$$-\boldsymbol{n} \cdot [-\frac{1}{\rho_c}(\nabla p_t - q_d)] = -\boldsymbol{n} \cdot a_0, \tag{16}$$

where a_0 is the normal acceleration and q_d is the external force term.

2.4. Acoustic Pressure and Audible Sound

Sound is measured by changes in air pressure. The louder a sound is, the larger the change in air pressure is. The change here is the change from normal atmospheric pressure or reference pressure to the pressure disturbance caused by the sound. Sound pressure is measured in the unit "pascals". A pascal (Pa) is equal to the force of one newton per square meter. The smallest sound pressure a human ear can hear is 20 μPa, which corresponds to zero dB. The sound pressure level (SPL) in dB can be calculated by

$$SPL = 20\, log_{10}\left(\frac{p}{p_{ref}}\right) \text{dB}, \tag{17}$$

where p_{ref} is the reference pressure 20 μPa in the case of audible sound calculations.

3. Results

The results of the numerical simulations are presented in this section. The simulations were done using COMSOL multiphysics software [24]. The specifications of the solid rotor induction motor are given in Table 1. The noise computation of an electric motor starts from the electromagnetic field computations, where the Maxwell equations are solved using finite element analysis. From the electromagnetic analysis, the forces of electromagnetic origin are taken into the mechanics domain to calculate the deformation and vibrations, where the forces are fed as an input to the solid mechanics calculations. Until the computation of vibrations, the FEM is used, and for the acoustic noise, the BEM is employed. In the first step, the electromagnetic simulation gives the magnetic forces due to Maxwell stress, and these forces are then fed to the structural computation as an input excitation. The supply frequency was 1008 Hz, and three time periods were simulated in the electromagnetic 2D computation. There were 19,750 linear triangular elements in the 2D mesh, as shown in Figure 3a, and the structural 3D mesh of the stator contained 104,931 tetrahedral elements, which is given in Figure 3b. Electromagnetic and structural mechanics time-stepping simulations were performed with a time-step length of 2.5 e^{-6} s for three time periods corresponding to a 0.003-s machine running time. The 2D electromagnetic computation took 17 min of CPU time to finish the simulation, and the 3D structural mechanics simulation took 4 h for each time period. A time-to-frequency FFT was done for the results of the third time period to transform the solution from a time to frequency domain. The acoustics computation to calculate the sound pressure level took 11 min of CPU time for each frequency. The magnetic flux density distribution across the cross-section of the motor is given in Figure 4.

Table 1. The solid rotor induction motor.

Specifications	Unit	Value
Power	kW	300
Voltage (line-to-line)	V	400
Rated speed	rpm	60,000
Number of poles		2
Stator outer diameter	mm	250
Stator inner diameter	mm	116

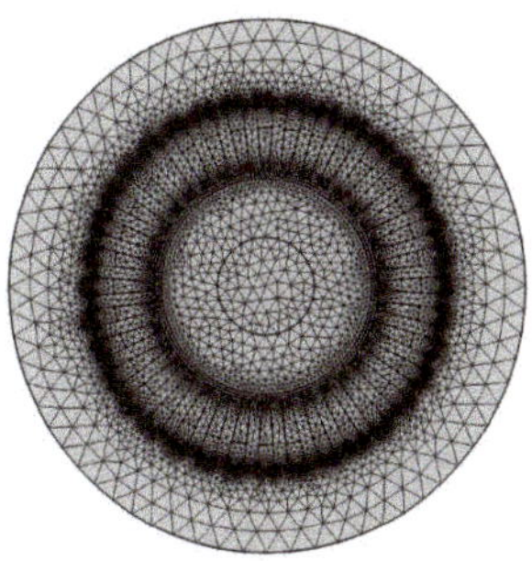
(**a**)

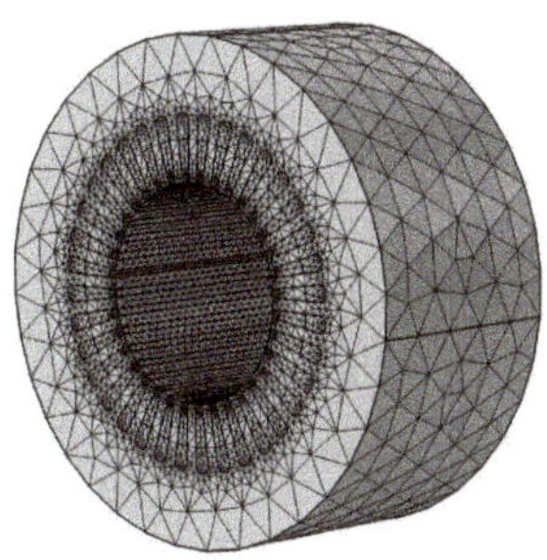
(**b**)

Figure 3. The finite element mesh of the motor: (**a**) the 2D triangular mesh of the rotor and stator and (**b**) the 3D tetrahedral mesh of the stator.

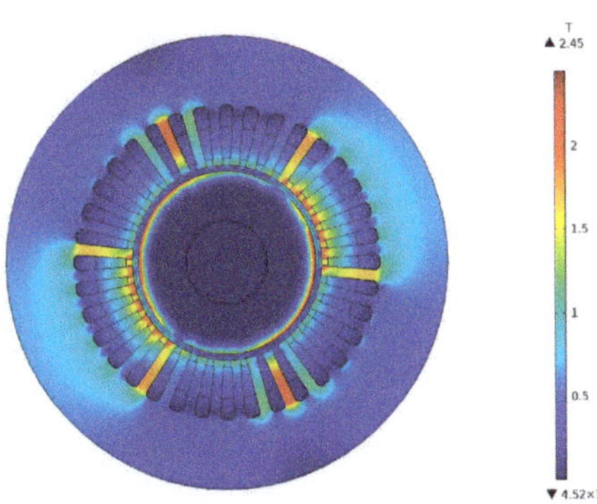

Figure 4. Magnetic flux density distribution in the motor cross-section from the electromagnetic finite element method (FEM).

In the structural simulation, displacement and acceleration of the stator body were computed. The magnetic forces were computed using Equation (2), and these forces were fed as the force term into Equation (3) to calculate the displacements. As a second stage of structural analysis, a Fourier analysis of the results was performed to discover the major frequency components in the deformation spectrum. The deformation pattern of the stator is depicted in Figure 5, and the Fourier analysis results of the displacement at a point on the stator outer surface are given in Figure 6, where the rotor rotational frequency is 1000 Hz, the twice-supply frequency, $2fs$ is 2016 Hz, and 4032 Hz corresponds to $2p^{*}2fs$, where p is the number of pole pairs.

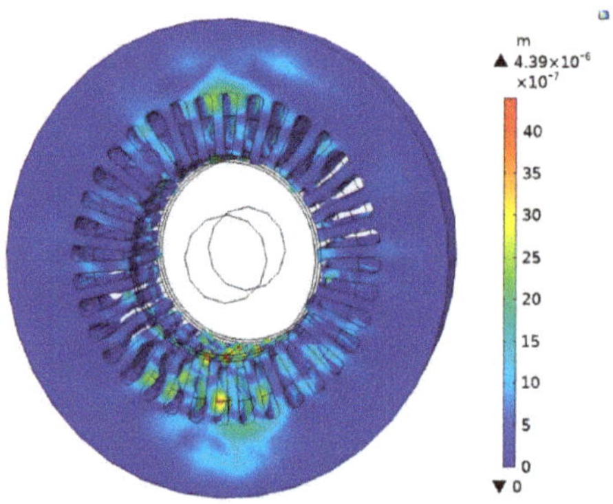

Figure 5. Deformation of the stator due to magnetic forces from the solid-mechanics FEM. Displacement scale factor: 5000.

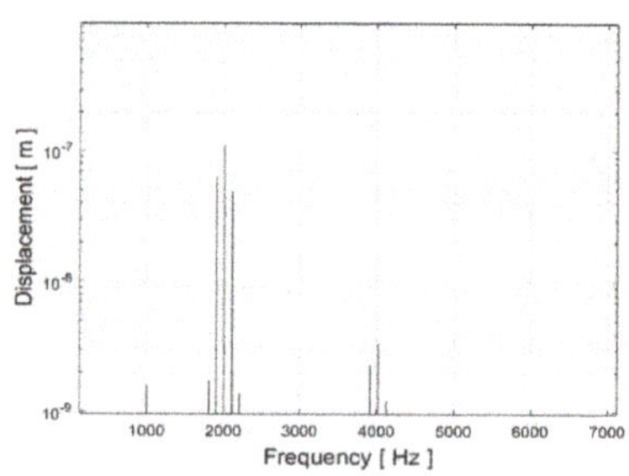

Figure 6. Frequency spectrum of the radial displacement of a point on the stator outer boundary.

In the final step of acoustics, the accelerations from the FE mechanics domain are imposed at the stator boundaries, as is theoretically explained in Equation (16). This specific feature couples acoustics with the structural analysis for an acoustic–structure interaction. For the acoustics computations, the air surrounding the motor is modeled as an infinite void, where no specific geometry or meshing is required. The BEM simulation calculates the acoustic pressure in the surrounding air of the motor and the resulting sound pressure level corresponding to different frequencies. Equation (15) is utilized in the calculation of sound pressure in the frequency domain. The acoustic pressure distribution outside the motor, corresponding to 2016 Hz of frequency, is shown in Figure 7. The sound pressure level produced by this acoustic pressure difference in the air is given in Figure 8. The same parameters for 4032 Hz are depicted in Figure 9a,b. Figure 10 shows the sound pressure level in far fields in different planes. The far-fields plots of the sound pressure level give a clear idea about the directivity of noise radiation in different planes around the motor at a particular frequency.

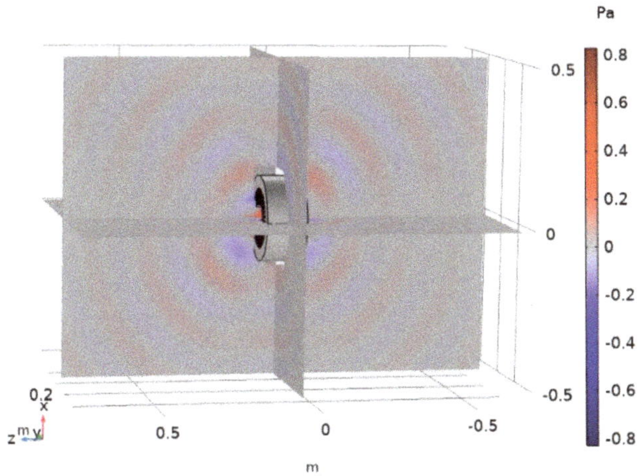

Figure 7. Acoustic pressure distribution in Pa outside the motor at 2016 Hz.

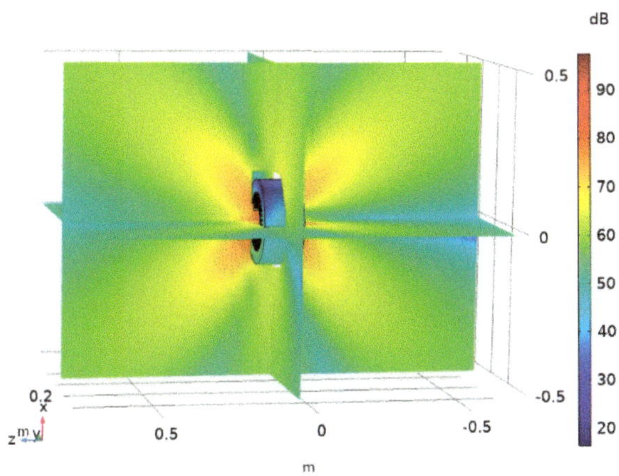

Figure 8. Sound pressure level in dB outside the motor at 2016 Hz.

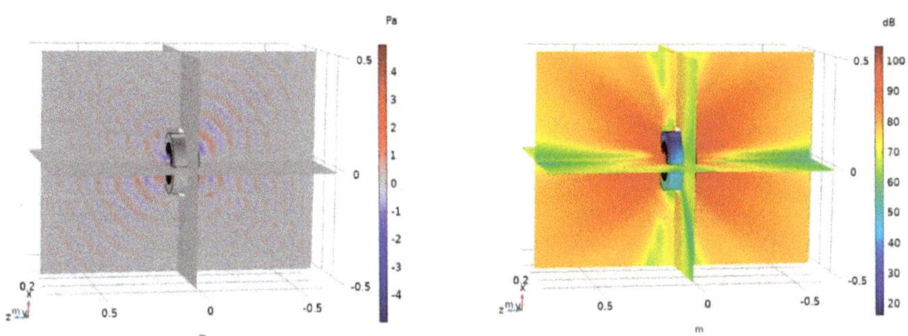

(**a**) (**b**)

Figure 9. (**a**) Acoustic pressure distribution in Pa outside the motor at 4032 Hz; (**b**) sound pressure level in dB outside the motor at 4032 Hz.

(a)

(b)

(c)

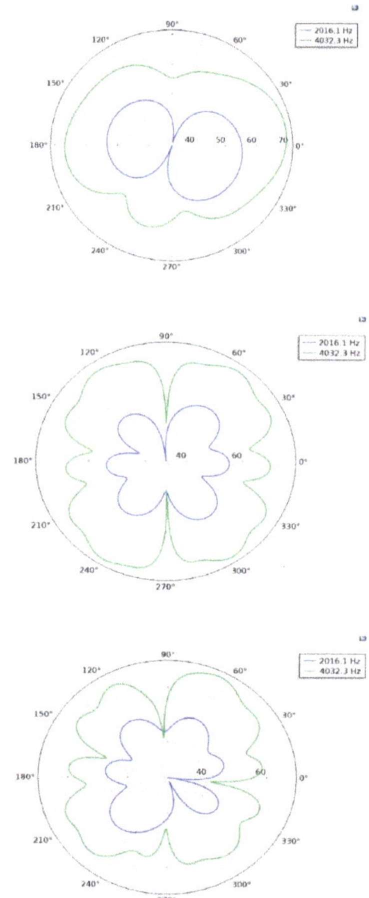

Figure 10. Far-field sound pressure level in dB at a distance of 1 m in the (**a**) *x*-*y* plane, (**b**) *x*-*z* plane, and (**c**) *y*-*z* plane.

4. Discussion

The results presented in this paper showcase how magnetic forces produce acoustic sound in an electrical motor. The emphasis of this study was mainly centered on a sound level computation through FEM–BEM coupling, and the results show that this multilevel approach combining two numerical methods and three different physics in a single finite element tool is an effective tactic for acoustic analysis of electrical motors. The structural mechanics results provide information about the deformation and vibration behavior of the machine and also the major frequency components in the vibration spectrum, as is shown in Figures 5 and 6. The computational results of the acoustic pressure variation occurring in the neighboring medium of the motor (Figures 7 and 9a) shed light on the pattern of sound pressure and how different frequencies cause dissimilar forms of pressure variations in terms of their directions and intensity, thereby varying the levels of sound (Figures 8 and 9b). The far-field sound pressure levels portrayed in Figure 10 offer an idea about how motor vibrations produce and direct sound in different directions.

One major drawback concerning numerical computations of acoustic noise is the relatively higher memory requirements and computational time required compared to analytical and semi-analytical models. The increased computational time of the FEM has been mentioned as a drawback of numerical methods in the literature, such as in References [7–9]. The use of the BEM in an acoustics domain fixes this issue to a considerable degree, without compromising the accuracy of the results. Although methods such as the permeance/mmf model [7,25] and the subdomain model [7,8,26] reduce the computational time, they possess some drawbacks in the modeling of complex geometries and in terms of the 3D effects of machines. In a multiphysical simulation environment, the use of the BEM instead of the FEM as the numerical method thus facilitates a competent framework for vibro-acoustic computations without compromising precision in detailed modeling, and it also has the benefit of reducing the computational time. If the FEM had been employed in the acoustic computations of the study presented in this paper, the entire surrounding air would have been meshed, and this would have caused increased memory requirements and computational time. A quantitative study comparing previous works could not be done because of the difference in motor types. However, qualitatively, the FEM–BEM approach has all of the benefits of numerical computations, especially in terms of accuracy and flexibility in modeling complex structures, and lessens the computational time significantly compared to a complete FEM model.

This paper does not include the measurement results of sound levels, and hence a future component of this study will be to conduct laboratory experiments to measure sound levels. In addition, to compare the measurement results to the simulations, the entire structure of the motor, including the frame and bearings, needs to be modeled in the future. Furthermore, computing and measuring the A-weighted sound levels corresponding to different frequencies need to be carried out in the second part of this study to precisely illustrate the sound pressure level in terms of human audibility [27]. In addition, in real-world problems where complex systems need to be simulated, materials cannot necessarily be connected: rather they are glued or clamped. Furthermore, the endplates and the frame of the motor could have an effect on the computed vibrations and sound pressures. Hence, the boundary layers in the structural FEM and acoustic BEM become more difficult to model. In those cases, accurate material models are required, and the model size could be increased. Both issues require measurements and iterative parameter adoption. However, the computational method presented in this paper facilitates a platform for vibro-acoustic studies by effectively modeling the acoustic noise of the motor.

5. Conclusions

The sound produced by an induction motor due to electromagnetic forces was successfully computed using a joint finite-element and boundary-element-based numerical computational technique. The energy conversion process in an electrical motor was modeled by showing how the electrical energy is converted into acoustic energy. The paper successfully illustrated how the proposed coupling methods and the FEM–BEM combination work effectively in acoustic studies of motors. The task of formulating a computational framework by combining theoretical equations in the right way and order, preparing numerical simulation stages, and coupling them to yield factual results in different stages and different physical domains was done in an original way in this study, as was presented in the paper. The proposed scheme is a vivid method describing how electromagnetic, mechanical, and acoustic domains can be analyzed and coupled in a numerical simulation platform. It provides an expedient and effectual tool for researchers in electrical machine and acoustics fields conducting magneto-mechanical and vibro-acoustics studies.

Author Contributions: Conceptualization, S.S.; methodology, S.S. and U.A.; software, S.S. and U.A.; validation, A.B., S.S., and U.A.; formal analysis, A.B., S.S., and U.A.; writing—original draft preparation, S.S.; writing—review and editing, S.S.; visualization, S.S.; supervision, A.B. All authors have read and agreed to the published version of the manuscript.

Funding: This research received no external funding.

Conflicts of Interest: The authors declare no conflicts of interest.

References

1. Vijayraghavan, P.; Krishnan, R. Noise in electric machines: A review. *IEEE Trans. Ind. Appl.* **1998**, *35*, 1007–1013. [CrossRef]
2. Belmans, R.J.M.; D'Hondt, L.; Vandenput, A.; Geysen, W. Analysis of the audible noise of three phase squirrel cage induction motors supplied by inverters. *IEEE Trans. Ind. Appl.* **1987**, *23*, 842–947. [CrossRef]
3. Belmans, R.J.M.; Verdyck, D.; Geysen, W.; Findlay, R.D. Electromechanical analysis of the audible noise of an inverter-fed squirrel-cage induction motor. *IEEE Trans. Ind. Appl.* **1991**, *27*, 539–544. [CrossRef]
4. Cameron, D.E.; Lang, J.H.; Umans, S.D. The origin and reduction of acoustic noise in doubly salient variable reluctance motors. *IEEE Trans. Ind. Appl.* **1992**, *28*, 1250–1255. [CrossRef]
5. Besnerais, J.L.; Lanfranchi, V.; Hecquet, M.; Brochet, P. Characterization and reduction of audible magnetic noise due to PWM supply in induction machines. *IEEE Trans. Ind. Electron.* **2010**, *57*, 1288–1295. [CrossRef]
6. Le Besnerais, J.; Lanfranchi, V.; Hecquet, M.; Brochet, P.; Friedrich, G. Acoustic noise of electromagnetic origin in a fractional-slot induction machine. *COMPEL-Int. J. Comput. Math. Electr. Electron. Eng.* **2008**, *27*, 1033–1052. [CrossRef]
7. Besnerais, J.L. Fast Prediction of Variable-Speed Acoustic Noise due to Magnetic Forces in Electrical Machines. In Proceedings of the XXII International Conference on Electrical Machines (ICEM), Lausanne, Switzerland, 4–7 September 2016; pp. 2259–2265.
8. Devillers, E.; Hecquet, M.; Besnerais, J.L.; Regniez, M. Tangential effects on magnetic vibrations and acoustic noise of induction machines using subdomain method and electromagnetic vibration synthesis. In Proceedings of the IEEE International Electric Machines and Drives Conference 2017 (IEMDC), Miami, FL, USA, 21–24 May 2017.
9. Devillers, E.; Hecquet, M.; Cimetiere, X.; Lecointe, J.P.; Le Besnerais, J.; Lubin, T. Experimental benchmark for magnetic noise and vibrations analysis in electrical machines. In Proceedings of the 23rd International Conference on Electrical Machines 2018 (ICEM), Alexandroupoli, Greece, 3–6 September 2018; pp. 745–751.
10. Fakam, M.; Verbeke, D.; Hecquet, H.; Lanfranchi, V.; Brochet, P.; Randria, A. Electromagnetic noise comparison between 'spm' and 'ipm' concentrated winding synchronous machine. In Proceedings of the ISEF 2011: XV International Symposium on Electromagnetic Fields in Mechatronics, Electrical and Electronic Engineering, Funchal, Madeira, 1–3 September 2011.
11. Islam, R.; Husain, I. Analytical model for predicting noise and vibration in permanent-magnet synchronous motors. *IEEE Trans. Ind. Appl.* **2010**, *46*, 2346–2354. [CrossRef]
12. Juhl, P.M. The Boundary Element Method for Sound Field Calculations. Ph.D. Thesis, The Acoustic Laboratory, Technical University of Denmark, Copenhagen, Denmark, 1993.
13. Henriquez, V.C.; Juhl, P.M. OpenBEM—An open source boundary element method software in acoustics. In Proceedings of the Internoise 2010, Odense, Denmark, 13–16 June 2010; pp. 1–10.
14. Wang, C.; Lai, J.C.S.; Astfalck, A. Sound power radiated from an inverter driven induction motor II: Numerical analysis. *IEE Proc.-Electr. Power Appl.* **2004**, *151*, 341–348. [CrossRef]
15. Herrin, D.W.; Martinus, F.; Wu, T.W.; Seybert, A.F. An Assessment of the High Frequency Boundary Element and Rayleigh Integral Approximations. *Appl. Acoust.* **2006**, *67*, 819–833. [CrossRef]
16. Roivainen, J. Unit-Wave Response-Based Modeling of Electromechanical Noise and Vibration of Electrical Machines. Ph.D. Thesis, Helsinki University of Technology, Espoo, Finland, 2009.
17. Neves, C.G.C.; Carlson, R.; Sadowski, N.; Bastos, J.P.A.; Soeiro, N.S.; Gerges, S.N.Y. Calculation of electromagnetic-mechanic-acoustic behavior of a switched reluctance motor. *IEEE Trans. Magn.* **2000**, *36*, 1360–1363.
18. Furlan, M.; Černigoj, A.; Boltežar, M. A coupled electromagnetic-mechanical-acoustic model of a DC electric motor. *COMPEL-Int. J. Comput. Math. Electr. Electron. Eng.* **2003**, *22*, 1155–1165. [CrossRef]
19. Schlensok, C.; Van Riesen, D.; Küest, T.; Henneberger, G. Acoustic simulation of an induction machine with squirrel-cage rotor. *COMPEL-Int. J. Comput. Math. Electr. Electron. Eng.* **2006**, *25*, 475–486. [CrossRef]
20. Deng, W.; Zuo, S.; Lin, F.; Wu, S. Investigation of vibration and noise characteristics in axial flux permanent magnet synchronous motor with different magnet shapes. *Proc. Meet. Acoust.* **2016**, *29*, 040005.
21. Järvenpää, S.; Huttunen, T.; Vanne, A.; Malinen, M.; Roivainen, J.; Ylä-Oijala, P. Fast boundary element simulation of noise emission from vibrating structures. *Raken. Mek. (J Struct. Mech.)* **2016**, *49*, 137–150.

22. Arkkio, A. Analysis of Induction motors Based on the Numerical Solution of the Magnetic Field and Circuit Equations. Ph.D. Thesis, University of Helsinki, Helsinki, Finland, 1987.
23. Belahcen, A. Magnetoelasticity, Magnetic Forces and MS in Electrical Machines. Doctoral Thesis, Helsinki University of Technology, Espoo, Finland, 2004.
24. COMSOL Multiphysics Version 5.4. Available online: http:www.comsol.com (accessed on 1 July 2019).
25. Timar, P.L. *Noise and Vibration of Electrical Machines*; Elsevier: Amsterdam, The Netherlands, 1989.
26. Boisson, J.; Louf, F.; Ojeda, J.; Mininger, X.; Gabsi, M. Analytical approach for magnetic and acoustic modeling of flux-switching permanent-magnet motors: Application to geometrical optimization. In Proceedings of the 17th International Conference on Electrical Machines and Systems (ICEMS), Hangzhou, China, 22–25 October 2014; Volume 3, pp. 3060–3066.
27. *Electroacoustics, Sound Level Meters-Part 1: Specifications*; IEC 61672-1; International Electrotechnical Commission: Geneva, Switzerland, 2013; p. 5.

energies

MDPI

Article

Mathematical Modelling of Drive System with an Elastic Coupling Based on Formal Analogy between the Transmission Shaft and the Electric Transmission Line

Andrzej Popenda [1,*], Marek Lis [1], Marcjan Nowak [1] and Krzysztof Blecharz [2]

1 Faculty of Electrical Engineering, Częstochowa University of Technology, 42-201 Częstochowa, Poland; marek.lis@pcz.pl (M.L.); marcjan.nowak@pcz.pl (M.N.)

2 Faculty of Electrical and Control Engineering, Department of Controlled Electric Drives and Energy Conversion, Gdańsk University of Technology, 80-233 Gdańsk, Poland; krzysztof.blechar20@pg.edu.pl

* Correspondence: popenda@el.pcz.czest.pl

Received: 14 January 2020; Accepted: 28 February 2020; Published: 4 March 2020

Abstract: In the paper, the kinematic structure of the transmission shaft between the driving motor and the working mechanism is studied. The analysis is based on electrical and mechanical similarities. The equivalent circuits, typical for electrical systems, are defined for the transmission shaft concerned. Modelling of the transmission shaft based on a formal analogy between the transmission shaft and the electric transmission line is also proposed. The results of a computer simulation and experimental test are presented. The results confirm the high conformity of the proposed mathematical model with the physical object.

Keywords: electric drive; transmission shaft; electric transmission line; electrical and mechanical similarities; kinematic structure; equivalent circuit; mathematical modelling

1. Introduction

Rapid development and diversity of electric drive systems require extensive theoretical and practical knowledge as well as the use of a wide range of theoretical and constructional solutions from the contemporary designers and users. Electric drive systems are used in all branches of industry; therefore, the trouble-free operation of these systems is of crucial importance. An analysis of the states of electric drive operation is often related to ensuring its safety. A key issue is to detect the mechanical resonance phenomenon or the phenomena close to resonance as the most unsafe ones for the system. Ignoring the analysis and the ongoing diagnostics of vibrations in electromechanical systems results in many failures. The example, including fatalities, is inter alia the failure of turbo generator set in the power plant Kostromska (Russia). In the commission view, the cause of the failure was ignoring torsional vibrations, which led to the resonance in the turbine and transmission shaft included in the turbo-generator set.

A very interesting case of the electric submersible pumps system with a multi-component system powered by a soft start system is shown in [1]. In this real case scenario, the incorrect control case led to resonance phenomena and a shaft breakdown. Therefore, comprehensive research is needed to analyse the operating states of electrical and electromechanical systems. One method to investigate the dynamic behaviour of electrical and electromechanical systems is an experimental test directly on the system. In many cases, the experimental tests on the system are not possible, due to technical reasons, costs of experiments or their duration. The alternative approach to the investigation into the system behaviour is the formulation of the model and carrying out the investigations into the model. The model of a system is the representation using mathematical relationships, a physical model or an analogue model. The physical model of the real-world system includes smaller-scale components having the same physical nature as a real-world system. In the physical model, there are phenomena described by the same dependencies, differing only in the order of magnitude. The analogue model includes components having a different physical nature than the real-world system, but these components are easier to implement. The analogue models are based on physical analogy. An electrical circuit may be defined as the analogue model of the mechanical system. The mathematical model is the set of mathematical relationships, on the basis of which the behaviour of a system can be predicted. The mathematical model is most often formulated as differential equations describing the operation of a system. Each physical model has a corresponding mathematical model.

The mathematical model of a mechanical object is usually a set of partial differential equations. They are hard to solve both analytically and numerically. Discrete models of systems include ordinary differential equations; therefore, they are used in practice most often. Real-world mechanical systems are usually nonlinear, where the nonlinearity is determined by material properties, clearances, nonlinear nature of dissipation forces and characteristics of elastic elements. A limited possibility of analysis of nonlinear differential equations leads towards the use of linear models or the linearization of those. There are a lot of physical systems, which may be represented by linear models with accuracy acceptable for practice. In the process of creating the models of physical systems, the equations of motion are formulated.

An electric motor, being part of an electric drive, is coupled with a working mechanism via a driving shaft that is an element of mechanical power transmission. Mechanical power transmissions can be a single-path or a multi-path and can also include gear trains and clutches [2]. The long driving shafts, defined as transmission shafts, are used first of all in the drive systems for the steel industry, mainly in the drive systems for rolling mills—the transmission shafts are over 10 m long and their diameters are of 0.5 to 0.8 m. Transmission shafts are also used in drive systems for polymerization reactors. The length of these shafts is from 4 to 7 m. Moreover, transmission shafts are used in hydro generator sets, ship drive systems, submarine drive systems, etc. Depending on the length and cross-section, transmission shafts can demonstrate different susceptibilities to the impact of the moment of torsion, as measured by a value of the angle of twist [2]. In the case of short mechanical coupling, the value of the angle of twist is insignificant and may be omitted by the assumption of rigid mechanical coupling, whereas in the case of longer mechanical coupling the value of the angle of twist cannot be ignored and such coupling should be considered as the elastic one [1–10].

In order to model the transmission shaft, classical methods based on the d'Alembert principle or Newton's second law as well as the variational method are used. The use of the variational method results in the distributed parameter model consisting of partial differential equations [3]. An example of a partial differential equation used to describe a long elastic element is the wave equation [11]. Partial differential equations may be solved analytically, but this is both tedious and time-consuming. Thus, these equations are usually transformed into ordinary differential equations by using finite differences [2]. The model of the long elastic element, based on the ordinary differential equations, corresponds with a multi-mass lumped parameter model, e.g., in [2,12,13], especially the two-mass model, e.g., in [6,8,14], which does not guarantee the accurate results of numerical analysis. The equations included in the multi-mass model are usually solved using numerical integration.

Research on stresses and oscillations of long drive shafts using various methods of analysis is undertaken in the literature [15–18]. In [18], the author presents a mechanical analysis of the propeller shaft on a warship in the event of a variable moment of shaft load. Modelling of the object is selected on a classic approach using a system of nonlinear equations. A similar approach, but using a different model and a different scale, with a three-mass system, is shown in [16]. The author shows comparative results between the analytical calculations of the two-mass system and the numerical calculations for the three-mass system. An extended approach to analysing the issue of three-mass and two-mass system including changes in the moment of inertia is shown in [18]. In all of the above described publications, the authors used a classical approach in different variations.

In this paper, the kinematic structure of the transmission shaft between the driving motor and the working mechanism was studied. The telegrapher's equations and their d'Alembert solution are proposed for the mathematical description of the transmission shaft. The proposed modelling is an alternative to the multi-mass model based on ordinary differential equations as well as a distributed parameter model based on partial differential equations. The advantage of the proposed model is its simplicity, due to the fact that the model is based on discrete equations, which do not require numerical integration in contrast to the models based on differential equations. The presented modelling is based on electrical and mechanical similarities. Identifying these similarities is very helpful for electricians in finding a relevant interpretation of mechanical systems, which is particularly important in the case of professionals dealing with electromechanical energy converters or drive systems [2]. The aim of the presented study was to achieve very high conformity of the mathematical model with the physical object, whilst keeping a short time for numerical calculations.

2. Methods of Modelling of Transmission Shaft on the basis of Electrical and Mechanical Similarities

The mass of the real transmission shaft (or any element) is distributed continuously. Representing such a transmission shaft by a model based on lumped parameters causes discrepancies in the results of the analysis in relation to accurate models. These discrepancies decrease with the number of points of concentration in the model. Dividing a transmission shaft with distributed mass into several shorter elements, described by lumped parameters such as mass, elasticity and damping of the i-th element, it is possible to obtain the results of computer simulation essentially not different from the results obtained if the shaft is divided into infinite number of elements, that corresponds with the wave model. The process of the abovementioned division is referred to as discretization of kinematic structure and corresponds with the multi-mass lumped parameter model of the transmission shaft.

In order to present the synthesis of the proposed models of the transmission shaft, the following assumptions have been adopted:

- transmission shaft is the elastic element,
- in the case of the lumped parameter model, the shaft is divided into a finite number of identical elements, the parameters of which (elasticity, mass and dissipation) are the same,
- in the case of the distributed parameter model, the shaft is divided into an infinite number of identical elements, the parameters of which are the same; the shaft is represented by specific (per unit of length) constant parameters: a specific moment of inertia J', specific coefficient of viscous friction inside the shaft D' and a specific coefficient of torsional susceptibility S'_c,
- circular motion is represented by angular velocities, whereas shaft deflection is omitted,

1. viscous friction inside the shaft is also represented by lumped damping parameters D_{1m} and D_{m1},

- angular velocity and the torsional moment are defined for each element of the shaft.

On the basis of the above-given assumptions, the kinematic structure of transmission shaft (Figure 1) transferred into m discrete elements was formulated [2], where $J_1, \ldots, J_m, C_{s,12}, \ldots, C_{s,m-1,m}, S_{c,12}, \ldots, S_{c,m-1,m}, D_{12}, \ldots, D_{m-1,m}$ are the moments of inertia, coefficients of torsional stiffness, coefficients of torsional susceptibility ($S_c = 1/C_s$) and coefficients of viscous friction inside the respective elements of divided transmission shaft; D_1, D_m are the coefficients of friction defined for bearings.

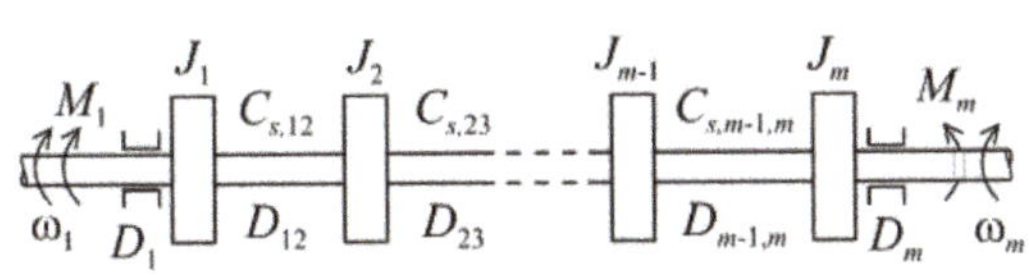

Figure 1. Kinematic structure of transmission shaft transferred into m discrete elements.

In accordance with the electromechanical energy conversion theory, the generalised coordinates are forces or torques in mechanical systems and voltages in electrical circuits, whereas the generalised velocities are the velocities or angular velocities in mechanical systems and currents in electrical systems. As a consequence, the parameters of transmission shaft are equivalent to those of the transmission line, i.e., the moment of inertia is equivalent to inductance, the torsional susceptibility is equivalent to capacitance, the coefficient of friction, if any, on the shaft surface is equivalent to the resistance and the coefficient of the viscous friction inside the shaft is equivalent to the inverse of the conductance. Adopting the abovementioned analogies and considering the specificity of transmission shaft, the equivalent circuits for the transmission shaft (Figure 2) has been created.

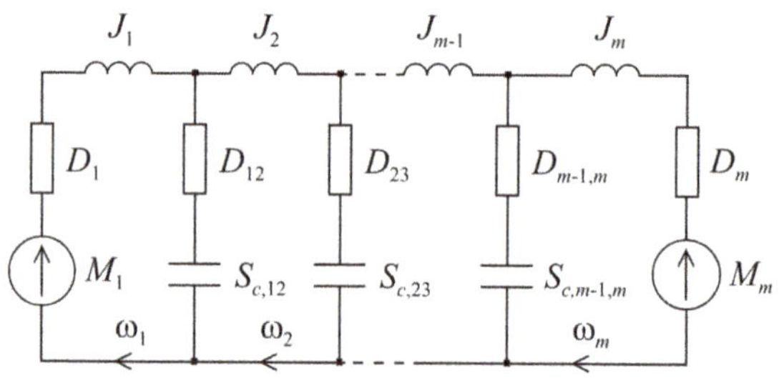

Figure 2. Equivalent circuit corresponding with the kinematic structure depicted in Figure 1.

On the basis of the above given equivalent circuit the m equations of moments and torques as well as $m-1$ equations of angular velocities, corresponding with the Kirchhoff's circuit laws typically applied in order to analyse electrical circuits, may be written for the mechanical coupling concerned [2]:

$$M_1 - D_1\omega_1 - \frac{d}{dt}(J_1\omega_1) - D_{12}(\omega_1 - \omega_2) - M_{c,12} = 0, \tag{1}$$

$$M_{c,k-1,k} + D_{k-1,k}(\omega_{k-1} - \omega_k) - D_{k,k+1}(\omega_k - \omega_{k+1}) - \frac{d}{dt}(J_k\omega_k) - M_{c,k,k+1} = 0 \text{for} \quad k = 2, \ldots, m-1, \tag{2}$$

$$M_{c,m-1,m} + D_{m-1,m}(\omega_{m-1} - \omega_m) - \frac{d}{dt}(J_m\omega_m) - D_m\omega_m - M_m = 0, \tag{3}$$

$$\omega_k - \omega_{k+1} = \frac{d}{dt}\left(S_{c,k,k+1}M_{c,k,k+1}\right) \quad \text{for} \quad k = 1, \ldots, m-1. \tag{4}$$

On the basis of the equivalent circuit (Figure 2) it can be concluded that the transmission shaft may be analysed in a similar way as the electric transmission line [2]. In order to formulate the distributed parameter model of transmission shaft, the following equivalent circuit of a shaft section, represented by the two-port system, can be used (Figure 3). The adopted length of the shaft section is Δx. It is assumed that the section Δx is short enough to use, in respect of this section, the dependencies adequate for the mathematical description of the lumped parameters model.

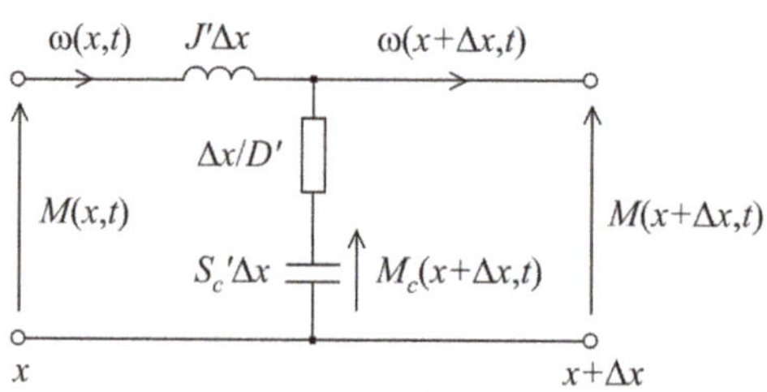

Figure 3. Equivalent circuit of a transmission shaft section.

The equivalent circuit of transmission shaft section (Figure 3) is similar to that of the transmission line section [19]. There is a formal analogy between the lossless transmission shaft and the lossless transmission line.

Balancing moments of torsion around the closed-loop and angular velocities at the point leads to the following dependencies [2]:

$$M(x,t) = J'\Delta x \frac{\partial \omega(x,t)}{\partial t} + M(x+\Delta x,t), \tag{5}$$

$$\omega(x,t) = S_c'\Delta x \frac{\partial M_c(x+\Delta x,t)}{\partial t} + \omega(x+\Delta x,t), \tag{6}$$

$$M(x+\Delta x,t) = D'S_c' \frac{\partial M_c(x+\Delta x,t)}{\partial t} + M_c(x+\Delta x,t), \tag{7}$$

where J', D', S_c' are specific (per unit of length) parameters of the shaft, i.e., specific moment of inertia, specific coefficient of friction and specific coefficient of torsional susceptibility, respectively, $S_c' = \rho/GJ'$, ρ is the mass density in kg/m^3 and G is the shear module in GPa. Transforming Equations (5), (6) and (7) as well as assuming $\Delta x \to 0$ the following equations can be obtained:

$$-\frac{\partial M(x,t)}{\partial x} = J'\frac{\partial \omega(x,t)}{\partial t}, \tag{8}$$

$$-\frac{\partial \omega(x,t)}{\partial x} = S_c'\frac{\partial M_c(x,t)}{\partial t}, \tag{9}$$

$$M_c(x,t) = M(x,t) + D'\frac{\partial \omega(x,t)}{\partial x}. \tag{10}$$

The further equations result from Equation (8) as well as Equations (9), (10) and (11):

$$-\frac{1}{J'}\frac{\partial^2 M(x,t)}{\partial x^2} = \frac{\partial^2 \omega(x,t)}{\partial x \partial t}, \tag{11}$$

$$-\frac{\partial \omega(x,t)}{\partial x} = S_c'\frac{\partial M(x,t)}{\partial t} - \frac{S_c'D'}{J'}\frac{\partial^2 M(x,t)}{\partial x^2}. \tag{12}$$

A lossless transmission shaft may also be considered, if $D' = 0$:

$$-\frac{\partial M(x,t)}{\partial x} = J'\frac{\partial \omega(x,t)}{\partial t}, \quad -\frac{\partial \omega(x,t)}{\partial x} = S_c'\frac{\partial M(x,t)}{\partial t}. \tag{13}$$

If the shaft parameters do not change along the axis, Equation (13) may be decoupled in respect of moment of torsion and angular velocity:

$$-\frac{\partial^2 M}{\partial x^2} = J'S_c'\frac{\partial^2 M}{\partial t^2}, \quad -\frac{\partial^2 \omega}{\partial x^2} = J'S_c'\frac{\partial^2 \omega}{\partial t^2}, \tag{14}$$

$$-\frac{\partial^2 M}{\partial x^2} = \frac{1}{v^2}\frac{\partial^2 M}{\partial t^2}, \quad -\frac{\partial^2 \omega}{\partial x^2} = \frac{1}{v^2}\frac{\partial^2 \omega}{\partial t^2}, v = \frac{1}{\sqrt{J'S_c'}} = \sqrt{\frac{G}{\rho}}, \tag{15}$$

where v is phase velocity in m/s. The general solution of the equations of type (15) was given by d'Alembert [19]:

$$M(x,t) = \frac{1}{2}\left[M_1\left(t+\frac{x}{v}\right) + M_1\left(t-\frac{x}{v}\right)\right] + \frac{v}{2}\int_{t-\frac{x}{v}}^{t+\frac{x}{v}} \frac{\partial M(0,\tau)}{\partial x} d\tau, \tag{16}$$

with the boundary conditions:

$$M_1(t) = M(x,t)\big|_{x=0}, \quad \frac{\partial M(0,t)}{\partial x} = \frac{\partial M(x,t)}{\partial x}\bigg|_{x=0}. \tag{17}$$

On the basis of the first Equation (13) and the Equation (16) the following Equations (18), (19) and (20) may be written:

$$\frac{\partial M(0,t)}{\partial x} = -J'\frac{\partial \omega(0,t)}{\partial t} = -J'\frac{\partial \omega_1(t)}{\partial t}, \tag{18}$$

$$\frac{v}{2}\int_{t-\frac{x}{v}}^{t+\frac{x}{v}} \frac{\partial M(0,\tau)}{\partial x} d\tau = -\frac{1}{2}\sqrt{\frac{J'}{S'_c}}\int_{t-\frac{x}{v}}^{t+\frac{x}{v}} d\omega_1(\tau) = -\frac{z_v}{2}\left[\omega_1\left(t+\frac{x}{v}\right) - \omega_1\left(t-\frac{x}{v}\right)\right], \tag{19}$$

$$M(x,t) = \frac{1}{2}\left[M_1\left(t+\frac{x}{v}\right) + M_1\left(t-\frac{x}{v}\right)\right] - \frac{z_v}{2}\left[\omega_1\left(t+\frac{x}{v}\right) - \omega_1\left(t-\frac{x}{v}\right)\right], \tag{20}$$

where z_v is wave impedance, $z_v = \sqrt{J'/S'_c} = vJ'$. The analogous formula for calculation of angular velocity may also be written as follows:

$$\omega(x,t) = \frac{1}{2}\left[\omega_1\left(t+\frac{x}{v}\right) + \omega_1\left(t-\frac{x}{v}\right)\right] - \frac{1}{2z_v}\left[M_1\left(t+\frac{x}{v}\right) - M_1\left(t-\frac{x}{v}\right)\right]. \tag{21}$$

On the basis of the Equations (20) and (21) the moments of torsion at the shaft beginning (input) and the shaft end (output) may be expressed as follows:

$$M_1(t) = z_v\omega_1(t) - z_v\omega_m\left(t-\frac{l}{v}\right) + M_m\left(t-\frac{l}{v}\right), \tag{22}$$

$$M_m(t) = -z_v\omega_m(t) + z_v\omega_1\left(t-\frac{l}{v}\right) + M_1\left(t-\frac{l}{v}\right), \tag{23}$$

where l is the shaft length. The abovementioned moments of torsion may also be expressed in the discrete form:

$$M_1(j) = z_v\omega_1(j) - z_v\omega_m(j-n) + M_m(j-n), \tag{24}$$

$$M_m(j) = -z_v\omega_m(j) + z_v\omega_1(j-n) + M_1(j-n), \tag{25}$$

with the initial conditions $M_1(0)$, $M_m(0)$, $\omega_1(0)$ and $\omega_m(0)$, where: n is the number of numerical calculation points, $n = l/vh$, h is the width of the step size between the points, $j = 0,1,\ldots$. The number n is also a discrete-time of moving the mechanical wave from one end of the shaft to the other. The continuous-time of moving the wave along the shaft axis, corresponding with number n, is given in the Equations (22) and (23) as l/v. The optimum number of numerical calculation points is the number, the increase of which does not cause significant differences in simulation results, whereas the decrease of this number causes significant differences in simulation results. The terms from the previous steps $\omega(j–n)$ and $M(j–n)$ in the Equations (24) and (25) may be determined using shift registers. For $j \leq n$:

$$M_1(j) = z_v\omega_1(j), \; M_m(j) = -z_v\omega_m(j). \tag{26}$$

The angular velocities at the beginning (input) and the end (output) of the transmission shaft may be calculated on the basis of the equations of motion for the rotors of both an electric motor and working machine:

$$J_e\frac{d\omega_1(t)}{dt} = M_e(t) - D_1\omega_1(t) - M_1(t), \tag{27}$$

$$J_L\frac{d\omega_m(t)}{dt} = M_m(t) - D_m\omega_m(t) - M_L(t), \tag{28}$$

where $M_e(t)$ and $M_L(t)$ are motor output torque and a load torque of working machine, J_e and J_L are moments of rotor inertia for motor and working machine, respectively, D_1 and D_m are coefficients of friction defined for bearings.

The model with two lumped damping parameters D_{1m} and D_{m1} may be adopted in order to represent the viscous friction inside the shaft. Then, the Equations (27) and (28) take the following form:

$$J_e \frac{d\omega_1(t)}{dt} = M_e(t) - D_1\omega_1(t) - D_{1m}(\omega_1(t) - \omega_m(t)) - M_1(t), \tag{29}$$

$$J_L \frac{d\omega_m(t)}{dt} = M_m(t) - D_m\omega_m(t) - D_{m1}(\omega_m(t) - \omega_1(t)) - M_L(t). \tag{30}$$

Such an extension of the model allows for maintaining the advantageous features resulting from the simplicity of numerical calculations.

3. Simulations and Experimental Results

Research Implementation Methods

To verify the proposed solution of the long driveshaft modelling algorithm by the use of an electric transmission line, simulation and experimental research were performed. In the first step, the algorithm computer program in C++ programming language was written by the research team. Open and free software was used to write the simulation program. The development of a simulation program independently enables the optimization of the algorithm code that is to be implemented in the future in an inverter with a digital signal processor.

The simulation contains the brushless dc motor (BLDC) model with a full control system, long driveshaft model according to Equations (24) and (25) and a direct current generator model. In the simulation, the parameters of 4 kW BLDC motor and steel transmission shaft of length 0.66 m and diameter 0.008 m, according to the real laboratory setup elements, were taken. Other parameters of the simulation model and reference units are shown in Table 1.

Table 1. Model parameters.

Parameter	Symbol	Value
Length of shaft	l	0.66 m
Diameter of shaft	d	0.008 m
Specific moment of inertia	J'	$3{,}18 \cdot 10^{-6}$ kg·m²/m
Specific coefficient of friction	D'	$7.3 \cdot 10^{-3}$ Nm²s
Specific coefficient of torsional susceptibility	S'_c	$32 \cdot 10^{-3}$ 1/Nm²
Wave impedance	z_v	0.01 Nms
Mass density	ρ	7900 kg/m³
Shear module	G	77.5 GPa
Phase velocity	v	3132 m/s
Motor output torque	M_e	N·m
Load torque	M_L	N·m
Moment of rotor inertia for motor	J_e	0.025 kg·m²
Moment of inertia for load	J_L	0.11 kg·m²
Moment of inertia for additional elements	J_a	0.079 kg·m²
Coefficients of friction for bearings	D_1; D_m	$1.6 \cdot 10^{-3}$ Nms
Lumped damping parameters representing viscous friction inside the shaft	D_{1m}, D_{m1}	0.022 Nms

In order to determine the optimum number of numerical calculation points, the numerical simulations of the transmission shaft behaviour were carried out, in which various numbers of numerical calculation points n were taken into account, respectively, 10, 15, 20 and 25 points. Figure 4 shows angular velocities for beginning (input) and end (output) of the transmission shaft for different calculations variants. The end of the transmission shaft was loaded by a rotating mass J_L = 0.11 kg·m^2 and a mechanical torque of rated value.

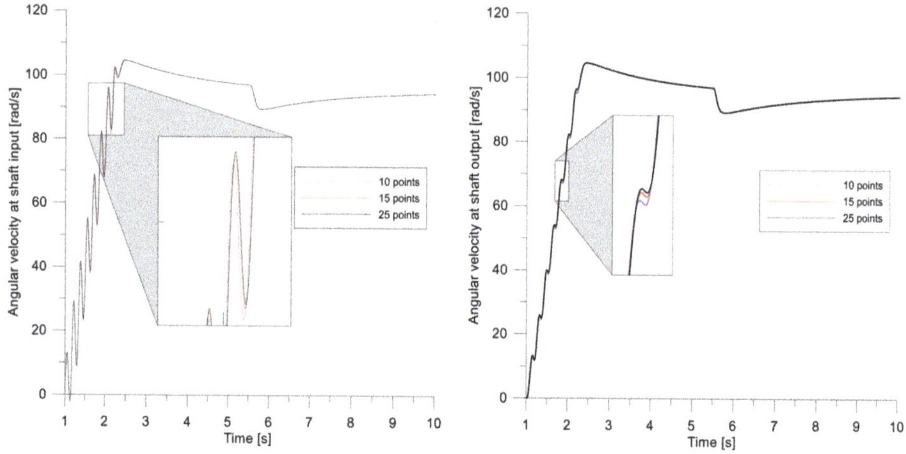

(a) (b)

Figure 4. Angular velocity at shaft input (**a**) and output (**b**) vs. time.

Figure 5; Figure 6 show respective the differences in angular velocity and angle of twist for the transmission shaft for different calculations variants.

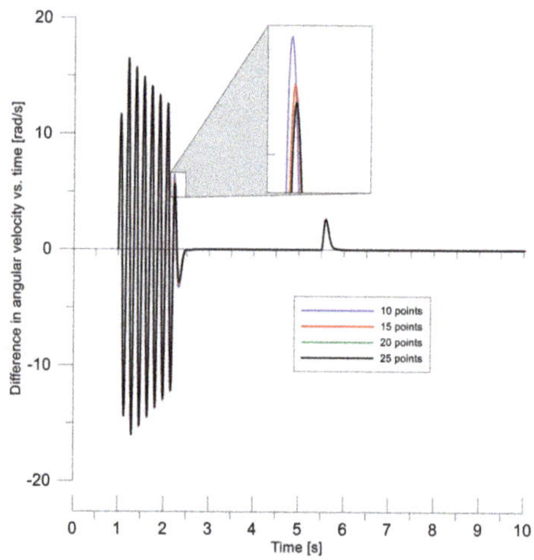

Figure 5. Difference in angular velocity vs. time.

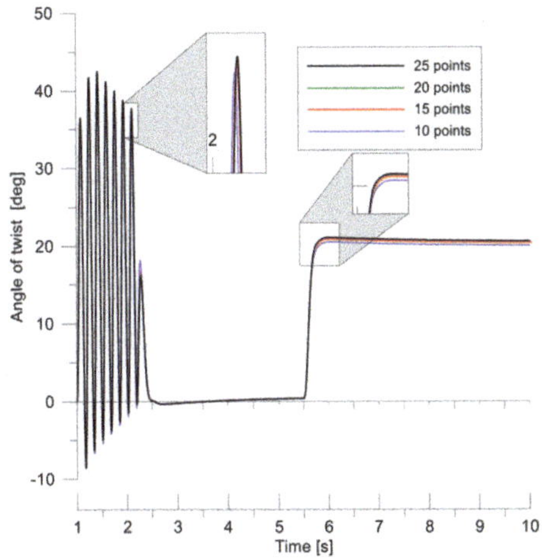

Figure 6. Angle of twist vs. time.

If a lower number of numerical calculation points is chosen (10 and 15) then the effect similar to the shortening of the shaft occurs (slightly less amplitude of oscillation and slightly less natural frequency as well as the decrease of steady angle of twist after 8 s from starting the system at the time $t = 1$ s).

For 20 and 25 points, the differences in amplitude of oscillation, natural frequency and steady angle of twist are not significant. It can be read from the graphs. For example, the difference in amplitude of oscillation and difference in steady angle of twist for 10 and 25 points, respectively, is in a range of 1 degree, which is around 5% in respect to the steady angle of twist, whereas for 20 and 25 points, respectively, this difference is a fraction of a percentage. The differences can disappear for an infinite number of points; therefore, a balance between the accuracy of numerical calculations and the speed of numerical calculations should be found.

In the next step, the simulation results were verified by the experiment. Experimental investigations were made using a test stand including a BLDC motor coupled with a DC generator by a steel long driveshaft. The parameters of the shaft (length, diameter) in the experimental setup was designed in accordance with the power of experimental setup components. This setup allows for connecting the shafts of various diameters, which allow for a good representation of the parameters of real-world systems. The laboratory setup is shown in Figure 7. In addition, in the figure, a simplified diagram of an experimental stand layout corresponding to the actual configuration was presented. The test stand is fully configurable; therefore, it is possible to mount additional steel discs at the shaft on the DC generator side that allow to change the moment of inertia of the load. Two rotary incremental encoders with a resolution of 3600 pulse/rev installed on the transmission shaft ends, are used to measure angular displacement and angular velocities. Hall effects transducers were used to measure motor phase currents. The measuring signals from the transducers are sent to a laboratory computer equipped with two multifunctional I/O devices PCIe-6351 from National Instruments.

Laboratory Virtual Instrument Engineering Workbench (LabVIEW) was used for data collection, whereas a computer program to visualise the measurement datasets was created by the authors. The test stand is fully automated and allows testing of the driveshaft with different geometries and made of different materials.

All presented time waveforms correspond to starting the motor and the operation under rated load after 5.5 s. The characteristic time waveforms for angular velocities at shaft input and output, the difference in angular velocity, angle of twist and phase current of BLDC motor, are shown in Figure 8, Figure 9, Figure 10, Figure 11, Figure 12.

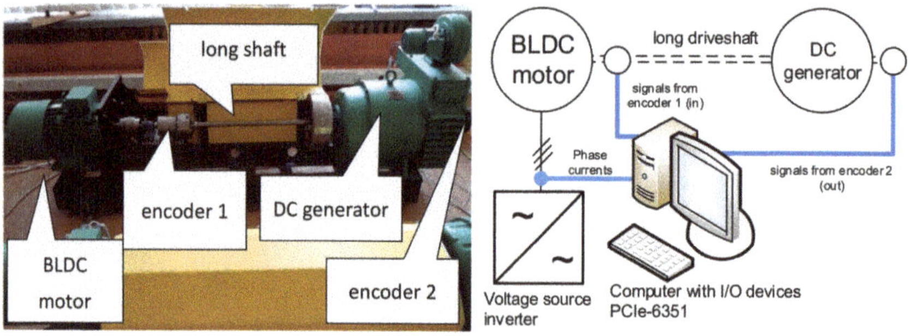

A photo of the investigate electric drive A simplified diagram of a laboratory stand

Figure 7. Stand for testing a drive system with elastic coupling.

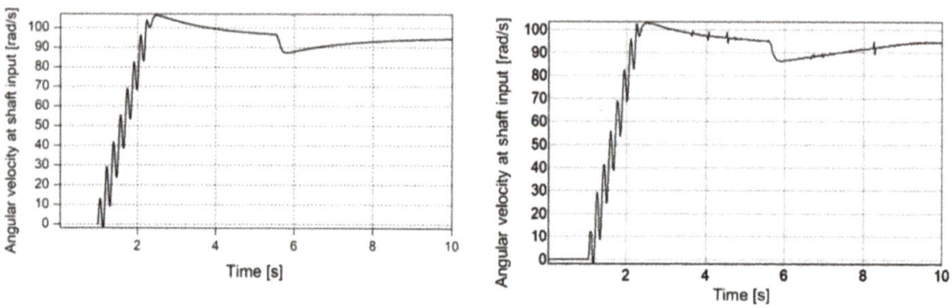

simulation experiment

Figure 8. Angular velocity at shaft input vs. time.

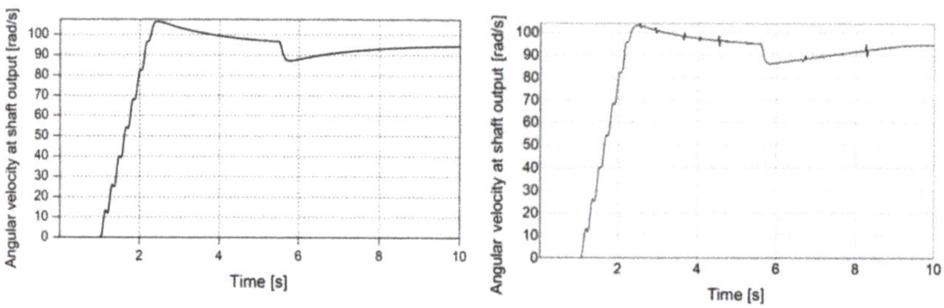

simulation experiment

Figure 9. Angular velocity at shaft output vs. time.

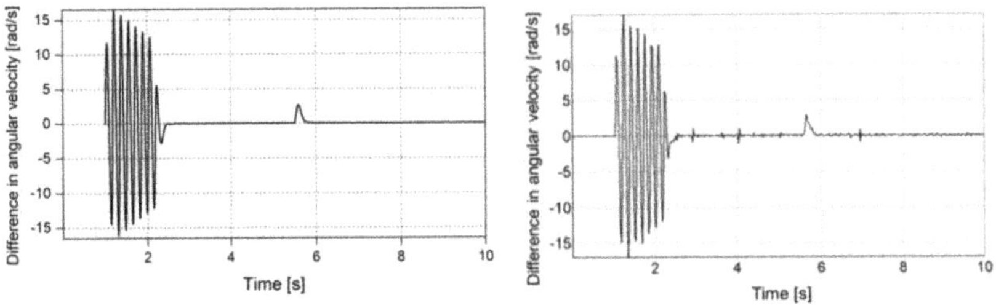

simulation experiment

Figure 10. Difference in angular velocity vs. time.

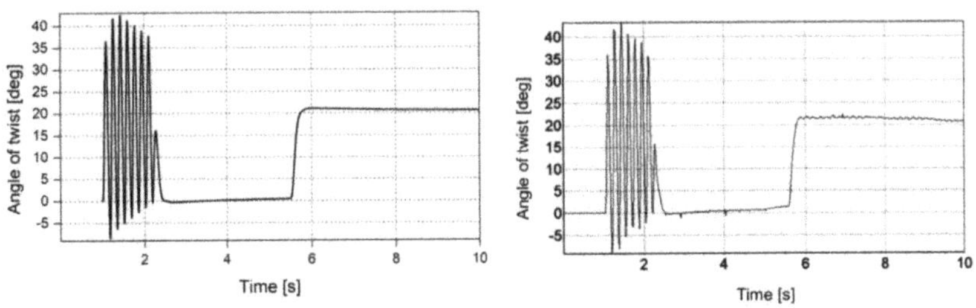

simulation experiment

Figure 11. Angle of twist vs. time.

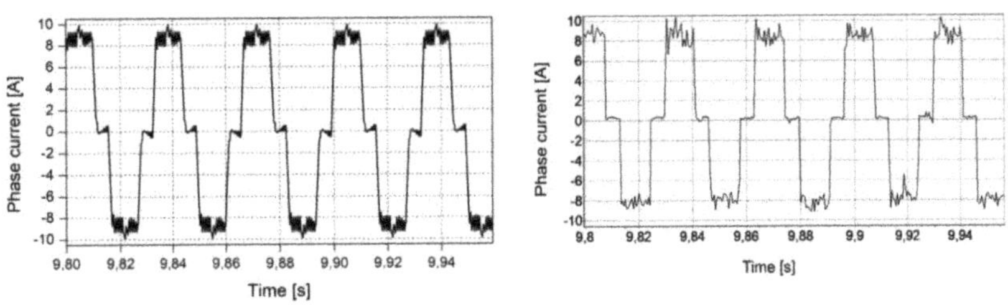

simulation experiment

Figure 12. Phase current of BLDC motor under load vs. time.

High similarity of both the experimental results confirms good accuracy of the proposed model.

4. Discussion

The results of the computer simulation and experimental tests presented in the paper deal with the modelling of the transmission shaft being a mechanical component of an electric drive. The high similarity of both experimental and simulation results confirms the high conformity of the proposed mathematical model, based on electrical and mechanical similarities with the physical object. This conformity is crucial from a practical point of view for design activities.

The telegrapher's equations and their d'Alembert solution are proposed in the article for the mathematical description of the transmission shaft. If the boundary conditions for the two points at the ends of the shaft are adopted, the solution is determined only for these points, without the need to track the processes within the shaft. If the shaft is part of a complex system, including an electric motor, power converter, mechanical coupling and working machine, then only the variables at the shaft ends can be taken into account, without the need to estimate the phenomena for any other value of shaft axis coordinate x. This assumption leads to the very simple equations of the discrete model of the lossless transmission shaft. This approach is known as Bergeron diagram method. The solution of the abovementioned equations does not require numerical integration, which can cause a problem with the stability of the integration method and it is often carried out in more than one stage. As a consequence, a shorter time for carrying out the numerical calculations is required if the proposed model is used than in the case of the models based on differential equations. Thus, high accuracy of the proposed model goes hand in hand with its simplicity and, as a consequence, with a short time of numerical calculations in contrast to the modelling based on numerical integration of differential equations or analytical solution of partial differential equations. For example, in the case of the multi-mass model of the transmission shaft divided into m discrete elements, $2m - 1$ ordinary differential equations have to be integrated numerically k times for each step, where k is the order of numerical integration method. The use of the proposed model, based on the d'Alembert solution for telegrapher's equations, requires for each step the single numerical calculation of $4n - 2$ discrete equations of type X [j + 1] = X [j] and 2 discrete equations given in the article as (24) and (25), where n depends on the shaft length, phase velocity and width of the step size in accordance with dependency given in the article. In order to ensure high accuracy of numerical integration of ordinary differential equations, which is comparable with the accuracy of numerical calculation of the discrete equations included in the proposed model, the width of the step size for the numerical integration should be tens of times shorter than for the numerical calculation of the abovementioned discrete equations. This means that using the proposed model leads to less numerical calculations in the same unit time than for the models based on differential equations.

5. Conclusions

Long elastic elements, especially transmission shafts, may be described mathematically by multi-mass lumped parameter models based on ordinary differential equations or distributed parameter models based on partial differential equations, which may be solved analytically. However, the analytical solution is both tedious and time-consuming. Thus, partial differential equations are usually transformed into ordinary differential equations by using finite differences. In both the abovementioned cases, the ordinary differential equations are usually integrated numerically.

In the paper, the kinematic structure of the transmission shaft between the driving motor and the working mechanism is studied. The analysis is based on electrical and mechanical similarities. The equations of moments and angular velocities, analogous to the Kirchhoff's law-based equations typically applied in the mathematical analysis of branched electrical circuits, are defined. Modelling of transmission shaft based on a formal analogy between the transmission shaft and the electric transmission line is also proposed. The high conformity of the proposed mathematical model with the physical object has been confirmed by the comparison of the results of computer simulation and experimental tests.

The advantage of the proposed model is its simplicity, due to the fact that the model bases on discrete equations, which do not require numerical integration in contrast to the models based on differential equations. The presented approach to modelling of the entire electromechanical system can be easily developed towards the system containing more than one elastic coupling, electric motor or working machine.

The proposed mathematical model may be used to analyse the impact of control and load on an electromechanical system as well as the failure states associated with or close to the resonance phenomenon. Simulation results may be used to design electric drives as well as to optimise electromechanical systems during operation.

The proposed transmission shaft model based on discrete equations is an open-loop model, but nothing precludes its use in a closed-loop control system in an advanced electric drive. This will be the subject of further research work of the authors team.

Author Contributions: Conceptualization, A.P. and K.B.; Methodology, A.P.; Software, A.P. and K.B.; Validation, A.P., M.L., M.N. and K.B.; Formal analysis, A.P. and K.B.; Investigation, A.P., M.L., M.N. and K.B.; Resources, A.P., M.L., M.N. and K.B.; Data curation, A.P., M.L., M.N. and K.B.; Writing—Original draft preparation, A.P. and K.B.; Writing—Review and editing, A.P. and K.B.; Visualization, A.P., M.L., M.N. and K.B.; Supervision, A.P. and K.B.; Project administration, A.P. and K.B.; Funding acquisition, M.L. All authors have read and agreed to the published version of the manuscript.

Funding: This research received no external funding.

Conflicts of Interest: The authors declare no conflict of interest.

Nomenclature of All Variables

$C_{s12}, \ldots, C_{s,m-1,m}$	are coefficients of torsional stiffness of the respective elements of divided transmission shaft,
D_1, D_m	are coefficients of friction defined for bearings,
$D_{12}, \ldots, D_{m-1,m}$	are coefficients of viscous friction of the respective elements of divided transmission shaft,
D_{1m}, D_{m1}	are the lumped damping parameters representing viscous friction inside the shaft
D'	is the specific (per unit of length) coefficient of viscous friction inside the shaft
G	is the shear module,
h	Is the width of the step size between the numerical calculation points,
$J_1, \ldots, J_{\rm m}$	are the moments of inertia of the respective elements of divided transmission shaft,
J_e, J_L	are the moments of rotor inertia for motor and working machine,
J'	is the specific moment of shaft inertia,
l	is the shaft length,
M, M_c	are the torsional moments applied to the shaft section,
$M_1, \ldots, M_{\rm m}$	are the torsional moments and torques applied to the respective elements of the divided transmission shaft,
m	is the number of discrete elements of divided transmission shaft,
n	is the number of numerical calculation points,
$S_{c,12}, \ldots, S_{c,m-1,m}$	are the coefficients of torsional susceptibility of the respective elements of the divided transmission shaft,
S_c'	is the specific coefficient of torsional susceptibility,
v	is the phase velocity,
x	is the shaft axis coordinate,
Δx	is the length of the shaft section,
ρ	is the mass density,
ω_1, ω_m	are the angular velocities at shaft ends,
$\omega_2, \ldots, \omega_{m-1}$	are the angular velocities of the respective elements of the divided transmission shaft.

References

1. Rabb, S.F.; Kahnamouei, J.T.; Liang, X.; Yang, J. Shaft Failure Analysis in Soft-Starter Fed Electrical Submersible Pump Systems. *IEEE Open J. Ind. Appl.* **2020**, *1*, 1–10. [CrossRef]
2. Popenda, A. Mathematical modelling of transmission shafts based on electrical and mechanical similarities. *Przegląd Elektrotechniczny* **2019**, *95*, 196–199. [CrossRef]
3. Czaban, A.; Lis, M. Mathematical Modelling of Transient States in a Drive System with a Long Elastic Element. *Przegląd Elektrotechniczny* **2012**, *88*, 167–170.
4. Rusek, A.; Czaban, A.; Lis, M.; Klatow, K. Mathematical model of an electromechanical system with a long elastic element. *Przegląd Elektrotechniczny* **2015**, *91*, 69–72. (In Polish)

5. Szabat, K.; Orłowska-Kowalska, T. Application of the Kalman filters to the high-performance drive system with elastic coupling. *IEEE Trans. Ind. Electron.* **2012**, *59*, 4226–4235. [CrossRef]
6. Serkies, P.; Szabat, K. Application of the MPC controller to the Position Control of the Two-Mass Drive System. *IEEE Trans. Ind. Electron.* **2013**, *60*, 3679–3688. [CrossRef]
7. Orłowska-Kowalska, T.; Szabat, K. Control of the drive system with stiff and elastic couplings using adaptive neuro-fuzzy approach. *IEEE Trans. Ind. Electron.* **2007**, *54*, 228–240. [CrossRef]
8. O'Sullivan, T.; Bingham, C.C.; Schofield, N. High-performance control of dual-inertia servo-drive systems using low-cost integrated SAW torque transducers. *IEEE Trans. Ind. Electron.* **2006**, *53*, 1226–1237. [CrossRef]
9. Li, W.; Hori, Y. Vibration suppression using single neuron-based PI fuzzy controller and fractional-order disturbance observer. *IEEE Trans. Ind. Electron.* **2007**, *54*, 117–126. [CrossRef]
10. Valenzuela, M.A.; Bentley, J.M.; Lorenz, R.D. Evaluation of torsional oscillations in paper machine sections. *IEEE Trans. Ind. Electron.* **2005**, *41*, 493–501.
11. Swanson, D.C. *Signal Processing for Intelligent Sensor Systems with MATLAB®*, 2nd ed.; CRC Press: Boca Raton, FL, USA, 2017; pp. 7–23.
12. Xu, J.; Su, X.; Peng, B. Numerical Analysis and Demonstration: Transmission Shaft Influence on Meshing Vibration in Driving and Driven Gears. *Shock Vib.* **2015**, *2015*. [CrossRef]
13. Łuczak, D. Mathematical model of multi-mass electric drive system with flexible connection. In *Proceedings of the 19th International Conference on Methods and Models in Automation and Robotics, Miedzyzdroje, Poland, 2–5 September 2014*; IEEE Publisher: New York, NY, USA, 2014; pp. 290–295. [CrossRef]
14. Muyeen, S.M.; Mohd, H.A.; Takahashi, R.; Murata, T.; Tamura, J.; Tomaki, Y.; Sakahara, A.; Sasano, E. Transient Stability Analysis of Wind Generator System with the Consideration of Multi-Mass Shaft Model. In *Proceedings of the 2005 International Conference on Power Electronics and Drives Systems, Kuala Lumpur, Malaysia, 28 November–1 December 2005*; IEEE Publisher: New York, NY, USA, 2005. [CrossRef]
15. Grządziela, A. Modelling of propeller shaft dynamics at pulse load. *Pol. Marit. Res.* **2008**, *15*, 52–58. [CrossRef]
16. Pielorz, A. Dynamic analysis of torsional discrete-continuous systems with position-dependent variable inertia. *J. Theor. Appl. Mech.* **2014**, *52*, 107–115.
17. Pielorz, A.; Skóra, M. Modeling of multimass systems torsionally deformed with variable inertia. *Int. J. Differ. Equ.* **2006**, *2006*. [CrossRef]
18. Pielorz, A.; Skóra, M. Analytical approach in torsional multi-mass discrete-continuous systems with variable inertia. *Meccanica* **2009**, *44*, 121. [CrossRef]
19. Rosołowski, E. *Computer Techniques for Analysis of Electromagnetic Transient States*, 1st ed.; Oficyna Wydawnicza Politechniki Wrocławskiej: Wrocław, Poland, 2009; pp. 23–31. (In Polish)

MDPI
St. Alban-Anlage 66
4052 Basel
Switzerland
Tel. +41 61 683 77 34
Fax +41 61 302 89 18
www.mdpi.com

Energies Editorial Office
E-mail: energies@mdpi.com
www.mdpi.com/journal/energies

www.ingramcontent.com/pod-product-compliance
Lightning Source LLC
Chambersburg PA
CBHW061549220926
51850CB00005B/288